EARTH SCIENCE

F. MARTIN BROWN
Wright-Ingraham Institute

WAYNE BAILEY
Jersey City State College

EARTH SCIENCE

GENERAL LEARNING PRESS

SCOTT, FORESMAN AND COMPANY

EARTH SCIENCE

F. MARTIN BROWN WAYNE BAILEY

CRITICAL READERS

KATHERINE LAING BLADH	University of Arizona at Tucson
WALTER N. DUFFETT	Eastern Illinois University
P. L. KEHLER	University of Arkansas at Little Rock
PAUL W. KIRST	Miami-Dade Community College
PAUL FRANK MILLER	Norwalk Community College
PATRICK J. STANLEY	Ball State University

Copyright 1978 Silver Burdett Company

All rights reserved. No part of this publication may be reproduced or transmitted in any form or by any means, electronic or mechanical, including photocopy, xerography, recording, or any information storage and retrieval system, without permission in writing from the publisher.

Manufactured in the United States of America
Published simultaneously in Canada
Library of Congress Catalog Card Number 77-014694
ISBN 0-382-18035-6
For information write to:
General Learning Press
250 James Street
Morristown, N. J. 07960

CONTENTS

CHAPTER ONE
THIS PLANET EARTH

The study of earth science — 4
Why study the earth?
The methods of science — 6
Investigating a problem • Systems of measurement
Earth science in history — 8
Aristotle • Werner • Hutton
Describing our planet — 11
Earth's addresses • Determining latitude and longitude • Just what time is it? • Duration of day and night • The seasons and solar energy • The zones of the earth
Matter and movement — 18
Kepler's discoveries • Newton and gravity • Conservation
Summary — 22
Review questions — 23

IN FOCUS:
Earth science: where lab and field work meet — 14
BOX:
The usefulness of oddballs — 19

CHAPTER TWO
THE EARTH'S MATERIAL

Starting at the beginning — 28
Atoms • Atomic structure • The nature of light • Chemical properties of atoms • A quick look at physics
Minerals — 33
Crystals • Identifying minerals
Defining and classifying rocks — 36
Igneous rocks • Sedimentary rocks • Metamorphic rock • The rock cycle
Summary — 52
Review questions — 53

IN FOCUS:
Modern prospecting for mineral resources — 44
BOX:
Turquoise jewelry — 47

CHAPTER THREE
CHANGING THE EARTH

Soil — 56
Influences on soil formation • The soil profile • Types of soil
Erosion — 59
Surface water • Groundwater • Mass wasting • Wind erosion
Glacial change — 77
Glacial formations • Glacial deposits • Continental glaciers
Sedimentation — 84
Sedimentary rocks
Fossil fuels — 87
Summary — 90
Review questions — 91

IN FOCUS:
The geologist as detective — 60
BOX:
The day the dam burst — 67

CHAPTER FOUR
MOUNTAIN BUILDING

A Look at mountains	94
Uplifting	95
The theory of isostasy • Evidence from gravity	
Faulting	100
Types of faults • Fault block mountains	
Folding	104
Recumbent folding • A closer look at the Appalachians	
Dome landforms	111
The Black Hills • Other kinds of domes	
Summary	116
Review questions	117

IN FOCUS: The mountain climbing experience — 108

BOX: Radioactive waste disposal: can we close Pandora's box? — 115

CHAPTER FIVE
THE EARTH'S INTERIOR AND EARTHQUAKES

The earth's interior	120
Density and volume of the earth • Heat	
Seismology	121
The seismograph	
The core of the earth	126
Theories of core composition	
The crust and the mantle	128
Temperature and pressure in the mantle • Upper mantle and crust	
The magnetism of the earth	132
Reversing magnetic poles	
The earth moves	134
What causes an earthquake? • Where do earthquakes occur? • The elastic rebound theory • Measuring and recording earthquakes • Where was the earthquake?	
Some famous earthquakes	140
Can earthquakes be predicted? • The Chinese example The Guatemalan example • The California example	
Summary	150
Review Questions	151

IN FOCUS: Can the next California quake be predicted? — 144

BOX: Experiencing an earthquake — 143

CHAPTER SIX
VOLCANOES AND IGNEOUS ACTIVITY

Magma formation	154
Why do rocks melt? • Composition of magma • Magmatic differentiation	
Intrusive rock masses	156
Dikes	
What is a volcano?	157
Volcanic eruptions • Fissure eruptions	
Studying volcanoes	164
Volcanoes and earthquake activity	
Other igneous activity	166
Summary	168
Review questions	169

IN FOCUS:	Volcano watch: predicting the next blow-up	162
BOX:	Tapping geothermal energy	155

CHAPTER SEVEN
THE CASE FOR CONTINENTAL DRIFT

The earth's crust in motion — 173
Mountain formation

A theory: continents are adrift — 174
The debate over continental drift • Continental drift and geologic features • More drift evidence from ice action • The crust: new levels of understanding • How the ocean floor grows • Underwater data: key to new concepts • The ocean floor dips into trenches • The convection cell hypothesis

The earth as a giant magnet — 191
The wandering poles • Magnetic records on the ocean floor

Summary — 196
Review questions — 197

IN FOCUS:	Drifting continents, changing climates	180
BOX:	Antarctic coal: can we get at it?	189

CHAPTER EIGHT
PLATE TECTONICS

A new view of the world — 200
Moving plates at the surface • Birth and death of the lithosphere • Plate features in the Pacific Ocean • Transform faults • Where plates come into existence • Island arcs • Raw material for plates • A collision that added to a continent

Evidence from the ocean floor — 212
Submersibles supply firsthand evidence • Sounding the depths • Bringing the bottom to the top • More tools for probing the sea floor

What is ahead in plate tectonics — 219
What new knowledge will yield

Summary — 224
Review questions — 225

IN FOCUS:	The ring of fire	220
BOX:	Surtsey: the birth of an island	209

CHAPTER NINE
THE ROCK RECORD

The problem of chronology — 228
The rock controversy • The geological column • Naming the rocks

Fossil evidence — 233
How fossils are formed • Where fossils are found

Dating the earth — 236
Calling cards from the past • Nuclear dating • Radiocarbon dating • Absolute and relative dating

The age of the earth — 244
Summary — 246
Review questions — 247

IN FOCUS:	Reconstructing the past from human coprolites	240
BOX:	Our dwindling fossil fuels	235

GEOLOGIC CHART History of the earth — 248

CHAPTER TEN
THE EARTH'S HISTORY

Before the beginning — 252
The beginnings of life • An open-call audition • The play's the thing

Act I: the Paleozoic Era — 254
Scene i.—Cambrian Period • Scene ii.—Ordovician Period • Scene iii.—Silurian Period • Scene iv.—Devonian Period • Scene v.—Carboniferous Period • Scene vi.—Permian Period

Act II: the Mesozoic Era — 259
Scene i.—Triassic Period • Scene ii.—Jurassic Period Scene iii.—Cretaceous Period

Act III: the Cenozoic Era — 266
Scene i.—Tertiary Period • Scene ii.—Quarternary Period • Epilogue: a review of the play

Summary — 272
Review questions — 273

IN FOCUS:
The hot-blooded dinosaurs — 262
BOX:
The unrepeatable Carboniferous Period — 267

CHAPTER ELEVEN
THE ATMOSPHERE

The nature of the atmosphere — 278
A capsule look at the atmosphere • Atmospheric gases

How gases behave — 281
Temperature and pressure • Kinetic theory of gases

The pressure of air — 285
The barometer • Pressure and altitude

Solar energy — 289
The nature of solar radiation • The energy budget of the atmosphere • Energy at work: the Tropical Zone • World use of solar energy

The structure of the atmosphere — 294
The troposphere • The stratosphere • The mesosphere • The thermosphere

The spinning earth — 298
The tropical heat belt • Angular velocity and linear velocity • The Coriolis effect

Summary — 304
Review questions — 305

IN FOCUS:
Is a new ice age on the way? — 296
BOX:
Are we destroying the ozone layer? — 301

CHAPTER TWELVE
WIND AND WEATHER

Water in the air — 308
Evaporation • Humidity • Condensation • Clouds and fog • Precipitation

Air masses — 313
The polar air cell • The battle between the air masses

Some interesting weather — 316
Thunderstorms • The New England northeaster • A Texas norther

Air pressures and winds — 319
Highs and lows of North America

	Storms	**321**
	Hurricanes—small-diameter cyclones • Tornadoes	
	Predicting the weather	**328**
	Analyzing the weather	
	Summary	**334**
	Review questions	**335**
IN FOCUS:	Killer tornadoes: the American nightmare	326
BOX:	"Eighteen hundred and froze to death"	329

CHAPTER THIRTEEN
THE OCEANS

	Structure of the ocean	**340**
	Layers of the ocean • Ocean temperatures • Salinity	
	Ocean currents	**343**
	Gravity and density • The force of wind • Deep circulation	
	Waves	**350**
	Kinds of waves • Causes of waves • Waves and storms • Tsunami	
	Tides	**361**
	Kinds of tides	
	Summary	**364**
	Review questions	**365**
IN FOCUS:	Exploring the last frontier	356
BOX:	Harnessing tides for power	359

CHAPTER FOURTEEN
OCEAN AND LAND

	The ocean bottom	**368**
	Ocean basins • Ridges and trenches	
	The continental margin	**376**
	Composition • Geographic features	
	The coastal plain	**380**
	Formation • Submergent coasts • Emergent coasts	
	The beach	**385**
	Wave and current action	
	Summary	**392**
	Review questions	**393**
IN FOCUS:	The coral reef: living geology	374
BOX:	The changing seashore	383

CHAPTER FIFTEEN
THE SOLAR SYSTEM

	An overview of the solar system	**398**
	Astronomical measurements	
	The sun	**402**
	General properties of the sun • The outer layers of the sun • The sun's rotation • Solar activity • Other surface phenomena	
	Planets and their orbits	**408**
	Bode's law • The Roche limit	
	The terrestrial planets	**412**
	Mercury • Venus • Mars • Pluto	
	The Jovian planets	**421**
	Jupiter • Saturn • Uranus • Neptune	
	The moon	**425**
	The motion of the moon • Phases of the moon • Eclipses	

• The moon's missing atmosphere • The lunar surface • Lunar geology • The interior of the moon • The origin of the moon

Visitors from space — 434
Meteors

Summary — 438
Review questions — 439

IN FOCUS BOX:
Solar activity and the earth — 410
The atmosphere of Venus — 415

CHAPTER SIXTEEN
THE UNIVERSE AND BEYOND

Observing stars — 442
Brightness • Measuring stellar distances • Luminosity • Colors of stars • Spectra of stars • Masses of stars • Putting it all together

Stellar structure and evolution — 451
The energy of stars • Source of the sun's energy • Formation of stars • Red giants • White dwarfs • Neutron stars, pulsars, and supernovas • A more extreme ending

Galaxies — 459
Our galaxy • The Doppler effect • Radio astronomy • Interstellar matter • External galaxies • Clusters of galaxies • Galactic violence

Cosmology — 476
Surveying the universe • Some cosmological ideas • The big bang theory • The steady state theory • The discovery of quasars • The 3K cosmic radiation • The ultimate question

Summary — 486
Review questions — 487

IN FOCUS BOX:
Hello out there! — 466
Keep away from black holes! — 458

CHAPTER SEVENTEEN
CURRENT ISSUES

A look at ecology — 492
Energy — 493
Energy pathways

Pollution — 496
Air pollution • Groundwater pollution • Ocean pollution • Thermal pollution • Light pollution • Radioactive wastes: a special problem

Resources — 503
Mineral resources • Biogeochemical resources • Conservation • Energy and economy • Energy resources • Energy in the future

Summary — 512
Review questions — 513

IN FOCUS BOX:
Is our technology destroying us? — 504
Turning garbage into fuel — 507

TERMS AND CONCEPTS — 514
INDEX — 520
CREDITS — 529

PREFACE

This is a particularly interesting time for earth science. It is now twenty years since the International Geophysical Year, 1957–1958, when nations cooperated with each other in a detailed and extensive study of the earth. Since that year several crucial discoveries have transformed our approach to understanding this planet. The way in which the seafloor spreads, the way in which continents drift apart, and the similarities and differences between the earth-like planets were all topics at quite a rudimentary stage even a few years ago. In addition, the impact of human civilization on the earth was barely understood.

In recent years we have begun to understand how plate tectonics changes the face of the earth. We know a great deal about severe storms, and ways are even being devised to prevent them. Spacecraft have visited all the planets up to Jupiter, and Saturn is due for a visit in 1979. Information gathered by these spacecraft has helped us to learn what features are common to the earth and other planets, and what features are unique to our world. This knowledge has led, in turn, to a greater appreciation of the complex factors that make the earth what it is.

In this book we have tried to interpret the vast amount of new material that has become available, in a way that can be understood by the average college student. We have attempted to be rigorous but readable—to make students aware of the quantitive nature of earth science but to spare them some of its mathematical complexities. To further aid students' understanding, we have made a special effort to acquire recent and striking photographs and to tie these photographs directly to the text.

Earth science has traditionally been separated into geology, meteorology, and oceanography, with astronomy providing a backdrop against which to examine the special nature of the earth. We have retained these divisions in our book. But if anything has been learned in recent years it is the unity of all earth studies. Each of the seemingly separate disciplines when pursued in depth leads to the others. Who would have suspected, forty years ago, that understanding the processes in the interiors of hot stars would tell us why the earth's composition is what it is? In this book the intricate interplay between the disciplines constituting earth science has been especially stressed.

We have also emphasised how earth science enters into our day-to-day lives. The effects of human beings on the environment and of the environment on human beings have been noted in every chapter. We have described, wherever possible, the steps we have taken to protect ourselves against the terrors of nature such as tornadoes, earthquakes, and volcanoes. We have also described the terrible, often irreversible, changes produced in the environment by uncontrolled technology, and its accompanying pollution.

Part I: Introduction. This chapter gives the student a general overview of earth science—its nature, branches, history, and place among other sciences. The methods of scientific inquiry, including how an investigator attacks a problem and how phenomena are measured, are also explored. This, in turn, leads to a discussion of the fundamental physical laws which govern the earth, the solar system, and the universe itself.

Part II: Geology. This section, made up of nine chapters, opens with an examination of earth materials. It begins with atomic structure and the chemical properties of atoms, and then moves on to a discussion of the nature and characteristics of minerals and rocks. Next follows an investigation of soil and the agents that shape the face of the earth—erosion, glacial change, and sedimentation. The processes connected with mountain building are discussed, followed by an examination of earthquakes and volcanic activity. Two entire chapters are devoted to continental drift and plate tectonics, covering the latest developments in this rapidly growing specialty. The final two chapters are concerned with earth history; the topics analyzed include chronology, fossils, and techniques of dating the earth, plus a review of the major periods in the planet's development.

Part III: Meteorology. The two chapters in this section deal with the earth's atmosphere and weather. The nature and structure of the atmosphere, and the laws that govern the behavior of atmospheric gases, are covered in full. Also examined are the elements that make up the weather—humidity, condensation and precipitation, clouds, air masses, and global air currents.

Part IV: Oceanography. The two chapters in this section are concerned with the composition and structure of the earth's ocean, and with the interactions between ocean and land. The topography of the ocean floor is discussed in depth. Next follows an examination of the major ocean currents and the forces that move them. Special attention is given to new methods of investigation and exploration in the field.

Part V: Astronomy. The two chapters in this section deal with the solar system and the universe. Beginning close to home with a discussion of the sun and the moon, the first chapter moves outward to a consideration of the other planets, with a close-up look at the possibility of life on Mars. The second chapter probes deeper into space, exploring the nature, structure, and evolution of the stars and galaxies. The chapter concludes with a discussion of current theories regarding the origin and possible end of the universe.

Part VI: The Environment. This final section deals with ecology—a topic that is becoming increasingly important in the modern world and is of particular interest to young people. The chapter discusses the nature of environments, ecosystems, and the laws that govern them, as well as the various problems created by the impact of human activity on the environment. The book ends, appropriately enough, with a look to the future—specifically, to future methods of meeting human energy needs.

Features
1. *In Focus.* Each chapter contains a special two-page feature designed to catch the student's attention and focus it on some intriguing aspect

of the earth sciences. Written in journalistic style, some of the features zero in on the latest theories and controversies in the field. Others are presented because the subject matter is attractive to the beginning student. Among the feature topics are: Is another ice age on the way? Can scientists predict earthquakes and volcanic eruptions? Were the dinosaurs cold-blooded—or hot-blooded? How do oceanographers plumb the ocean depths? Is our civilized way of life destroying us?

2. *Boxes*. In addition to the "In Focus" features, each chapter includes a briefer feature which highlights some interesting aspect of a topic related to the chapter. For example, the box in Chapter One looks at "the usefulness of oddballs"—specifically, at an organization which believes that the earth is flat. The chapter on earthquakes includes an intriguing box that tells what it is like to be caught in the middle of an earthquake. And the final chapter on environment contains a box describing how some cities are generating extra power by burning waste as a fuel.

3. *Summaries and Review Questions*. At the end of each chapter is a succinct summation of its principal ideas, designed to refresh the student's memory. In conjunction with this important study aid are review questions which test the student's recall and comprehension of the material.

4. *Glossary Keyed to the Text*. At the end of the book is an extensive glossary which not only provides the student with a concise definition of the key terms presented in the book, but also cites the chapter in which each term is first introduced.

5. *Color Portfolios*. Four portfolios are a special feature of the book, each containing eight pages of color photographs. The photos were chosen both for their visual impact and for the way they illustrate important aspects of the earth sciences. The portfolio devoted to astronomy is particularly noteworthy, and is designed to impart some of the immediacy of this highly visual discipline. Photographs of Mars taken by the recent Viking mission are nothing short of breathtaking.

Acknowledgements. It is impossible to cite by name all the persons whose labors are essential to creating a book of this scope. Appreciation is extended to all those who contributed their talents to various parts of the manuscript, reviewed and criticized chapters, and attended to the thousand details of art and production that make the difference between an inviting and a dreary-looking final product. We would especially like to thank the reviewers whose names are listed on the copyright page. Their criticisms and suggestions were invaluable to us in the preparation of the final manuscript. They are not, however, responsible for any errors of fact or interpretation that may remain. Special thanks also go to Sharon Rule (Project Editor); Dilip Balamore (Managing Editor); Sarah M. Parker (Assistant Editor); Daniel Liberatore (Editor for feature material); Eileen Ann Max (Production Director); and Ellen Klugherz (Photo Researcher). The art work was done by SoHo Studios Inc., and by Gloria Priam, whose meticulous and imaginative solutions to the problems that arose proved invaluable to the successful completion of the book.

PART 1

INTRODUCTION

CHAPTER ONE

THIS PLANET EARTH

"Daddy, where did the world come from?" "Well, once upon a time there was only the great sea, the goddess Tiamat, and Apsu her mate. And from Tiamat and Apsu were born all the other gods. But the gods disturbed Tiamat's rest with their noise and feasting, so she sought to destroy them. Then Marduk, the son of Ea, the greatest of the gods, rose up and struck Tiamat. With his mighty sword he killed her, dividing her body in two. From half of Tiamat's body Marduk made the heavens, and from half he made the earth. From her eyes he made to flow the great rivers, the Tigris and the Euphrates. Then on the earth Marduk placed trees and green plants and all beasts that creep and run and fly; in the waters he placed fish and all strange creatures of the deep. And he set men in the earth to till it, to serve the gods and honor them with sacrifices and offerings and songs." Some such answer you might have gotten to your question if you were a Babylonian child of the second millenium B. C. On the other hand, if you were a Dogon of West Africa you would be told about the cosmic egg, and the human being inside it who spoiled the creator's plan for a perfect world by breaking out of the shell too soon. And if you were an Iroquois Indian you might learn of the woman sent from heaven who conceived by a great mud turtle and bore a daughter. The daughter in turn bore two sons, one good and one evil, who together with their grandmother created the earth.

What all this goes to show is that people have been asking about the origins of the earth for a very long time. Where did it come from? How was it made? Why is it plagued with imperfections? Every tribe and culture has asked these questions. Usually their answers were not meant to be taken literally. They tell of things that happened in a mythic time, not a historic past. Their purpose is more ethical than scientific: "What are we human beings here for and what are we

this planet earth

supposed to be doing with the world we live in?"

It was left to the nosy Westerners to start asking literal-minded questions about how the earth came to be and—even more typically Western—when. A natural place to look for information was their own creation story—the Genesis story in the Jewish–Christian Bible. So, in 1654, the scholarly Anglican Archbishop James Ussher, of Armagh in Ireland, sat down with the book of Genesis to make some very careful calculations. When he had finished, he announced the precise year in which the world had been created—4004 B.C. A later scholar refined the date even further: the creation had taken place on October 26 at 9 o'clock in the morning.

Well, the Archbishop was off by about four and a half billion years, according to modern estimates. But we need not laugh at him, because he was doing what scientists at their best have always done—taking the best information he had available and using it to help him reason his way to new insights, new conclusions, and perhaps a new round of fruitful questions. Today's scientists work with the same basic tools—factual data, intellectual honesty, and the power of logical thinking.

the study of earth science

What we know about the earth can be divided into three closely related sciences: **geology,** the study of the solid earth; **oceanography,** the study of the oceans; and **meteorology,** the study of the earth's atmosphere and weather patterns. Since the earth is a part of the universe, its study also involves the science of **astronomy.**

Natural forces are responsible for much of the earth's most spectacular scenery. This is a view of Glacier National Park, a popular tourist attraction.

this planet earth
why study the earth?

Let us look at some of the practical contributions made by the various fields of earth science. Geology is the branch about which we know most. Most of our knowledge has come from study of the thin outer layer or crust of the earth, a great deal of which can be observed directly. Our understanding of the rest of the 6,400 kilometers (4,000 miles) down to the center of the earth is based on indirect observations. By combining the two, geologists are gaining a clear picture of the earth's internal structure and composition. Among the most important of their contributions is the growing ability to predict earthquakes and volcanic eruptions.

Meteorologists, too, make important practical contributions. Their understanding of what happens to the atmosphere as it is heated or cooled is the basis for current methods of weather prediction. Today, for example, the approach of a severe storm — hurricane, typhoon, or tornado — can often be predicted in advance. Only a few years ago, such storms usually occurred with no warning whatsoever.

Oceanographers have given us a much better understanding of the interrelationship among the seas, the land, and the earth's atmosphere. Slowly, we are learning how to make use of this knowl-

At left, artificial mineral crystals created in the laboratory. At right, satellite photos show the growth and movement of a hurricane over a period of three days.

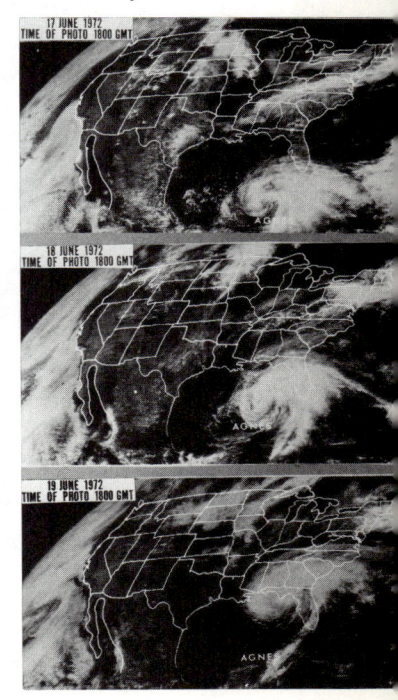

this planet earth

edge. For example, many new food products are now being obtained from the sea, and are providing food for the populations of both industrial and developing nations. Soon, it is hoped, we will develop new and economical methods for recovering valuable mineral resources from the ocean floor.

Although astronomy is the study of the heavens rather than of the earth, here, too, important contributions have been made. A correlation can now be seen between periods of solar activity and long-term changes in the earth's climate. Soon we may be able to predict coming years of drought or of unusual cold, a valuable aid to agricultural planning. Artificial satellites now orbiting the earth are equipped to monitor changes in the content of the atmosphere, such as those which arise from industrial pollution of the air.

the methods of science

Scientists learn by making observations, studying them, and drawing general conclusions. The training of a scientist includes a great many things that will sharpen powers of observation, develop a logical approach to problems, combine objectivity with a sense of caution, and give each individual an appreciation for the background of his or her own particular field. Contrary to much popular misinformation about scientists and their work, philosophical consideration of a problem is just as important as tinkering in the laboratory.

The study of earth science, like that of all the other sciences, is conducted according to the **scientific method**—an awesome-sounding concept, but one which has a beautiful simplicity. Reasoning according to the scientific method is almost the same as any other sort of logical reasoning, in which we move step by step until we reach an objective conclusion. The only difference is that scientific reasoning involves an unusual amount of attention to the accuracy of each step. Step A and Step B, for example, may be correct; but if Step C is wrong, then Step D and all the steps to follow will also be wrong.

Scientific study of a given phenomenon begins with making observations about it and then drawing a general conclusion. Once scientists have studied a phenomenon thoroughly and believe they fully understand it, they try to discover the underlying law that governs it. Some of the most complex phenomena in nature can often be explained in terms of a simple law.

Once a law is formulated, its predictions must be tested. Often the testing can be done under laboratory conditions. If not, the experimenters must make many more observations and tests of the phenomenon under natural conditions. An idea that stands the test

investigating a problem

This satellite is used to identify possible locations of valuable natural resources.

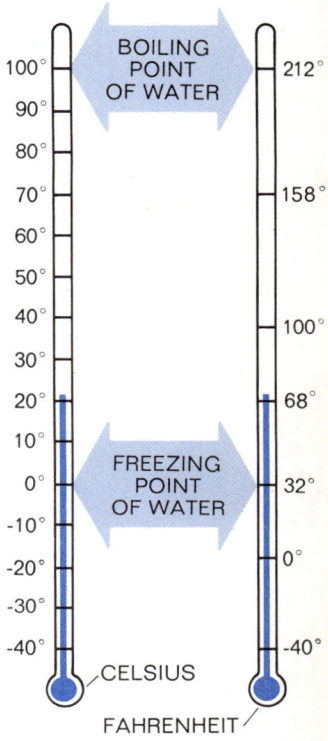

this planet earth

At left, astronomers examine a meteorite that fell in Oklahoma in 1970. At right, a comparison between the Celsius and Fahrenheit temperature scales.

of laboratory experiments or further observation becomes a working **hypothesis,** and serves as a tool for further exploration of related problems. If the hypothesis later proves to be universally applicable, it becomes known as a **law.**

Throughout history many systems of measurement have been devised. Today two systems are in general use. The United States and certain African nations use the **English system,** based on units such as the foot, pound, and quart. The **metric system,** or **International System of Units,** is used in most other countries, and is now being adopted by the United States. Scientific work in all countries is carried out in the metric system, which is based on units such as the meter, gram, and liter.

In this book our emphasis is on the international system. When the measurement is commonly made in either system, we may give both measurements. Table 1-1 shows how to convert measurements in the English system into their international equivalents, and vice versa. As you can see from the table, one of the simplicities of the international system is that any unit is larger or smaller than the next by a multiple of 10.

systems of measurement

this planet earth

Units of length	
1 centimeter	10 millimeters
1 meter	100 centimeters
1 kilometer	1,000 meters

Units of volume	
1 liter	1,000 milliliters
1 milliliter	1 cubic centimeter

Units of mass	
1 gram	1,000 milligrams
1 kilogram	1,000 grams

| Units of conversion ||
English system	International system
0.39 inch	1.0 centimeter
1.0 inch	2.54 centimeters
39.37 inches	1.0 meter
0.62 mile	1.0 kilometer
1.0 mile	1.6 kilometers
1.0 quart	0.94 liter
1.06 quarts	1.0 liter
1.0 pound	454.0 grams
1.0 pound	0.45 kilogram
2.2 pounds	1.0 kilogram

Table 1-1 Metric conversions

temperature scales

The temperature of an object can be expressed in several ways. In the United States we usually speak of temperature in degrees **Fahrenheit**. According to this scale, normal body temperature is 98.6°F, pure water freezes at 32°F and boils at 212°F, and so on. In Europe, and in an increasing number of other places, the **Celsius** (or centigrade) scale is used. In this system, pure water freezes at 0°C and boils at 100°C. The Celsius scale is generally used in scientific investigations.

earth science in history

Although science attempts to be entirely objective, its progress has sometimes been greatly modified, and even impeded, by the personalities of investigators. As we begin our study of earth science it is appropriate to discuss some of the people who made significant contributions.

Aristotle

Aristotle, the well-known Greek philosopher who lived during the fourth century B.C., did much to advance our understanding of the earth. However, all of his hypotheses were arrived at simply by thinking along philosophical lines; he did not test any of them by experimentation. Not all of his teachings were correct, but his reputation as a teacher was so great that many of his followers later refused to admit that he might have made some mistakes. The end result was that in some cases Aristotle's incorrect explanations of natural phenomena retarded earth science more than they advanced it.

Werner

Abraham Werner was a great teacher of the eighteenth century, the last of those who used Aristotelian methods. Like Aristotle, however, he did not test his hypotheses. In the course of his teaching he made many statements about the earth's structure that were based on theories he had neither tested nor proved. For instance, he taught that the earth was once covered entirely by water, from which rocks were supposedly formed, producing mountains, valleys, and plains. Then something happened—Werner did not say just what—which caused the water to drain away, leaving continents, islands, and oceans.

Even though Werner had little or no evidence for his theory, his students accepted it as though it had been proved. In those days teachers, whether good or bad, were regarded as unquestionable authorities. No one dared to ask Werner where the water went when it drained off the earth, or how water could change into rocks—nor did Werner ever offer any explanation for these points.

Hutton

Among the few people who doubted Werner's teachings was a Scotsman named James Hutton. As a trained physician who later became a farmer, Hutton was in close contact with soil and rocks, so it was logical for him to wonder how they had been formed. Unlike Werner, he arrived at his ideas through observation. He watched the action of streams and ocean waves, and became interested in the way sand and pebbles were carried by moving water. As a result of these observations, Hutton began to arrive at some new ideas about how valleys and beaches were formed.

In 1795, at the urging of friends with whom he had discussed his theories, Hutton published a book entitled *A Theory of the Earth, with Proofs and Illustrations*. As the title implies, Hutton's theories were based on actual physical evidence. And they were so different from Werner's teachings that the publication of his book started a controversy that divided the scientists of that time into two groups.

When we compare Werner and Hutton, we see that it is very difficult to pin down Werner's opinions. Since he published very little, we must depend on what his students published. How much of this is pure Werner and how much Werner-plus-student, no one can be sure. Hutton, on the other hand, set down this ideas in very definite form.

Hutton's methods

Hutton believed that observations of what is happening in

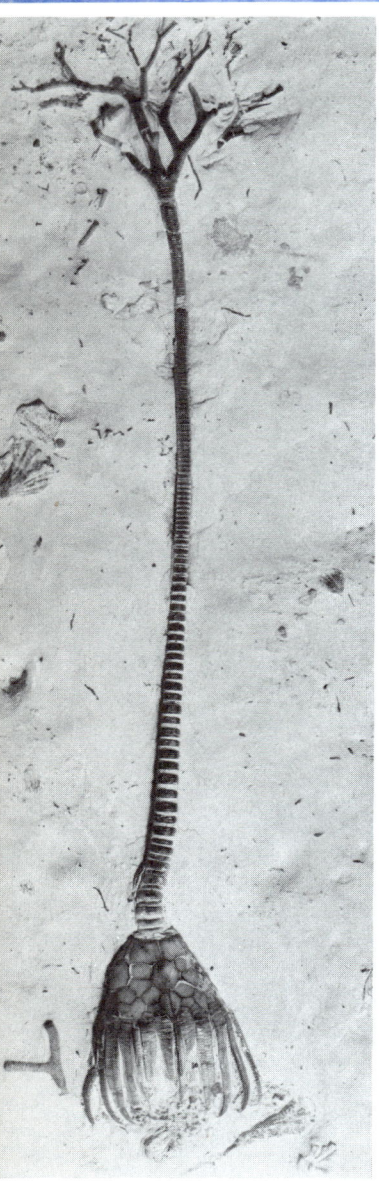

this planet earth

the present should be applied to what happened in the past. This basic idea that Hutton contributed is known as the principle of **uniformitarianism.** The philosophy of uniformitarianism can be stated very simply: "The present is the key to the past." In essence, this says that the laws that govern the physical world today are those that also governed it in the past.

As we have stated, Hutton's theories were based on direct observation rather than pure philosophizing. Take, for example, his discovery that land moves up and down. In a shallow bay, he observed the trunks of dead trees standing upright under the water. It was obvious to him that the land must have sunk in relation to the sea. In other places, far above the present water line, Hutton found seashells in rocks. Again, he reasoned that these must be the remains of animals that had lived in the sea; the land here must have risen.

Hutton's conclusion was that the land was not always stationary, but that it could move up and down in relation to the sea. When it rose, rocks that had originally been formed at the ocean bottom were lifted above the sea. In these rocks were the shells of sea animals that had been buried in sediments, which later were changed to sedimentary rocks before the land rose.

Hutton's theory that rocks were formed from sand and mud on the ocean floor gave a big boost to biological investigations of the time. For years, curious objects had been found in rocks— objects that looked like seashells, animal bones, and pieces of plants. Werner's followers believed that these were simply odd rock formations. Those who accepted Hutton's principle of uniformitarianism, however, agreed with him that they must be the remains of real animals and plants. Because these objects were dug out of rocks,

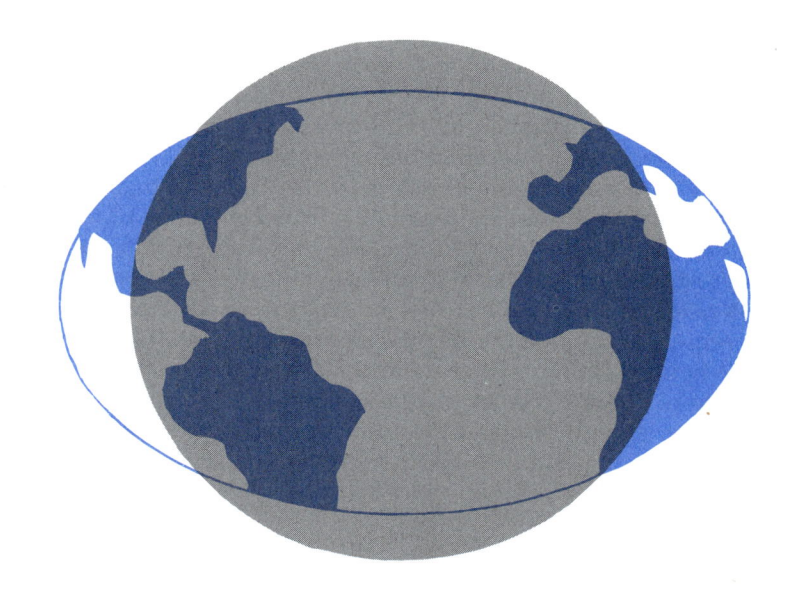

Above, the fossil of a crinoid or sea lily, preserved in sedimentary rock. Hutton observed many fossils like this one. At right, the shape of the earth. It is an oblate spheroid rather than a perfect sphere.

they became known as **fossils,** from the Latin *fossilis* meaning "dug up." Later, it was discovered that fossils could be used to determine the age of the rocks and to trace the evolution of life forms.

describing our planet

Perhaps the best way to begin our study of the earth is to examine it as a whole. Most people, of course, know that the earth is spherical. Its diameter is approximately 12,700 kilometers (8,000 miles). For some special purposes, however, it is not quite accurate to call the earth a sphere. Its true shape is quite complicated, because its interior is not uniform.

The exact shape of the earth is the shape of the sea level, not involving the shapes of mountains, valleys, or other surface features. This shape is called an **oblate spheroid.** (Oblate simply means flattened.) We can construct an oblate spheroid by taking an ellipse and rotating it around its shorter axis.

About 29 percent of the earth's surface is covered by the continents, averaging 840 meters (2,750 feet) above sea level. The rest of the earth is covered by the oceans, which have an average depth of 3,810 meters (12,500 feet). The polar radius of the earth is 6,357 kilometers (km), and the equatorial radius is 6,387 km. Thus we see that the earth bulges slightly at the equator.

In our study of the earth we must have a set of references that will enable us to locate a specific spot on the earth's surface. The early Greek map makers, living about 400 B.C., solved this problem by dividing their maps into many small rectangles, using one set of parallel lines running east and west and a second set running north and south. The Greeks, like other civilizations of the time, believed that the earth was flat, and also that it was wider from east to west than from north to south. The east–west lines were called **latitude,** and the north–south lines were called **longitude.** Today these names are still in use, although we are applying them to a spherical earth.

Latitude and longitude are measured in number of **degrees** away from a reference line. Imagine that you are standing at the center of the earth, facing directly toward the equator. The angle by which you must raise or lower your eyes to look at a place above or below the equator is the latitude of that place. Latitude lines therefore indicate angles north and south of the equator. Since the lines are all parallel to the equator, they are called the **parallels of latitude.** The parallels are numbered from 0° at the equator to 90° north (N) at the North Pole and 90° south (S) at the South Pole.

Our reference for longitude is a line called the **prime meridian,**

earth's addresses

Parallels of latitude run east and west. These imaginary lines represent angular distances north and south of the equator.

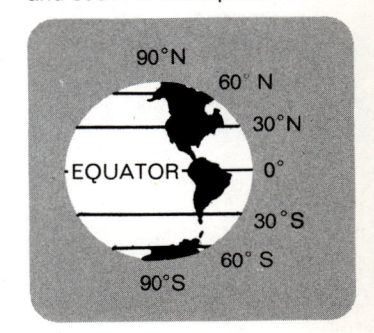

this planet earth

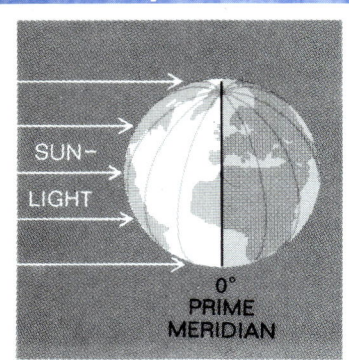

Lines of longitude, unlike parallels of latitude, run north and south from the Poles and are perpendicular to the equator.

which passes through the Royal Astronomical Observatory in Greenwich, England. From your vantage point at the center of the earth, suppose you now face directly toward the prime meridian. The angle at which you must turn your head to look at another place is the longitude of that place.

Each line of longitude, called a meridian, passes through the North and South Poles and is perpendicular to the equator. Meridians indicate positions east and west of Greenwich. They are not parallel to each other, but converge as they approach the poles. Each meridian is directly under the sun when it is noon at that longitude. (In fact, the word "meridian" comes from the Latin word for "midday.") The meridians are numbered in degrees east (E) or west (W) from 0° at Greenwich to 180° in the Pacific Ocean.

Each degree is divided into 60 minutes, written as 60'. A minute is further subdivided into 60 seconds, written as 60''. By using the number of degrees from the equator and from the prime meridian, any place on the earth's surface can be located and given a precise "address" or geographical coordinate.

determining latitude and longitude

We can find the latitude of any location by observing the stars. In the Northern Hemisphere, Polaris—the North Star at the end of the "handle" of the Little Dipper—is always almost directly over the North Pole. We can estimate our latitude by measuring the angle formed by the North Star, our own position, and the horizon.

Getting a fair determination of longitude requires an accurate clock or chronometer. The chronometer shows the time at the prime meridian, called **Greenwich Mean Time (GMT)** or **Universal Time (UT).** First we note the GMT for the instant the sun crosses the meridian we are upon. The difference between GMT and our time allows us to calculate the longitude of our meridian. For example: at noon local time, when the sun crosses our unknown longitude, the GMT or prime meridian time reads 1 P.M. The difference in time between the prime meridian and our longitude is therefore one hour. Since the sun appears to travel 15 degrees west every hour, our longitude is thus 15 degrees west of the prime meridian.

just what time is it?

Problems can arise if the local time is linked directly to the passage of the sun. Suppose we use the sun's crossing of a meridian as the signal for midday, twelve noon. Practically every city and town in the United States would have a different time showing on its clocks—and in fact, this actually was the case only a hundred years ago.

With every city and town going on its own time, no great harm was done so long as the fastest way to travel was by horse and buggy. But when railroads began crossing the country, and railway schedules were necessary, the situation became very confused. Unless some way could be found to standardize clock time within a giv-

this planet earth

en region, it would be impossible to say just when a train would arrive at a given place. In 1880, therefore, the United States was divided into four **standard time zones,** each about 15 degrees of longitude wide. Within each zone, all clocks were set to exactly the same time, or **civil time.** Since each 15 degrees of longitude represents one hour, travelers moving between time zones thus had to change their watches to conform to the civil time for the new zone.

Clock or civil time moves from east to west. In the accompanying figure, we see the arrangement of meridians as viewed from above the North Pole. Notice that when it is noon Eastern Standard Time on the 75th west meridian (75°W longitude) it is one hour earlier, 11 A.M. Central Standard Time, on the 90th west meridian (90°W longitude). In one hour it will be 1 P.M. Eastern Standard Time and noon Central Standard Time. That is, the sun will appear to have moved 15 degrees westward.

For the long-distance traveler, all this shifting of times can raise an interesting problem. As I move west from, say, Chicago, I set my watch back one hour each time I pass into a new time zone. But suppose I go westward all the way around the earth. By the time I get back to Chicago, I will have set my watch back 24 hours. In effect, I will have "lost" an entire day. If Chicago has experienced 20 days while I was gone, I will have experienced only 19. And if I told my family in Chicago to meet me at the airport on October 2, I may find that they were out there looking for me the day before I arrived. If, on the other hand, I travel eastward, the reverse will happen. I will "gain" an entire day, experiencing 21 days while the folks at home go through 20.

One way to solve such problems would be to agree on some system of automatically moving one day ahead on the calendar as we go around the earth toward the west, and one day back as we go

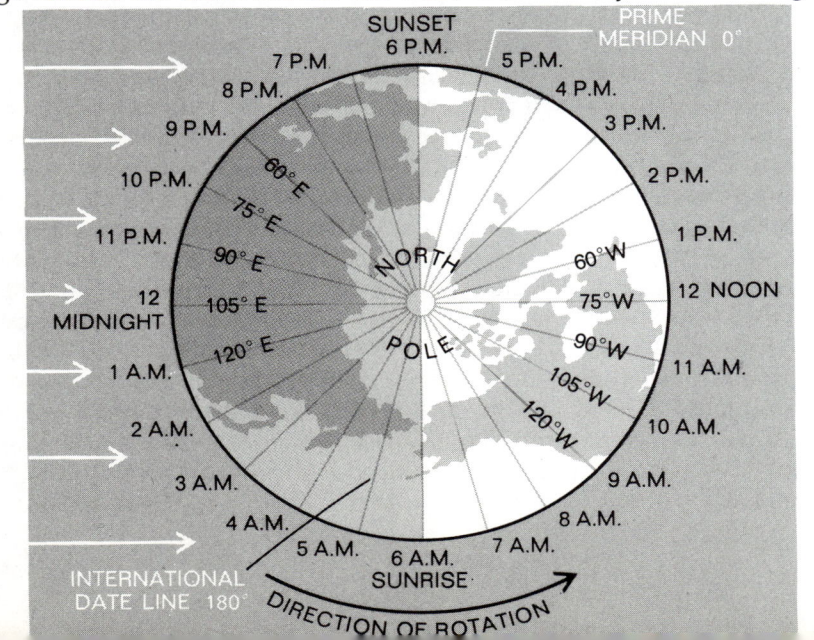

The standard times zones of the earth. Both the day and the date change at the International Date Line.

If it is Tuesday the 15th of July in Alaska, what are the day and the date in Japan?

IN FOCUS:
EARTH SCIENCE: WHERE LAB AND FIELD WORK MEET

If you are contemplating a career choice and are drawn to ivy-towered scholarly pursuits or the rarified atmosphere of remote labs and libraries, earth science is probably not for you. The various branches of earth science are unique in that they require unusually close contact between the investigator and the subject under investigation. Danger, in fact, is often part of the earth scientist's work.

Earth science uses concepts and tools from many disciplines. These include astronomy, cosmology, biological sciences, social sciences such as anthropology, and even many aspects of history and geography. But many careers also involve a number of "shirt-sleeve" skills, both basic and advanced. Let's take a look at some of the possibilities.

If you enjoy working outdoors under a wide variety of conditions, are reasonably strong, and can drive a truck, you might become a geological observer—helper. Your duties would include driving an equipment truck to worksites; unloading dynamite and seismometers and carrying them to designated locations; and connecting electrical cables between the equipment and the truck. Next, you would load shotholes with dynamite, and tamp the charges.

As a seismograph shooter on the same job, you would detonate dynamite set in the shotholes, producing seismic waves that can be recorded and interpreted to reveal subsurface rock formations. You would also be responsible for moving, storing, and maintaining inventories of the explosives, a job that requires a license from the Federal Government. You would, in addition, repair and service electrical wiring and other shooting equipment.

On a different level of work in geology, you might be a mineralogist, developing data and theories about the origin and possible uses of minerals. You would identify and classify samples, examining them microscopically to determine their physical characteristics. Next, you would perform various physical and chemical tests to determine their composition and crystal structure.

If you decide to specialize in studying the topographic features, rocks, and sediments of the ocean bottom, you would be technically classified as a marine geologist or geological oceanographer. Your laboratory might be a submersible research vessel that could descend to a thousand meters or more below the surface.

Is your future in the stars? As an astrophysicist you would observe and interpret celestial phenomena, and relate your research to scientific knowledge in several fields. In fact, you might even become an astronaut. A number of qualified scientists took part in the highly successful 56-day mission of Skylab, and many others will be needed for the upcoming Space Shuttle program.

If your particular interest is in celestial mechanics, you might be concerned with studying the motions of objects in relation to gravitational attraction. Or, as an astrometrist, you would specialize in measuring the coordinates of celestial bodies, to determine where the earth is in relation to the rest of the universe.

Are you a weather buff? Then you might consider a career in meteorology. You might specialize in weather prediction, perhaps being based at one of the severe storm warning centers. Or, you might be more concerned with research and development in the rapidly-growing field of weather modification, conducting experiments in such things as seeding of clouds with silver iodide to produce rain.

It should be clear by now that most of these pursuits involve close contact with earth, land, and sea. For example, the best way to study the internal structure and activity of a hurricane is to get inside the hurricane; this is now being done by scientists who fly specially-constructed hurricane-hunter planes into the eye of the storm, where they can make close observations.

Most earth scientists take such risks in stride. If you are passionately interested in coral reef structure, for instance, and the only way to get a closer look is by diving beneath a reef in scuba gear, dive you will, despite the presence of stinging jellyfish, poisonous sea snakes, or an occasional prowling octopus. If you are interested in cave limestones, a horde of bats that may possibly carry rabies will not deter you from your examination of a particularly interesting stalactite. And if a fascinating sedimentary rock formation happens to lie near the top of an unusually inaccessible mountain, you will find a way to get there to look at it—pack mule, helicopter, or even by crawling on your hands and knees for a hundred yards or so up a sheer mountain face. Earth science, perhaps more than any of the other sciences, is a labor of love.

around to the east. This would not change the number of days travelers experience, but it would at least keep the travelers and the homestayers—or the travelers in opposite directions—at the same place on the calendar, and enable scientists to be sure they are talking about the same day as their colleagues halfway around the earth. Such a system has in fact been established, and the dividing line has been set at the 180th meridian. As we cross this meridian, both the date and the name of the day change. For example, if we leave San Francisco on Monday the 10th and fly west, when we cross the 180th meridian we will suddenly find ourselves in Tuesday the 11th—even if it is only 2 o'clock in the afternoon. But if we leave Japan on Thursday the 13th and fly east, we will cross the meridian into Wednesday the 12th. Hence the 180th meridian is called the **International Date Line.**

duration of day and night

We usually think of the sun as rising in the east and setting in the west. The precise direction of the rising and setting sun, however, varies considerably over the course of a year. In the Northern Hemisphere, for example, the sun rises to the north of east and sets to the north of west, from late March until late September. During this period, the period of daylight is longer than the period of night. For the rest of the year in this hemisphere, the sun rises and sets to the south of east or west. The longest day of the year is about June 21, the **summer solstice,** and the shortest day is around December 21, the **winter solstice.**

the seasons and solar energy

The earth makes one complete orbit around the sun in 365.24 days—one earth year. During that time, we pass through the seasons of winter, spring, summer, and fall. The earth's axis of rotation is tilted, or **inclined,** to the plane of the earth's orbit around the sun. But the north end of the axis points in the same direction all the time—almost directly toward Polaris, the North Star.

On June 21 the Northern Hemisphere is tilted 23.5 degrees toward the sun, as shown in the accompanying figure. This makes the sun appear to be directly overhead at noon at 23.5°N (or 23°30') latitude, the parallel known as the Tropic of Cancer. On December 21, when the earth has moved to the opposite side of its orbit, the Southern Hemisphere is tilted toward the sun, which appears directly overhead at 23.5°S latitude, the Tropic of Capricorn. Thus, the sun appears to travel 47 degrees southward between June 21 and December 21, and 47 degrees northward during the next six months. The seasons of the year are caused by this apparent traveling of the sun and the inclination of the earth's axis.

Summer in the Northern Hemisphere begins on about June 21, when the Northern Hemisphere receives its maximum radiation from the sun. At this time the Southern Hemisphere is receiving its minimum solar radiation, and winter begins there.

On March 21 and September 22 the sun appears at its **zenith,**

this planet earth

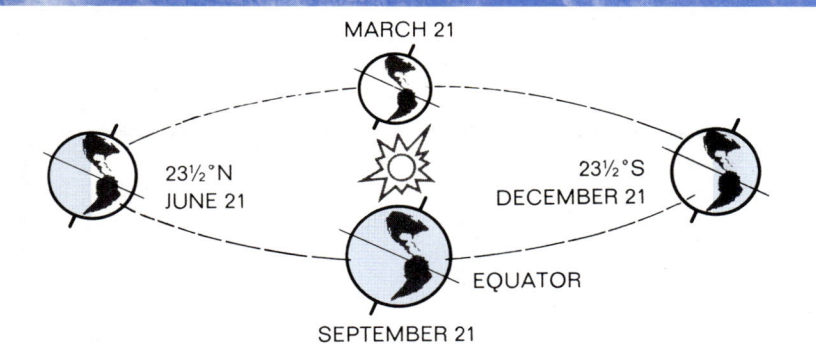

Four positions of the earth in its orbit around the sun. Would there be seasons if the earth's axis were perpendicular to the plane of its orbit?

or directly overhead, at noon over the equator. On these days there are exactly 12 hours of daylight and 12 hours of darkness everywhere in the world. These two days are called the **equinoxes,** a Latin word meaning "equal nights." March 21 is the **vernal** equinox and September 22 is the **autumnal** equinox.

The major difference between summer and winter is the temperature, which, in turn, affects the weather. The amount of solar energy received at a given place on earth is strongly determined by the angle at which the sun's rays strike the earth, and by the duration of sunlight. For example, at noon on June 21 the sun appears as high in the sky as it ever does in the Northern Hemisphere. At a position of 23.5°N, a beam of sunlight one meter square would illuminate exactly one square meter of ground that lies perpendicular to the sun's rays. A flagpole at that point would cast no shadow. On that day the sun shines for almost 13.5 hours. Each square meter of ground would receive 19,200 calories of energy from the sun each minute, if there were no atmosphere to interfere.

Six months later, at noon on December 21, the sun is at zenith at 23.5°S. At 23.5°N, a sunbeam now strikes the ground a full 47 degrees away from the perpendicular. The incoming energy from a one-meter-square sunbeam is thus spread over more than one square meter of area. That area is consequently warmed less than it was by the zenith sun on June 21.

the zones of the earth

During the summer, there is a region in the Northern Hemisphere where the sun sometimes does not set at all. At the North Pole, the sun does not go below the horizon between March 21 and September 22. At the southernmost parallel of this region, there is only one day when the sun does not set: the day of the summer solstice. In between the North Pole and this southernmost parallel, the number of nightless days increases as we travel north.

The regions where, on at least one day of the year, the sun does not set are called the **Polar Zones.** The parallels that mark the edges of the polar zones nearest to the equator are the **Arctic Circle** in the north and the **Antarctic Circle** in the south.

this planet earth

In the **Tropical Zone** on either side of the equator—between the Tropic of Cancer and the Tropic of Capricorn—the sun reaches zenith at noon on at least one day of the year. North and south of the

The temperature zones of the earth. This figure shows the earth's inclination toward the sun on June 21, the beginning of summer in the Northern Hemisphere. What is happening in the Southern Hemisphere?

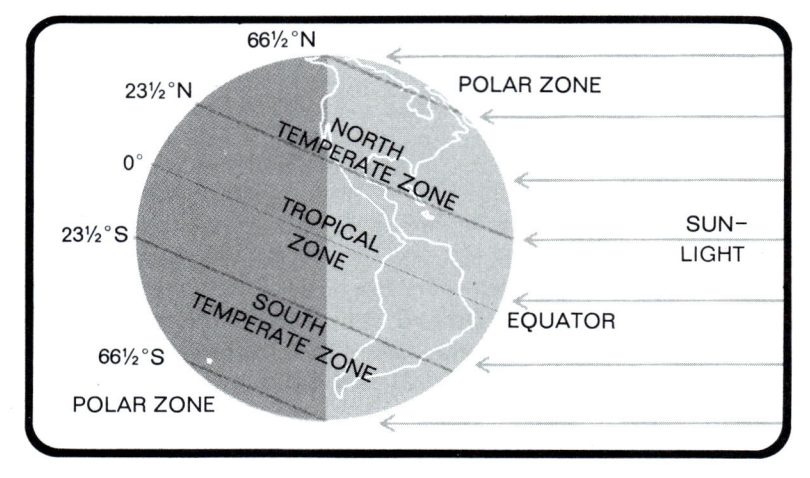

Tropical Zone are the **Temperate Zones,** where the sun never reaches zenith and where it shines every day, although it may often be obscured by clouds.

matter and movement

At the beginning of the sixteenth century, Nicolaus Copernicus, a Polish astronomer, came to the conclusion that all the planets, including the earth, revolve around the sun in circles.

Kepler's discoveries

Copernicus' theory that all planets circle the sun was largely ignored for nearly a hundred years. Then, at the beginning of the seventeenth century, the German astronomer Johannes Kepler set about calculating the orbit of Mars, which he assumed was a circle. He based his calculations on the remarkably accurate observations of his teacher, a Danish astronomer named Tycho Brahe. The essence of Kepler's work was an attempt to predict the future positions of Mars. After long and careful calculation, he found that his predictions were wrong.

Kepler had great faith in Tycho, so he therefore reexamined his own methods. Suddenly, with the sort of insight that is so often a factor in scientific discoveries, he guessed the answer. The orbit of Mars was an ellipse, not a circle. Kepler later discovered that all the planets, including the earth, travel in elliptical orbits around the sun. This is now known as **Kepler's first law of planetary motion.**

Kepler's second discovery was that Mars moves faster when

THE USEFULNESS OF ODDBALLS

"Well, of *course* the earth is round, you dummy! Everybody knows that! Hey, you guys, get this! This screwball says how do we know the earth ain't flat!" Derisive laughter all around.

Yes, "everybody knows" the earth is round. But then, once upon a time, "everybody knew" the sun went around the earth. It was only after a few oddballs started asking embarrassing questions that what everybody knew turned out to be wrong, and the human race learned a bit more about the real nature of its galactic environment.

Very few people today would seriously doubt that the earth is round, but some do. The International Flat Earth Research Society, with its headquarters in California, maintains that the "spinning ball theory" has never really been proved. Instead, its members believe, the earth is a flat disk, infinite in size.

Note the word "research" in the name. The society's stated purpose is to seek provable knowledge about the earth, ignoring theory as something imaginary. In a sense this is not too far from what conventional science tries to do—that is, start with facts and work out a theory that fits them. What we might ask, though, is whether the Flat Earth researchers are really doing what they claim to do. Are they not really—perhaps without realizing it—starting with a theory and choosing only those facts which can be made to support it?

This, however, is a risk that lies in wait for all scientists. Anyone who gets enthusiastic about an idea is in danger of seeing only the facts that agree with it. This is why a scientific theory is never taken as proved until its opponents have had plenty of chance to try to disprove it. And it is one reason why "oddball" groups such as the Flat Earth Society are valuable. Even when their theories are totally wrong, they sometimes come up with real facts that the "establishment" scientists need to face and deal with—facts that otherwise might be overlooked or quietly ignored.

And sometimes, of course, the oddballs turn out to be right. The notion of continental drift, for instance, was considered quite silly by the establishment when it was first proposed only a few decades ago. But now it is generally accepted—in fact, this book contains a whole chapter on it. Not only has it been accepted, but its acceptance has enabled scientists to make better sense out of a lot of miscellaneous information they already had. So a wise scientist will never laugh too hard at people who disagree with what "everybody knows." They might have something to teach him.

this planet earth

it is closer to the sun than when it is farther away. Kepler stated his discovery thus: in any unit of time, the area swept by an imaginary line joining the sun to Mars is constant, regardless of whether Mars is moving slower or faster than average. Other astronomers found the same thing to be true of other planets and of the earth—moon system. This principle is now called **Kepler's second law.**

Next, Kepler discovered that those planets far from the sun not only have a longer distance to travel but their velocity is slower than the velocities of planets with smaller orbits. Kepler demonstrated that the **period** of each planet—the amount of time it takes to complete one orbit—is related to the size of that orbit. This is now known as **Kepler's third law.**

Newton and gravity

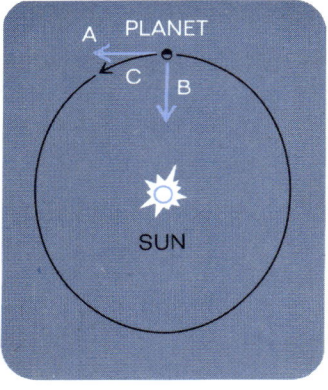

During the Great Plague that struck London in 1664, a young man named Isaac Newton went to live on his mother's farm for the duration of the epidemic. According to his biographer, Newton saw an apple fall from a tree and wondered if the same force that had acted on the apple also acted on the moon. It took him 15 years to arrive at an answer. When he had done so, he proposed his **theory of universal gravitation,** which withstood all tests for 300 years.

Newton, who had studied Kepler's laws, reasoned that an object in space should move in a straight line unless something interferes with it. He was convinced that something interferes with the planets, because their orbits are elliptical. All bodies, he knew, have **mass**—that is, they contain a certain amount of matter. Mass, in turn, possesses **inertia,** the tendency to resist change in motion. A mass remains at rest or travels with constant velocity in a straight line, unless a force acts on it.

Newton came to the conclusion that the force which interferes with the straight-line flight of a planet is the same force that causes an apple to fall. This is **gravity,** the force of attraction that exists between each object in the universe and every other object.

If we consider two objects, each object exerts a gravitational force on the other. This force depends upon the masses of the two objects and the distance between them. Or, stated more precisely: the gravitational force between two objects is measured by the product of their masses divided by the square of the distance between them. If the distance between two masses is increased n times, the force between them decreases n^2 times. This is called the **inverse square law of gravity.**

If the gravitational attraction of the sun were not present, the planets would move in straight lines rather than following elliptical orbits.

conservation

A conservation law states that certain quantities are **conserved**—that is, they remain constant. Thus, when a number is calculated both before and after an experiment, the number will remain the same if it is a conserved quantity. Conservation laws are very useful, since we do not need to know the intermediate details of a process in order to calculate a conserved quantity.

this planet earth

conservation of mass

The law of conservation of mass states that matter cannot be created or destroyed by chemical means. This idea began to develop when an English chemist, Joseph Priestley, extracted oxygen from the air by heating mercury in air to form a red powder called mercuric oxide. When he heated the powder further, it decomposed back into mercury and oxygen.

Priestley journeyed to Paris to discuss this discovery with the French chemist Antoine Lavoisier, who was studying the problem of combustion in an effort to discover why and how things burn. Lavoisier demonstrated that oxides are combinations of metals and oxygen. The work of Lavoisier and Priestley led to the formulation of the idea that matter is indestructible but that it can be changed in composition. Joseph Louis Proust stated this as the **law of conservation of mass** around 1800.

conservation of energy

During the first half of the nineteenth century, physicists explored the properties of the known forms of energy—heat, light, electrical, and mechanical—and discovered that all forms of energy can be converted into heat energy. During the middle of the century, relationships between the other forms of energy were also established. Eventually all the information was put together mathematically by James Clerk Maxwell, a Scottish physicist and mathematician. He demonstrated that energy, like matter, is conserved. Energy may be changed from one form to another, but in the process none is lost and none is created. Another powerful tool was put into the hands of scientists when this **law of conservation of energy** was proven.

mass and energy

The twentieth century has seen an important modification of the conservation laws in one respect. We now know that mass can be converted into energy, and that, conversely, energy can be converted into mass. The basis for this understanding is the work done by Albert Einstein, a German physicist. In 1905, Einstein published his Special Theory of Relativity, in which he demonstrated that space and time are parts of a four-dimensional **spacetime**.

Einstein further asserted that energy itself has mass. He suggested that if mass were destroyed completely a great deal of energy would be released. A large city could be kept functioning for an entire day by converting just one gram of matter into energy.

In 1939, Einstein suggested to President Franklin Roosevelt that it might be possible to release enormous amounts of energy from the radioactive element uranium. The result of that suggestion was the creation of the first atomic bomb. In this device a small mass of uranium was converted to energy, resulting in a huge explosion.

What happens to the laws of conservation? It has become necessary to reformulate them as a single **law of conservation of mass and energy.**

this planet earth

SUMMARY

Earth science is made up of the separate disciplines of geology, meteorology, oceanography, and astronomy. Above all, it is both a practical and a useful science.

Scientific study of a problem involves making observations about it and then reaching a general conclusion by means of scientific reasoning. The scientific method requires that concepts be clearly defined and theories carefully tested. An idea that stands the test of experiment becames a hypothesis. If it proves to be universally applicable, it becomes a law.

Aristotle, Werner, and Hutton, were all important figures in the history of earth science. Both Aristotle and Werner did not test their hypotheses. Hutton, however, attempted to verify every theory he suggested, and published his results. Among his important contributions was the principle of uniformitarianism.

The shape of the earth is an oblate spheroid. Any point on the earth can be located according to its latitude and longitude. Latitude is its angular distance north or south of the equator. Longitude is its angular distance east or west of the prime meridian. Latitude in the Northern Hemisphere can be roughly estimated by observing the altitude of the North Star. Measurement of longitude, however, requires an accurate chronometer.

For ages, it was believed that the earth was the center of the universe, and that the universe moved around it. The investigations of Copernicus and Kepler, however, led to a new understanding. We now know that the earth and the other planets orbit the sun.

Kepler, basing his work on Tycho's careful observations, proposed three laws that govern planetary motion. Newton explained how masses interact with forces, and how they attract each other. His theory of gravitation provided a deeper understanding of Kepler's laws.

Complicated situations are often best explained in terms of conservation laws. Momentum, for example, cannot be created or destroyed. It was once believed that energy and mass were also conserved independently. However, the work done in this century by Einstein has demonstrated that mass can be converted into energy, and vice versa.

REVIEW QUESTIONS

1. Describe the shape of the earth. Why isn't it a sphere?

2. What are latitude and longitude? How are they determined?

3. Why is earth science studied? Name the divisions of earth science.

4. Describe the achievements of Hutton. How did his methods differ from those of Werner?

5. Why do seasons occur?

6. Name and describe the different temperature zones of the earth.

7. What is uniformitarianism?

8. Briefly describe the scientific method.

9. What is gravity? How did Newton's law of gravity explain the orbits of the planets?

10. List Kepler's three laws of planetary motion.

11. If it is 12 noon at Greenwich, what time is it in New York, at a longitude of 75° West?

12. What is the International Date Line?

13. Describe at least two conservation laws.

14. How was the law of conservation of mass discovered?

15. What do you think would happen if the mass of your textbook were suddenly converted into pure energy?

PART 2

GEOLOGY

CHAPTER TWO

THE EARTH'S MATERIALS

King Lear stood on the stormswept heights. The Romans found a fort already standing when they arrived, strengthened it, and added two lighthouses to supplement it. William the Conqueror made the area a strategic objective after he landed in 1066. It held off a French invasion in the time of King John. Henry II built the massive castle keep that still stands today, and the whole area was extensively fortified during the Napoleonic Wars at the start of the nineteenth century. In World War II huge guns boomed from the walls, lofting shells across twenty miles of water to German-occupied France, and an answering rain of shells and bombs came back. The shattered remnants of a British force escaping from the death trap of Dunkirk were landed here. From earliest years, this southeasternmost corner of England, within sight of the continent of Europe on a clear day, has figured dramatically in history. The White Cliffs of Dover—a bastion of resistance by British tribesmen against the encroaching Romans, a heart-lifting symbol of hope and freedom in the dark days of Nazi might.

But the years in which humans have known the White Cliffs contain only a small fraction of the history that the rocks themselves disclose to one who knows how to interpret them. The chalk of which they are formed is made up of the skeletons of uncountable billions of microscopic sea creatures, deposited when the broad seas of the Cretaceous Period covered most of Europe from a hundred and thirty-five to seventy million years ago. Larger fossils of fascinating ancient creatures are embedded in the fine-grained, limey rock. Since the chalk was laid down at a rate of something less than an inch in every thousand years, the 400-foot height of the cliffs is an awesome witness to the lengths of time that have gone into the building of this planet we short-lived, self-important human beings scurry around on.

Of course, the ancient chalk deposits are not found only

the earth's materials

at Dover. They once covered all of southern and eastern England, and they appear at the surface now in various places. A line of low chalk hills marks their furthest reach, beyond which the old pre-Cretaceous rocks of the western highlands stood above the waters of the encroaching sea. Later, massive geological forces folded and bent the hundreds of feet of chalk. London sits above a downfold in the chalk, which was later filled in by sands and clays. South of London a huge upward fold ran east and west. The Dover cliffs are part of the remnants of the northern side of that fold, now known as the North Downs. The southern side—the South Downs—ends in a matching set of cliffs, even higher, around Beachy Head near the seaside resort of Brighton. In between, the arched top of the fold has long since been eroded away, clear down to the underlying clays and sandstones. The eroded rock face gives geologists a handy cross-section view of quite a few million years of the earth's past.

So the British tribesmen who shouted catcalls from the clifftops at Caesar's troops sailing past below, and the English soldiers who manned the great guns to defend England against the dark threat of Hitler's mad drive to rule the world, were standing on top of far more history than they knew. In the perspective of those millions of years, what we humans call history sometimes looks like very small stuff indeed.

When we begin to study our planet in detail, it is appropriate to start with the substances that make up the solid earth, since geology is the oldest and most highly developed earth science. But before we can look at the rocks and minerals that compose the earth's crust, we need to know something about the structure of matter in general. This will help us understand why particular rocks and minerals have particular qualities and properties.

starting at the beginning

The simplest substances that we can isolate by chemical means are the **elements.** At least 88 elements occur in nature and have so far been isolated, and 16 others have been produced artificially in the laboratory. At room temperature, most of the elements are solid, but eleven of them are gases, and two others, bromine and mercury, are liquid.

But there are many more substances on the earth than those 88 natural elements. These are **compounds,** made up of two or more elements that have united. Hundreds of thousands of compounds have been identified so far.

atoms

The idea that all matter is made up of fundamental tiny units

the earth's materials

Table 2-1. The ten most abundant elements in the earth's crust.

Element	Percent
Oxygen	46.6%
Silicon	27.7%
Aluminum	8.1%
Iron	5.0%
Calcium	3.6%
Sodium	2.8%
Potassium	2.6%
Magnesium	2.1%
Titanium	0.44%
Hydrogen	0.14%

dates back to the early Greek philosophers. One of them, Democritus, suggested that matter was composed of indivisible entities called **atoms.** Aristotle later argued that everything was made of four elements: Air, Fire, Water, and Earth. This belief persisted until the nineteenth century. Then, chemists discovered that it was convenient to think of chemical reactions as taking place between various kinds of atoms.

When elements combine to form compounds, the atoms of the elements link up to form **molecules,** which are the smallest recognizable parts of a compound. Suppose we take water as an example. Water is a compound made up of two elements, hydrogen (H) and oxygen (O). Two atoms of hydrogen link up with each oxygen atom to form a water molecule, written as H_2O. We will see a little later just how these linkages occur.

atomic structure

For many years scientists believed that atoms were indeed indivisible. Then, early in this century, it was discovered that each atom contained even more fundamental particles. The most important of these are **electrons, protons,** and **neutrons.**

Electrons have a very small mass. About 10^{27} of them would weigh only one gram. (This is so large a number that it would take us many billions of years to count them all.) Electrons carry a negative electrical charge.

Protons are about 1,836 times as massive as electrons. They carry a positive electrical charge, equal but opposite to the charge carried by the electron. Like charges repel each other, and unlike charges attract each other. Thus electrons repel electrons, protons repel protons, but electrons and protons attract each other.

Neutrons are still larger, about 1,838 times as massive as electrons. Neutrons have no electrical charge, and only a limited lifetime when free in space. At the end of its lifetime, a neutron **decays** or breaks down into a proton and an electron.

Neutrons normally exist without decaying only when they are bound to other neutrons and protons. The nucleus of an atom consists of neutrons and protons bound together, and both particles are referred to as **nucleons.**

Hydrogen is the simplest element. Its atom contains one proton and no neutrons in the nucleus, which is orbited by one electron.
The nucleus of the oxygen atom contains eight protons and eight neutrons. It is orbited by eight electrons.

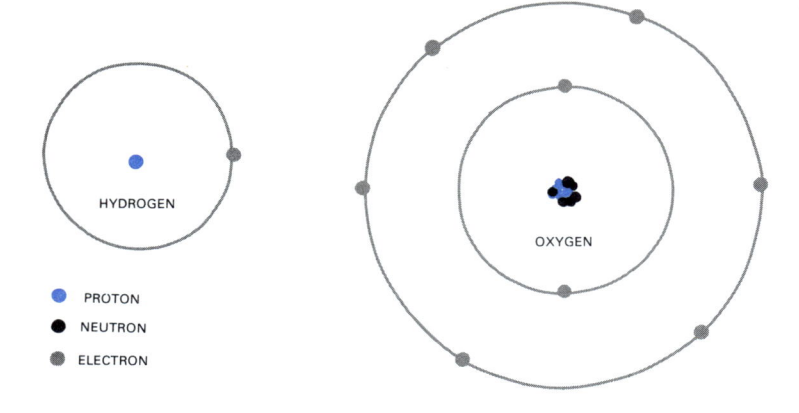

HYDROGEN

OXYGEN

● PROTON
● NEUTRON
● ELECTRON

Our present knowledge of atomic structure suggests that each atom possesses a positively charged central nucleus about a trillionth of a centimeter across. Negatively charged electrons orbit the nucleus at various distances. On the whole, atoms are neutral, with as many orbiting electrons as there are protons in the nucleus.

Atoms can exist in various energy states. The different states correspond to different configurations of the electrons about the nucleus. An atom changing from one configuration to another either absorbs or emits energy. It cannot be in any random state of energy, but only in certain particular states characteristic of that atom. This is similar to the situation of a man in an elevator who can get off at any floor but not in between floors.

The lowest energy state for an atom is called its **ground state.** An atom in the ground state can absorb energy and become **excited,** but in time it will fall back into the ground state and give up some energy. If an atom gains so much energy that one or more electrons escape it completely, it is said to be **ionized;** it has become an **ion,** and possesses a positive charge. Positive ions usually capture electrons, and thus return to being atoms. Only tiny amounts of energy are involved in these changes. The energy absorbed or emitted by an atom is usually in the form of light.

the nature of light

The precise nature of light energy is complicated, although a modern theory exists which explains it very well. This theory can only be expressed exactly within a mathematical framework, so we need not describe it here. Instead, we will discuss light in simple terms, remembering that words cannot allow us complete accuracy. Explained in words, light is something like a wave; but it is also rather like a collection of particles.

In a vacuum, light travels along straight lines at a speed of 300,000 km/sec (186,000 mi/sec). This high velocity makes it seem to travel instantaneously. As it travels, the electric and magnetic properties of space change along its path. When we plot these changes on a graph, a wavy pattern emerges. For this reason we refer to light as an electromagnetic **wave,** although its path is really straight.

A wave possesses velocity, wavelength, and frequency. The

wavelength is the distance between one crest of the wave and the next crest. The color of light is determined by its wavelength. For instance, the wavelength of red light is longer than the wavelength of blue light. We measure light in **nanometers (nm).** A nanometer is equal to 1/100,000,000 of a centimeter. Red light has a wavelength of 700 nm, green light about 550 nm, and blue light about 400 nm. Electromagnetic waves longer than about 750 nm are called **infra-red radiation.** They cannot be seen by human eyes, although certain insects can detect them. Waves shorter than about 400 nm are also invisible, and are known as **ultra-violet radiation.** The shortest waves are called **X-rays** or **gamma-rays.**

The **frequency** of a wave is the number of waves that pass an observer every second. Since light travels at 300,000 km/sec, the frequency is the same as the number of waves that can be fitted into 300,000 kilometers. Fewer long waves can fit into this distance than shorter waves. So if the wavelength is long, the frequency is low, and vice versa.

Atoms of each element emit radiation at certain wavelengths. We can thus analyze the radiation emitted by a sample of gas and determine what elements are in it. In general, atoms absorb the same wavelengths that they emit. Molecules behave in the same way, but their energy states are more numerous than those of atoms.

chemical properties of atoms

The simplest atom is that of hydrogen. It consists of one proton with one electron in an orbit around it. The energy states of the hydrogen atom are very simple. A neutron can combine with the central proton to produce **deuterium,** or heavy hydrogen. If two neutrons combine with the central proton, we get **tritium,** or heaviest hydrogen. Chemically, these three substances are almost identical.

Helium has a nucleus of two protons combined with either one or two neutrons. Two electrons in orbit balance the charge of the protons. **Lithium** has three protons and three to six neutrons in its nucleus. Three electrons orbit in its two **electron shells.**

Atoms with the same number of protons and electrons may differ in the number of neutrons they possess. When atoms interact chemically, only the electron shells are involved. The number of electrons in the shell thus determines the chemical properties of an atom. This **atomic number** distinguishes the atoms of one element from those of another. Atoms of a given element always have the same atomic number. However, they may have different masses because their nuclei may contain differing numbers of neutrons. Atoms with differing masses but the same atomic number are called **isotopes.**

When we look at the chemical properties of elements, we find that the elements are divided into several groups. The **inert gases** are extremely unreactive and form few compounds. This property is probably related to the structure of their outer electron shells, which are exceptionally stable.

the earth's materials

The **alkaline metals** are all extremely reactive. Their atoms have one more electron than needed for a stable structure. So their high reactivity results from their willingness to give up this electron and achieve stability.

The **halogens,** on the other hand, have one *less* electron than they need for stability. They are highly reactive because they are eager to grab an electron from another atom in order to become stable.

The **alkaline earth** elements have two electrons too many to be stable. Another group which includes oxygen, sulfur, and selenium, is two electrons short of stability.

chemical bonding

Atoms are stable only when their electron shells have inert gas structures. They try to achieve this structure by linking up with other atoms. This link is called a **chemical bond.** This bonding is so effective that only the inert gases exist as single atoms. All other atoms are usually bound to atoms of their own or a different sort.

There are several types of chemical bonds. We can understand **ionic bonding** by examining common table salt, or sodium chloride (NaCl). Sodium is an alkaline metal willing to give up an electron and achieve the stable structure of neon, an inert gas. Chlorine is a halogen eager to grab an electron and achieve the stable structure of argon, another inert gas. What happens, then, is that chlorine accepts an electron from sodium. Since chlorine has one too many electrons, it is negatively charged; sodium, with one electron too few, is positively charged. The difference in their charges attracts them to each other.

The water molecule, H_2O, is a good example of **covalent bonding** in which atoms achieve stability by sharing electrons. Oxygen lacks two electrons for stability. It can achieve stability by sharing the

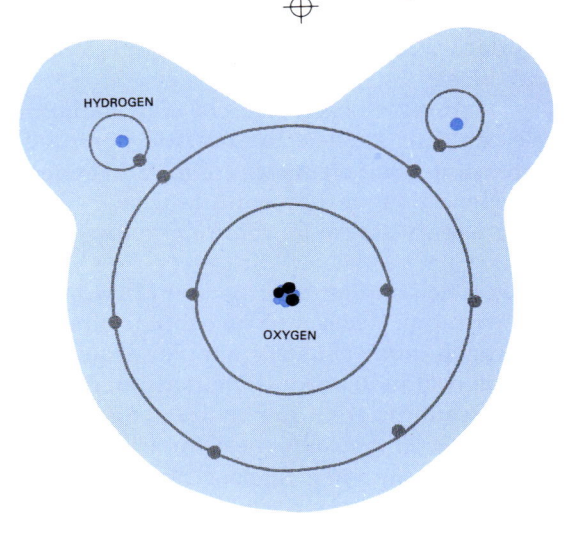

Crystals of sodium chloride (common table salt) are created when sodium atoms form ionic bonds with chlorine atoms.
Molecules of water (H_2O) are formed by covalent bonding of two hydrogen atoms and one oxygen atom.

the earth's materials

electrons of two hydrogen atoms.

Carbon atoms can form four covalent bonds at the same time. They link up very easily with other carbon atoms. They are thus able to form very complicated molecules, often in long chains. This property of carbon makes it the basis for organic life as we know it. Millions of carbon compounds are known, and **organic chemistry** is devoted to the study of these compounds.

a quick look at physics

As we get deeper into our discussion of earth science, we will need to understand some of the simple basic ideas of physics. As we saw in chapter 1, the **mass** of an object is the amount of matter it contains. In ordinary conversation we often use the words *mass* and *weight* interchangeably. In physics, however, the two terms have different meanings. Mass is a measurement of how much matter an object contains. **Weight,** on the other hand, expresses the force of gravitational attraction between an object and the nearest large astronomical body.

Let's take an example of this. Suppose we have, on earth, a lump of iron weighing exactly 1 kilogram (kg). That lump of iron contains a certain amount of matter. If we take the lump to the surface of the moon, it will still contain exactly the same amount of matter. However, its weight will change, because the moon's gravitational attraction is only one-sixth as strong as that of the earth. The lump of iron, on the moon, will weigh only $1/6$ of a kilogram.

A little later in the book we will discuss rocks and other materials in terms of their **density.** Simply stated, density refers to mass per unit volume—the amount of matter a material contains within a given amount of occupied space. We might ask, for example, whether teakwood is denser than balsa wood—that is, which material contains more mass? We cannot determine this unless we compare samples of each wood, both samples the same size and therefore occupying the same volume of space. So we compare 1 cubic meter of balsa wood to 1 cubic meter of teakwood. The teakwood is heavier, and this tells us that it is more dense—that is, it contains more matter than the block of balsa wood the same size. Now, we might have compared a block of teakwood only $1/2$ of a cubic meter in size to 1 cubic meter of balsa wood. The teakwood would still be heavier than the balsa—but this would not be a measurement of their comparative densities, since the two samples compared did not occupy the same volume.

minerals

Minerals are substances that have a definite chemical composition.

the earth's materials

Quartz, also known as rock crystal, is a mineral composed entirely of silicon and oxygen. It has a hardness of 7 on the Mohs scale.

crystals

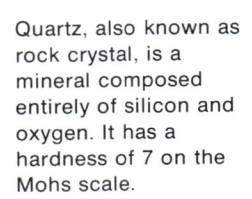

Usually they are produced by natural inorganic processes, but in some cases they may once have been part of living organisms. The study of **mineralogy** is concerned with identification of minerals and the classfication of their structures.

The atoms in most minerals are arranged in a precise order. The minerals are thus **crystalline**. If this internal order affects the external shape of a mineral, we have a **crystal**. Some minerals do not have a definite ordering of atoms, but are **amorphous** instead.

In some cases, the same chemical substance can have different internal structures. For example, both diamond and graphite are composed of carbon atoms. In diamond, the atoms are arranged in a pyramidal shape, making it the hardest substance known. In graphite, however, the atoms are arranged in layers. The atoms within each layer are strongly bound, but the layers themselves are weakly bound to one another and can slide over one another easily. This makes graphite soft and slippery.

Today crystal structure is understood in terms of how the atoms composing the crystal are arranged. But long before such understanding was possible, scientists had already determined that there are basically only six different shapes of crystals. They described the different shapes by examining the line around which the crystal was symmetric in some manner. These lines are called **axes** (singular: **axis**).

Isometric crystals have three perpendicular axes of equal length. In **tetragonal** crystals one of these axes has a different length. In **orthorhombic** crystals all three axes are of different lengths.

Only two of the three unequal axes meet at right angles in

the earth's materials

Isometric   **Tetragonal**   **Hexagonal**

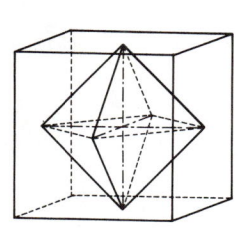

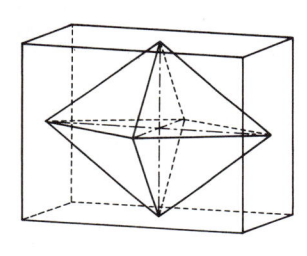

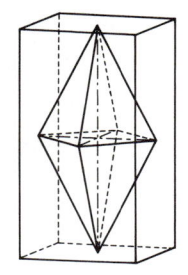

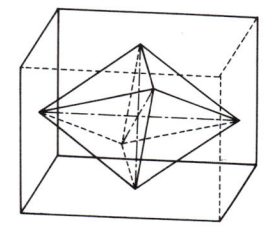

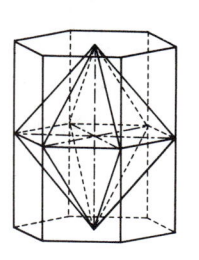

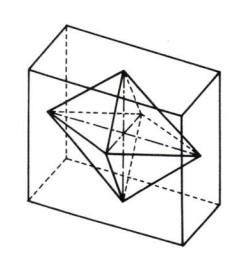

Orthorhombic   **Monoclinic**   **Triclinic**

There are six basic types of crystal structure. Most minerals are crystalline and fit into one of these categories.

monoclinic crystals. In **triclinic** crystals, all three unequal axes meet in oblique angles. **Hexagonal** crystals have three axes of equal length meeting at 120°; a fourth longer or shorter axis is perpendicular to these three.

The principal minerals that make up the earth are **silicates.** Their usual structure is the silicon—oxygen tetrahedon. Each small silicon atom is surrounded by four large oxygen atoms. Atoms of other elements attach themselves to two of these oxygen atoms.

identifying minerals

Most minerals can be identified using a combination of the following methods. The metallic minerals are particularly easy to identify by their **color.** For example, **azurite,** a copper carbonate, is blue; **malachite,** a copper carbonate with a different structure, is green. Minerals can also be identified by the color of the **streak** or powder they leave when they are scratched across an unglazed porcelain plate.

Closely associated with the color of a mineral is its **luster.** Luster describes the way light is reflected from the surface of a mineral. Minerals may be metallic or non-metallic in luster. Some fibrous minerals, such as asbestos, have a **silky** luster. Talc has a **pearly** luster. Still other minerals may have a **greasy** or **vitreous** (glass-like) luster. Cerussite and anglesite, both compounds of lead, have an **adamantine** or diamond-like luster.

the earth's materials

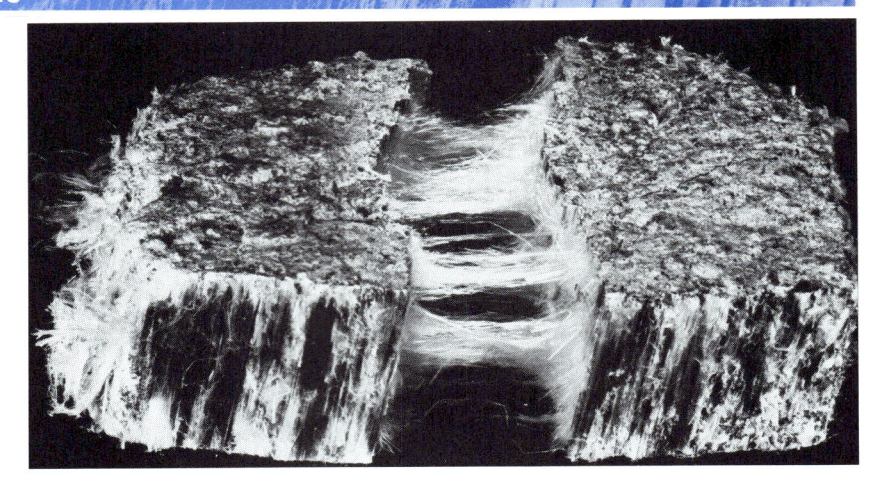

Serpentine is a soft mineral (about 2½ on the Mohs scale) that displays silky fibers.

Minerals can also be identified by their **hardness.** In 1822 Friedrich Mohs set up a scale of relative hardness, ranging from 1 to 10, in which 1 is the softest. A mineral with a higher number can scratch a mineral with a lower number. We should note that this scale does not give precise degrees of hardness; it simply ranks the hardness of minerals relative to each other.

A number of minerals display a property called **cleavage**—that is, they tend to break along predictable planes at points of weak bonding. A chunk of a mineral that possesses cleavage will break in such a way that all the smaller pieces produced have the same shape as the parent chunk. Other minerals do not show cleavage when they are broken. Instead, they **fracture,** usually into irregularly shaped pieces. Both cleavage and fracture are often helpful in identifying a given mineral.

defining and classifying rocks

What, exactly is a rock? This question is not as easy to answer as we might think. Unlike minerals, most rocks do not have a uniform chemical makeup, although some do. Most rocks are composed of two or more different minerals, but a few contain only one. We ordinarily think of rocks as being made up of substances that have never been alive; yet some rocks are formed from the skeletons of long-dead animals or the tissues of ancient plants. As a rule of thumb, we can say that a rock is usually an aggregate of naturally-occurring mineral substances—a collection of mineral particles that have been brought together and formed into a solid mass by some natural process.

the earth's materials

A crystal of yttrium iron garnet. The garnets are hard crystals that can be found in a variety of colors.

Fortunately, rocks are easier to recognize than they are to define. For the most part, we can look at an object and know whether it is a rock or not. But to make our study of rocks useful, we need to classify the rocks, fitting them into categories that tell us something about their character. The branch of earth science that deals with this matter — the origin, composition, and classification of rocks — is known as **petrology**.

What is the most satisfactory way to classify rocks? Several possibilities come to mind. Rocks might be classified according to the size and shape of the particles, or **grains,** that compose them. However, there are too many variations in grain size to make this practical. They could be classified on the basis of the minerals they contain; but very different rocks may contain the same minerals. Or they could be classified by their chemical makeup; but since this requires a laboratory, it is not very helpful to workers out in the field.

After we rule out grain size, mineral content, and chemical makeup (at least for now — we will come back to some of them later)

the earth's materials

there remain two other criteria that turn out to be useful for our purpose. The first of these is the geological processes by which the different rocks were formed. Some were formed by the cooling and hardening of hot liquid rock substances, or **magma,** from deep beneath the surface of the earth. These are called **igneous** (made by fire) rocks, because of the tremendously high temperatures at which they originated. Other rocks were formed from layers of **sediment**—particles of various substances that accumulated on the bed of a body of water. Such rocks are called **sedimentary** rocks. Finally, there are **metamorphic** rocks, such as marble. These are rocks that were once sedimentary or igneous, but which were changed (the word "metamorphic" means "changed form") by intense heat and pressure in the depths of the earth.

The other useful criterion for classification is the **texture** of the rock. Some rocks are made up of tightly fitted crystals that formed as the magma was cooling or the sediment was being deposited. These are known as **crystalline** rocks. Granite and marble are both crystalline. Others are composed of worn and broken fragments of preexisting rocks, and are said to have a **clastic** texture. (The word "clastic" comes from a Greek word meaning "to break.")

igneous rocks

Liquid magma contains a rich mixture of elements. In basaltic magma, which is in some ways the most important, oxygen and silicon are the most abundant elements. Next most common are usually aluminum and iron, while still others are present in smaller quantities. Each of the minerals that make up igneous rock contains some of these elements, but none contains all of them. How are the elements sorted out, so that a single piece of rock contains separate crystals of three, or four, or even more different minerals?

Most igneous rocks are silicates. In some silicate minerals, each molecule is formed with a single silica tetrahedron. In others, the tetrahedrons are joined together in long chains, flat sheets, or structures linked tightly in three dimensions. These more complicated molecules can form only at lower temperatures in the magma. That is, when hot magma begins to cool, the first crystals to form will be those with single-tetrahedron structures. At a somewhat lower temperature, chain-type minerals will start to crystallize, and so on. The last minerals to crystallize will be those with a complicated three-dimensional structure. In this way, rocks containing different minerals can be formed out of a single body of magma.

Bowen's reaction series

The researcher N. L. Bowen studied this sequence of crystallization in the 1920s, and made some interesting discoveries. Of course, he could not study magma in a volcano or under the earth's crust. Instead, he melted minerals in the laboratory to create an artificial magma, known as a **silicate melt.** The melt contained elements in about the same proportions as might occur in a typical magma. After heating his melt, Bowen let it cool gradually, observing what hap-

the earth's materials

Hornblende, a silicate with an elaborate chain structure, is formed during the discontinuous reaction series. Hornblende is also known as amphibole.

pened as it did so.

As Bowen had expected, the first crystals to appear were **olivine.** This is a greenish mineral with a single-tetrahedron structure. For a little while the olivine crystals kept growing. Then, as the melt became cooler, the olivine disappeared! In its place, crystals of **pyroxene** began to form. Pyroxene is a chain-structure silicate. Like olivine, it contains iron or magnesium, but it can also contain calcium and sometimes aluminum. Apparently what had happened was that, since the olivine crystals were still mixed with the melt, they had reacted with it as soon as it was cool enough to permit chain structures to form. They had picked up some new atoms, lost some old ones, and linked themselves into pyroxene chains.

As the melt cooled further, the pyroxene crystals disappeared, and **hornblende,** which has a more elaborate chain structure, appeared. Then the hornblende also vanished, and was replaced by **biotite mica,** which has a sheet structure.

This sequence is known as a **reaction series.** At each stage, a chemical reaction takes place between the crystals and the melt, producing a new type of crystal. Because, each time, one kind of crystal disappears and is replaced by a crystal with a different structure, it is known as a **discontinuous** reaction series.

However, we have not yet gotten to three-dimensional structures. These, Bowen found, form a completely separate reaction series, which behaves somewhat differently. The first mineral to form in this series was a type of **feldspar,** containing silicon, aluminum, and calcium. As the melt cooled, there was no sudden change; the feldspar did not disappear. Instead, some of the calcium and aluminum was gradually removed from the structure, and atoms of sodium and silicon were taken up in their place. Thus the new mineral was just a slightly different kind of feldspar. Because the structure remains basically the same, this series is known as a **continuous** reaction series.

The two series can take place at the same time in a single melt or magma. This is possible because, as we said above, silicon and oxygen are far more abundant in a typical magma than are any other elements. The minerals of the discontinuous olivine-to-biotite series are all mafic minerals, which contain iron and magnesium. There are relatively small amounts of these elements in a typical magma—not nearly enough to use up all of the silicon and oxygen. So, even though the mafic minerals have smaller molecules and therefore usually begin to form sooner—that is, at higher temperatures—there is usually a plentiful supply of silicon, oxygen, and other elements left over. Feldspars do not need iron or magnesium, so they can form from this left-over material when the proper temperature is reached.

At an even lower temperature, other minerals form, mostly another type of feldspar and, finally, **quartz,** which is pure silicon and oxygen. These behave somewhat differently, however, and are not considered part of either reaction series.

All this is very neat, but it seems to leave us with a problem. We have accounted for the formation of quite a number of minerals—but we are left at the end of our two series with only a few of them. All the others have disappeared on the way down the temperature scale. Yet all these minerals are found in rocks. Somehow they have escaped being destroyed. How might this have happened?

In the laboratory, it happens if the crystals of each material are removed from the melt before they have a chance to change. Once they are out of contact with the remaining liquid, the chemical reactions that change them cannot take place. Olivine thus remains olivine; it does not change to pyroxene as it cools. Pyroxene may still form in the remaining melt, if enough of the proper ingredients are still present. However, it forms directly from the liquid, without any "help" from the olivine. The same holds true for the other minerals. Could something of the same sort happen in a natural magma?

As a matter of fact, it might. Mineral crystals differ in density from the liquid magma. The early-forming mafic minerals such as olivine and pyroxene are denser than the magma, so they would tend to sink downward and form a layer at the bottom. If this layer were thick enough, some of the crystals would be covered by others. They would no longer be in contact with the liquid magma, and would

the earth's materials

Quartz is important to the communications industry. These are artificial quartz crystals, created in the laboratory for use in electronic equipment.

not react with it as the temperature dropped. Many feldspars, in contrast, are less dense than the magma. They would tend to float to the top, and the same thing might happen. Even minerals that did not sink or float might form clusters of crystals. The crystals in the center of the cluster would be protected against further reactions.

The processes that occur in nature are undoubtedly more complicated than this, but they must make use of the same general principles. We can assume that the great variety of minerals in present-day igneous rock came into being in some similar way.

the earth's materials

types of igneous rock

When magma cools slowly, deep within the earth's crust, it stays for a long time at the proper temperature for crystallizing a particular mineral. Therefore, the crystals have time to grow fairly large, and the resulting rock is said to be **coarse-grained.** Such deep formed rocks are called **intrusive** or **plutonic.** (The second name comes from Pluto, the classical god of the underworld.) On the other hand, sometimes magma is poured out on the surface, usually as volcanic lava. Then it cools quickly, so it forms small crystals. The resulting rock is **fine-grained,** and is called **extrusive** or **volcanic** rock. (Volcanoes take their name from Vulcan, who was supposed to be the blacksmith of the gods.)

On the basis of crystal size and mineral content, we can distinguish four main types of igneous rock:

1. **Granitic rocks:** coarse-grained, intrusive rocks composed mainly of **felsic** minerals (that is, minerals that contain mainly feldspar and silica), usually light-colored. Granite, which contains mostly quartz and feldspar, is the best-known example.
2. **Rhyolitic rocks:** fine-grained, extrusive rocks composed of the same minerals as the granitic kind. Related to them and to the basaltic rocks are the andesites, which are common in the volcanic regions around the Pacific Ocean.
3. **Gabbroic rocks:** coarse-grained, intrusive rocks with a high proportion of mafic (iron–magnesium) minerals and calcium-type feldspars, usually dark-colored and heavier than felsic rocks. Related to them are the **ultramafic** rocks, containing at least 90 percent mafic minerals. One of these, peridotite, is thought to be the main component of the earth's mantle, deep beneath the crust.
4. **Basaltic rocks:** fine-grained, extrusive rocks composed chiefly of mafic minerals. Basalt, commonly found on the ocean bottom, is the most common example.

In addition to these, certain other types of igneous rock may form under special conditions.

Many of the earth's more rugged-appearing rock formations are made of granitic rock. This intrusive granite became exposed when erosion and weathering destroyed the sedimentary rock above it.

the earth's materials
sedimentary rocks

Anyone who has climbed mountains knows that bare rock surfaces often have loose fragments of stone that can slip dangerously under the climber's hands or feet. Another danger is areas of "rotten stone" that may break or crumble when stepped on. Both of these hazards result from a complex of processes known as **weathering.** The weathering of rock is extremely important to life. One of its products is soil—the earth in which all of our land plants grow.

the destruction of old rock

Weathering is of two basic types: mechanical and chemical. **Mechanical weathering** simply breaks up the rock; it does not change its chemical makeup. A common type of mechanical weathering takes place when water freezes in a crack in the rock. Water expands when it freezes, so the growing ice crystals actually push against the rock on either side. Gradually the crack deepens and widens, and eventually a piece—a chip or a boulder—may break free of the parent rock.

Sharp changes in temperature may also cause physical weathering. Rock expands when it is heated, and contracts, or shrinks, when it cools. Sometimes a sunwarmed rock surface may expand so much, or so fast, that it splits away from the cooler rock below. Or a sudden change of temperature may cool the surface rapidly, while the rock below is still warm. The cooling surface contracts faster than the warm subsurface, and the two sections split apart. The fact that the rock is made up of a variety of minerals helps the process along. Different minerals expand and contract at different rates at the same temperatures. This creates stresses within the rock that can lead to splitting.

In **chemical weathering,** the rock minerals are actually changed. Some elements are taken out, new ones are put in, the molecules are taken apart and rebuilt in different structures. Nearly always this requires the presence of water.

Because of the way its molecules are constructed, water tends to break apart the molecules of many other substances—that is, to dissolve them. We see this happen with salt and sugar every day. Minerals on the surface of a rock may be dissolved by rainwater and carried away, a process called **leaching.** Other minerals may actually "soak up" the water, or parts of it, in another process known as **hydrolysis.** Often these two processes occur together.

The chemical weathering effects of water are usually increased by another factor. When the water is carrying dissolved elements, many of these elements are in the form known as ions. Ions carry an electrochemical charge, so they are chemically active. That is, they are likely to react with other substances they meet. Consequently, they may cause chemical changes in rock much faster than plain water would do it. For instance, water may pick up ions of sulfur from rocks or from industrial wastes in the air. In combination with the water, sulfur forms sulfuric acid, which has a powerful destructive effect on

IN FOCUS:
modern prospecting for mineral resources

The mere mention of the word "prospecting" often summons up an image of the grizzled, dusty desert rat outfitted only with pick, shovel, and burro, roaming the barren Western countryside in his quest for treasure. He had only to use his eyes to spot the gold or silver lying at his feet. This type of prospecting, common a century ago, has been replaced by much more sophisticated methods. Faced with rapidly dwindling resources, governments and huge corporations involved in modern prospecting have turned to geology, geophysics, geochemistry, and the space sciences to help them locate precious earth materials.

It goes without saying that the services of the geologist are essential to prospecting, because so many mineral resources are found in conjunction with particular rocks and geological formations. Zinc, copper, and lead, for example, are found in igneous rocks, whereas oil and gas occur in old sedimentary deposits which trap and hold them. Coal lies in younger sedimentary strata and in outcrops—beds of rock that poke up through the earth's surface. Sedimentary basins are also sources of salt, potash, and phosphate. Finally, gold, silver, and nickel frequently appear in mountainous regions.

What are some of the new methods for locating these resources? From the science of geophysics—the application of physics to earth studies—we know that the various earth materials have different physical characteristics. Most metallic ores, for instance, have densities and magnetic and gravitational properties different from those of the strata where they are found. This enables scientists to use magnetic, gravitational, or seismic methods to ferret out these ores.

The quickest and most economical way to determine if a large area contains valuable resources is to survey it from the air. Variations in the earth's magnetic field, often produced by magnetized rocks, can be detected by an airborne magnetometer. Because of their iron content, igneous and metamorphic rocks are more strongly magnetic than sedimentary rocks. Many air surveys attempt to determine the extent of igneous and metamorphic formations and to pinpoint the magnetic rocks lying in them. Metallic ores found in such rocks include magnetite, limonite, and iron sulfide, as well as some manganese and chromite. Magnetic surveys also determine the size and shape of sedimentary basins which may contain fossil fuels.

Probably the commonest geophysical method for finding oil and natural gas is seismic reflection. In this technique, earthquake-like seismic waves are generated by explosives or other means. As the waves travel through the earth, seismometers measure the time it takes them to return to the surface after they are reflected from the boundaries between different strata. Seismic waves travel at different known velocities through different materials. They generally move fastest through dense igneous rock, and slow down considerably when they pass through loose materials such as clay, sand, or shale. Consequently, from the picture of underlying structures drawn by wave reflection, prospectors can pick out those strata which are likely to contain hydrocarbons.

One of the most interesting aspects of modern prospecting is the use of biogeochemistry to locate hidden resources. Biogeochemical techniques detect any changes in the chemical composition of water, soil, air, or plants which may have been caused by metallic ores. Presently, these methods are most widely used with plants. Geobotanists know, for example, that some plants grow only where certain groups of elements appear in the soil in large quantities. Thus, lodgepole pine grows in soil containing copper, zinc, iron, gold, boron, manganese, and silver. Not surprisingly, such plants are called indicator plants, because they alert a prospector to be on the lookout for certain minerals.

Perhaps the most promising development in prospecting has been the use of remote sensors—from aircraft and satellites—to locate mineral resources. Photographs and infra-red and radar imagery have revealed faults, fractures, and other structures which proved to contain resources. Photographs from NASA's Earth Resources Technology Satellite, for example, have been enhanced by computer to bring out certain contrasting details which help identify surface signs of mineral deposits. Some materials give themselves away by the color of the earth overlying them. Minerals show up in computer-enhanced imagery as green or brownish green, whereas underground water pockets—a valuable resource in arid regions—appear reddish brown.

The magnetometer and seismometer have also been employed in deep-sea prospecting. Since many ocean prospecting firms are interested only in mineral nodules lying on the ocean floor, submerged television cameras are often used to locate and determine the extent of such deposits. More recently, some prospectors have turned to an electrical device that monitors sea water for the presence of metallic ores which easily detectible electrical properties.

Future solutions to the problem of dwindling natural resources may lie in our ability to develop techniques for getting at some of them—Antarctic coal or deep-sea oil, for example—which have so far been inaccessible. Many believe that the rapidly growing space and computer technologies will eventually supply at least some of these techniques.

the earth's materials

detrital sedimentary rock

most substances. We have had considerable evidence of this in recent years. There is so much sulfur in the air in some places that rainwater falling through it becomes quite acid. When this acid falls on brick and stone buildings, it weathers them very fast.

Still another form of chemical weathering is **oxidation.** This is the tendency of some elements to combine with oxygen. The most familiar such combination is iron oxide, or rust. Oxidation can take place without water, but it occurs more rapidly when water is present.

Weathering produces, fundamentally, two things: (1) broken particles of rock and (2) dissolved minerals. By various processes and in various combinations, these are turned into sedimentary rock.

Usually the rock particles are gradually sorted, by wind and water, into layers of more or less the same size and material. This happens through the process of **sedimentation,** which we will discuss in detail in chapter 3. Thus gravel may form a layer along the bed of a stream, and quartz sand may be piled up by waves along a beach. Silt may settle to the bottom of a lake.

Rock that forms from these sediments is known as **detrital sedimentary rock.** (Detritus is any accumulation of loose fragments.) Such rock always has a clastic texture. Sometimes it is made up entirely of sediments that have been pressed tightly together by the weight of other layers on top. More often it is **cemented** together by mineral crystals that form in the spaces between the detrital particles.

These crystals come from the other part of the weathered-away igneous rock—the dissolved, leached-out minerals. Water circulates through the layers of sediment and dissolves certain minerals out of them. Under proper conditions, the minerals are later **precipitated**— that is, their ions separate from the water and combine again into solid crystals.

Detrital rocks are usually classified by the size of the particles,

This sulfur was mined in Sicily. When combined with water, sulfur forms sulfuric acid, a corrosive agent that causes rock to weather very quickly.

TURQUOISE JEWELRY

Turquoise jewelry is currently quite fashionable. Rings, necklaces, earrings, and bracelets containing this attractive phosphate mineral can be purchased in many places—jewelry stores, chic boutiques, even from street vendors. A stunning exercise in turquoise mosaic can be seen in the so-called Holy Toad, a religious emblem crafted by the Zuni Indians of the American Southwest. This appealing figure consists of a sea shell shaped like a toad and encrusted with garnets and tiny pieces of tuquoise.

Turquoise usually appears in arid regions, either in small veins or in large, kidney-shaped deposits in sedimentary or volcanic rock. The color of the mineral ranges from a dramatic sky blue (the most valued) to dark green (considered the least desirable). The most sought-after variety of turquoise comes from Neyshabur, Iran. Other sources include North Africa, Australia, and the American Southwest, where turquoise has always played an important role in the religious imagery of the American Indian.

Because of its softness, turquoise is difficult to fashion. Great care must be taken in cutting, grinding, and polishing it. When turquoise is found, it must first be separated from the rock in which it is embedded, first with a hammer and chisel and then with a diamond saw. Next, a grinding wheel is usually employed to round it into a dome shape known as a cabochon. Finally, the scratches are sanded out with a damp carborundum cloth, and the gem is then polished with a leather buffer coated with cerium oxide. The stone is now ready to place in a metal setting—silver is traditional—or it may be bored through to become a pendant or part of a necklace.

the earth's materials

or grains, they contain. Within this grouping, they can also be classified by their mineral content. Rock composed of extremely tiny grains (less than $1/16$ mm in diameter) deposited in very thin, easily split layers is commonly called **shale.** Shales contain a high percentage of clay minerals derived from the weathering of feldspar, and this contributes to their distinctive structure. They also tend to contain more organic material than other detrital rocks. Sometimes shales are distinguished from **siltstone** and **mudstone,** which are composed of similar-sized particles but do not split in the same characteristic way.

Sandstone includes rocks composed of somewhat larger particles ($1/16$-2mm). Quartz is the most common mineral producing grains of this size. When a sandstone is composed mainly of quartz grains it is called **quartz sandstone.** If it contains considerable feldspar as well, it is known as **arkose sandstone.** A dark sandstone, containing angular fragments of various kinds, is called **graywacke.**

Conglomerate is made up of larger fragments (over 2 mm) that have been smoothed and rounded by erosion (see chapter 3). If fragments of this size are sharp and angular, the rock formed from them is called **breccia.** Most conglomerates are cemented together by quartz, the most resistant and durable of the common minerals.

Another sort of detrital rock contains fragments of the remains of living organisms. Many small sea creatures consume the dissolved minerals in the water and use them to form their shells and skeletons. Generally these are composed mainly of silica or of calcium carbonate. After the organisms die, the hard remains collect on the sea bottom. There they are weathered and broken by the action of the water. Finally they are cemented together, just as rock detritus is. The most common of these rocks are various kinds of **limestone.**

Examine the figure above and then refer to the paragraph at right. Is this an example of conglomerate or breccia?

chemical sedimentary rock

Some sedimentary rocks contain no detritus at all. They are made up entirely of minerals that have been precipitated out of water. Because they are formed in this way, they have a crystalline texture. One such rock is **halite** or rock salt, which crystallizes when very salty water evaporates.

Halite dissolves so easily in water that it is normally found only in dry desert areas, or in extremely salty water such as that of the Great Salt Lake or the Dead Sea. A more common chemical sedimentary rock is **limestone.** Basically, precipitated limestone is simply calcium carbonate cement, or related compounds, without any detritus mixed in. It is often formed where water flows over layers of detrital limestone, leaching out some of the minerals as it goes. A spectacular example is **stalactites,** those long stony "spears" that hang from the roofs of limestone caves. Chemical limestone can even be precipitated out of water that has flowed over concrete. Stalactites—small ones, but real—have been seen hanging from the concrete roofs of tunnels in the New York City subway system.

the earth's materials

sedimentary rocks and the earth's history

Because of the way in which sedimentary rocks are created, the newer, younger layers, or **strata,** always form on top of the older layers. By studying these strata, scientists can learn a great deal about them and about the environment in which they were formed—its climate, the size of its oceans, the plants and animals that lived in them and so on. This study is the science of **stratigraphy.** As a simple example, suppose stratigraphers are examining a cliff in which several strata of sedimentary rock are exposed. Ordinarily, they can assume that the lowest stratum is the oldest, and the top stratum is the youngest. This is known as the **principle of superposition.** ("Superposition" means "putting on top of.")

Knowing the relative age of rock strata (that is, their age in relation to each other) enables us to tell other things. Sedimentary rocks often contain fossils—petrified remains of plants and animals that were buried in the sediment before it hardened. By knowing which layers certain fossils were found in, the scientist can tell something about the age of the fossil. Also, the character of the rock itself can tell us much. Different types of rock in different layers often reveal changes in the climate or other conditions of the region.

metamorphic rock

Rocks on the surface of the earth weather away and become soil. Meanwhile, those beneath the surface may undergo other changes. Sometimes a layer of igneous or sedimentary rock becomes more and more deeply buried. Other layers are piled on top of it, weighing it down under ever-growing pressure. At the same time, the rock tends to heat up, both because of the pressure and because of the activity of radioactive substances within it. The rock may become hot enough to melt, turning once again into magma; when it eventually hardens, it will form igneous rock.

But sometimes the rock does not melt. Instead, it undergoes certain chemical and physical changes, caused by the heat and pressure, while still remaining solid. These changes are called **metamorphism,** and the types of rock they produce are metamorphic rock.

What are the changes that occur in metamorphism? At an early stage, any water and gases in the rock pores are squeezed out by the pressure. As the rock gets hotter, chemical reactions change some of the mineral compounds into new ones. These new minerals are usually very dense and hard. They include the precious stones used in fine jewelry—rubies, garnets, sapphires, and so on. Hardest of all are **diamonds,** which are crystallized at very great depth—a hundred kilometers or more beneath the surface.

Another change happens because pressure is usually greater in one direction than another. The crystals of the rock are squeezed, stretched, and lined up in one direction rather than another. This is called **orientation** of the crystals. In some rocks it produces a very clear layered appearance known as **foliation.** Foliated rocks will split easily along the plane of orientation.

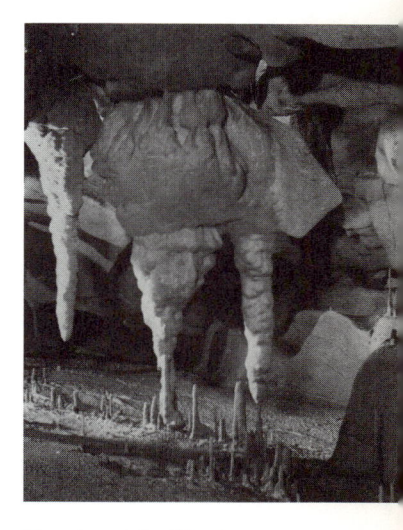

Massive stalactites hang from the roof of a portion of the Carlsbad Caverns. Smaller stalagmites rise from the floor of the cave.

the earth's materials

Manhattan schist is very common in the New York City area. The small garnet crystals embedded in the schist are often readily visible to the casual observer.

types of metamorphic rock

So many minerals are formed by metamorphism (there are over fifty common ones) that it would be confusing to classify the rocks by their mineral type. A simpler system is to classify them by their crystal size and by whether or not they are foliated.

There are four main types of foliated metamorphic rock. **Slate** is fine-grained and splits easily into thin, smooth plates with a polished-looking surface. It is usually formed from shale and similar sedimentary rocks. **Phyllite** is a slate that has undergone further change. Some large crystals, usually of mica, have formed among the small ones. These large crystals are visibly stretched and flattened in the direction of orientation. **Schist** is coarser-grained, with very clear, thin planes of orientation. **Gneiss** is similar to schist, but coarser, with alternating light and dark bands, and not quite so evenly foliated. Both schist and gneiss are sometimes formed from granitic rock, but schist may also be formed from shale.

Some metamorphic rocks are **nonfoliated.** Many of them seem to have been formed at lower temperatures and pressures than the foliated rocks. Their crystals are not oriented strongly in one direction, so they do not tend to split in layers. The most common nonfoliated rocks are **marble,** formed from limestone, and **quartzite,** formed from quartz-rich sandstone.

the rock cycle

More than half the United States has a surface of sedimentary

the earth's materials

rocks, although they may be covered by soil. Most areas of sedimentary rock were once under water. The surface of the rest of the United States is made up of crystalline rocks. Much of the eastern part of the United States consists of metamorphic rocks, and considerable areas in the western part of the United States consist of igneous rocks. Large areas of Canada's surface rock are metamorphic, covered by a very thin layer of soil. In the central part of Canada, as in the United States, there are very thick layers of sedimentary rocks directly under the soil.

In many places where the surface rocks are crystalline, there is good evidence that sedimentary rocks once covered the present surface rocks. The layers of sedimentary rocks that have disappeared were reduced to debris and washed back into the sea, forming new sedimentary rocks.

It is probable that in some cases the heat deep in the earth was so intense that metamorphic rocks melted and then cooled, forming new igneous rocks. In that way, what originally had been part of an igneous rock passed through a full cycle and again became part of an igneous rock. This cycle is called the **rock cycle**.

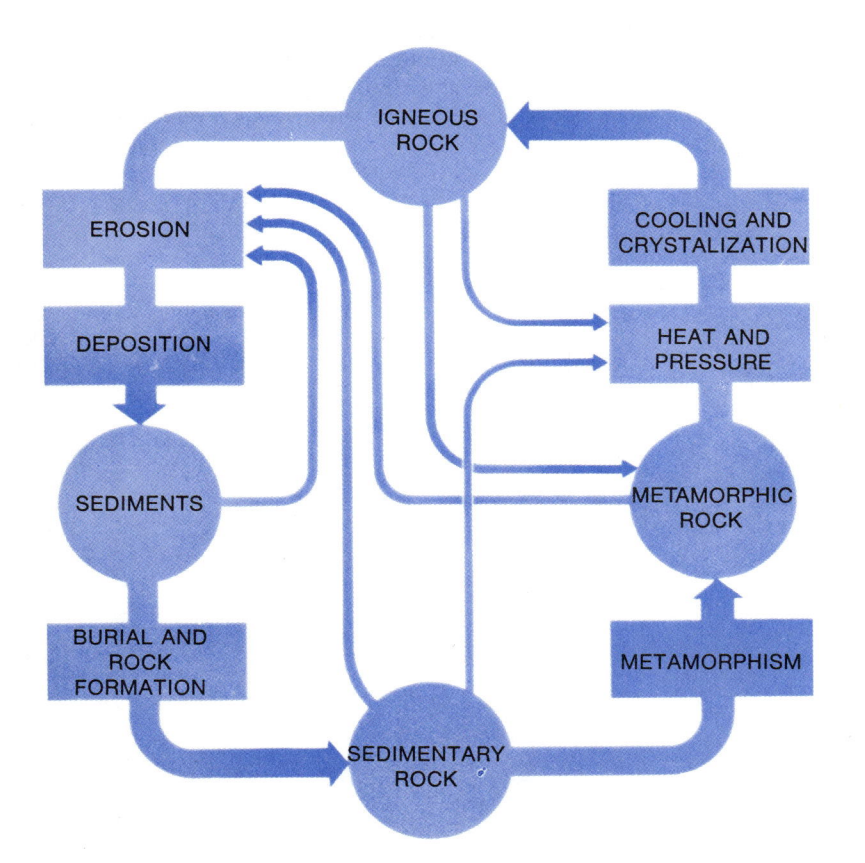

A giant crystal, the "Patricia" emerald, was mined in Colombia, an important source of these valuable minerals. Emerald crystals of this size are quite unusual.

Following the rock cycle, a rock may be changed from one form to another, eventually returning to its original form once more.

the earth's materials

SUMMARY

The simplest natural substances are the elements. But most substances are compounds, combinations of two or more elements. Atoms are the fundamental unit of matter. The most important atomic particles are electrons, protons, and neutrons. When atoms link together by bonding, they form molecules, the smallest recognizable parts of a compound.

Light is an electromagnetic wave which travels at 300,000 km/sec. Light possesses velocity, wavelength, and frequency.

The elements are divided into several groups, including the inert gases, the alkaline metals, the halogens, and the alkaline earth elements.

Each mineral has a definite chemical composition. Most minerals have a crystalline structure, but a few are amorphous. In some cases, the same chemical substance can have different internal structures.

Six basic crystal structures exist. Crystals may be isometric, tetragonal, orthorhombic, monoclinic, triclinic, or hexagonal. The structure depends to some extent on how the atoms are packed together.

Minerals may be identified on the basis of several properties. Among these are color, streak, luster, hardness, and cleavage.

Rocks are classified as igneous, sedimentary, or metamorphic. Igneous rocks were formed by volcanic processes. Sedimentary rocks were formed from deposits of solid materials on the beds of bodies of water. Metamorphic rocks result from the effects of pressure and/or heat on igneous and sedimentary rocks.

Sedimentary rocks are particularly useful for studying the earth's history. According to the principle of superposition, the lowest stratum of sedimentary rock is the oldest, and the upper layers become progressively younger. Sedimentary rocks often contain the fossilized remains of plants and animals.

the earth's materials

REVIEW QUESTIONS

1. Distinguish between an element and a compound.
2. Name and describe the three most important atomic particles.
3. Name and define three properties of light waves.
4. Distinguish between ionic bonding and covalent bonding.
5. What is a crystalline mineral?
6. Name and describe the three classifications of rocks.
7. Explain the principle of superposition.
8. How are metamorphic rocks formed?
9. Describe Bowen's reaction series.
10. What is a molecule?
11. List at least four simple ways of identifying a mineral.
12. Describe the four main types of igneous rock.
13. Describe the general structure of an atom.
14. Name and describe at least three groups of elements.
15. Distinguish between mechanical and chemical weathering.
16. What is the velocity of light?
17. List the six basic types of crystals.
18. What do we call the lowest energy state of an atom?

CHAPTER THREE

CHANGING THE EARTH

Long before the descendants of Europeans set foot there, the northwest coast of North America supported a prosperous Indian culture. Its people lived by hunting, fishing, and gathering; their villages were close to the water's edge, in sheltered bays and coves. Rainfall was abundant; forests covered the steep hill slopes that ran down almost to the shore. In some places migrating whales and seals cruised near enough to be hunted by daring paddlers for their meat, blubber, skin, and oil. There was plenty of wood for building; cedar bark, dried and split, was skillfully woven into baskets. During the winter there was time to hold great ceremonies, or carve ingenious masks, combs, bowls, clubs, and other articles of ritual and practical use. The villagers worked hard, and the men risked their lives each time they paddled out in their canoes to seek the whales, but they lived well.

  It was early June, sometime in the last years before the first traders arrived in their strange ships. On what is now the Olympic Peninsula of Washington, the rains had been heavy, and the ground was deeply soaked. But in one village, daily life went on much as it had for some 2,000 years. The hunters were out. Someone brought from the forest a basket of fresh-cut cedar bark and set it aside to be dried and split the following day.

  That night the village died. Perhaps there was a slight earthquake. In any case, the rain-soaked clay hillside behind it gave way, and tons of mud poured down the steep slope. No one caught in its path can have escaped. Houses were knocked askew and buried, with all their contents, as much as eight to twelve feet deep.

  Within a few years the village was rebuilt. Its location was good, and the people of the Makah tribe remembered that this was their ancestral land. Under the name of Ozette, it lasted until a generation ago. Perhaps the old disaster was

changing the earth

forgotten. But under the mud the smashed houses, the carved bowls and combs, the harpoons and fishhooks, the neatly rolled cedar bark — even a few green leaves someone had dropped on the floor that last evening — all lay hidden, later to be found and excavated by twentieth-century archaeologists. The massive mudflow, bringing death and destruction to one generation, preserved the record of a culture for people of another age.

Such incidents must have been common in the centuries of human history. They serve to remind us forcefully that this seemingly solid earth is actually in a state of constant change. The planet's generally rough-hewn shape is the result of great internal forces, which we will discuss in the chapters to follow. But forces that operate on the surface of the earth — water, wind, and glacial action — are responsible for polishing it into the familiar landscapes we see.

In the last chapter we discussed weathering, the chemical and mechanical processes that break solid rock down into small pieces and often change its very composition. The ultimate product of weathering is soil. It takes a long time to produce soil; thousands of years of weathering are often necessary to build a single inch.

soil

Geologists generally use the term **regolith** to refer to all the loose surface material that lies above the solid bedrock. This material may include glacial deposits and sediments as well as true soil. The distinguishing feature of true soil is that it contains not only minerals but organic components as well. Soil, unlike the rest of the regolith, is capable of nourishing plant life.

The organic material in soil, known as **humus,** is produced by the biological activity of living things. It may include the decayed remains of plants and animals as well as the wastes they produced while living. Humus furnishes nutrients for plants and bacteria. The amount of humus in a particular type of soil is usually a gauge of the soil's **fertility,** its ability to support plant life. Soil that is rich in humus is generally very fertile.

Along with organic and inorganic solids, soil also contains water and air, which circulate in the spaces or **pores** between the soil particles. The porosity of a soil affects its fertility. A soil that is too porous will allow water to escape from it quickly, either by evaporation or by percolating down into the regolith. A very non-porous soil, on the other hand, resists penetration by water and air. Ideally, about half the volume of a soil should consist of pore spaces, which not only allow a reasonable amount of water and air to circulate but also enable the roots of plants to spread downward easily.

changing the earth

influences on soil formation

Many factors influence the type of soil formed by weathering processes. Most geologists believe that climate is the most important of these. Warm, humid climates, for instance, permit a greater amount of chemical weathering to take place. In cold, dry climates weathering is predominantly mechanical. Chemical weathering usually produces a heavier layer of soil within a given time than mechanical weathering can.

At one time it was believed that the type of soil formed was strongly dependent on the type of **parent rock** which was subjected to weathering, and that different types of parent rock produced different types of soil. Now, however, geologists believe that no matter what the type of parent rock—igneous, metamorphic, or sedimentary—in a particular climate, it will break down into approximately the same type of soil if the weathering processes continue long enough.

As this suggests, time is another important factor in the formation of soil. Generally speaking, the longer the weathering processes continue the heavier the layer of soil that is formed. As weathering goes on, the soil's resemblance to the parent rock decreases.

We mentioned earlier that the amount of organic matter in a soil partly determines its degree of fertility. Thus, the animal and plant life in a region are also important factors in soil formation. Earthworms and burrowing insects are particularly important to soil formation in temperate, humid climates. Not only do their movements stir up the soil and allow more water and air to penetrate, but they also consume organic material and dispose of it in the form of wastes that enrich the soil.

the soil profile

When we dig down into the soil we usually find that its composition varies. In most soils we can distinguish several different layers or **horizons,** which together make up the **soil profile.**

Starting from the top, we find the **surface soil** or **A horizon.** This layer is also known as **topsoil.** The upper part of the surface soil is usually rich in humus, and so appears darker than the lower layers. As water penetrates the surface soil, it leaches out some of the soluble material and carries it downward.

The next layer is the **subsoil** or **B horizon.** It is also known as the **zone of accumulation,** because it contains the materials that were leached out from the surface soil. Although the subsoil is thus enriched, it contains less organic matter than the surface soil. Together, the surface soil and the subsoil are often called the **solum** or true soil.

The next layer is called the **C horizon.** The C horizon is a transitional layer between the solum and the bedrock. Material in this layer has undergone only partial weathering. It is made up of coarser fragments than the two upper horizons, and contains very little organic matter. Below it lies the solid, unweathered bedrock. When the three horizons are very distinct and well-developed, we say that the soil is **mature.**

changing the earth

Humus

Sandy loam (less humus)

Loosely packed clay

Well packed clay

Partially weathered soft rock

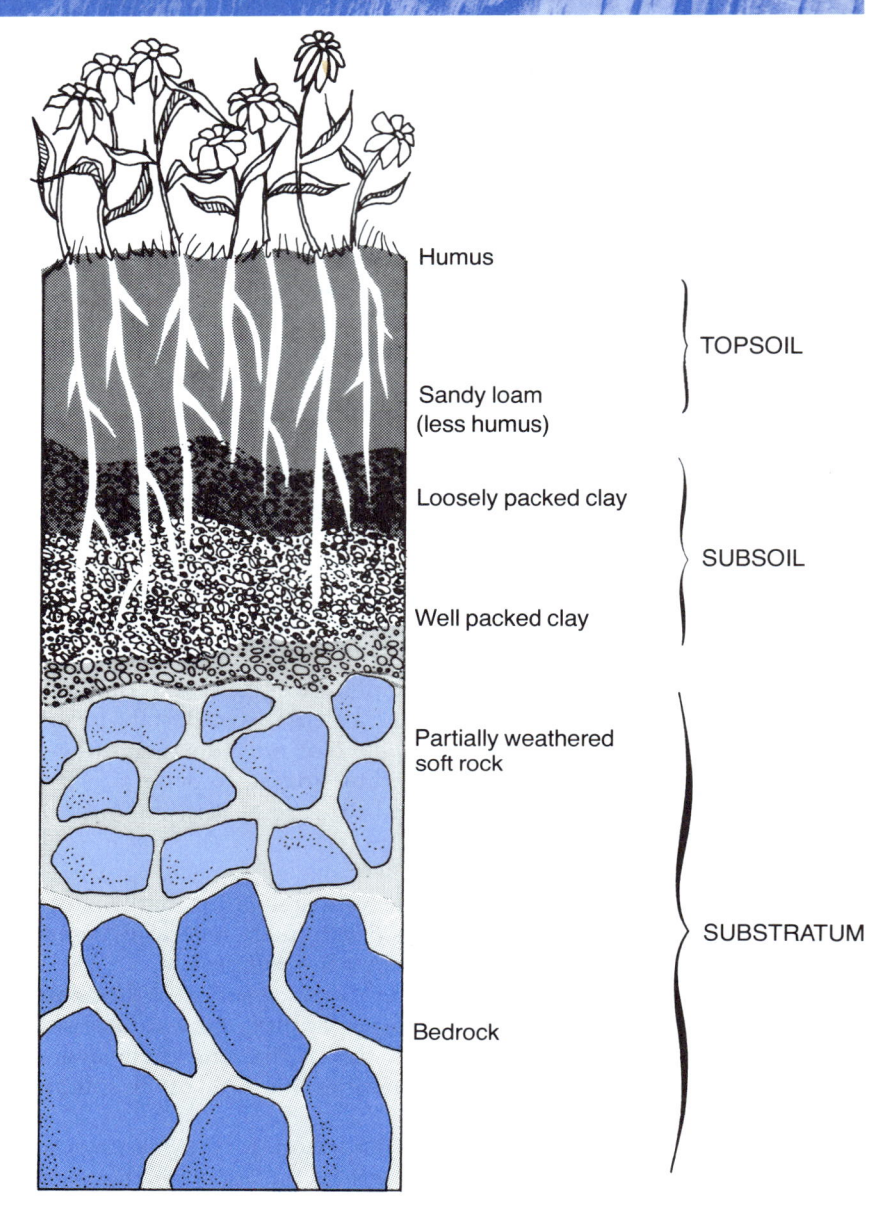

Bedrock

TOPSOIL

SUBSOIL

SUBSTRATUM

Soil profile in a temperate humid region. The topsoil rich in organic matter rests on clay, which in turn rests on partially weathered rock. Unweathered bedrock is found at the bottom. The soil profile differs from place to place, according to the climate.

types of soil

Since soil is largely created by weathering, it is not surprising that its composition varies according to climate. Different parts of the world with similar climates are likely to have similar soils. The rain forests of Africa and South America have similar soils, and so do the subarctic zones of North America and the Eurasian land mass. The classification of soils is a very complex subject. At least ten general types of soil and forty subtypes are known. Here, of course, we can only discuss a few of the more common types of soil.

changing the earth

At least two major soil types are found in North America. In the humid eastern and northwestern parts of the continent, we find **pedalfer.** This soil takes its name from the Greek prefix *ped-*, meaning "soil," and the suffixes *-al* and *-fer*, which represent the chemical symbols for aluminum and iron. Pedalfers contain an accumulation of iron oxides and aluminum-rich clay in the subsoil.

Pedocals are the typical soils in the western half of the United States and northern Mexico, regions with a generally dry, temperate climate. *Pedo-* again stands for "soil," and *-cal* for calcium. These soils contain a heavy concentration of calcium carbonate.

In moist, tropical climates a variety of **laterites** can be found. These soils begin with much the same composition as pedalfers, but the different climatic conditions change them. During much of the year they are saturated with water, which filters downward and carries away silica and other soluble substances. Most subsoils contain clay and iron oxides, but in laterites this material is replaced by a concentration of iron oxides and hydrous aluminum oxide. Lateritic soils take their name from the Latin word for brick, and in fact are so hard when dried that they are sometimes used for making bricks. They have proved to be a major obstacle to agriculture in the tropics. Since the heat and humidity of these regions encourage a high level of bacterial activity, vegetal matter decays very fast and laterites therefore have a low content of humus. This, along with the fact that they dry to such hardness that they are difficult to cultivate, makes them unsuitable for agricultural use.

In most arctic regions soils are relatively immature, because chemical changes take place slowly at low temperatures. In some areas **tundra soil** has developed. Tundra soils have a shallow surface layer that thaws during warm weather. But the average temperature of these regions is so low that the deeper layers remain permanently frozen, and are known as **permafrost.** Like the laterites, these arctic soils have little agricultural value. Since vegetation is sparse in such climates, arctic soils contain only tiny amounts of organic material, and the presence of permafrost below the surface permits only shallow cultivation.

erosion

Once the processes of weathering have broken up solid bedrock into smaller particles, those particles can be moved from one place to another by water, glacial action, or wind. This is known as **erosion**—the removal, transport, and deposition of parts of the regolith. Water is by far the most important agent of erosion. Wind and glacial movement play a part only in certain regions.

IN FOCUS:
the geologist as detective

One autumn day in 1904, a seamstress was strangled to death with her own scarf in a bean field near Frankfort, Germany. The police had apprehended a suspect, but lacked sufficient evidence to charge him with the crime. Finally, they decided to call in one Georg Popp, a chemist, microscopist, and geologist, to examine their sole lead—a dirty handkerchief found at the scene of the crime. Popp analyzed the nasal mucus on the cloth and found that it contained snuff, coal dust, and grains of the mineral hornblende. The police knew that their suspect used snuff. They also knew that he worked at a nearby granite pit as well as at a gasworks that burned coal.

This information prompted Popp to examine scrapings from beneath the suspect's fingernails. Analyzed under the microscope, the scrapings contained telltale coal and hornblende. Popp then found that soil from the man's trousers contained minerals that matched those in the soil at the murder scene. In the face of this evidence, the murderer confessed.

This neat bit of sleuthing has been related by geologist Raymond Murray. As Murray points out, Popp put his detective skills to work in many other criminal cases. For his contributions, he deserves to be called the father of forensic geology—the branch of geology which uses earth materials, including rocks, soil, fossils, and minerals, as evidence in criminal proceedings.

The beginning of this intriguing science has been traced by Murray to the British author Sir Arthur Conan Doyle, creator of the world's most famous fictional detective, Sherlock Holmes. Holmes, according to his assistant Doctor Watson, could immediately determine the exact part of London where any sample of soil had originated. An amateur gumshoe himself, Sir Arthur helped obtain a pardon for a man imprisoned for mutilating and slaughtering farm animals, by showing that the soil on the convicted man's shoes did not match the soil at the scene of the crime.

Today, forensic geology has moved far beyond the one-man methods of Holmes and Popp. Earth materials are examined in thousands of criminal investigations each year. The Federal Bureau of Investigation, for example, employs earth scientists to analyze geological evidence in its sophisticated Washington laboratories. Indeed, the F.B.I. is considered a leader in the field of forensic geology. Important forensic geology centers are also located in several other states, as well as in Canada and Great Britain.

From a legal standpoint, geological materials fall into the category of physical evidence known as class evidence. This type of material, which includes such things as hair, blood, and paint, can originate from a wide variety of sources. By contrast, individual evidence—fingerprints and spent ammunition, for example—can come from only a single source. Since the usefulness of a particular item of class evidence depends upon how common it is, geologic evidence is most useful because it is nearly impossible for a person to walk through a park, a field, or even a city street without picking up some geological material on his body and clothing. The different types and combinations of rocks, minerals, and soils are virtually unlimited. Igneous and metamorphic rocks contain a wide variety of minerals. Besides appearing in different types and quantities, such minerals also are found in different sizes and textures. Weathering processes add to the diversity by breaking minerals down and forming them into entirely new combinations. Most important, all of these materials can be precisely identified.

Synthetic mineral products—laundry powder, wallboard, masonry, and insulating materials, to name a few—can also help the forensic geologist in criminal cases. Murray relates a particularly interesting case. A man arrested for a minor crime appeared at the police station with what looked like a bad case of dandruff. A sharp-eyed detective took a closer look at the man's hair and saw that the fallout was not what it seemed to be. Microscopic analysis revealed that the mysterious material was a dust composed of diatoms—tiny fossilized algae from the diatomaceous earth that is often used as an insulating material in safes. When such safes are drilled, blown, or pried open, the insulation loosens and spills into the immediate vicinity, frequently settling on the clothing of anyone nearby. This knowledge led the police to compare the diatoms on the suspect with those in the insulation of a safe that had been burgled the day before. The diatom species from the two sources matched, and the suspect was charged with the crime.

Forensic geology is attracting a growing number of earth scientists who wish to use the skills and tools of their profession to further the administration of justice. As more and more crimes are solved with the help of these modern-day Sherlocks, geology will undoubtedly play an increasingly important role in crime detection and investigation.

changing the earth
surface water

Because the land is never perfectly level, water always flows from one point to another on the surface. **Surface water** is one of the most powerful agents of erosion.

Any water moving across the land in a channel is called a stream, regardless of its size. Streams perform three kinds of work as they flow: they erode material, transport it, and deposit it. Although all parts of a stream transport material, some parts may perform more erosion than deposition, and in other parts the reverse is true.

A **stream system** is composed of all the streams that unite and eventually reach the ocean as a major river. The land from which a stream or stream system collects water is called its **watershed.** A watershed can be part of a single mountainside, or it can include millions of square kilometers of land. The watersheds of two neighboring stream systems are separated by a **divide,** a highland from which water flows in two different directions.

Many streams flow together to form a mighty river. The watershed of a large stream system may extend over millions of square kilometers.

The steepest parts of a stream system are generally near the **headwaters** where individual streams begin. The flattest parts of the system are near the **mouth,** where the stream empties into a body of still water. As a stream erodes its bed, the average slope of the land over which it flows becomes less steep.

changing the earth

If nothing interfered, erosion would eventually reduce the slope of a stream until it was almost level. Water would then flow slowly all over the land. Moving so slowly, it would not be able to erode its bed any further. This lowest level to which a stream can cut is called its **base level.** In practice, it is roughly the level of whatever body of quiet water the stream is entering—pond, lake, or ocean. No major stream system ever completely reaches base level. Geologic changes, such as uplift of land, prevent this from happening. The lower part of the Mississippi River system and the last few kilometers of the Colorado River system are almost at base level.

Oxbow lakes and complex meandering structure of White River, Des Arc. Prairie County, Arkansas.

The steepness of a stream's course is defined by how much change in elevation occurs per unit of distance—the **stream gradient.** A stream that flows over steep ground has a high gradient; it may drop as much as 30 meters in a kilometer. A slow-moving, nearly level stream may drop less than a third of a meter over the same distance. Streams with high gradients flow faster than those with low gradients. The faster a stream flows, the more material it can erode and transport.

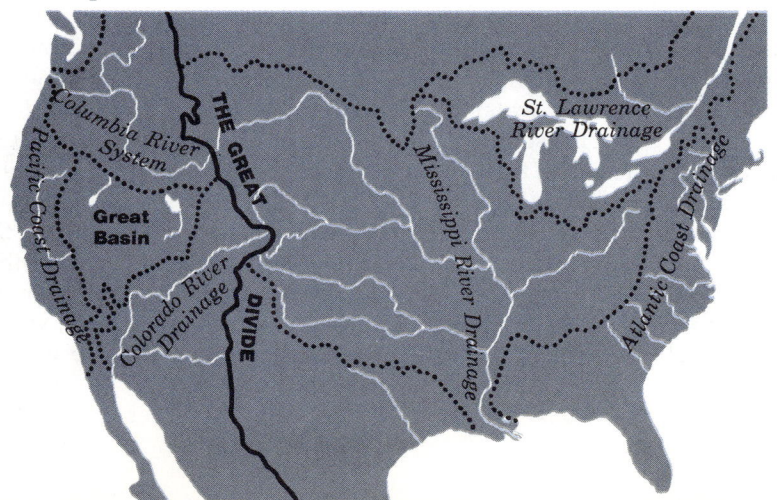

The major stream systems in the United States. Each consists of several smaller systems that join together. The Mississippi River system, the greatest of the stream systems, has a watershed occupying half of the country.

changing the earth
formation of a stream system

A stream system begins when water runs off a high area in a definite path, or **gully.** If water follows a particular gully for any length of time, erosion deepens the gully into a channel. Although the water is flowing downhill, erosion at the highest point of the gully extends the channel uphill. As this **headward erosion** continues, the stream channel collects water from a larger and larger area.

No natural stream flows in a straight line for any great distance. It must swing around obstacles—boulders, fallen trees, whole hills. As its gradient becomes lower, the processes of erosion tend to exaggerate, rather than straighten, the curves produced by these obstacles. Eventually this action produces a series of winding curves, or **meanders,** flowing through a broad area of level land known as a **floodplain.**

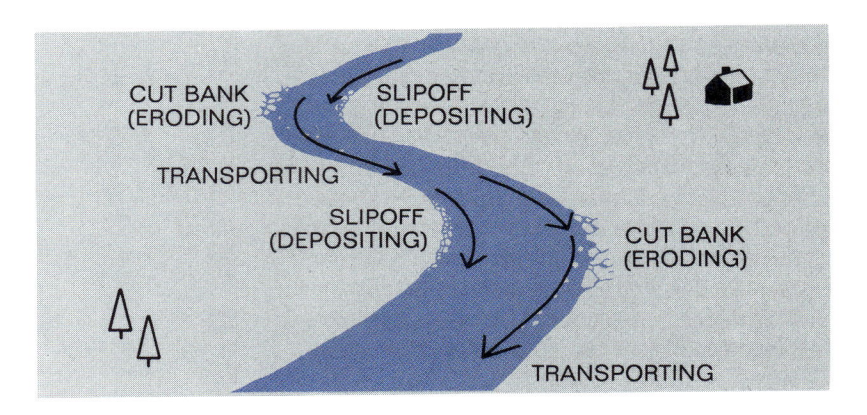

The erosion of the riverbank takes place at the cutbank; the deposition of sand and stones takes place at the slipoff.

How does this happen? To begin with, as the gradient becomes lower and water movement slows, the stream tends to cut into the sides of its channel rather than the bottom. In addition, as we shall see a little later, the velocity of flowing water varies within different parts of a stream. Among other differences, it is greater along the outside of a curve than at the inside. (It is easy to visualize this if we bear in mind that the water on the outside has farther to go to get around the curve.) Because it flows faster, it erodes more material from the bank on that side. Then, when it reaches a curve in the opposite direction, this water is on the inside of the curve, so it slows down and deposits some of the sediment it is carrying. By such means, over the years, the curve becomes more pronounced, and also gradually moves downstream.

Sometimes, as a result of heavy rains or melting of snow in the headwater regions, so much water flows into a stream channel that the stream rises above its banks and **floods,** spreading out over the surrounding land. Often the floodwater cuts across meanders, which tends to shorten the stream's length without changing the vertical

changing the earth

distance that the stream has dropped. The stream's gradient is thus increased. The higher gradient increases the velocity, and therefore the eroding power, of the flooding stream.

In mountainous country a sudden heavy rainfall may cause a stream that is normally only a few inches deep to swell to a torrent several feet deep, causing a **flash flood.** The velocity of water during a flash flood is enough to move large boulders. Flash floods often occur with little or no warning, and can do great damage.

When a river flows into an ocean or lake, its speed decreases as it comes closer to base level, until it is barely moving. This may occur at the mouth of the river, or many miles offshore. For example, when the sediment-loaded water of the Mississippi reaches the Gulf of Mexico, it slows down. The coarser and heavier particles settle to the bottom first, and then the smaller, lighter particles, forming large, muddy deposits on the ocean bottom. These are called **deltas,** because many of them are shaped like the Greek letter △.

Some rivers have no deltas. They do not flow directly into the oceans, but into **estuaries.** There are several types of estuaries; some are actually drowned river valleys. All estuaries, however, are partially enclosed bodies of water in which fresh water from a river mixes with the salty water of the sea. Estuaries are interesting to study because they display a gradual change from freshwater life to marine life. The Hudson River in New York and the Amazon in Brazil have estuaries rather than deltas.

Meandering river. Notice the several oxbow lakes.

Debris left over after a flood suggests that human attempts to control natural forces sometime suffer setbacks.

changing the earth

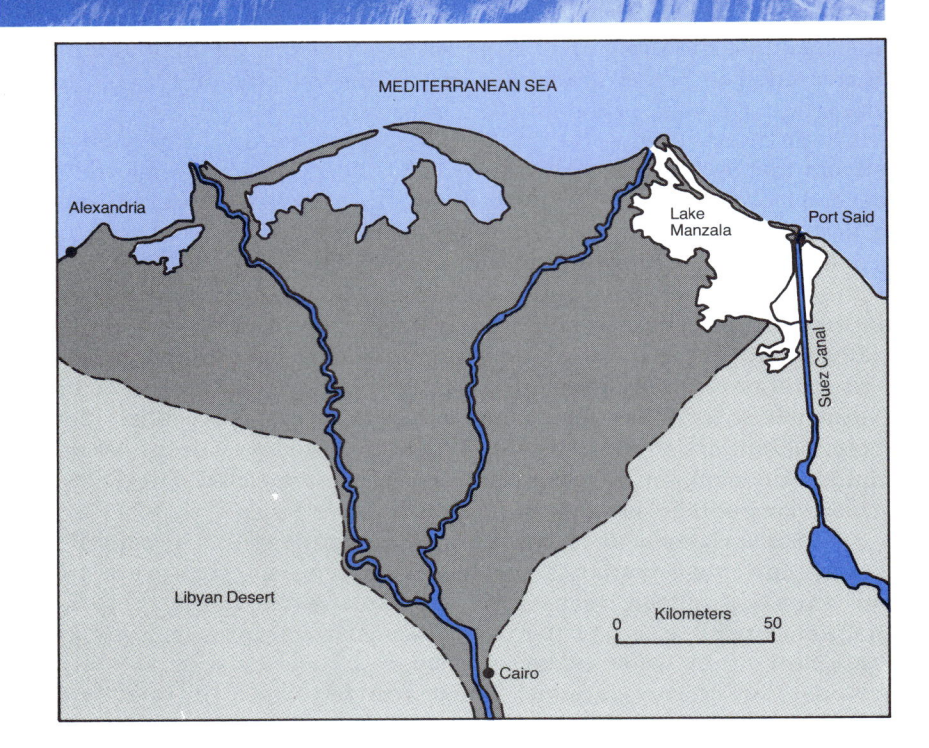

66

The Nile Delta extends over a million square kilometers. Civilization sprang up early in this fertile green region surrounded by hostile desert.

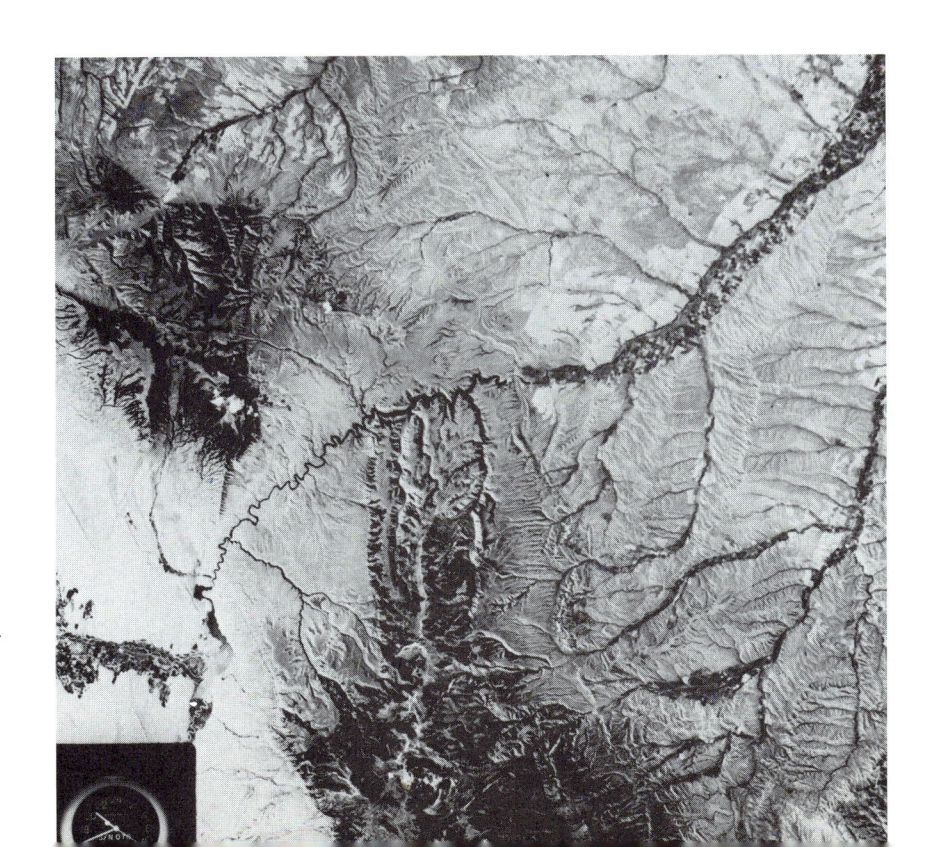

Approximately 9000 square kilometers of north central Wyoming and southern Montana, photographed by the U.S. spacecraft *Skylab* from orbit around the earth. The photo shows the Big Horn River flowing northward and crossing the Big Horn Mountains.

THE DAY THE DAM BURST

At three minutes before noon on June 5, 1976, Idaho's new Teton Dam broke, releasing 80 billion gallons of water. A wall of water 20 feet high roared down the Teton River canyon, drowning 300 square miles of fertile farmland in the Snake River Plain. When the waters receded a week later, eleven people were dead, 3,000 were homeless, and 13,000 cattle had drowned. The topsoil had been stripped from 100,000 acres of farmland, and one billion dollars worth of property had been damaged. A crude hand-written sign, found propped against a telephone pole on a silt-covered street, proclaimed local sentiment about the disaster: "Wanted: damn engineer—dead or alive."

The saddest thing about this catastrophe is that it never should have happened. Earth scientists had warned of serious geological problems some time before construction of the dam got underway. Indeed, in a lawsuit initiated by an environmental group in 1973 to halt the project, an engineering geologist testified that the Teton Canyon's rhyolitic rock was riddled with thousands of fissures and caverns. Water from the dam would seep into this porous lava, make its way around the dam's earth fill, and erode the structure. The dam builder—the United States Bureau of Reclamation—countered this testimony by stating that 520,00 cubic feet of grout—a sand and cement filler pumped into the rocks under pressure—had sealed the most serious cracks.

Apparently the grout did not do the trick. The 300-foot-high dam, which stretched 3,000 feet across the Teton canyon, sprang a leak in mid-morning of June 5, 1976. Bulldozers climbed the sloping face of the dam and attempted to plug the holes with huge boulders, but at 11:57 A.M. the water broke through the dam's north face, carrying half of the $70 million structure with it. The federal government has earmarked some $2 million to help reimburse the victims of the disaster, and may end up paying even more as the result of lawsuits now being filed. The sole bright spot of this unnecessary tragedy is that the dam broke when it did. Had it ruptured during the night when people were asleep, thousands of lives might have been lost.

changing the earth

how streams work

If we observe a stream flowing through a very smooth channel, several things are immediately noticeable. The surface of the stream is quite smooth, especially near the middle. Twigs or leaves floating near the middle move more rapidly than those near the edges, where the water swirls in little eddies.

The smooth central flow is called **laminar** (layered) **flow.** It occurs with little friction between layers of water. The slower movement at the edges of the stream is called **turbulent flow.** In turbulent flow there is much more friction, and the water thus travels downstream more slowly. The eddies in turbulent water are caused by friction between the water and the bed and banks of the stream. This friction erodes particles of soil and rock, and the stream carries them along.

The material carried by a stream is called its **load.** Streams carry their loads in several ways. Very fine particles are carried suspended in the moving water. Materials that dissolve in water are carried in solution. The heavier, larger particles of the load slide or roll along the bottom, and are known as the **bed load** of the stream.

groundwater

Another important agent of erosion is **groundwater,** which moves through soil or porous rock beneath the earth's surface. Groundwater is distributed in such thin layers or tiny drops that it is greatly slowed by friction. It may take weeks, months, or years for it to move a few kilometers. Groundwater thus uses its energy very slowly, and does not have the power to move large amounts of material.

The speed with which groundwater moves downward depends on the size and shape of the particles through which it is moving, and on the amount of space between them. Coarse gravel, for example, is said to be a **permeable** material; the large spaces between the particles allow water to move with relatively high velocity. Clay, on the other hand, often acts as an **impermeable** surface; its particles are separated by such small spaces that very little water can move through it. Soils containing a great deal of clay usually have very poor drainage.

The top of the water-soaked layer of soil underground is called the **water table.** Below the water table the soil and rock are always saturated with water. Water moving downward through the regolith after a heavy rainfall may raise the level of the water table. Conversely, the water table sinks lower during prolonged dry spells because of the natural drainage of the groundwater. The soil above the water table may be moist because water from the water table creeps upward a short distance. Many desert plants are able to survive because this moisture reaches their roots.

The layer of soil or rock through which groundwater flows is called an **aquifer,** a general name for any water-bearing layer. The aquifer extends downward as far as the soil is permeable to water. Sooner or later, though, the water reaches an impermeable layer, often of clay or shale, known as an **aquiclude** (from Latin *claudere*, "to close"). If the aquiclude is level, the water may be blocked from any further movement. It will collect there, and the water table will rise—perhaps eventually reaching the surface and creating a swamp or bog. Often, though, the aquiclude slopes, more or less as does the surface of the ground above. In this case the groundwater will flow slowly down the slope of the aquiclude.

The movement of groundwater is very evident when we observe a spring, where flowing groundwater comes to the surface. A **spring** occurs where the downward slope of the land meets the top of a sloping aquiclude. The water above the aquiclude flows outward and down the side of the surface slope.

Occasionally, an aquifer lies between two aquicludes. The water trapped between the two impermeable layers cannot escape unless one of the aquicludes ends or is broken. When the water in such an aquifer is under very high pressure an **artesian spring** may form. This occurs when the water rises to the surface through a break in the overlying aquiclude. **Artesian wells** may also be drilled through the upper aquiclude to take advantage of the pressure.

Where there is not enough pressure to force the water to the surface, it must be pumped up. Wells either dug or drilled to a level below the water table tap the groundwater that is flowing closest to the surface. Wells of this type may run dry if the level of the water table drops far enough.

When an aquiclude lies above the water table, and is topped by another aquifer, a **perched water table** may be formed. This is a "false" water table located some distance above the real one. It may form natural springs and can be tapped for wells, just as the real one can.

water action in limestone regions

Since limestone is soluble in water, water affects the land in limestone regions in a unique way. As rain falls through the atmosphere, it dissolves carbon dioxide. The water and carbon dioxide combine chemically to form a weak acid called carbonic acid. On the ground, carbonic acid in turn reacts with the limestone, which is solid calcium carbonate. A new compound, calcium bicarbonate, is thus formed. Calcium bicarbonate is also soluble in water and is carried away by the moving groundwater, thus gradually dissolving the limestone formation.

The processes of weathering enlarge any fractures or **joints** in the bedrock. Water enters these joints and widens them by dissolving the limestone. In many limestone regions there are no surface streams, because rainwater that falls on the surface finds its way underground through the joints. There it dissolves the calcium carbonate and forms underground tunnels. As the groundwater pene-

changing the earth

trates deeper, these tunnels are emptied. Eventually they may form a system of caves or caverns.

Water can often be seen dropping from joints in cavern ceilings. As drops of water fall and evaporate, dissolved carbon dioxide is released back into the air. Carbonic acid is no longer present to keep the calcium bicarbonate in solution. The solution therefore breaks up, and the calcium carbonate is deposited again as a form of solid limestone called **travertine**. Stalactites and stalagmites, which we discussed in the last chapter, are formed by the gradual deposition of travertine.

Sometimes in a limestone region so much rock is dissolved along joints that deep saucer-shaped depressions called **sinkholes** are formed. Sinkholes vary in size from a single meter to many meters across, and are often filled with water to form small lakes. Occa-

Stalactites and stalagmites in a Kentucky cave.

changing the earth

Tourists view stalactites and stalagmites.

sionally the roof of an underground tunnel near the surface becomes so thin that it gradually caves in, forming a long depression called a **collapse valley.** Regions in which groundwater has shaped the landscape by dissolution of limestone are known as **karst regions,** named for the Karst Valley in Yugoslavia which is typical of such landscapes.

As a karst region ages, sinkholes enlarge and join one another, collapse valleys form, and eventually most of the limestone is removed. Streams appear in the lowlands, flowing over nonlimestone rocks. Between the streams, steep-sided low hills are all that remain of the original limestone. Different stages in the development of a karst landscape can be seen near Mammoth Cave in Kentucky.

Next to surface water action, **mass wasting** is usually considered the most important form of erosional activity. Mass wasting is land movement over a large area. It is produced by the pull of gravity and affected by the water content of the material, and is generally

mass wasting

changing the earth

creep

classified according to whether the movement takes place quickly or slowly.

The most widespread form of slow mass wasting is **creep,** which includes very slow, imperceptible movements that affect large areas of soil or broken rock debris. Wherever slopes are covered with soil or regolith, these movements are constantly going on. However, they are rarely noticed by the casual observer, because the rate of movement amounts to only a few centimeters, or occasionally as much as a meter, over periods of many years.

As creep takes place, the layers of the regolith move downward at different rates of speed. Ordinarily, the upper layer moves more quickly. As a result, trees, posts, and even buildings on the slope tilt in the direction of the downslope movement, because they are rooted in the slower-moving lower layer of the regolith. Because of the slowness with which it occurs, creep is an unspectacular form of mass wasting, but it nonetheless transports an enormous amount of material.

rapid movements

The rapid forms of mass wasting are far more dramatic. Perhaps the most familiar is the **landslide,** in which enormous amounts of rock, soil, or both move down a steep slope at great speed, often burying every structure at the bottom of the slope. Related to landslides is the **avalanche** of the alpine regions, which involves a similar rapid downhill movement of rock and glacial ice or snow.

Mud flows are rapid movements of water-laden soil. They usually occur in mountain regions, and are also associated with volcanic eruptions, which we will discuss in chapter 6. These great rivers of mud are powerful enough to move huge boulders, automobiles, and even houses that lie in their path. **Rock falls** are caused by the movement of nearly waterless rock fragments breaking away from the sides of steep cliffs. The accumulations of fallen rock at the bottom of the cliff are called **talus.**

wind erosion

Erosion by wind is largely limited to regions in which the earth's surface is covered by fine, loose particles. It is most easily observed in deserts, where sand is not held down by a cover of vegetation. Wind erosion modifies the land forms already shaped by water erosion.

Wind can move dust, silt, fine sand, and, under very strong wind conditions, even moderately coarse sand, in a process called **deflation.** The finer material is picked up and carried in suspension. Material of slightly larger size rolls and creeps along the surface.

Deflation produces **deflation hollows** or **blowouts,** depressions where material has been blown away. Small hollows are found in sand dunes, and relatively large deflation hollows are formed in loose materials. These hollows are at most a few hundred meters deep, and several square kilometers in area. They are not usually obvious to an observer on the ground, but often show up clearly in topographic maps.

The wind helps to carve some unusual rock formations in desert areas. It is assisted by differential weathering, in which certain materials within a rock mass weather and break up more rapidly than others, and then are removed by the wind or washed away by rain. Particles of sand and silt carried in suspension by the wind also cut away the softer portions of rock masses. This action is called **abrasion** or **sandblasting,** and the formations it produces are known as **ventifacts** ("made by wind"). The abrasive action of windborne material often produces intricately carved pedestal rocks, notches, and alcoves in desert cliffs.

Landscape arch in Utah.

Sometimes grains of loose surface material are too heavy to be picked up by the wind, but the wind causes them to jump from one place to another in a process called **saltation** (from the latin *saltare,* "to jump"). One grain strikes a second grain and knocks it into the air, where it travels a short distance before landing. Most wind-

"Fond lover never wilt thou kiss." The words of the poet Keats surely apply to the "Two Lovers," a famous peak overlooking Svolvær harbor in Lofoten Isles, Norway.

Balanced Rock in Utah was produced by sand-blasting.

changing the earth

75

Delicate Arch in Arches National Park, Utah, is an example of the unusual effects that sandblasting can produce over millions of years.

blown material moves within a few centimeters, or at most within a meter or so, of ground level. Dust and very fine sand, though, are sometimes carried high in the air, and may be blown hundreds or thousands of kilometers.

wind deposits

The most impressive and noticeable deposits of the wind are **sand dunes,** which can be seen not only in deserts but also along sandy ocean coasts. The popular image of a desert is a wasteland of sand, but actually sand covers only a small fraction of most deserts. Only about one-seventh of the huge Sahara desert is covered by sand dunes; the rest is stony or rock-floored. About 35 percent of the Arabian desert and 10 percent of the desert area in the United States are covered by sand dunes.

Dunes vary from a few meters high and a few meters across to many kilometers in length and heights of more than 100 meters. They come in a great variety of shapes, but the most familiar are **crescent**

changing the earth

dunes, also called **barchans.** These dunes are formed where there is a moderate supply of sand and a prevailing wind. The points of the crescent lie downwind from the prevailing wind—that is, in the direction toward which the wind is blowing. Prevailing winds also move the sand of such dunes forward, while preserving their shape. The wind carries material up the windward slope and over the top of the dune, where it then settles on the leeward side which is sheltered from the wind. Moving dunes become fixed and stationary if a vegetation cover grows on them, but if the cover is removed they may start moving again. As they creep across the country the dunes often overrun forests and buildings, which later reappear when the dunes move on.

Sometimes wind-blown silt accumulates in sheltered places. There it forms a soft, fine, buff-colored material called **loess.** Loess is a permeable, unlayered sediment consisting primarily of microscopic particles of quartz and calcite. Great deposits of loess are found in northern China, resulting from the long accumulation of dust swept from the desert basins of central Asia. Extensive deposits are also found in the central parts of the United States.

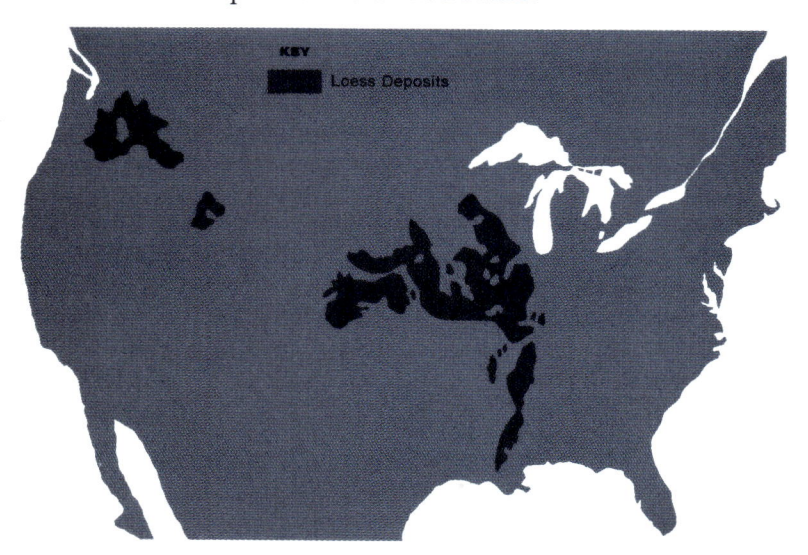

Sand dunes.

Loess deposits in the United States. The loess probably originated elsewhere and was transferred to its present location by winds.

Soil scientists believe that much of the loess in the world was originally laid down in deposits which washed out from glaciers (we will discuss this subject in the next section). When the deposits dried out, the wind then picked up the fine silt and deposited it where it is now found.

Important loess deposits lie in the Mississippi River valley, and corn is extensively cultivated on the loess plains in states such as Iowa and Illinois, where there is enough rainfall. Wheat is grown farther west on the loess plains of Kansas and Nebraska and in eastern Washington.

Sleeping Bear Dunes National Lakeshore in Michigan is noted for its beaches, sand dunes, forests, and lakes. It is still in the process of formation.

The importance of loess as a world agricultural resource cannot be easily overestimated. Loess plains and plateaus have developed rich, black soils especially suited to cultivation of grains. The highly productive plains of the Ukraine, the Argentine pampas, and the rich grain region of northern China are underlain by loess.

Some loess must have been deposited during great **dust storms.** Dust storms on a lesser scale continue to occur in many regions. They are generated where ground surfaces have been stripped of protective plant growth by cultivation or overgrazing, or where no vegetation cover exists because of an extremely dry climate. In dry seasons over such regions, strong, turbulent winds lift great quantities of dust into the air. A dust storm appears to the observer as an immense, dark, moving cloud, extending from the ground to heights of a thousand meters or even more. Within the dust cloud, deep gloom or even total darkness prevails. Visibility is cut to a few meters, and a fine choking dust penetrates everywhere.

Soil scientists have estimated that as much as 4,000 tons of dust can be suspended in a cubic kilometer of air. At that rate, a dust storm 300 or 400 kilometers in diameter might be carrying more than 500 million tons of dust—enough to make a hill 30 meters high and nearly 3 kilometers in diameter at the base.

During dry periods, dust storms occur in the American Midwest and Southwest. During the severe drought of the 1930s, these storms were so severe that the region including western Kansas, Oklahoma, and the Texas panhandle was nicknamed the Dust Bowl. Much of the fertile topsoil was blown away, creating a major agricultural disaster.

glacial change

Glaciers are large natural accumulations of ice, which move slowly over the land under the force of gravity and their own enormous weight. In high mountains glaciers form in long, narrow valleys and are called **valley glaciers** or **alpine glaciers.** In arctic regions, prevailing temperatures are so low that ice can accumulate over broad areas. This type of ice mass is called a **continental glacier, ice sheet,** or **icecap.** The sheer weight of the ice exerts tremendous pressure on the underlying terrain, so that as it moves downslope, a glacier is a powerful agent of change.

Ice itself is an important erosional force. If the weather is warm, glacial ice melts during the day and fills the joints in the bedrock with water. When that water freezes as temperatures drop after nightfall it expands, forcing the joints apart and sometimes shattering the rock. Fragments of rock are often frozen into the ice, and thus become part of the glacier.

After many years, a mountain glacier will remove enough rock from the area where it begins to make a depression shaped like a

Sperry Glacier on the west side of the Continental Divide. This glacier has been melting steadily. Its present area of about one square kilometer is only a third of its extent in 1901.

half-bowl. Where a glacier has disappeared, it looks as though a giant ice-cream dipper has scooped out a piece of the mountain. This steep-walled basin in which a glacier originates is called a **cirque.** A small lake known as a **tarn** is sometimes left in a vacated cirque.

Usually more than one cirque will form on a mountain. High ridges called **arêtes** are left between the cirques. If at least three cirques and arêtes form around the top of a mountain, the peak has a very rugged, pointed appearance. Glaciated mountains of this type are called **horns;** the most famous example is the Matterhorn, in Switzerland.

Rivers in mountain valleys are rarely more than 3 or 4 meters deep, and so do not extend high up the valley walls. Where streams flow at a constant velocity, the walls slope sharply to the valley floor, forming a V-shaped cross section. Valley glaciers, on the other hand, may be as much as several hundred meters thick. They extend far up the valley wall, and their erosive action constantly deepens and

changing the earth

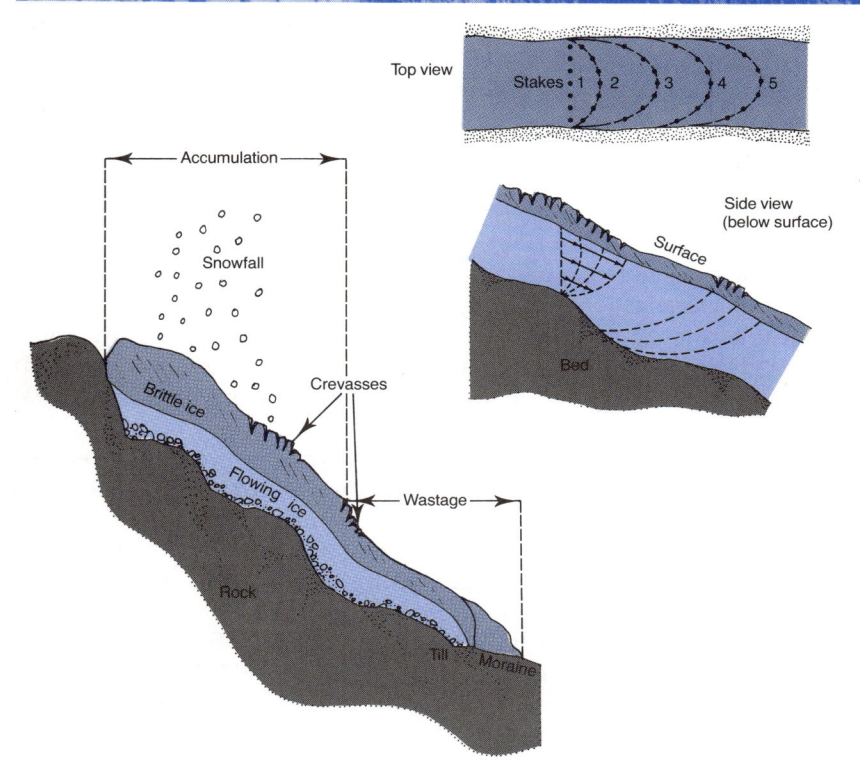

Pressure of ice causes glacier to flow. The flow is faster at the center than at the edges. Stakes driven in a straight line across a glacier gradually take up positions along an arc as a result. The surface of the glacier develops great cracks as it adjusts to the flowing ice underneath.

broadens the channel through which they flow. The result is a **glacial trough** which is U-shaped in cross section.

Just as rivers have tributary streams, large mountain glaciers also have tributaries. In valleys from which glaciers have disappeared, U-shaped **tributary valleys** often enter the main valley from the side. Some tributary valleys are at the same level as the main valley they join. In many cases, however, the floor of the valley of the tributary glacier enters the main glacial valley high up on its walls, because the weight of the tributary glacier was not great enough to cut its valley as deep as the main valley. These features are therefore called **hanging valleys.** Streams flowing out of them often tumble downward in spectacular waterfalls.

Glacial troughs differ in shape from river valleys. Rivers generally cut deeper than glaciers.

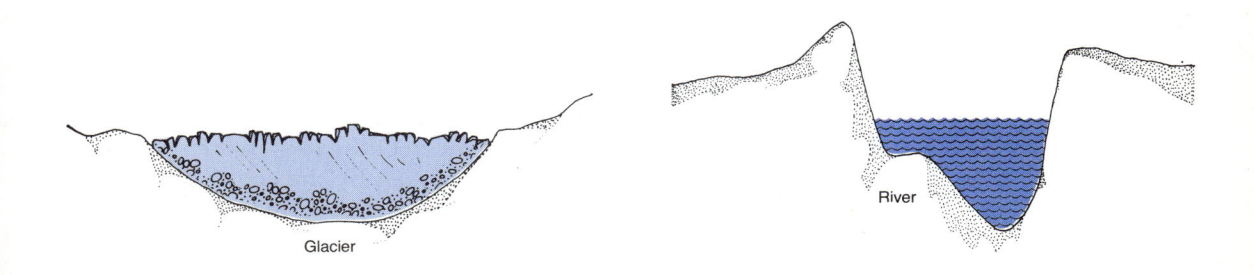

changing the earth

glacial formations

As a glacier moves down a valley, fragments of rock that are frozen into the ice scratch and scrape the bedrock. If the bedrock is soft the glacier digs into it quite rapidly. Where the bedrock is hard, however, the glacier can only abrade and polish it, leaving marks. Small marks that are barely more than scratches are called **striae,** and may also be found on the flattened faces of stones carried along by the glacier. In some places long, deep grooves are cut into the bedrock. Striae and glacial grooves are some of the clues which enable geologists to recognize an area that once was glaciated.

Glaciers last only as long as the climate is cold enough to produce at least as much ice as will melt in a year. On its downhill journey, a glacier eventually arrives at a place where ice melts faster than it is produced or than it can move down from upper parts of the glacier. This is the end or snout of a glacier.

Mountain glaciers follow periodic cycles of growth or decline. In cold, wet years they grow rapidly and move further downslope. In years that are warm and dry enough to melt all the new snow as well as some of the packed snow of previous years, the snout of the glacier melts, and the glacier is said to be retreating. The last time climatic conditions were just right for sustained growth of new glaciers was between about 1000 A.D. and 1500 A.D. Since then, warm periods have been alternating with cooler periods. Many scientists believe that we are now entering a cold period, since recent photographs taken from satellites show that glaciers have been advancing during the last ten years.

glacial deposits

When a glacier retreats, the debris that was frozen into it is deposited as the ice melts. When a stream deposits its load, it drops the heaviest particles first and the lightest particles last, but glacial deposition is quite different. Because of the way a glacier carries its load, particles are not sorted when the ice melts. Thus, the particles dropped by a glacier are made up of everything from finely ground-up rock, called **rock flour,** to huge boulders or **erratics**. This unsorted material is collectively known as **till.**

The piles of glacial till dropped by a glacier are called **moraines,** and the various types of moraine are named according to the position they had in the glacier. Piles of debris left at the snout of a glacier are called **terminal** or **end moraines.** Those left along the sides of the ice are known as **lateral moraines.** Debris dropped from the bottom of the ice along the ground forms **ground moraines.** Material carried in the middle of a glacier that formed when tributaries joined is dropped to create a **medial moraine.**

When a glacier is growing, the ice presses against the cirque and valley walls. At the same time, much of the rock in the upper part of the cirque is being cracked and broken by the action of freezing water, which causes masses of rock to crumble and fall. When the glacier disappears, there is no longer anything to hold the rock walls

Terminal moraine of Clements Glacier, a retreating glacier located in the cirque of a U-shaped valley from the Ice Age. The glacier no longer exists.

in place except the strength of the rock itself. Further cascades of frost-broken stones collapse into the valley. In some places millions of tons have fallen, forming great piles of debris called **talus slopes** which sometimes extend far down the valley.

If a glacial trough is formed in an area of soft rock, the ice may erode a basin, which becomes a lake when the ice melts. Other lakes may form behind the dam created when ice and moraine block the mouth of a tributary valley. In time, such lakes often drain by cutting through the moraine dam.

In the cold climate of high latitudes, mountain glaciers may descend to sea level and even push out into the ocean. So long as the ice is much thicker than the depth of the water, the glacier is heavy enough to rest on the sea bottom, and it deepens its trough underwater. But as it pushes out into deeper water, the buoyancy of the ice raises the snout off the bottom, and huge blocks of ice finally break off from the snout and become icebergs. If a glacier as it approaches the ocean cuts a trough deeper than the sea level, the sea invades the trough after the glacier disappears, producing a long, narrow steep-walled bay. Such a flooded glacial valley is called by its Norwegian name, **fjord.** Many fine examples of fjords are found along the Alaskan coast.

continental glaciers

So far we have been considering types of erosion and deposition caused by valley glaciers. The effects produced by continental ice sheets are very different.

At one time or another, glacial ice covered most of the land north of the Missouri and Ohio rivers and between the Appalachian Mountains and the northern Rockies. Ice flowed across New York, northern New Jersey, and New England, and onto what is now ocean bottom. The water that formed the huge glacial ice sheets came from

This fjord in Norway was cut by a valley glacier during a recent glacial period. When the glacier retreated the sea flowed in to take its place.

View of Geiranger in Norway's fjord country, as seen from Flydal Canyon. Oceangoing ships, such as the one in the picture, often travel up and down the the larger fjords.

changing the earth

An isolated kame in eastern Wisconsin. The material in a kame is stratified and sorted.

the oceans, lowering the level of the sea. Scientists have found indications that during this glacial period the sea level was as much as 100 meters lower than at present.

The lowering of sea level by this amount would result in marked changes in some places. For example, the water that separates Siberia from Alaska is less than 100 meters deep. What is now the Bering Strait must have been dry during the ice age, forming a land bridge that connected Asia with North America. The British Isles were attached to Europe during this period, and the islands of Sumatra, Java, and Borneo were part of the mainland of southeastern Asia. Farther east, Australia and New Guinea formed a single landmass.

The continental ice sheets did not leave cirques and glacial troughs behind. The terrain they once covered is gently rolling country. But they did leave piles of glacial till similar to the ground moraines of valley glaciers. Great blocks of ice were trapped in these moraines. When the ice melted, the remaining debris settled, forming depressions called **kettles** in the moraines. Some kettles are now lakes and ponds, while others are dry. Moraines which are marked with kettles are called **knob-and-kettle moraines.**

Material of glacial origin that is carried away from the melting ice by streams is called **fluvioglacial outwash** (from the Latin *fluvius*, "river"). This material is partially sorted by stream action, but it also contains pebbles and boulders shaped by the glaciers. When the continental glaciers melted, outwash was deposited by the meltwaters on the side of the moraine away from the ice. On the side facing the ice, lakes formed behind the moraines as the ice melted. When these drained, old lake basins lined with stratified sediments were left behind.

In some places glacial debris piled up under the ice. Near the melting end of the ice, cracks and holes extended from the surface to the bottom of the ice. Meltwater moving across the surface washed debris into these holes, and at their bottoms it piled up into little hills called **kames.** In other places meltwater flowed in streams under the ice. These streams constructed snake-like, winding ridges of clay, sand, and gravel, which are known as **eskers.**

Glacial hills called **drumlins** can be found in many parts of the northern United States. They are largely composed of till, and are believed to have formed under the moving glaciers. Drumlins have an elliptical outline, steep at one end and gently sloped at the other. The long dimension of a drumlin is oriented parallel to the direction of the glacier's flow. Drumlins usually are found in clusters, or swarms.

sedimentation

We have been considering the agents of erosion which pick up debris, transport it, and eventually deposit it. As we saw in the last chapter, these deposits are called sediments, and the process of deposition is called sedimentation.

There are various types of sediments. Some are materials that were once dissolved in the water. Substances in solution may remain dissolved indefinitely without being deposited. The saltiness of the ocean is chiefly due to the presence of such substances, dissolved out of rocks by rain, surface water, groundwater, and seawater itself. But under certain conditions part of the dissolved material may return to a solid state and form sediments. For example, salt can precipitate out of seawater when the solution becomes so concentrated that the water cannot dissolve and carry any more salt. In other cases, these materials may be left behind when all the water evaporates.

The most common sediments, however, are not dissolved in the water, but are merely carried along by it. They are pieces of rock that have been ground either finely or coarsely by the processes of weathering and erosion. As the water slows down it possesses less energy, and its capacity to carry material decreases. The stream thus drops rock fragments too heavy to carry. As we might expect, it lays down the densest materials first, with the result that the sedimentary materials are winnowed or sorted out. The denser debris is left in one place, while the less dense material is carried farther, to be deposited in another area. A similar sorting takes place in sediments carried by the wind.

In contrast, as we have seen, glaciers do not sort out their deposits. Unless meltwater forms a stream that carries the outwash and sorts it along the way, glacial deposits are a jumbled mass of unsorted till.

Weathering, which we discussed in chapter 2, is the ultimate source of the materials that compose sediments. The nature of the source material determines what substances are found in the sediment. Mountains made of rocks containing only calcite, for example, could never be a source area for a sediment composed of quartz sand.

Nor could a terrain made of quartz particles produce a sediment composed of calcite particles.

As might be expected, the agents of sedimentation leave their sediments in very different environments. Wind and ice usually deposit their loads in relatively dry places, while streams leave deposits along their edges, or in the lakes or oceans into which they empty.

Sediments which are deposited in the ocean are called **marine** sediments. All others are classified as nonmarine or **terrestrial** sediments. Most sediments being deposited today are marine. As we will see in chapters 9 and 10, extensive portions of the present land areas were once submerged beneath shallow seas. Nearly all of the sedimentary deposits found on land today, such as the various types of limestones, are marine deposits from those ancient seas.

The closer the place of deposition is to the source area, the less the particles of sediment will be modified from their original condition. The longer they are carried by their agents of transportation, the more the particles are worn. Sediment found on the continental shelf, just off the coast, is very different from that found in the deeper parts of the ocean basins. Shelf sediment particles tend to be quite rounded, as a result of constant stirring by waves. Those in quieter, deeper waters are less affected by wave movement, and they include some relatively unstable minerals that cannot survive in the agitated shelf regions.

As we discussed in chapter 2, sedimentary materials often form new layers of rock after they are deposited. Rocks formed in this fashion from the fragments of pre-existing rocks are known as clastic sedimentary rocks. Four basic steps are involved in their formation. The source rock must be broken down into fragments by weathering or erosion; the products of breakdown must be transported; the material must be deposited as sediment; and the sediment must be changed from a loose aggregate of materials to a solid rock. These four steps are (1) breakdown; (2) transportation; (3) deposition; and (4) consolidation. We have discussed the first three steps already; now let us consider the fourth.

For a loose sediment to be transformed into a rock, it must undergo consolidation or **lithification.** This is a very different process from the formation of igneous rock, which you will remember results from the cooling of molten materials. Lithification is usually more complex.

Nearly always, it begins when a layer of sediment is compacted or crushed by pressure from the additional layers of heavy sediment accumulating above it. Layers of fine-grained sediments may be compressed to less than half their original thickness. Compaction is usually accompanied by **desiccation** or loss of water, as water in the pore spaces is forced out. The pressure of compaction can pro-

sedimentary rocks

rock formation

changing the earth

sedimentary patterns

duce a variety of solution and precipitation effects within the sediment.

Cementation is an important process in the lithification of sediments made up of relatively large grains. It occurs when a mineral substance is introduced into pore spaces by a solution moving through the ground. The cementing material may be derived from the same sedimentary layer, or it may originate in adjacent layers or far-distant regions.

How long does it take for a sedimentary deposit to turn into rock? This can vary widely. Some sediments are lithified almost immediately after deposition. The deposits around geysers in Yellowstone Park, for example, become rock as soon as the water they contain evaporates. Rocks of this kind in Yellowstone have preserved the tracks of a buffalo and the moccasin prints of an Indian pursuing it; the chase took place less than a thousand years ago. By contrast, some deposits of sediment remain soft for tens of millions of years. Just inland from the New Jersey sea coast are deposits some 80 million years old. After all this time, ancient sea shells can still be pulled out of the soft, largely uncemented sand and clay.

Sedimentary rocks are not just large masses of undifferentiated stone. They display various patterns, or **sedimentary structures,** that result from the ways they were formed. The most obvious and characteristic structure is the layering which we call **bedding** or **stratification.** The distribution of the layers, and their internal characteristics, reveal the geological history of the terrain in which they are found. Furthermore, if we can get to the bottom of a series of layers, we nearly always find that they rest on a bed, or basement, of crystalline rock, either igneous or metamorphic. We find this true in the Grand Canyon, for example. All of these features allow stratigra-

Eroded mesa land. Erosion reveals cliffs of sedimentary material, mostly shale and clay.

phers to draw some conclusions about the relative ages of the rock layers and the conditions under which they were laid down.

The individual beds or strata of sedimentary rock are most perfectly developed in marine sediments, where a single stratum can often stretch for hundreds of square kilometers. Sedimentary strata may have almost any thickness; most are less than about a third of a meter thick, but some may reach ten times that thickness, and still others are paper thin. The term **lamina** is sometimes applied to a very thin stratum.

A sequence of strata—ranging from a few to thousands—may make up a recognizable unit because of shared characteristics. Such a group of strata is called a **formation.** Sequences of strata are sometimes exposed to view by erosion or by artificially-made cuts. These sequences are known as **sections.** Usually the layers of sediment found within a section are laid down in horizontal beds. Sometimes, however, one or more strata will lie at an obvious angle to the general orientation of the section. This is known as **cross-bedding.** Cross-bedding is usually found where deposition took place quite rapidly by water moving along a steep incline. It is commonly seen in sedimentary rocks that formed in deltas along the margin of lakes or oceans.

Mud cracks result when a deposit of mud or silt dries out and cracks. The cracks may be filled, covered, and preserved by the next layer of sediment. Mud cracks are usually seen in sedimentary rocks that formed in such shallow water that the bottom was exposed during low tide. Rocks formed under these conditions may also display **ripple marks** caused by wave action.

Clastic sedimentary rocks that originated in deeper water may display **graded bedding,** a gradation in grain size from top to bottom of a stratum. In shallow water, clastic sediments are subjected to constant current action that usually sorts out different grain sizes. The resulting sediment in any one area thus has a relatively uniform texture. But when particles of different sizes are dumped together into quiet, deeper water, the larger particles settle faster and the finer material sinks more slowly. Sandstone formed in shallow water is thus usually composed of grains of fairly uniform size. Sandstone formed in deeper water, on the other hand, most commonly contains grains of various sizes, coarser at the bottom of the stratum than at the top.

fossil fuels

One type of sedimentary deposit has a special economic importance. This is the **hydrocarbon compound,** which is composed of atoms of hydrogen and carbon. Some examples of these compounds are peat and the various kinds of coal; petroleum is also a

changing the earth

hydrocarbon compound formed under somewhat different conditions. Hydrocarbon compounds, because they are the major source of energy for industrial societies, are also known as **fossil fuels.**

All of the hydrocarbon compounds are organic; that is, they derive from living things. Most of the substances contained in living matter, or secreted by it, are compounds of carbon. This element is considered the most characteristic of the elements that occur in plants and animals. Without carbon, there would be no life on earth. Minerals derived from living things are said to be **biogenic.**

In early geological times, a large part of the earth was covered with marshes and swamps, which supported a very luxuriant vegeta-

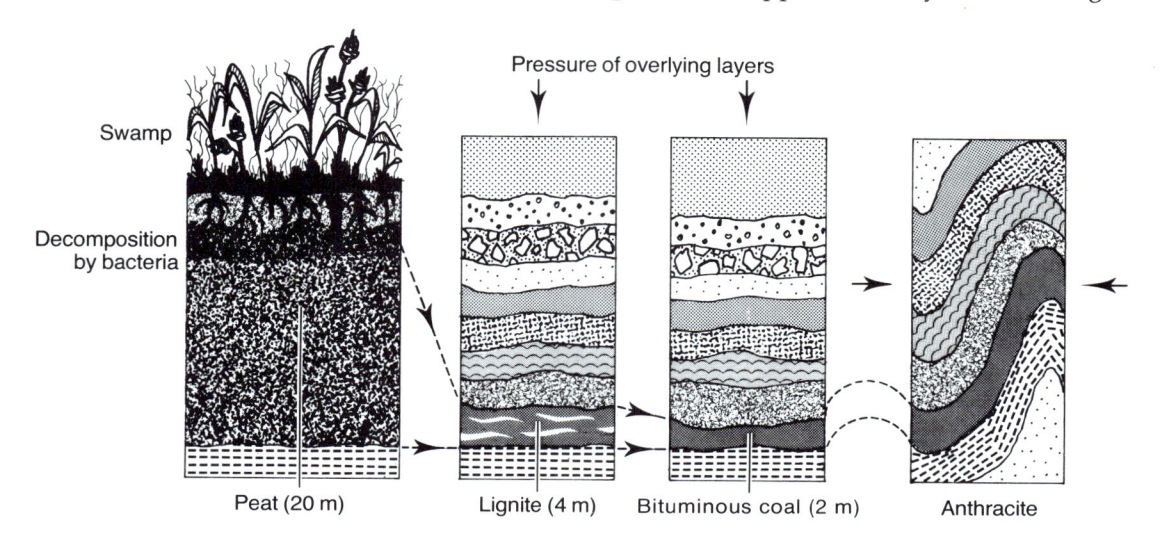

Plant remains not decomposed by bacteria become peat. Over millions of years, peat is compressed into lignite, a brown coal. Further compression produces the familiar bituminous coal. In some cases this process goes one step further, producing jet-black, lustrous anthracite.

tion. When conditions later changed, great masses of vegetation were buried under sediment and water. Very little oxygen could reach this organic material, so it decayed only partially, and was changed very slowly by chemical reactions.

Coal is a biogenic sedimentary rock derived from the accumulation and compaction of plant matter. Coal beds begin as thick, wet layers of **peat,** a spongy mass of vegetal matter. Peat bogs are still being formed today. Plant matter under water forms a sort of cushion on which other plants grow. These bogs often appear to be solid ground, but if a person walks over them the whole mass moves. In the course of time, extensive deposits of peat are formed. In northern Europe, blocks of peat are dried and used as fuel; some of the peat now being burned may be millions of years old.

In some regions a thick layer of peat was built up and was then covered by layers of mud or other sediment when a change occurred in the environment, often as a result of flooding by the sea. As the overlying material slowly accumulated, the peat was compacted under the growing weight. It underwent chemical changes, giving off carbon dioxide and other gases, and became **lignite,** a soft form of brown coal. As compaction continued, the lignite was squeezed to a small fraction of its original thickness, and became hard and black, the rock we call **bituminous coal.** When bituminous coal is burned, the remaining gases, not driven off during compaction, are finally able to escape.

By far the greatest quantities of coal were formed during the period of geological time known as the Pennsylvanian, which began about 320 million years ago. The Pennsylvanian coals of West Virginia, Pennsylvania, and Illinois occur in beds layered one above the other through thousands of feet of sedimentary section. This indicates that the Pennsylvanian period was a time of repeated flooding of the low-lying continents by sea water, each inundation being followed by a withdrawal of the sea. On the swampy ground left behind grew the primitive forests which eventually became the next bed.

In a few regions, such as eastern Pennsylvania, a very hard and especially desirable coal called **anthracite** is found. Anthracite was formed when strata of bituminous coal were folded and greatly compressed. Anthracite contains fewer remaining gases than bituminous coal, because they were driven out by the heat or great pressure to which it was subjected. It therefore burns with less smoke, but it is very difficult to ignite.

Petroleum or oil is another hydrocarbon compound, derived from deposits of vegetal matter and possibly the remains of single-celled animal organisms as well. Petroleum is found in several types of underground traps, usually layers of permeable sand or sandstone capped by impermeable shales or clays. These traps hold the liquid petroleum in pools, just as groundwater is trapped in a layer of permeable rock. Like groundwater, petroleum is commonly extracted from the ground by wells.

During the formation of petroleum, the decaying organic matter gives off gases, but these do not escape into the air. Instead, they are trapped underground with the oil. These gases are known as **natural gas,** and like coal and oil they are a valuable fuel. The pressure of natural gas is often employed to force trapped petroleum up the well to the surface.

Because of their immense importance as a source of energy, many earth scientists fear that we are depleting our fossil fuel resources much too quickly. We will discuss this problem in detail in chapter 17.

changing the earth

SUMMARY

Soil is finely divided, loose surface debris that contains organic materials. It is produced from solid rock by chemical and mechanical weathering—wind and water action, chemical change, and the action of living things. Areas with similar climates have similar soils. Laterites develop in moist, tropical climates. Pedalfers are found in moist temperate climates, and pedocals in drier climates. Immature tundra soil is often found in arctic regions. Humus is the organic component of soil.

Erosion is the removal, transport, and deposition of the loose materials on the earth's surface. Mass wasting includes slow, imperceptible creep and more rapid movements such as landslides, rock falls, and mud flows.

Surface water is an important agent of erosion. Streams unite with other streams to form stream systems, which eventually flow into bodies of still water. High-gradient streams flow faster than low-gradient streams. Low-gradient streams are most likely to flood during the spring when snow has melted.

Groundwater is a slower agent of erosion. It flows through aquifers, and may occasionally be forced to the surface by pressure. Groundwater is responsible for the formation of caves and sinkholes, which are typical of a karst landscape.

Wind erosion and deposition are particularly evident in desert areas, where wind action may form hollows or dunes. Differential weathering may produce odd-looking ventifacts. Accumulations of loess are formed from wind-blown silt, creating very fertile agricultural land. Dust carried by wind may produce severe dust storms.

Valley glaciers and continental glaciers are also agents of erosion. They grow during relatively cold periods and retreat during warmer ones. The material they deposit is usually not sorted according to size. The prehistoric glacial ice sheets drew their water from the oceans, lowering sea level by about 100 meters. Land bridges connected many islands and continents during this period.

The process of sedimentation lays down the densest water-borne materials first, and the less-dense materials last. Sediments are transformed into rock by lithification. The four steps in sedimentation are breakdown, transportation, deposition, and consolidation. Sediments may be either terrestrial or marine. Most present-day sedimentary deposits on land were laid down by ancient seas. The most characteristic sedimentary structure is called stratification.

Hydrocarbons are considered a form of biogenic sedimentary rock, and are the earth's most important source of fuel. They were formed by chemical changes in partially decayed vegetable matter. These compounds include peat, lignite, bituminous coal, anthracite, petroleum, and natural gas.

changing the earth

REVIEW QUESTIONS

1. What is soil? What do we call the process that forms soil?
2. Trace the formation and structure of a typical stream system flowing into the ocean.
3. Name and describe four important types of soil.
4. What are ventifacts? What process creates them?
5. Distinguish between the two major types of glacier.
6. Why are laterites relatively low in fertility?
7. Describe some limestone formations created by groundwater.
8. What is clastic rock?
9. Name and describe at least three types of moraine.
10. How are hydrocarbon compounds formed?
11. Where does turbulent flow occur? Why?
12. Describe the two main categories of mass wasting.
13. How can we distinguish between a rock deposited by a glacier and one deposited by moving water?
14. What are the four steps in the process of sedimentation?
15. Name and describe at least four hydrocarbon compounds.
16. What is the soil profile?
17. Distinguish between deflation and saltation.
18. How does flooding vary according to the type of stream in which it occurs?
19. What is the difference between the way a stream drops its load and the way a glacier drops its load?
20. Name and describe several common sedimentary structures.
21. Why are deposits of loess so important? How are they formed?
22. Describe three ways in which a stream carries its load.
23. What is the water table?
24. Most sedimentary deposits found on land today belong to what type?

CHAPTER FOUR

MOUNTAIN BUILDING

"If anything can go wrong, it will go wrong." That bit of wry wisdom is generally called Murphy's Law, but Alfred Brandt must sometimes have thought it should be named for himself. Certainly he had expected problems when he undertook to drill a tunnel under the Alps, but—so *many* problems?

The Simplon Pass, in southern Switzerland on the Italian border, had been an important route across Europe's highest mountains since the thirteenth century. A road built by Napoleon soon after 1800 had opened it to wheeled traffic. But by the latter years of the nineteenth century, railroads were the most important means of land transportation. And there was no way in the world to build a railroad up the steep slopes and over the top of the pass at 2,000 meters above sea level. Even if there were, the pass was blocked by snow for nearly half of every year. If railroads from the rest of Europe were going to get into Italy, they would have to go, not over the mountains, but under.

Several short tunnels had been built by the time Brandt and his German firm tackled the Simplon in the 1890s. Brandt knew it would be a difficult project. It was to be 20 kilometers long, and at its deepest point it would be more than 2,000 meters below the mountain peaks. The pressure would be tremendous, and so would the heat. Under such conditions, anything that went wrong would probably be worse than expected.

It was. To begin with, the mass of the surrounding rock was so great that its gravitational field interfered with the surveying instruments, making it hard to keep the tunnel on course. When the drills cut into underground streams, the water was under so much pressure that it flowed out at incredible speeds. As much as 13,000 gallons per minute sometimes had to be drained out of the tunnel. Sometimes, it was true, the streams could be blocked; one very watery

mountain building

stretch was conquered by lining the growing end of the tunnel with steel frames half a meter thick. Then there were the rocks themselves. These, like the water, responded violently when pressure on one side of them was suddenly relieved by the opening up of the tunnel space. Weak stretches of rock bent, swelled into the tunnel, and had to be heavily braced. Other rocks, more dangerously, exploded right off the walls in shrapnel-like bursts, threatening workmen's lives—all of this in temperatures that frequently equaled those in the hottest deserts.

The problems were solved, and the first tube was opened in 1906. The second, delayed by World War I, was not opened until 1921. Long before that Brandt had died—worn out, perhaps, by the troubles that dogged the project. But he left behind him an extraordinary engineering feat, and what is still the longest railroad tunnel in the world.

a look at mountains

The earth's surface is not stretched smooth like the skin of a balloon. It is covered with ridges and wrinkles and other isolated irregularities, which are called **relief features.** The tallest of these is Mount Everest, nearly 9 kilometers high; the deepest are the ocean trenches,

An aerial view of Greece shows a wide variety of relief features. As seen from this height, portions of earth's terrain appear so rugged that a visitor from another planet might be tempted to conclude that no one lives here!

mountain building

some of which go down more than 10 km below sea level. Since about 70 percent of the earth's surface is covered with water, most of the relief features are not readily observable.

Geologists believe that at some time during their history most continental regions were mountain masses. In the last chapter we saw how mountains are continously eroded by the steady action of wind and water. The longer this goes on, of course, the more eroded they become.

The highest mountains in the United States, other than in Alaska, are between 4 and 5 kilometers above sea level. If these mountains are being worn away at the rate of about $\frac{1}{3}$ of a meter in 1,000 years, it would take about 15 million years of weathering and stream action to wear them down to just above sea level.

Suppose we assume that the only time mountains could have been formed was when the earth itself was formed. If all mountains had been built at that time, there should not be any mountains today, because erosion would long ago have obliterated them. Since there obviously are mountains, however, we must find a way to account for them. In fact, mountains have been formed at many different times in the past. Somewhere on the earth mountains are always growing.

In this chapter we shall see how mountains are formed. The process of mountain-making is called **orogeny,** and the mountain ranges, groups of which appear on the earth's curved surface as arcs, are known as **orogenic belts.**

uplifting

From the time of Hutton or earlier, geologists have been interested in marine sediments exposed on the land. Some rocks appeared to have been lifted from the oceans without much change, but others seemed to have been folded into huge wrinkles to form the mountains. It was not until the last century that geologists began to understand how this folding could have taken place.

the theory of isostasy

Scientists believe that under the crust of the earth there is a deep layer of rock which, for now, we may call the subcrust. The subcrust is under enormous pressure from the weight of the outer crust, the oceans, and the atmosphere above it.

This subcrust is denser than the outer crust which rests upon it. And under the tremendous pressure of the overlying weight, it is somewhat plastic. That is not to say that it is liquid; in fact, it is probably rigid, and would shatter if we struck it a sharp blow. But it is able to adjust to unequal forces over long periods of time. We can compare it to cold tar, which shatters easily if struck, but which will slowly flow downhill if left undisturbed for a long time.

mountain building

The Seven Sisters mountains, in northern Norway near the Russian border, are one of the country's many scenic attractions.

Because wood is less dense than water, a wooden block will always float with part of its thickness out of the water. If a portion of the block is cut off, the buoyancy of the denser water will cause the block to rise.

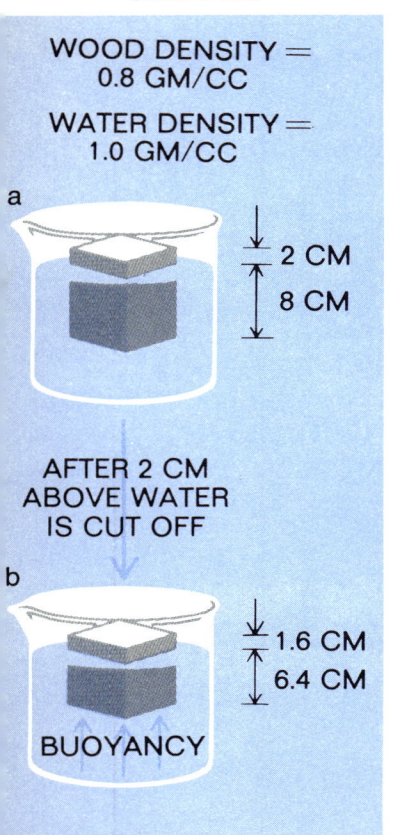

WOOD DENSITY = 0.8 GM/CC

WATER DENSITY = 1.0 GM/CC

a

2 CM
8 CM

AFTER 2 CM ABOVE WATER IS CUT OFF

b

1.6 CM
6.4 CM

BUOYANCY

The outer crust rests upon this denser plastic subcrust much as a block of ice rests upon denser water. In fact, it "floats" on the subcrust in much the same way as ice floats on water.

Two objects of different densities will sink different distances into the water. For instance, a block of light balsa wood will not sink as far as a block of heavy mahogany of the same size, because the mahogany must displace more water to balance its own greater density. On the other hand, suppose we have two objects of the same density but different size—such as a thick and a thin ice cube. Because their densities are the same, both ice cubes will displace amounts of water equal to the same percentage of their size. But the thick ice cube, just because it is bigger, has more total mass, so it has to displace a greater total amount of water in order to reach a balance. Therefore it will sink to a greater depth than the thin cube. However, it will also stick up farther, because both cubes will have the same percentage of their total masses out of the water.

Something of the same sort appears to be true of the earth's floating crust. Some parts of it are much thicker than others, so they penetrate much deeper into the plastic subcrust. And the deepest parts of all are also the thickest—that is, the mountains.

Just as with the thick and thin ice cubes in water, this difference in depth is a matter of displacement. Since mountains are thicker than the low-lying portions of continents, they must displace a greater portion of the subcrust. Like large icebergs which must have a correspondingly large submerged portion to displace water and keep the whole block afloat, so mountains have a large submerged portion that makes them buoyant.

The matter is somewhat complicated (but not seriously) by the fact that the crust itself is not all of one density. It consists of two

mountain building

Ice is less dense than water. Icebergs such as this one must accordingly have a large submerged portion to keep the entire block afloat. Only the "tip of the iceberg" appears above the water's surface.

general layers, which correspond roughly to the categories of felsic and mafic rock which we discussed in chapter 2. The lower layer, consisting mostly of basaltic rock, is called **sima** (from **si**lica and **ma**gnesium, two of its chief components). The upper layer, which is mainly granitic rock, is called **sial** (from **si**lica and **al**uminum). The sial is less dense than the sima. It forms the upper portion of the continental crust, shading gradually downward into a lower layer of sima. Ocean crust, in contrast, is almost entirely sima.

The interesting part of this for our present purposes is that the extra thickness and depth of mountains is almost entirely accounted for by sial. Underneath the mountains, the relatively light sial extends far down into depths usually occupied by sima or even by the subcrust. It can do this only because of the weight of the high mountains above, pressing it down. If this weight were removed, the deep mountain roots would be pushed up to the same level as the rest of the crust.

Just as the plastic subcrust is depressed in some areas by the weight of the continents and mountains upon it, so in other areas, where the weight is removed, it thrusts up, in the same way displaced water thrusts up around an object floating in it. Thus the total pressure of crust and subcrust is equally distributed. This equal distribution is called **isostasy** ("equal standing").

This equilibrium, however, is constantly being disturbed, because the weight of the earth's crust is continually being redistributed. Erosion is one of the chief causes of redistribution, as wind and water wear away mountains and deposit the debris at lower ele-

mountain building

vations. Equilibrium is also disturbed by the formation and disappearance of huge continental glaciers. When ice sheets form on the continents, they add weight; and meantime, the withdrawal of water from the oceans to form them subtracts weight from the ocean floors.

The plastic subcrust responds to these changes in pressure with a very slow flow. This process is called **isostatic adjustment.** Where the landmasses lose weight through erosion or the melting of

CRUST DENSITY ABOUT 2.8 GRAMS PER CUBIC CENTIMETER

ISOBARIC LEVEL

PLASTIC UNDERLYING ROCKS DENSITY ABOUT 3.3 GRAMS PER CUBIC CENTIMETER

The mountains of the earth float on the plastic subcrust in much the same way that a block of wood or an iceberg floats in water. As erosion wears away a mountain, the portion that depresses the subcrust becomes correspondingly less.

ice sheets, less subcrust needs to be displaced to support them. Accordingly, it flows in beneath them, thrusting them up. Where landmasses become heavier, they displace more of the plastic subcrust, which flows away and thrusts up where the weight is less.

When great ice sheets formed and overloaded the land beneath them, these regions sagged. But now that the ice sheets have disappeared, the land is slowly rising. In parts of Scandinavia, this **uplift** is occurring at rates of as much as 100 cm per century. Similarly, when a mountainous region loses weight through erosion, it also rises a little. But when the sea bottom receives the material removed from the mountains, it tends to sink a little, because of the increased weight of the debris.

evidence from gravity

Gravity is one of those things that everyone knows about but that most of us might have some trouble defining. It is most conveniently described as the force of attraction between two masses. When you step on the bathroom scale to check your weight, what you are really measuring is the gravitational effect of the earth's mass on your mass. If the earth had less mass than it does, you would weigh less than you do, even though your mass would remain the same.

What is more surprising is that you would not even weigh the same amount everywhere on earth. Standing at sea level on the equator, you would be a little lighter than if you stood at sea level near the North Pole. At the top of a high mountain you would weigh slightly less than at sea level. Why? How can gravity differ at different places on the earth? Scientists have been working on that question for more than two centuries.

In 1735, Pierre Bouguer, a French mathematician, was taking part in a surveying expedition to Ecuador, high in the Andes Mountains of South America. His work required very precise time measurements, so he was using a pendulum clock that he knew to be highly accurate. To his consternation, he found that the clock was running slow. Trying to figure out why, he wondered whether the height above sea level had anything to do with it. After all, it was at sea level that the clock had originally been set to run accurately. So he carried the clock up to the top of nearby Mount Pichincha, which is about 4,800 meters high. Sure enough, it ran even slower here than it had at the city of Quito, about 2,900 meters high.

The principle on which a pendulum clock works is that its period—the time it takes for the pendulum to make one swing—is always the same for a pendulum of a given length. But what determines that period is gravity. A stronger force of gravity will pull the pendulum down more quickly from the top of its swing, shortening the period. A weaker gravity, since it does not pull so hard, allows a longer period. If each swing of the pendulum moves the clock hands one second's distance, a longer pendulum period means a longer "second," and thus a slower clock.

Further measurements by Bouguer and other scientists have shown us other respects in which gravity differs from one place to another. As was noted above, a person weighs less at the equator than at the North Pole. That is, gravity varies with latitude, because the earth is not a true sphere. Because its rotation causes it to bulge slightly at the equator, and to flatten at the poles, a person standing at the equator is actually farther from the center of the earth. As more and more measurements were made, scientists were able to calculate what the sea-level gravity ought to be at every latitude. This theoretical gravity value is known as **standard gravity.**

Many measurements, however, cannot be made at sea level. Often they are made on plateaus or mountains. In order to know whether the measurements match what theoretically ought to be the gravity for these places, certain corrections have to be made. One correction allows for the higher altitude, which, as we have seen, decreases the pull of gravity. Another correction takes account of the extra mass of the mountain or plateau between sea level and the altitude where the measurement is made. This extra mass, of course, tends to increase the pull of gravity. Still other corrections allow for the gravitational effects of even higher mountains or deep valleys nearby—things which might make the gravitational pull a bit lopsided. When all these corrections are made, the final figure represents what the true gravity would be if one could stand at sea level directly under the spot where the measurements were made. Presumably, this figure ought to be the same as the standard gravity for the same

mountain building

latitude. In fact, though, it generally is not. The difference between the two figures is called the **Bouguer anomaly,** and it gives us one of our clues to things going on beneath the surface of the earth.

Measurements of the Bouguer anomaly have been made in many places. Similar calculations have been made for the moon and Mars.

The Bouguer anomaly presents us with a puzzle—how can it happen? Recall the theory of isostasy. According to this theory, the earth's crust is floating on the denser plastic subcrust. Each part of the crust displaces just enough of the subcrust to support its own mass and weight. This means, logically, that from any point on the earth's surface down to the center of the earth, the total weight and mass should be the same.

What the Bouguer anomaly seems to tell us is that this is not so. A negative Bouguer anomaly where gravity is less than would be expected for the latitude implies a shortage of mass compared with what is needed for isostatic balance. A positive Bouguer anomaly—gravity stronger than expected—implies an excess of mass.

Scientists have found that positive and negative anomalies do not occur randomly. Positive anomalies are likely to be found where a good deal of geological activity is going on—volcanic eruptions, earthquakes, and mountain building. Negative anomalies often occur in regions where uplift is believed to be taking place. This makes sense if we bear in mind that isostatic adjustment is a very slow process. Volcanic activity and movements of the earth's crust may pile up new mountains faster than the subcrust can be displaced to balance them. This would produce an excess of mass, and therefore a positive anomaly such as we find in regions of young mountains. On the other hand, the great continental glaciers melted away much faster than the subcrust could flow in to replace their missing weight and uplift the crust. This would produce a temporary shortage of mass—and a negative anomaly—until uplift is completed. Again, some mountains, such as the Sierra Nevada of California, have a negative Bouguer anomaly. This could mean that erosion is removing material from the surface of the mountains faster than uplift and the flow of the subcrust can replace it underneath.

faulting

Isostatic adjustment and localized tensions within the earth's crust can exert enormous pressure. Sometimes breaks occur in the brittle surface rocks as a result of unequal stresses. These breaks are known as **faults,** and the process is called **faulting.**

mountain building

Faulting is accompanied by a slippage or displacement parallel to the plane of breakage. Faults often run a long distance horizontally. The visible part of a fault at the surface is called the **fault trace.** Fault traces sometimes extend along the ground for more than 100 km. Little is known about the vertical extent of faults, but recent studies suggest that some are relatively shallow and that others may extend downward for at least a thousand meters.

In our discussion of faults, an important point to remember is that faulting always involves a change in position of one side of the fault, relative to the other. This may be achieved by movement of either side, or of both sides. This movement is the factor which distinguishes between a fault and a fracture in the earth's crust.

types of faults

Faults are classified according to the nature and the relative direction of the displacement of the rock along the plane of breakage, or **fault plane.** A **normal fault** is created by tensional forces that tend to pull the two sides of the fault apart. The fault plane is inclined at a steep angle, and the slippage is vertical. As shown on the next page, one side of the fault drops below the other. We say that this side of the fault is **downthrown** relative to the other, which is **upthrown.**

This movement exposes the steep face of the upthrown side of the fault. This face is called the **fault scarp,** and its height gives us an approximate idea of how much vertical displacement has taken place. Fault scarps range in height from a meter or so to as much as a thousand meters. On rare occasions they may be as much as 100 to 200 km long. Fault scarps may form imposing barriers across which it is difficult to build roads and railroads. The great Hurricane Ledge of southern Utah is a feature of this type. In places it presents a steep wall more than 800 meters high.

Normal faults are the most common, but several other varieties also exist. A **reverse fault** is created by compressional forces that push the two sides of the fault toward each other. The fault plane inclines in such a way that one side of the fault rides up over the other. We would expect that the fault scarp thus formed would have a steep overhang, but since faulting is a slow process except on the rare occasions when it accompanies a severe earthquake, the forces of erosion usually keep pace with the rate of uplift. In the figure, we see that landsliding has not only destroyed the scarp overhang but has actually caused the scarp to incline in the opposite direction. For this reason it is often difficult to distinguish quickly between the scarp of a normal fault and that of a reverse fault.

An **overthrust fault** resembles a reverse fault in some ways, since its movement too is caused by compressional forces. However, the fault plane of an overthrust fault lies almost horizontally, so that one slice of rock rides up over the ground surface next to it. A thrust slice may be from a few hundred to more than a thousand meters thick, and as much as 80 km wide.

mountain building

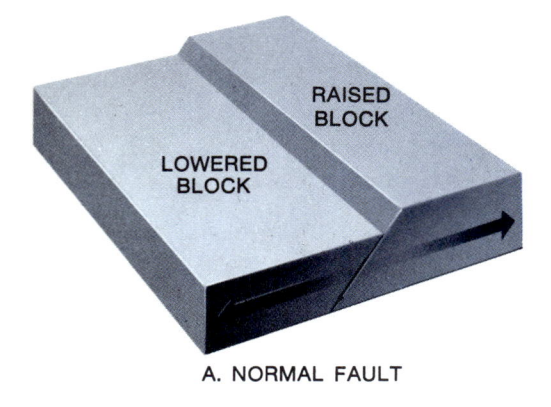

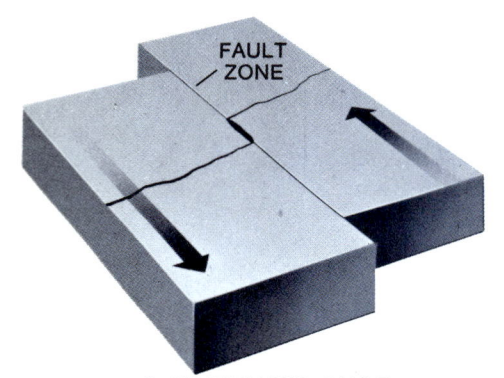

A. NORMAL FAULT — LOWERED BLOCK, RAISED BLOCK

B. REVERSE FAULT — LANDSLIDING, UP, DOWN

C. HORIZONTAL FAULT — FAULT ZONE

D. OVERTHRUST FAULT — THRUST PLANE

Four types of faults are shown here. Arrows indicate the direction of movement in each type. Note that the scarp of the reverse fault has been partially destroyed by landsliding.

In a **strike-slip** or horizontal fault, there is little vertical movement. Since the slippage is mostly horizontal, no scarp is usually formed by strike-slip faulting. These faults appear as thin lines on the surface of the earth. In some cases the fault trace is marked by a narrow trench or **rift.** Sometimes a stream will follow the fault trace for a short distance.

Faults are usually not isolated features, but generally occur in multiple arrangements. Normal faults, in particular, are frequently found in a parallel series. The series is parallel because the faults were produced by the same or similar stresses affecting the earth's crust at different points.

Such a series of parallel faults gives rise to a characteristic pattern of rock structure. A narrow block sometimes drops down between two normal faults, forming a valley with scarps facing each other on both sides. This structure is called a **graben** (German for "ditch"). Sometimes just the opposite occurs. Instead of being downthrust with respect to adjacent blocks on either side, a block may be upthrust. It then appears as an elevated mass called a **horst,** with scarps on both sides. Grabens appear as conspicuous trenches in the topography, with straight, parallel walls. Horsts resemble blocklike plateaus or mountains.

mountain building

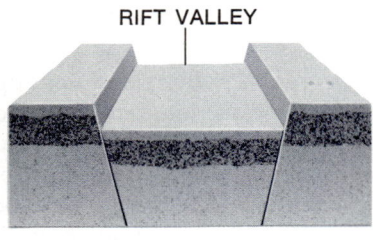

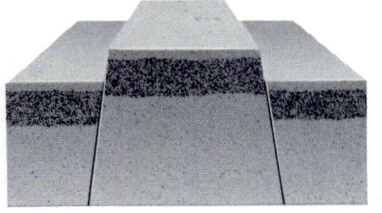

103

GRABEN   HORST

RIFT VALLEY

At left, the downthrusting of a block between two normal faults has formed a graben. The horst at right was created by upthrusting of the block in the middle.

Faults and fault systems are under intensive study by geologists. As we will see in the next chapter, they are associated with earthquakes, and a greater understanding of the faulting process may give rise to better methods of earthquake prediction.

The Grand Canyon has been cut to a depth of nearly 2 km by the Colorado River, through layers of sedimentary and crystalline rock. The reddish color of many of the strata indicates that arid conditions prevailed during much of the time when these rocks were forming.

fault block mountains

The scale of faulting may be large or relatively small. In regions where normal faulting occurs on a grand scale, with displacements up to several thousands of meters, huge mountain masses known as **fault block mountains** may be formed. Many of the mountain ranges in the western United States, such as the Teton Range in Wyoming, are fine examples of fault block mountains. Some of these blocks are a hundred or more kilometers long and tens of kilometers wide.

Several features are always present in fault block mountains. First, the faults along the edges of the blocks tend to be more or less parallel. Second, the blocks do not rise straight up. Usually one side of a block will rise faster and higher than the other, thus placing the highest point of the mountain range "off center"—nearer to one side of the block than the other. Since the blocks rise slowly, stream erosion has time to modify the steeper face of the uplifted block, shaping

mountain building

104

it into triangular facets. Such landforms can be seen in the Wasatch Mountains in Utah.

The process of uplift takes place very slowly, slowly enough that major rivers can flow across the uplifted regions and cut deep canyons through the uplifted blocks. Smaller streams, however, are usually diverted from their old courses as uplift occurs, and must establish new paths or disappear. Very often, because the rising block is tilted, their direction of flow is reversed. This leaves a **notch** on the summit, representing the old stream channel.

From what we know about the ability of streams to carve channels into rock, it appears that the rate of uplift of a fault block mountain must range from about a third of a meter to more than two meters over a period of a thousand years. The Arkansas River has cut

Mount Rundle is a large fault block mountain. Note that it appears to be slightly "off-center." Such mountains are known as tilted fault block mountains.

the Royal Gorge to a depth of 330 meters through an uplifted block of granitic rock. The Colorado River has cut the Grand Canyon to a depth of almost 2 kilometers in mostly sedimentary rock. Both of these rivers have been eroding their present channels through uplifted blocks for a least a million years.

folding

Where the earth's crust is less brittle, the pressure of uplifting causes the rock to bend instead of breaking. This is especially typical of sedimentary rock. The bending process, known as **folding,** may form small wrinkles or huge mountains.

mountain building

The fold mountains we see today are the remains of the most resistant sedimentary rocks. Some of these mountains have been formed from the rounded parts of an upward fold, or **anticline.** Others were fomed from downward folds, known as **synclines.** ("Cline" simply means "slope.")

As folding takes place, the slope of the terrain changes. Sedimentary strata that were stable as long as the land remained relatively flat lose their stability as the land begins to slope. Depending on the type of sediment, the thickness of the strata, and the amount of water they hold, they may begin to move gradually downhill. As they move, the layers of sediments tend to form wrinkles and folds as well as overthrust faults. This process, known as **gravity sliding,** can take place even when the slope of the land is very gentle.

In mountain ranges that have been formed by folding—such as the Appalachians in the United States—the summits of the individual mountains are often very nearly the same height. To explain this, one hypothesis suggests that at one time in the past the entire region was eroded to an almost flat plain, which was then uplifted and carved again by streams. The first part of this process is called **peneplanation,** and the flat region thus produced is a **peneplain.** As we have said, the Appalachians are folded mountains, but peneplanation can affect other types of mountains also.

If a peneplain is uplifted, stream erosion will cut away the softer rock and leave the more resistant rock standing as mountains. Because the tops of these mountains were once part of a level peneplain, according to the hypothesis, they are all about the same elevation above sea level.

Occasionally, as peneplanation occurs, an area of rock particularly resistant to erosion will remain sticking up out of the flat plain. These isolated hills or small mountains are called **monadnocks,** so-

At left, crystalline limestone and shale have been subjected to pronounced folding. This formation is located near Mckenzie, Maryland.

Sheep Mountain, at right, is an example of an anticline. What other types of formations can you identify in this picture?

mountain building

Extensive folding has occurred near Red Deer River in Alberta, Canada. The anticline shown is a point of contact between the Belly River and Bear Paw formations.

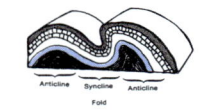

An artist's conception of an anticline and a syncline.

called because one of the most striking examples exists at Monadnock, New Hampshire.

When a peneplain is uplifted, the gradients of the streams that cross it increase, and the area is said to be **rejuvenated.** Old, slow-moving streams become swift-moving streams with great eroding power. These streams cut away the softer rock, leaving the more resistant rock standing as mountains. Because the tops of these mountains were once part of a level peneplain, they are all at about the same elevation above sea level.

Streams that cannot erode their beds as fast as the land is uplifted become dammed by the rising land. Water is trapped behind the natural dam until it finally overflows into a new course. The old course of the river may remain at the summit of the ridge as a **wind gap.** On the other hand, streams that can erode their beds as fast as the land is being uplifted are able to maintain their old courses. They cut deep notches, or **water gaps,** through the uplifted hard strata. Many wind and water gaps can be seen throughout the Appalachian Mountains.

recumbent folding

The Appalachian Mountains of Georgia and Tennessee differ in some respects from the northern part of the Appalachian range. In the north, the folds of the mountains are simple and symmetrical. In Tennessee, however, the folds tend to be uneven and lopsided. Researchers now believe that this **asymmetric folding** was caused by an excess of pressure from one side of the original pile of sediments. The Canadian Rockies contain many folds of this kind, but the most pic-

mountain building

turesque examples are found in the Alps of central Europe. In the Alps, some folds are tipped so far over to the side that they lie almost horizontally. Such complex folding is known as **recumbent folding,** and each fold is called a **nappe.**

Erosion has cut deep valleys into some of the Alpine nappes, forming "windows" through which the sequence of sedimentary rocks can be studied. When geologists first studied the strata deeper and deeper in the nappes, they reached a point where the sequence of strata repeated itself in reverse. This peculiar discovery gave geologists the clue to how the nappes had been formed.

The surface of contact between the folded sedimentary rock of the nappe and the underlying crystalline bedrock is called the **thrust plane.** Somewhere in the side opposite to where pressure was being applied to the sediments, the sedimentary rocks caught fast to the underlying crystalline rocks. The slowly flowing rocks began to pile up at the point on the thrust plane where friction was stopping their motion. The material at the top, however, continued to move onward

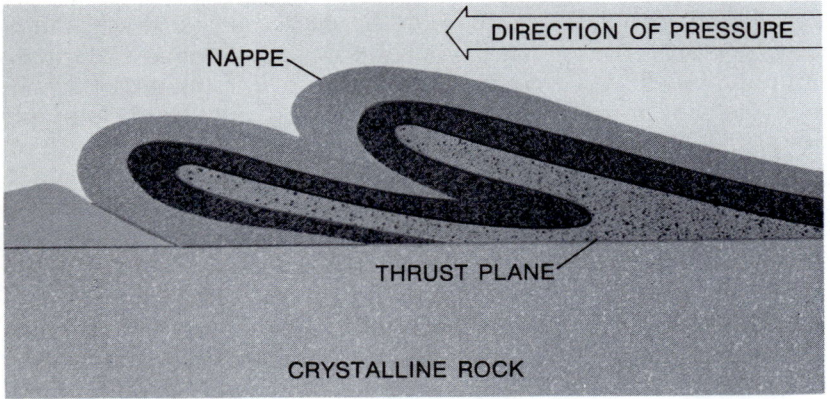

The sequence of rock layers within a series of nappes is shown here. Pressure has caused the rocks to become plastic, so that the top of each fold flows beyond the point at which the sedimentary rocks snagged on the underlying crystalline layer.

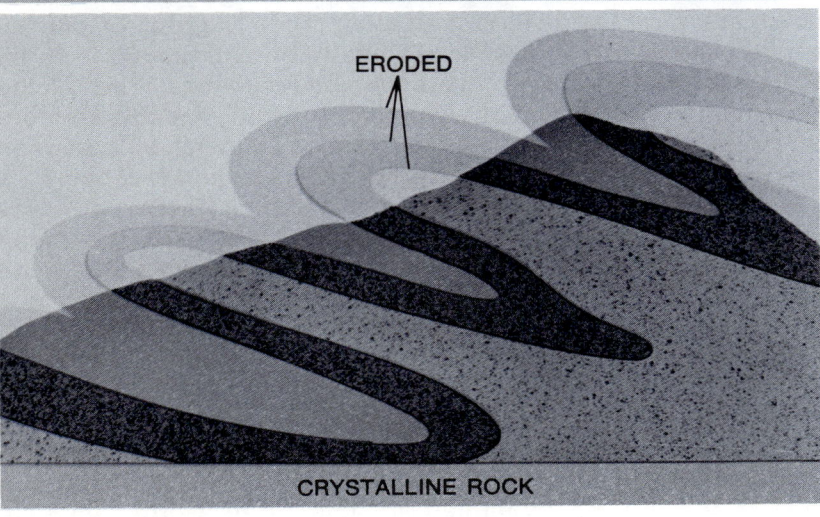

Eroded areas in this series of nappe folds make it possible for geologists to recognize several repeating sequences of sedimentary strata. This sketch also shows the strata before they were eroded.

IN FOCUS
the mountain climbing experience

One of the young child's first impulses, as it sets out to explore the world, is to climb and seek the highest point. The desire to scale the heights and reach for the peaks seems to be a universal human trait. Archaeological evidence shows that mountain climbing is no novelty in human history. Climbing irons or ice creepers, which can be attached to boots to give better footing on ice, have been found dating back as far as the fifth century B.C. The alpenstock, a wooden staff used by climbers, was a standard item in the mountain traveler's kit by the time of the Renaissance.

The first recorded ascent to a mountain peak in North America occurred in 1522. Cortes, exploring Mexico with his expeditionary party, found that he was running short of gunpowder. He offered a reward to anyone who could bring a supply of sulfur down from the crater of a nearby volcano. Accordingly, two of his men scaled Mt. Popocatepetl and were lowered into its crater. After filling sacks with as much sulfur as they could carry, they returned to collect their reward from Cortes.

In 1806 Lt. Zebulon Pike became the first of a line of explorers intent on climbing mountains in the American West. Pike had neither the time nor the resources to scale the highest peak, which now bears his name. This feat was accomplished in 1820 by Edwin James, who thus completed the first known high mountain ascent in this country. From a technical standpoint, James' ascent was no more than rough hiking. Certainly it was not in the same class as the ascent of Mont Blanc in the Alps, for instance, where much more skill and expertise in negotiating glacial ice is required.

Mountain climbing for the sake of pure sport began in the Alps more than 200 years ago. The first ascent of Mont Blanc was characterized by a great deal of sophistication and attention to detail. The climbers gathered at the village of Chamonix. They were already familiar with game trails in the area, so with the aid of a telescope they methodically and leisurely selected what appeared to be the best route. The whole expedition, in fact, seems to have been a very gentlemanly affair.

The establishment of mountain climbing as a sport in North America was greatly influenced by events taking place in the Alps. Until the 1850s, mountaineering there had been largely a Continental European affair. Then, a group of outside amateurs, mostly British, moved in to try for the unclimbed summits. The Alpine Club was formed in 1857, and this London-based organization did much to foster and publicize climbing as an amateur sport.

The "impossible" Matterhorn was first scaled in 1865. By 1880, few major European peaks remained unclimbed. But instead of bringing mountaineering to a halt, this started a new wave of interest in the sport. A search began for new routes on previously-climbed peaks in Europe. Attention next shifted to unclimbed mountains elsewhere in the world—the Andes, the Caucasus, the Himalayas, and the untamed mountains of North America.

World War I had brought the development of new types of climbing gear. During the 1930s pitons came into general use, despite the objections of some climbers. These "purists" complained that the metal spikes hammered into the rock face made climbing no more difficult than ascending a ladder, and that they threatened to disfigure the beauty of natural rock formations. Still, the popularity of the sport continued to rise, and the best-known German and Italian climbers became national celebrities during the late 1930s.

World War II brought further refinement and improvement of climbing equipment. Hempen rope was replaced by more durable nylon. There were now several kinds of pitons, and they were being made of lightweight aluminum.

The most significant recent change in the North American mountaineering scene has been the enormous increase in the number of people who are taking up the sport. But there may be trouble brewing in the paradise of the great outdoors. Wilderness areas are being threatened by crowds of people and the resulting environmental damage. The necessary regulation of access to these areas has become a barrier to serious but free-spirited climbers.

Will this lead to a decline of interest in the sport? No, but the focus of interest seems to be changing. A new breed of climber has appeared, including people of both sexes and all ages. These people might almost be classified as mountain cleaners rather than mountain climbers. Recently, for instance, a group of students from Evergreen College in Olympia, Washington, scaled Mount Everest on the first litter-picking expedition up the world's tallest mountain. The Outdoor Program of Oregon University organizes similar clean-up trips, regularly sprucing up Mount McKinley in Alaska.

This is not to say that the fundamental human urge behind mountain climbing—the desire to reach the top—is dead. In 1977, a 27-year-old toymaker and amateur mountain climber scaled the 110-story South Tower of Manhattan's World Trade Center. He made the vertical ascent using nylon rope and special metal clamps of his own design, which fitted into the metal runners that serve as tracks for window-washing equipment as it moves up the side of the building. His successful ascent was greeted with mixed reactions. At the top he received a police summons for "reckless endangerment," and once down again he received a citation from the mayor of New York. But after the excitement had died down, the climber was gratified to learn that the New York City Fire Department is looking into the possibility of using his invention to fight deadly "unreachable" fires in modern high-rise buildings.

mountain building

a closer look at the Appalachians

for a while, past the point where the material at the bottom was snagged on the underlying crystalline rocks. Thus, the top of the fold gradually moved ahead of the bottom, producing the structure of the nappes.

The man who taught us most about the sedimentary rocks in the eastern part of the United States was James Hall. In 1836, Hall was hired as an assistant in geology by the newly-formed Natural History Survey of New York. He was soon assigned to work in the western part of the state.

To most geologists, the layers upon layers of sedimentary rocks found in western New York were uninteresting and unimportant. However, Hall was challenged by his assignment, and found these rocks fascinating because of the many fossils they contained. He spent most of the next sixty years of his life studying rock formations in the eastern United States, including those in the Appalachian Mountains.

While Hall was working in New York, Henry Rogers and Wil-

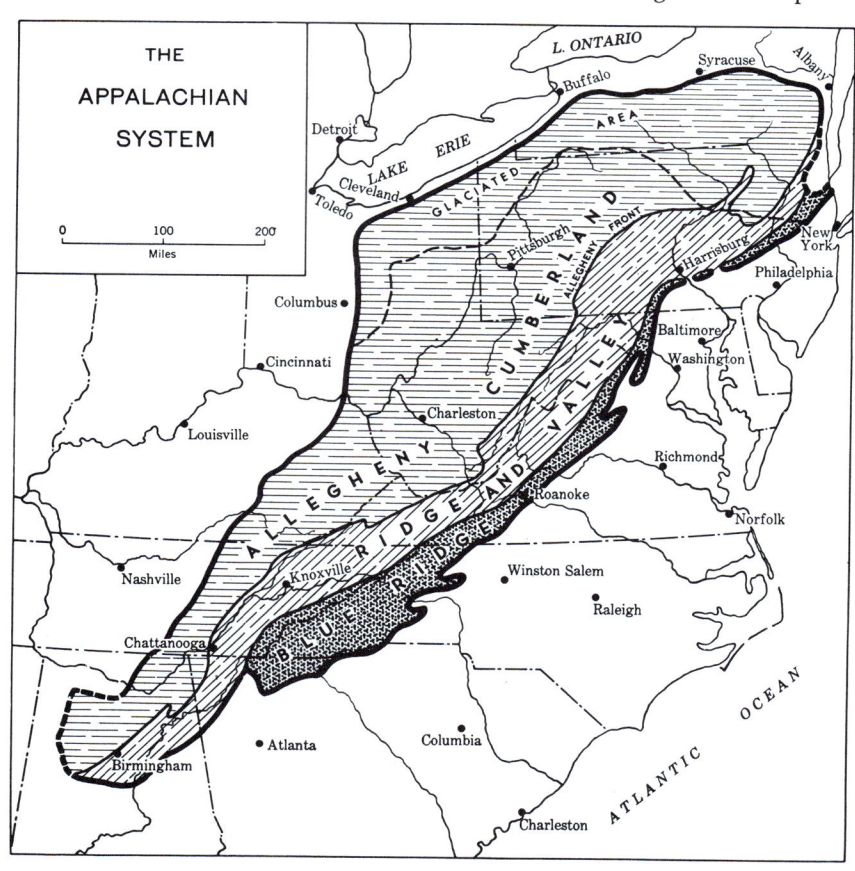

Map showing the extent and principal subdivisions of the Appalachian hill and mountain system.

liam Rogers, sons of a well-to-do doctor in Philadelphia, began to study the mountains of Pennsylvania and Virginia. Eventually, they reached the conclusion that these mountains had been carved by rivers from huge folds of sedimentary rock. The work done by the Rogers brothers and by Hall forms the basis for our present understanding of the Appalachians and all other folded mountains.

It seems probable that the Appalachian Mountains have been reduced to a peneplain several times during their long history. Each time this occurred, it was followed by a new uplift of the land and rejuvenation of the streams. A new cycle of erosion then commenced.

A recent hypothesis suggests that at least part of the Appalachian range was once an island arc, similar to the present-day Aleutian Island chain. We will discuss this hypothesis in detail in chapter 8.

dome landforms

In the 1870s, G. K. Gilbert, a young geologist newly hired by the United States Geological Survey, was assigned to study a region in the western United States. One of the areas he explored was the Henry Mountains in Utah. He soon realized that these mountains showed no signs of having been folded or carved from an uplifted fault block. Rather, the crust seemed to have been uplifted into an elliptical dome about 80 km long, and then eroded.

Dome mountains, such as those Gilbert discovered, can be considered a third mountain type, along with folded and fault block mountains. Not all dome mountains are formed in precisely the same way. The Henry Mountains were uplifted by the intrusion of hot magma from below into a crack between sedimentary layers. The pressure of the upthrusting magma forced the overlying rock layers upward. Eventually the magma hardened into a dome-shaped body of igneous rock called a **laccolith.**

Other dome mountains are pushed up, not by laccoliths, but by much deeper and more widespread forces. These forces are probably associated with isostatic adjustment, or with the building of major mountain ranges of other types. These domes are usually larger than the laccolithic domes, and in them not only the sedimentary layers but even the crystalline basement rocks have been uplifted. The Black Hills of South Dakota and Wyoming are carved from a dome of this kind.

Whichever way it is formed, a dome has a typical structure. On the top and sides are a series of sedimentary layers of varying hardness. Beneath them lies much harder crystalline rock—either a young laccolith or ancient basement rock. Naturally, the sedimentary

mountain building

the Black Hills

rock will erode more easily than the crystalline rock. This sets the stage for the development of an interesting set of landforms.

The dome from which the Black Hills have been eroded straddles the border between Wyoming and South Dakota. It is an almost perfect ellipse, about 160 km long from north to south and about half that wide from east to west.

Erosion has removed most of the sedimentary rock that once

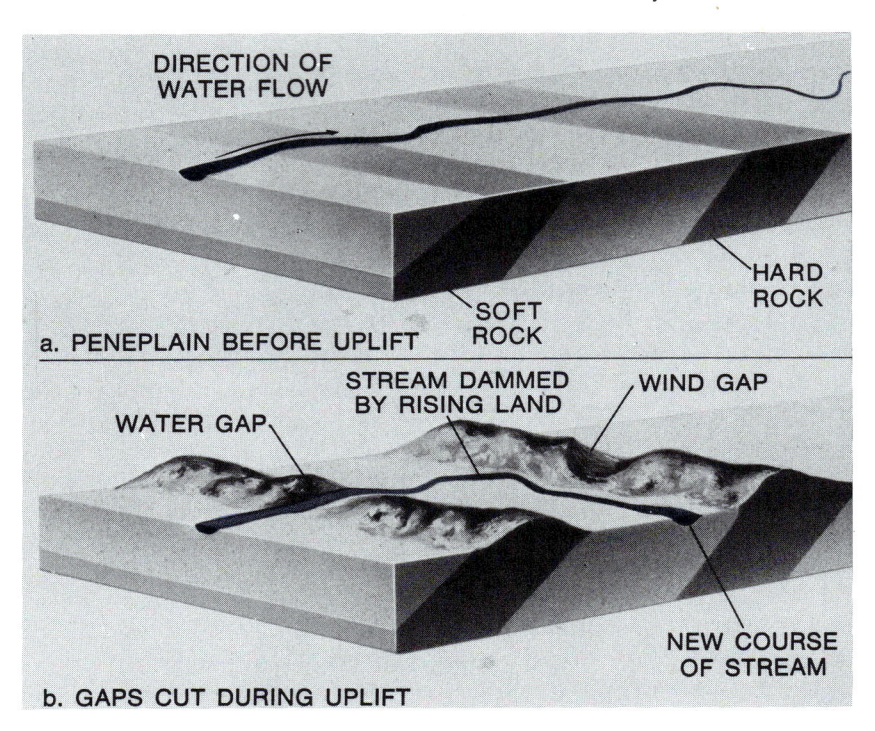

a. PENEPLAIN BEFORE UPLIFT — DIRECTION OF WATER FLOW, SOFT ROCK, HARD ROCK

b. GAPS CUT DURING UPLIFT — WATER GAP, STREAM DAMMED BY RISING LAND, WIND GAP, NEW COURSE OF STREAM

The Appalachians have probably been uplifted and then reduced to a peneplain several times during their history. The various types of gaps cut during the process of uplift are shown here. The cycle of peneplanation may take from 50 to 100 million years.

covered its eastern half. The western half is still capped by a thick layer of resistant limestone. Where the limestone is gone, very ancient crystalline rock has been exposed. This rock is much older than the sedimentary layers, which tells us that it is basement rock and not a laccolith.

Geologists do not know why the sedimentary cover on the eastern half eroded more completely than that on the western half. Perhaps the rocks of the eastern part were more sharply upthrust and therefore tended to crack and break more than those in the west.

Surrounding the Black Hills is a ridge of fairly hard sandstone called the Dakota Formation. This is actually the eroded remains of the topmost sedimentary layer that once covered the dome. Such encircling ridges are characteristic of eroded dome landforms. Usually the side facing the dome is steep and heavily eroded. If the outer side—the surface of the old sedimentary layer—slopes more gently,

mountain building

the ridge is known as a **cuesta.** If both slopes are steep, the ridge is called a **hogback,** since its jagged line, seen from a distance, suggests the spine of a razorback hog.

Cuestas and hogbacks are not limited to dome mountains. They are common wherever hard sedimentary strata slope upward with softer strata beneath them. Occasionally, as we shall see, they can form from the hard strata alone.

Between the sandstone ridge and the dome of the Black Hills is a broad valley, eroded from a layer of soft red shale under the sandstone. Because of its color, it is known as the Red Valley. Like cuestas and hogbacks, these **subsequent** (following) **valleys** are typical of dome landforms, though not limited to them. When the hard sandstone layer is first eroded away from the top of the dome, it and the newly exposed second layer are at the same level. But if the second layer is softer, it now erodes faster than the first, creating a hollow in the center of the forming ridge. As both surfaces erode outward and down the sides of the dome, the hollow becomes a circular (or elliptical) groove. Streams flowing down from the top of the dome or of the ridge turn and flow along the groove, deepening it still further into a valley. It is because the valley is thus formed, in a sense, after the ridge that it is called a subsequent valley.

In Heligoland, an exposed salt dome indicates that the region was once covered by an inland sea. Salt domes are often associated with deposits of natural gas and petroleum.

Beneath the red shale is a third layer, the hard limestone that still lies directly above the old crystalline rock in the western part of the Black Hills. To the east, as we saw, the limestone has eroded. But because there is no soft layer beneath it, the eroded edge has not formed a ridge. Instead, streams flowing down the dome have cut it into more or less triangular upturned slabs, still resting against the crystalline rock. Because of their shape, these slabs are known as **flatirons.**

Streams have eroded valleys into the crystalline core, leaving behind huge peaks of granitic rock. One of these, Mount Rushmore, is famous for the gigantic sculptures of presidents carved on its face. One or two of these granitic peaks reach as high as the uneroded limestone of the western region. This suggests that, before erosion, the eastern half of the dome may have been higher and steeper than the western—which could have been why it eroded more completely.

Streams have also cut valleys into the tough limestone. The most spectacular of these is Spearfish Canyon, over 300 meters deep, at the northern end of the dome. Along the broad reaches of the western limestone plateau, several low hogbacks have been cut into the limestone itself. The limestone plateau slopes very gently downward, making the western and eastern sides of the dome very different. By comparing the two sides, geologists have learned much about the way rugged mountains can be produced from a smoothly rounded dome.

other kinds of domes

On the coastal plains of Louisiana and Texas, and on into Mexico, are a great many low, domed structures. At one time these were thought to be monadnocks, but now we know they have a different explanation. They have been pushed up by rising columns of halite, or rock salt, and are therefore known as **salt domes.**

At certain times in geological history, thick layers of salt were deposited on the floors of shallow lagoons or drying seas. Later they were covered by successive layers of other sedimentary rocks, eventually becoming deeply buried.

Most rocks, of course, would just stay there, lying as they were deposited, unless they were moved by an earthquake or other such event. But salt differs from most other sedimentary rocks in two important ways. It is lighter, and it becomes plastic under pressure. Consequently, salt in a deeply buried layer, if given a chance, will flow slowly; and it will tend to flow upward. It gets this chance when, for some reason, the overlying layers press down less heavily at one spot than elsewhere. Under such conditions, a narrow **salt plug** may slowly push its way upward for thousands of meters, heaving up the overlying rock layers and thrusting them aside as it rises. The plug remains connected to the buried layer.

Somewhat surprisingly, the greatest commercial value of these salt domes is not as sources of salt, though they are mined for this. More important, oil and natural gas deposits are often found around them. These seem to be trapped in the sedimentary layers that are pushed up and sealed off as the plug rises through them.

Other valuable substances are formed from impurities in the salt itself. At the top of the rising plug, some of the salt may be dissolved by circulating groundwater, leaving a "cap" of insoluble minerals. One of these minerals, anhydrite, may later be chemically changed into gypsum, calcite, and sulfur. The sulfur deposits overlying the Gulf Coast salt domes have already yielded millions of tons of pure, high-quality sulfur, and more millions of tons are believed to be still in the ground.

RADIOACTIVE WASTE DISPOSAL: CAN WE CLOSE PANDORA'S BOX?

According to Greek mythology, Pandora was the first woman. She was entrusted by the gods with a box that held all the ills that could plague mankind. The gods warned her never to open it; but, unable to contain her curiosity, she did so, thus releasing the problems that afflict humanity. This may be a valid analogy for the dilemma facing our energy-short world today. Frightening as the prospect of nuclear warfare may be, an even greater problem is connected with the use of nuclear energy, even for peaceable purposes. That problem lies in finding a safe and reliable way to dispose of the radioactive waste products of nuclear reactors.

Civilian nuclear power plants have so far produced about 600,000 gallons of high-level radioactive waste—but this is chicken feed compared to the estimated 80 million gallons of waste produced by the American nuclear weapons program. Somehow, we must find a way to seal off at least part of this material for a period as long as 250,000 years—long enough for the deadly radioactivity to dissipate.

At several federal reservations, large quantities of these lethal wastes are currently being stored above ground in huge holding tanks, some of which have already leaked as corrosion destroys the tank walls. Of course, this is only a temporary solution.

In the United States, the Energy Research and Development Administration is looking into the possibility of storing radioactive wastes underground in Rocky Mountain granitic formations, or in a salt deposit near Carlsbad, New Mexico. Britain, ranking second to the United States in nuclear generating capacity, stores highly radioactive waste in stainless steel tanks within concrete silos. Low-level waste is periodically gathered together, cast in concrete, and dumped into a trench of the Atlantic Ocean. In West Germany, radioactive waste is sealed in glass, put in steel containers, and stored in abandoned salt mines.

Looking to the future, many scientists are turning their attention to the bowels of the earth as the ultimate dumping pit. The Japanese, who are particularly anxious about their growing nuclear waste problem, have recently suggested a plan in which wastes would be deposited in the deep ocean trenches of the Pacific. Tectonic movement of the earth's crustal plates, it is hoped, would then drive the wastes down deep into the earth's mantle, where they would gradually lose their radioactivity over geological ages.

SUMMARY

At one time during their history most continental regions were probably mountain masses, which were eventually worn away by erosion. Mountains are still forming, in a process called orogeny.

The earth's crust is depressed in some places by the weight of the continents and mountains. In other places the crust thrusts upward so that pressure is equally distributed. Isostasy refers to the equilibrium in the earth's crust maintained by such isostatic adjustments that equalize pressure.

Isostatic adjustment may cause the base of an eroded mountain range to rise and form a new range. This is known as uplifting. Deep-sea fossils found on mountain tops provide evidence for uplifting. Measurements of the variations in gravity at different altitudes and latitudes give further supporting evidence. A variation in gravity caused by a variation in the mass of the earth at a given point is known as the Bouguer anomaly, which may be either negative or positive.

Breaks in the earth's crustal rocks caused by stress are called faults. Faulting is accompanied by a displacement along the plane of breakage. The horizontal extent of a fault at the surface is known as the fault trace. A normal fault results from up-and-down slippage, resulting in an upthrown fault scarp. In a reverse fault, one side of the fault rises up over the other. Strike-slip faults move sideways, and usually form no scarp. Faults most commonly occur in a parallel series. A block between two faults may drop to form a graben, or rise to form a horst.

Fault block mountains are the result of normal faulting on a large scale. In such mountains, the faults along the edges of the block tend to be parallel. Usually one side rises faster and higher than the other. Stream erosion shapes the facets of the uplifted block into triangles. Faulted mountain blocks may be either tilted or lifted.

When uplifted rocks bend instead of breaking, folded mountains are formed. Anticlinal mountains are the result of an upward fold, and synclinal mountains have been formed from downward folds. The Appalachian Mountains were formed by folding.

Peneplanation is the process in which a mountain region is eroded to a flat plain. When a peneplain is uplifted again, the streams that cross it are rejuvenated. They may form either wind gaps or water gaps.

Dome mountains are formed when magma accumulates between sedimentary strata, forcing the material above to bulge upward. The Black Hills of South Dakota are the result of dome formation. Salt and sulfur domes are found on the coastal plains surrounding the Gulf of Mexico.

mountain building

REVIEW QUESTIONS

1. What do we call the process of mountain building?
2. Name three things that cause a variation in the force of the earth's gravity.
3. What three features are always present in block mountains?
4. Describe the process of faulting.
5. How are dome mountains formed?
6. Describe the formation and appearance of a graben and a horst.
7. What causes a negative Bouguer anomaly? Where is it usually observed?
8. What is isostasy?
9. What is folding? What two types of mountains does it form?
10. Name and describe at least three types of faults.
11. Describe the structure of a mountain formed by recumbent folding.
12. How is a peneplain formed?
13. What is a fault scarp? Do all faults form scarps?
14. What are relief features?
15. Describe Bouguer's observations about gravity.
16. What is a rift valley?
17. Explain the difference between a wind gap and a water gap.
18. What is the effect of the weight of glacial ice on the earth's crust?
19. Describe some of the evidence for uplifting.

CHAPTER FIVE

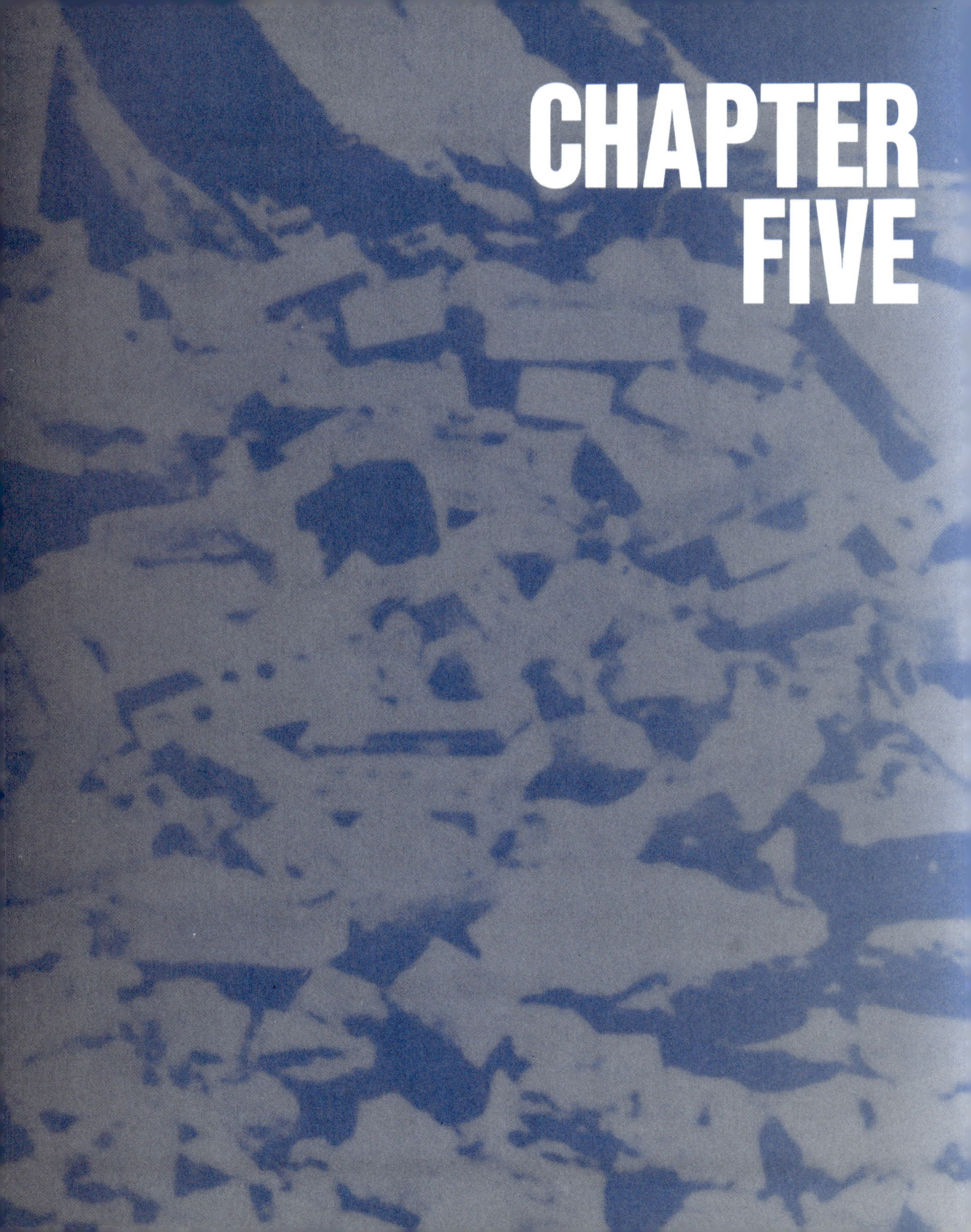

THE EARTH'S INTERIOR AND EARTHQUAKES

Late in the afternoon of March 27, 1964, the streets of downtown Anchorage, Alaska were quiet, almost empty. It was Good Friday, and most people had gone home early. At 5:36, with no warning at all, Anchorage was shaken by the strongest earthquake ever recorded on the North American continent. Buildings collapsed, pavements buckled, and portions of the Alaskan coastline crumbled into the sea. A seismic sea wave generated by the earthquake swept ashore, completely destroying at least one small town and, at Kodiak, depositing many of the ships of the fishing fleet on the streets of the city's business district. Slightly more than an hour later, that same wave flooded the town of Crescent City, California, leaving twelve persons dead.

    Fewer than 200 people died as the result of the Good Friday earthquake. But the toll might have been far higher if the quake had happened earlier in the day when crowds of people were concentrated in downtown areas, where the worst damage occurred. Unlike most natural disasters—floods, hurricanes, volcanic eruptions—earthquakes strike with little or no warning. Since the beginning of recorded history attempts have been made to devise a way of predicting them. Now, with modern scientific technology, accurate earthquake prediction may soon become a reality.

the earth's interior and earthquakes

the earth's interior

The devastation caused by the great San Francisco earthquake of 1906 is shown in this news photo.

Scientists agree that the key to understanding the causes of earthquakes lies in understanding the internal structure of the earth. Many of the basic facts about the earth were discovered long before the present century. Its circumference was determined over 2,000 years ago; more recently, scientists were also able to measure its mass. Space-age science, with the aid of sophisticated equipment, has refined this earlier work.

density and volume of the earth

An important measurement for understanding the structure of the earth's interior is density. As you will remember, density is the amount of mass in a substance for each unit of volume—what we commonly refer to as weight. To find the average density of the earth, we divide mass by volume. The volume of the earth has been calculated to be 1.08×10^{21} cubic meters. Its mass is known to be 5.96×10^{27} grams. The average density of the earth is therefore 5.52 grams per cubic centimeter (g/cm³).

The density of a material is often expressed in terms of **specific gravity,** the ratio of the material's density to that of pure water. Pure water has a density of 1 g/cm³. Since the earth's average density is 5.52 times greater than this, we say that the average specific gravity of the earth is 5.52.

But this is only an average. If we divide the earth into regions, we find some interesting variations. From actual measurements of rocks near the earth's surface, we know that they have an average specific gravity of roughly 2.8, only about half that of the earth as a

the earth's interior and earthquakes

whole. Evidently, somewhere within the earth there must be materials whose density is greater than the average density of the whole planet and much greater than that of the crustal rocks found near the surface. And, in fact, scientific investigations indicate that the core of the earth must be composed of extremely dense materials.

heat

Another important feature of the earth's interior is its heat. Without interior heat, the planet might never have developed as we know it — it would be dead and cold, like the moon. As it is, evidence of the dynamic power of heat is everywhere, in the geological traces of melting and volcanic activity. We also know, from experience in mines and with holes drilled deep into the earth's surface, that the temperature increases with depth.

Some of this heat is probably left over from the earth's formation. Some is generated at great depths by the tremendous pressure of the overlying layers of rock. Finally — especially in the outer layers — some is produced by the decay of radioactive materials. Even now, there is enough radioactive material in the upper mantle — the region just below the crust — to have raised its temperature by 20°C in 100 million years. Since the store of radioactive elements was formerly much greater, the heat generated during the earth's entire 4.6-billion-year history should have been enough to boil and vaporize the rock itself, even at the tremendous pressures deep in the core. Why has this not happened? For one thing, we may suspect that radioactive elements are not so abundant at great depths as they are near the surface. This would mean that the temperature would not rise as fast as great depths as it does in the region where we can measure it directly, just below the surface.

Furthermore, heat does not remain in one place. It flows, along a **thermal gradient** from a hotter to a cooler medium. In the case of the earth, this means that temperature steadily decreases as geothermal heat flows out from the interior toward the surface and finally radiates into space. If heat did not escape in this way, the earth's material would continue to grow hotter. But if the lost heat were not replaced, the planet would eventually become too cool to support life. Whatever heat is lost by heat flow must be balanced by internal heat production if the temperature of the earth is to remain fairly stable. In this respect, as in so many ways, the earth seems to have achieved something close to equilibrium.

seismology

One of the chief means by which scientists have gained their knowledge of the earth's interior is the study of earthquakes. Laboratory study and analysis of the physical properties of various types of rock,

the earth's interior and earthquakes

microscopic evaluation of molecular structures, and computer modeling have all contributed to our knowledge of the earth's structure and composition. But our most important source of information comes from the branch of geophysics known as **seismology,** which is concerned with the study of earthquakes and their attendant **seismic waves.**

As early as 136 A.D. the Chinese were using a measuring device that could indicate the direction of an earthquake from the observer's location. In 1731, Nicolas Cirillo invented a device which enabled him to calculate the relationship between distance and the magnitude of the ground wave or surface shake in Naples. British seismologist John Milne invented the first modern **seismograph** in 1897, and this is considered the date at which the modern science of seismology really begins.

the seismograph

A simple seismograph consists of a heavy pendulum with a marking device, and a rotating recording drum, both attached to bedrock. These two parts are not connected. The inertia of the pendulum prevents it from moving during an earthquake. But the re-

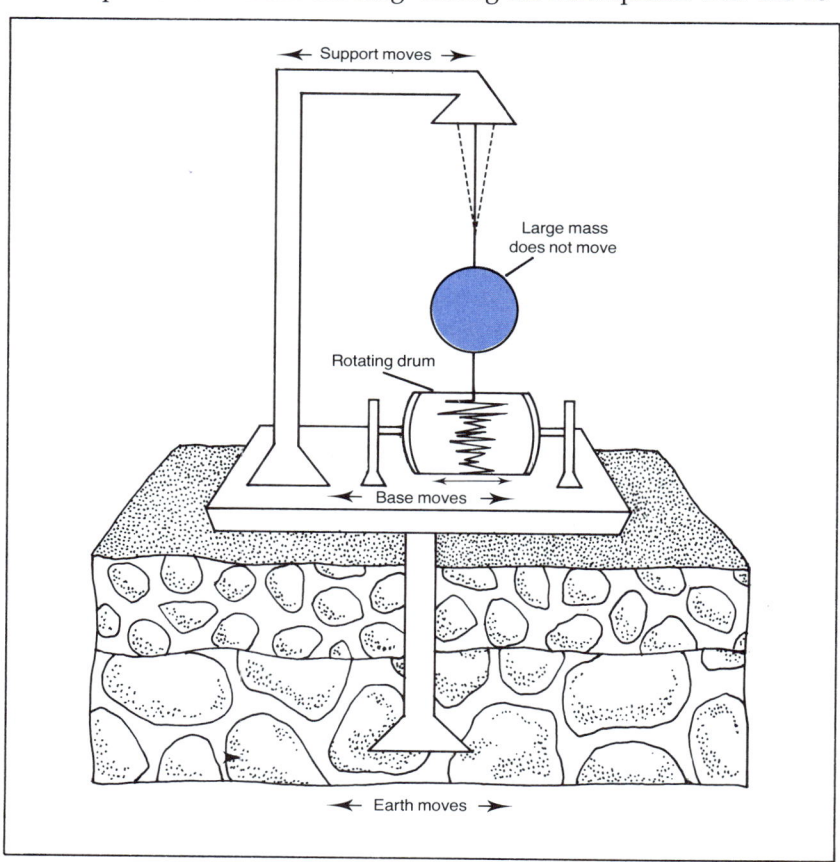

How a seismograph works. Arrows show the direction of movement. The marking device attached to the motionless mass records the seismic waves on the rotating drum.

the earth's interior and earthquakes

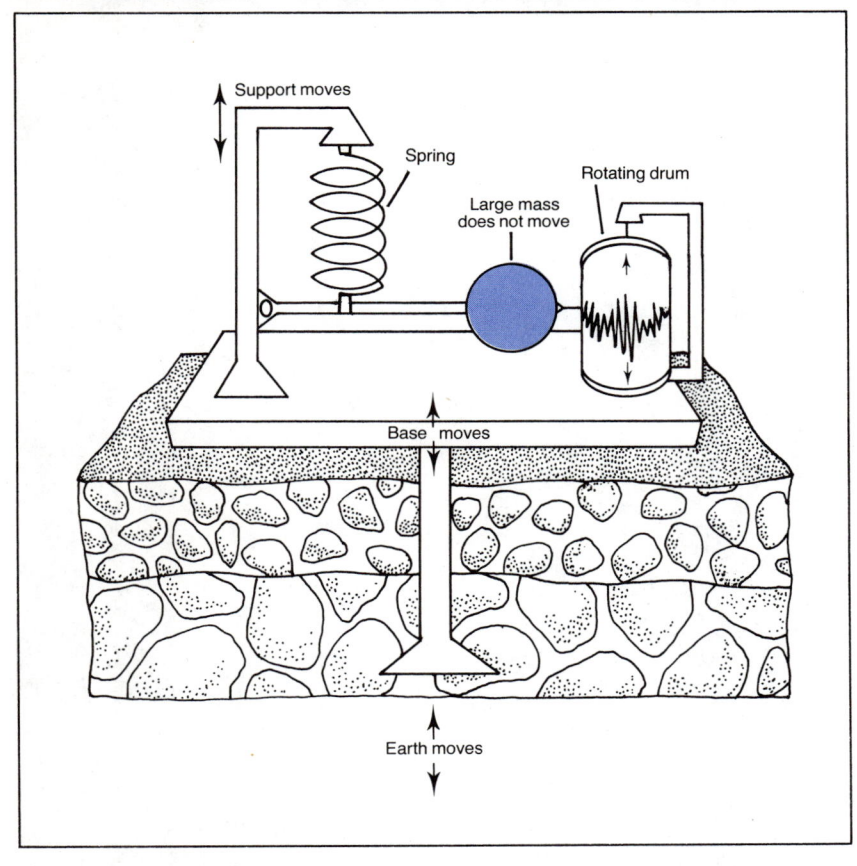

A second type of seismograph, in which the rotating drum is positioned vertically.

cording drum, being rigidly fixed, moves when the rock moves. Therefore, the marking device on the motionless pendulum marks on the recording drum the movements of the rock. Such a record is called a **seismogram.**

A laboratory equipped to record earthquakes has at least three seismographs, to pick up vibrations traveling in different directions. One records the vibrations along a north–south line; another, the vibrations along an east–west line. The third instrument, arranged in a slightly different way, records vertical movements of the earth's crust.

The seismograms of an earthquake form a complex pattern of lines. When there is no earthquake, the line is almost straight, although there is usually some background vibration caused by traffic or nearby machines. The first indication of a distant earthquake is a little wiggle in the line. This slight motion continues for a few minutes; then there is another, somewhat larger wiggle. For the next few minutes the wiggles continue at moderate size, with occasional slight peaks. The amplitude of the wiggles then increases greatly, and this increased amplitude continues for some time.

recording waves

This seismograph regularly detects tiny earthquakes along a series of faults in the northeastern United States.

what the waves tell us

Let us look at the simplest kind of record, as shown in the accompanying figure. The first wiggles are the **P-waves** or **primary waves** (from *primus*, meaning "first"). The second set are the **S-waves** or **secondary waves** (from *secundus*, meaning "second"). The large-amplitude wiggles are called **surface waves.** P-waves and S-waves travel through the earth, but surface waves travel around the outside of the earth, and are the cause of most earthquake damage.

The point below the earth's surface at which an earthquake originates is called the **focus.** The point on the earth's surface directly above the focus is known as the earthquake's **epicenter** (from the Greek *epi*, "on" or "over"). Seismic waves of all three types leave the point of focus at the same time, but they travel at different rates of speed.

A wave propagates itself by transmitting energy from particle to particle of the medium it is moving through. The P-waves trans-

the earth's interior and earthquakes

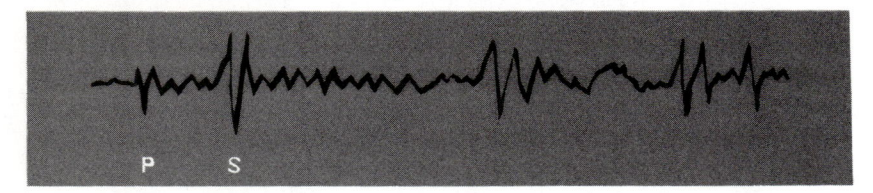

This greatly simplified figure shows P-waves and S-waves as they might appear in a seismogram.

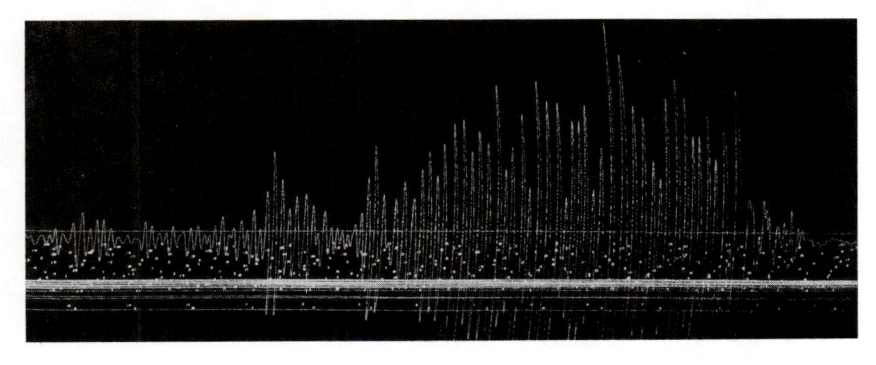

A typical seismogram, showing the seismic wave record of a major Japanese earthquake that occurred in 1939.

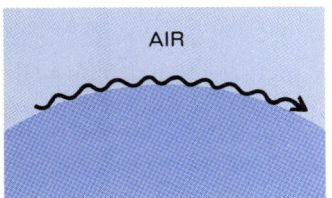

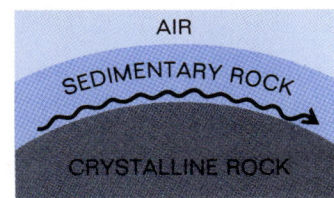

Surface waves travel through the sedimentary rocks near the earth's surface and along the interface between the sedimentary and crystalline layers.

mit their energy in a **longitudinal** direction; they can be compared to a long line of dominoes in which the first domino falls, and knocks down the next one, which in turn knocks down the next, and so on. P-waves can travel through solids, liquids, and gases. Their speed is greater through a very rigid medium. For instance, P-waves travel faster through dense basalt than they do through crumbly shale.

S-waves travel in a different way. Their energy is transmitted at right angles to their direction of movement, or **transversely.** They resemble the waves produced when a rope is held steady at one end and shaken rapidly up and down at the other end. Transverse waves can travel only through a solid medium, and they travel more slowly than longitudinal waves.

By observing the speeds of seismic waves and how those speeds vary, scientists can determine some of the physical properties of the material the wave is passing through, and thus gain a clearer idea of the earth's inner structure. Table 5-1 shows how much variation has been discovered.

Surface waves are complicated waves of long wavelength. Because they travel only through the crust of the earth, they provide little information about the inner structure. They can, however, tell us

the earth's interior and earthquakes

something about the crust—in particular, about abrupt changes in the kinds of materials through which they pass. Such a sudden change in material, such as glacial till next to dense granite, is called a **discontinuity.** It may provide useful clues to events in the earth's past. A surface where two discontinuous materials meet is called an **interface.**

Table 5-1. Variation in speed of earthquake waves in the earth's crust.

Wave type	Speed near the surface	Speed deep in the crust
P	5.5–6.0 km/sec	6.5–7.0 km/sec
S	3.3–3.7 km/sec	3.8–3.9 km/sec
Surface	2.0 km/sec	No wave

the core of the earth

On the basis of seismic wave data, we may picture the earth as a series of concentric layers: the core, the mantle, and the crust. Waves from an earthquake follow a curved path through these layers, indicating that they are being refracted by materials of varying densities and rigidity. This differential refraction produces the **shadow zone**—a zone between 103° and 143° from the epicenter of an earthquake where few, if any, of its shock waves are recorded on a seismograph. The shadow zone is caused by deflecting of the P-waves at the interface between the mantle and the core.

The **core** is the part of the earth deeper than 2,900 km. It has a radius of about 3,470 km. Seismic waves received between 143° and 180° from the epicenter have traveled through the core. Careful study of the seismograms of these waves reveals some interesting facts. First, no S-waves appear at all. This suggests that the core is either liquid or semiliquid, since S-waves can travel only through solid material. Furthermore, if the core were solid, P-waves that pass through the center of the earth should reach a seismograph about 16 minutes after leaving the focus. Their actual travel time, however, turns out to be 20 minutes—four minutes longer than expected.

This longer travel time implies that at least part of the core is liquid. But is all of it? To determine this, we must remember that the more dense and rigid a material, the faster P-waves will travel through it. Researchers have discovered that, although P-waves slow down as they enter the core, they actually speed up to a higher-than-normal velocity as they pass through the inner part of it. The core, then, must be made up of two components, a liquid outer part and a dense, solid center.

the earth's interior and earthquakes

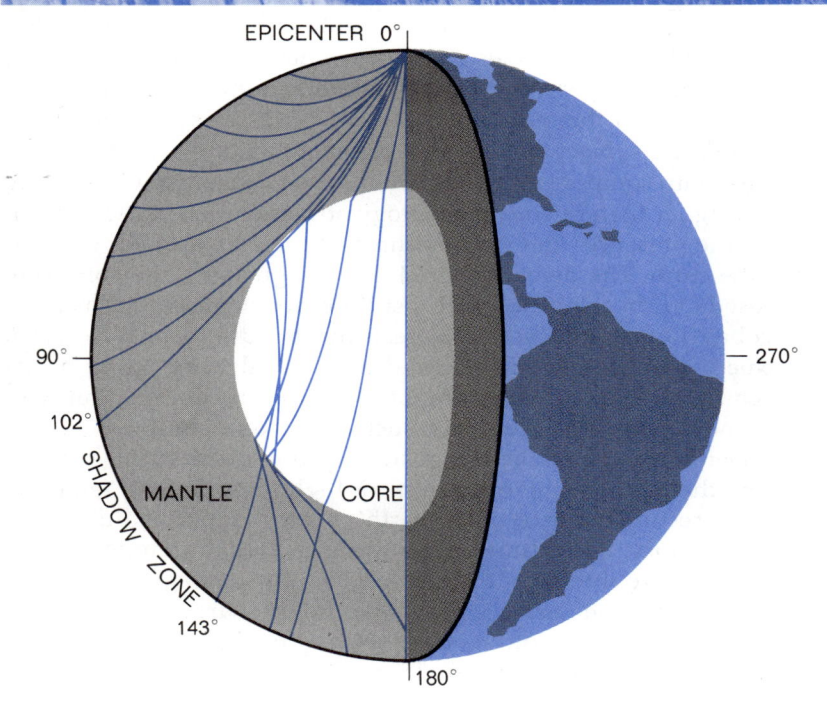

The shadow zone of the earth, in which P-waves are deflected as they reach the interface between the mantle and the core.

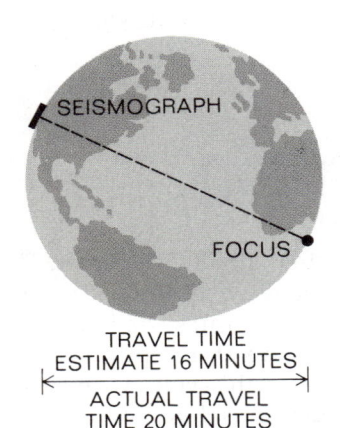

P-waves slow as they pass through the core, and S-waves are stopped by it, indicating that part of the core is fluid.

No one really knows how the core was first formed. However, many scientists believe that during its formative period the earth's interior became molten, a phenomenon called the **iron catastrophe.** As a result, a vast reorganization of the planet's entire body took place. According to this hypothesis, the earth's original composition was much the same throughout. But in the iron catastrophe, molten iron and associated heavy elements sank to the center of the earth to form the core. Lighter materials floated to the surface to form what we now know as the mantle and the crust.

the earth's interior and earthquakes

theories of core composition

The solid inner core is most probably composed chiefly of iron and nickel. Only these metals are heavy enough to explain the core's density, which jumps from 5.5 g/cm³ at the bottom of the mantle layer to 9.5 g/cm³ in the outer core. But the actual composition of the core is still the subject of some disagreement among experts. If iron were the main component, then the density of the core ought to be 20 percent higher than seismic technology now suggests, particularly in view of the pressures and temperatures that must exist near the center of the earth. The density should be even greater if the core were composed of iron combined with a still-denser element such as nickel or cobalt. One recent theory suggests that the discrepancy between the expected density of the core and its actual density might be accounted for if the core were composed of iron in combination with one or more less dense elements, such as silicon or sulfur.

Determination of the composition of the core has been complicated by the fact that the conditions of high pressure and temperature that presumably exist at the earth's center cannot be duplicated in the laboratory. It is estimated, for instance, that the temperature at the mantle–core interface is about 4,200°C. At the beginning of the inner core it may be nearly 6,000°C. Yet the pressure here is so great that iron (if the core is iron) remains solid, even though at the earth's surface it melts at about 1,500°C. Theoretically, at least, it is possible that metals in either solid or liquid form might possess entirely different properties and behave in entirely different ways under such conditions.

the crust and the mantle

Above the core lie the **mantle** and then, finally, the **crust**. The crust, solid and thick as it seems to us, is really a very thin skin, perhaps 40 km thick at most. The mantle below it is a thick shell of hot rock, extending downward to surround the core at a depth of 2,900 km. It accounts for about 83 percent of the earth's volume, nearly half its radius, and some 67 percent of its mass. It is thought to have an average composition corresponding to that of peridotite. Its density increases from roughly 3.5 g/cm³ near the surface to about 5.5 g/cm³ near the core.

The crust and its thin film of atmosphere and ocean are derived from the mantle. Melting of various components of the mantle during our planet's history produced magmas that rose to the surface and solidified. Besides contributing new rock to the crust, they gave off water vapor and other gases that added to the oceans and atmosphere.

Three distinct parts of the mantle have been identified

the earth's interior and earthquakes

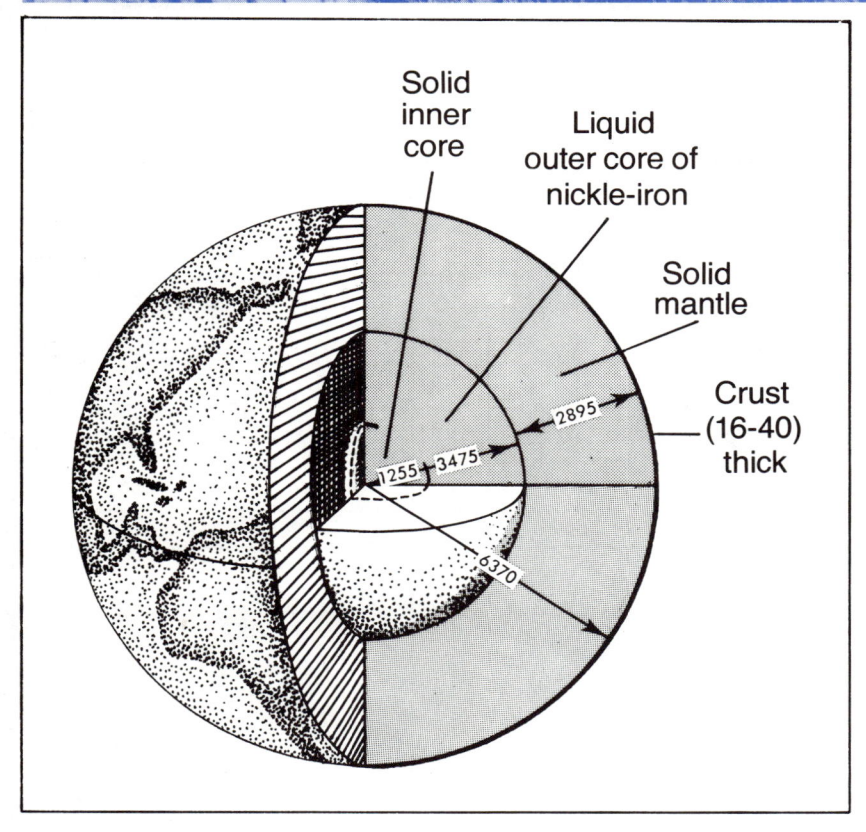

Solid inner core
Liquid outer core of nickle-iron
Solid mantle
Crust (16-40) thick
2895
1255 3475
6370

A cross-section of the earth, showing the relative thickness of the various layers.

A model of the earth's interior, showing the density of the crust, mantle, inner core, and outer core.

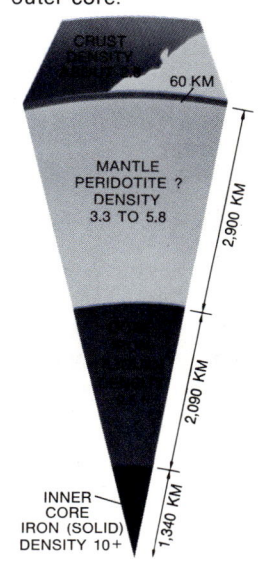

CRUST DENSITY — 60 KM
MANTLE PERIDOTITE ? DENSITY 3.3 TO 5.8 — 2,900 KM
— 2,090 KM
INNER CORE IRON (SOLID) DENSITY 10+ — 1,340 KM

through studies of seismic properties and density. The **upper mantle** extends from the bottom of the crust to a depth of almost 400 km. The mantle **transition zone** extends below this to about 650 km. The **lower mantle** reaches down from here to where the core begins, at about 2,900 km.

The figure on the next page charts the earth's depth and plots temperature distribution and melting points. Areas of both average and high heat flow are represented by thermal gradients. You will notice that the slope of both curves changes as depth and pressure increase. The accepted pressure gradient in the mantle is 400 atmospheres per kilometer (1 atm = atmospheric pressure at sea level, or about 15 pounds per square inch). It is estimated that the temperature at the mantle–core interface would have to be greater than 5,000°C to melt the mantle material. Since it appears to be only about 4,200°C, the mantle must be solid.

The data for this chart have been provided by a combination of theoretical determinations, laboratory experiments on the melting points of silicates and metals at different pressures, and measurements of heat flow. Again, it is important to bear in mind that we

temperature and pressure in the mantle

the earth's interior and earthquakes

cannot directly study what takes place at these great depths, or even reproduce all the necessary conditions in a laboratory. Laboratory equipment is now capable of producing pressures of up to 200,000 atm, which corresponds to the pressure believed to be found at a depth of 600 km. Beyond this depth our thinking is really educated

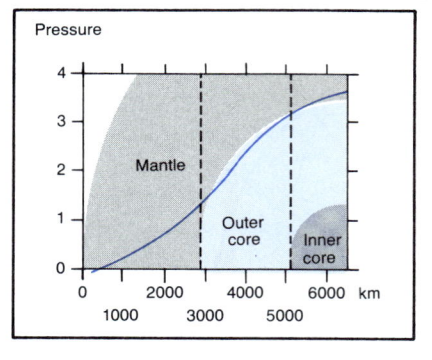

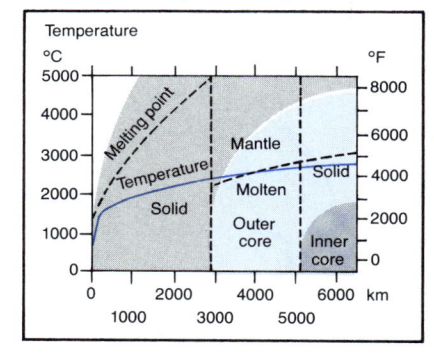

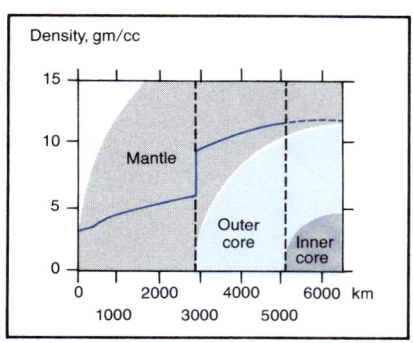

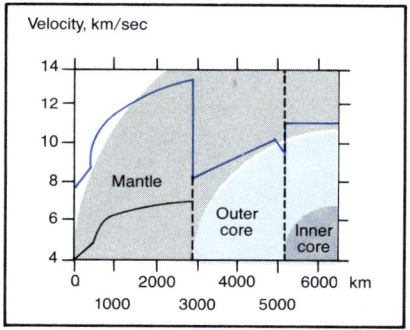

Properties such as temperature and pressure differ from layer to layer of the earth.

guesswork. Because we know many of the principles involved, the chances are that the guesses may be very good. Still, there is always the chance that factors beyond our knowledge are at work.

upper mantle and crust

The outer part of the earth, down to about 400 kilometers, has three distinctive subdivisions: the lithosphere, the asthenosphere, and the low-velocity zone.

The outer 60 kilometers comprises the **lithosphere,** which includes the crust and part of the upper mantle. The lithosphere acts as an insulating layer between the rest of the upper mantle and the surface, which differ greatly in temperature. Temperatures in the lithosphere generally range below 1,000°C.

The boundary between the upper mantle and the crust is marked by two sudden changes. One is that the crust contains distinctly more aluminum, sodium, and potassium, and less magnesium and iron than the upper mantle immediately below it. The other change is marked by a phenomenon discovered in 1909 by the Croatian scientist Andrija Mohorovičić. He found that at depths of 10 to 35 km P-waves suddenly increase in velocity from about 6.7 to 8 km/sec. The place where this occurs is called the **Moho** or **M-discontinuity.** Today it is recognized as the line of demarcation between the crust and the mantle.

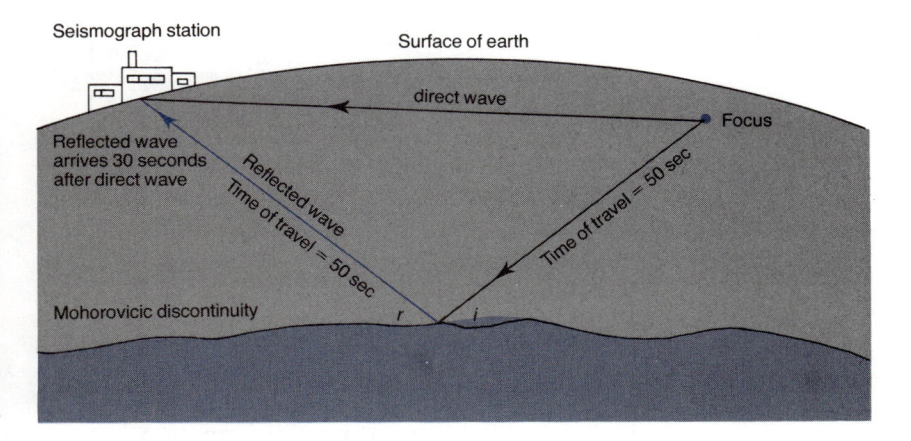

The Moho is the interface between the crust and the mantle. P-waves reflected by the Moho take a longer route from the focus to the seismograph at which they are recorded.

The **asthenosphere** is a region of the upper mantle 60 to 400 km below the surface. It consists of rocks that show some evidence of melting. Near the bottom of the asthenosphere, seismic waves that should travel faster than in the zone above actually travel six percent slower. The existence of this so-called **low velocity zone** provides a plausible explanation for the theory of plate tectonics, which we will discuss in chapter 8. Low velocity of waves implies that the material in this region is plastic and thus acts as a mobile surface on which the rest of the upper mantle rides.

composition of the crust

The earth's outer skin or crust varies in thickness from about 40 km under great mountain ranges such as the Himalayas to only a few kilometers under some parts of the ocean floor. Crust is classified as either continental crust or oceanic crust.

The upper continental crust generally consists of exposed sedimentary, metamorphic and granitic rock. Beneath this are layers of metamorphosed rock that is also strongly granitic in character. This material, as we saw in chapter 4, is called sial, because of its silica and aluminum content. These layers occasionally extend down to mantle depths. Together, the topmost part of the upper mantle and the continental crust form parts of **lithospheric plates**—quasi-rigid plates that appear to float independently on the asthenosphere below.

The oceanic crust is largely composed of basaltic rock or sima. The study of oceanic crust has proved very challenging. The average depth of the ocean is 4,000 meters, and the great pressure at such depths places severe limitations on the methods of exploration. Still, modern technology has helped to expand our knowledge of the oceanic crust. Devices have been developed that can drill into the ocean floor and bring up a plug or core of material for laboratory

the earth's interior and earthquakes

The distribution of sial and sima in the earth's crust. The study of seismograph records suggests that the less-dense sial is largely confined to the continents, and that a layer of sima lies at the bottom of the crust throughout the earth.

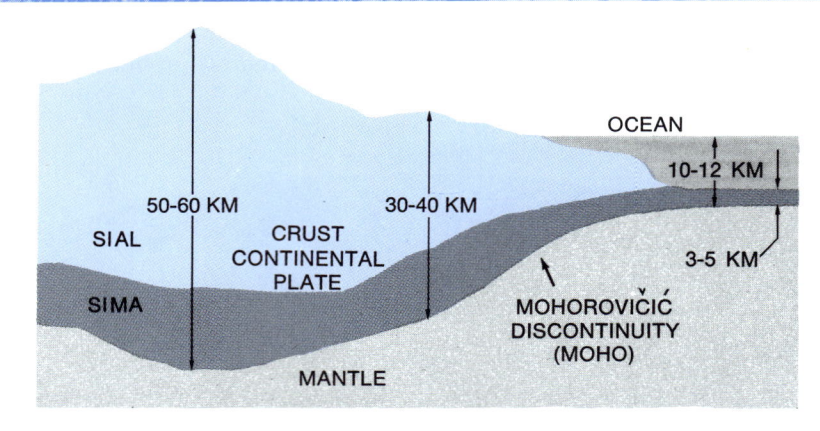

analysis. Some new undersea vessels are capable of diving to depths of more than 3,000 meters, carrying scientific equipment or even the scientists themselves. Echo sounding, or **sonar,** and analysis of seismic wave velocity have added to our store of information.

Combined evidence from all of this work has given us a picture of a continuous range of undersea mountains some 65,000 km in length, running through all the world's oceans. A hypothesis has been developed that these mid-ocean ridges are in fact a system of parallel ridges centered on one continuous gash or **rift** in the crust, from which the ocean floor is spreading out. The earth's crust is now considered to be made up of a series of quasi-rigid plates, which grow outward from sections of this rift where molten rock wells up. This process will be discussed at length in chapters 7 and 8.

the magnetism of the earth

When an electric current passes through a wire, a magnetic field is formed around the wire. The process can also be reversed: if a loop of wire is moved through a magnetic field, an electric current is produced in the wire. Any magnetic field has two poles, and the magnetic force forms lines between them, as shown in the accompanying figure.

As we know, the earth has a magnetic field, with poles near—but not precisely at—the geographic north and south poles. For convenience, we refer to the magnetic poles also as north and south. A compass needle points approximately north and south because one end of it "seeks" the north magnetic pole and the other end "seeks" the south magnetic pole. What is happening, in fact, is that the needle is aligning itself parallel to the earth's lines of magnetic force.

A number of substances in the earth are affected by magnetic

the earth's interior and earthquakes

force, notably iron and its ores. We have all seen iron filings, nails, and so on attracted and held by a hand magnet. One ore, magnetite, is so strongly affected that it sometimes becomes a magnet itself—the lodestone of old poems and tales. The response of minerals to magnetic force becomes, as we shall see shortly, a useful tool for studying certain things in the earth's history.

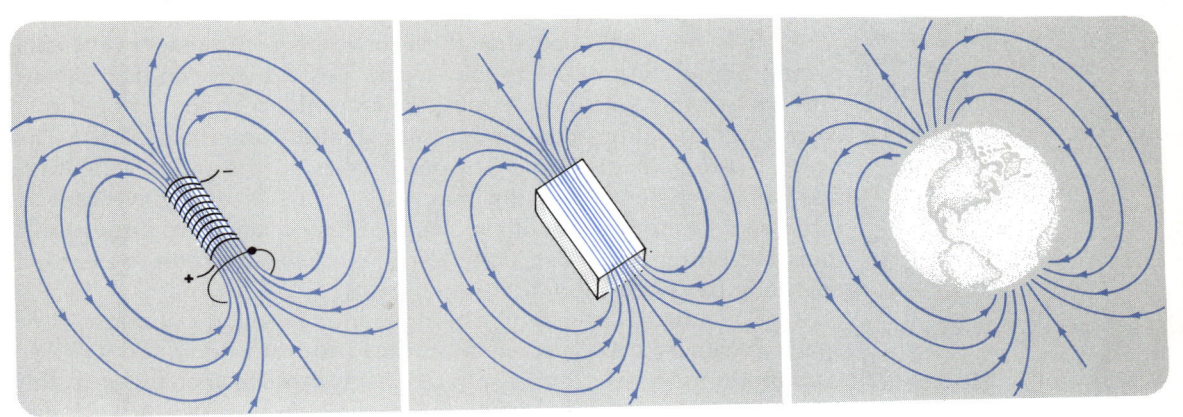

Scientists generally agree that the earth's magnetism is due to its partially-molten metallic core, which, as we have already seen, appears to be largely composed of iron. Several hypotheses have been advanced to explain how the magnetic field is generated.

The magnetic field produced by a current flowing through a wire may be related to the flow of electrons from atom to atom on the surface of the wire. Some scientists have reasoned that since at least part of the earth's core is fluid, there must be **convection currents** in it. Convection is the transfer of heat from place to place by a patterned flow of liquid. Such a flow, like the flow of electrons along a wire, might generate a magnetic field in the core. This hypothesis, however, has not been proved, and cannot be tested using present technology.

A second and similar hypothesis proposes that the magnetic field may be caused by a flow of electrons or ions in **convection cells** in the earth's mantle. These cells would be self-contained regions possessing convection currents. We may someday be able to test this hypothesis, since it is at least conceivable that we may be able to drill into the mantle, whereas we are never likely to penetrate directly to the core.

A wire carrying an electrical current produces a magnetic field. Permanent magnets and lodestones are magnetic because of the way in which electrical currents flow within their atoms. The earth's magnetic field is probably caused by convection currents which produce electrical currents.

When a magnetized needle is laid on a cork and the cork is floated in water, the north pole of the needle points toward the north and the south pole points toward the south. In much the same way, when a basaltic lava cools, the particles of iron-containing minerals in it become aligned parallel with the earth's magnetic field. As the lava hardens, a record of the magnetic field is thus preserved in the rock.

reversing magnetic poles

Recent studies of such **fossil magnetism** in rocks have led to a startling discovery. Far from being constant, the north and south magnetic poles have reversed their positions a number of times during the earth's history. Evidence for this discovery is the fact that rock specimens found in the same location but of different ages may display opposite magnetic polarity. The only explanation for this is that the magnetic field of the earth itself has undergone reversal.

It is now believed that there have been nine such reversals during about the last 3.5 million years. All evidence indicates that these magnetic reversals have occurred rapidly, perhaps over periods of only a few thousand years, and at short intervals of time as measured in geologic terms. The geophysicists who are investigating this subject believe that the magnetic reversals may have caused rapid changes in the rate of evolution of life on earth. To understand their reasoning, we must look at still another feature of the earth's magnetic field—cosmic radiation.

The earth is constantly being bombarded by radiation from space. We know from laboratory studies and from study of the survivors of atomic blasts that heavy radiation can cause changes in the part of living cells that carries information about heredity. Under normal circumstances, the earth's magnetic field traps much of the cosmic radiation in a belt in the upper atmosphere. But scientists believe that during the periods of magnetic pole reversal, the magnetic field disappeared almost completely. This allowed much more of the cosmic radiation to reach the earth's surface. As living cells were affected by the radiation, the hereditary information they carried was changed. Many mutations occurred, speeding up the normal rate of evolution, changing some species and destroying others. This theory has not been proved as yet, but future studies of the subject may yield some interesting results.

The reversal of the magnetic poles has also given strong support to the hypothesis that the ocean bottoms are spreading. This question will be discussed in greater detail in chapters 7 and 8.

the earth moves

We are used to thinking of the earth beneath us as the most solid and secure thing in our existence. Thus, we are particularly frightened when the ground trembles and trees and buildings shake. Yet earthquakes are very common. The relatively few that are strong enough to be felt are only a small fraction of the thousands that occur each year. The majority are so mild that no one notices them, and they can be recorded only by very sensitive instruments. The eight or ten that cause severe damage in any one year are the exceptions.

the earth's interior and earthquakes

what causes an earthquake?

An earthquake happens when something causes the rocks of the earth's crust to vibrate. Most major earthquakes occur because two large masses of rock slip a little along a fault. When vibrations are produced as these rock masses rub against one another, the earthquake is called a **tectonic earthquake.** It is this kind that we shall study in this chapter.

There are several other causes of earthquakes. One of them is volcanic action. For a long time, in fact, scientists believed that all earthquakes were associated with volcanoes, but now we know that this is not true. Furthermore, most of the earthquakes that do occur when a volcano is in eruption are tectonic, and not directly caused by the eruption.

Some earthquakes associated with volcanoes, however, are not tectonic. The gases dissolved in the hot magma expand with explosive violence as the magma rises toward the surface of the earth. Such explosions shake the rocks, causing earthquakes. Even without an explosion, the rush of volcanic gases through underground passages and out of the volcano's vent may cause the rocks to vibrate. This vibration is very much like the flapping of a flag in a strong wind. The collapse of an underground chamber in a volcano may also cause an earthquake.

It is possible to produce earthquakes by setting off huge explosions, such as underground nuclear tests. Many other human activities can cause vibrations of the earth that are felt for a short distance. A train rushing by, for example, may cause a micro-earthquake.

where do earthquakes occur?

No part of the earth is entirely free of earthquakes. Large areas of the continents and the oceans experience occasional mild shakes that pass unnoticed. Earthquakes that we do notice are called **sensible** earthquakes; that is, we can sense them. Sensible earthquakes almost always occur where the earth's crust is unstable, so they are most common where mountains are growing or where volcanoes are active. This locates the most violent earthquakes around the perimeter of the Pacific Ocean, in the Mediterranean Sea and eastward to the Himalaya Mountains, and along the mid-ocean ridges.

In North America the major earthquake area is the western coast, from southern Mexico to the western end of the Aleutian Islands. The Rocky Mountain region receives few sensible shocks. Occasionally, however, a damaging earthquake occurs in the vicinity of Yellowstone National Park, the site of hot springs that indicate underground volcanic activity.

The Midwest is almost without earthquakes. Along the East Coast only a few are noticed but many are recorded. Probably the great majority of earthquakes east of the Rocky Mountains are associated with isostatic rebound, which we discussed in the last chapter. This is the adjustment of the crust to the loss of weight of the glaciers that once covered most of Canada and a third of the United States.

the earth's interior and earthquakes

The major earthquake belts of the world. Circles indicate regions of mountain building or volcanic activity; lines indicate the location of mid-ocean ridges.

the elastic rebound theory

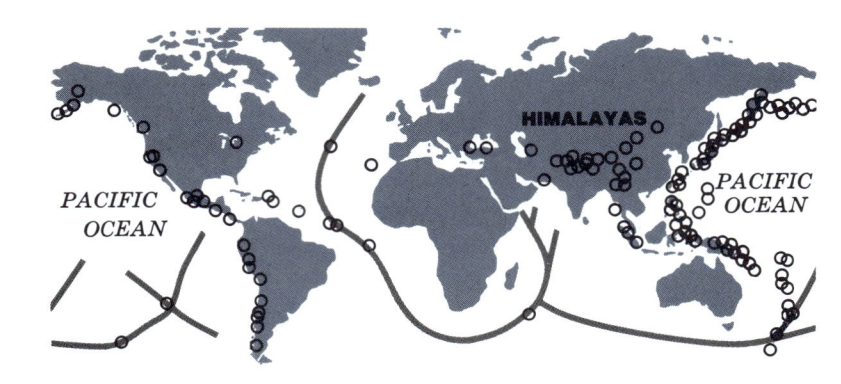

Observation of what happens to the rocks at the surface of the earth during an earthquake clearly indicates that many earthquakes are caused by movements along faults. One of the most famous is the San Andreas fault, shown in the accompanying figure.

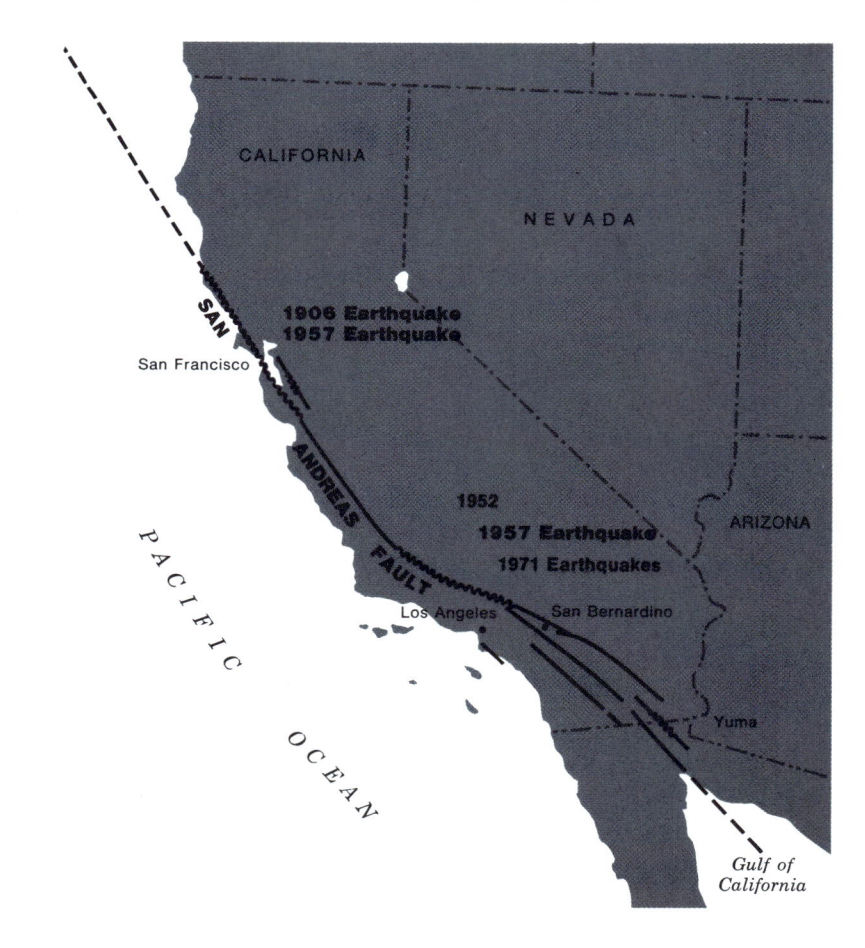

The San Andreas fault extends from the northern Pacific Ocean through much of California and into Mexico. The map shows locations and dates of major earthquakes along the fault.

the earth's interior and earthquakes

The San Andreas fault trace can be clearly seen in this aerial photograph of Riverside County, California.

The San Francisco earthquake of 1906 probably did more to stimulate research into the causes of earthquakes than did any previous earthquake. Cities a hundred kilometers south of San Francisco were seriously damaged. All these cities lie within the great San Andreas fault zone. The surface of the earth was torn apart for about 350 km along that fault. On the Marin Peninsula, northwest of San Francisco, the land on the western side of the fault zone moved about 7 meters northward with respect to the land on the eastern side.

Detailed studies of the movement along the San Andreas fault suggested that the western block had bent northward, slowly, over a period of from 60 to 100 years. The rough surfaces of the western and eastern blocks of the fault pressed against each other so forcefully that no movement occurred at the fault.

Instead, a great strain developed along the eastern margin of the western block. The surface of the block was bent (warped) into a shallow S-curve. When the stress that produced warping was sufficient to overcome friction along the fault, the edge of the western block snapped northward. Stages of this series of actions are shown on the next page.

the earth's interior and earthquakes

When the earthquake occurred, the strain was relieved almost instantaneously. However, the grinding of the two faces of the fault against each other caused the rock of the two blocks to vibrate. These vibrations moved through the blocks, shaking everything that rested on them—trees, buildings, water, animals, people.

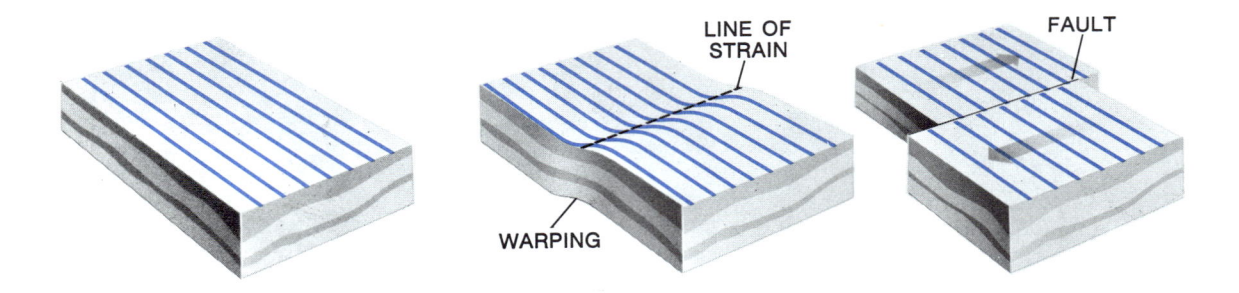

LINE OF STRAIN

FAULT

WARPING

The sudden release of strain that has built up over a long period of time will result in an earthquake.

This landslide followed a major earthquake. Damage caused by such slides is often greater than damage caused directly by the earthquake itself.

The ability of a material to rebound, or return to its original shape after a stress has been applied and removed, is called **elasticity**. In an earthquake, the stress disappears once the jammed rock faces slip and shift position. The elasticity of the rocks than causes each of the blocks to rebound to its original shape. For this reason the modern theory of earthquakes is called the **elastic rebound theory**.

measuring and recording earthquakes

It has taken a long time for scientists to devise an objective, universally applicable way to measure the forces involved in an earthquake. The first system was based on a scale of intensity invented in 1902 by Giuseppe Mercalli, an Italian student of earthquakes. The scale ranges from I to XII. Intensities I to V are felt by increasing percentages of the population. All people feel the shock of intensity VI; objects are moved, and some plaster falls from ceilings and walls.

the earth's interior and earthquakes

Intensity XII is marked by total destruction. With this crude scale, each earthquake is judged locally; the same earthquake might rate as intensity VII in one town and intensity III a few miles away. The energy actually released by the shock cannot be accurately measured.

Charles F. Richter, an American seismologist, later invented a scale that is more useful for scientific purposes, because it is related to the energy released by the earthquake, starting with a magnitude of 1 for the mildest quakes. It is based on seismographic records. The magnitude is determined by the distance the line of the seismogram is deflected. An earthquake of magnitude 2 makes a deflection 10 times as great as an earthquake of magnitude 1, and so on. But the energy difference is even greater than this suggests. Each magnitude number, in fact, represents about 60 times the energy released by an earthquake one number lower. Thus, an earthquake of magnitude 5 releases 60 times more energy than an earthquake of magnitude 4, and 360 times the energy of a quake of magnitude 3.

where was the earthquake?

Usually, the first news about a distant earthquake disaster is a vague announcement. The news will report a seismologist saying something like this: "A severe earthquake occurred 2,000 miles from this station, probably in the Aleutian Islands." Later in the day, a more precise location for the earthquake is announced. Determining the exact location involves several steps.

In theory, there should be no difficulty in estimating the distance between the site of an earthquake and the seismograph that records it. We assume that P-waves and S-waves start from the earthquake site at the same time. From the seismograph record we can measure the difference in their time of arrival. If we know the respective velocities with which the two kinds of waves travel, we should be able to estimate the time the earthquake occurred. Unfortunately, however, the waves pass through many different kinds of rock before they arrive at the seismograph, and this causes variations in their velocity. The result is that estimates based on travel time alone are rarely accurate. Additional information is needed.

Seismologists have painstakingly gathered all the available records about accurately located earthquakes. This information includes when the earthquake occurred, when its waves reached each seismograph, and the distance from the earthquake to each instrument. They have assembled this data into travel-time tables for each seismograph station. The information for one of these tables is plotted in graph form in the accompanying figure. Each station uses its own set of tables. The travel-time graph shows the relationship between time and distance for each of the many kinds of earthquake waves. Our graph is simplified and shows only travel times for surface waves, P-waves, and S-waves.

Locating the epicenter of an earthquake requires seismograms of waves received by at least three seismograph stations. Each station is used as the center of a circle; the radius of the circle is the same as

the earth's interior and earthquakes

that station's distance from the earthquake according to the travel-time table. When a circle is drawn for each of the three stations, the circles will intersect to form a small triangle. The epicenter of the quake lies somewhere within the triangle. Even more accurate determinations can be made by using records from more than three seismograph stations.

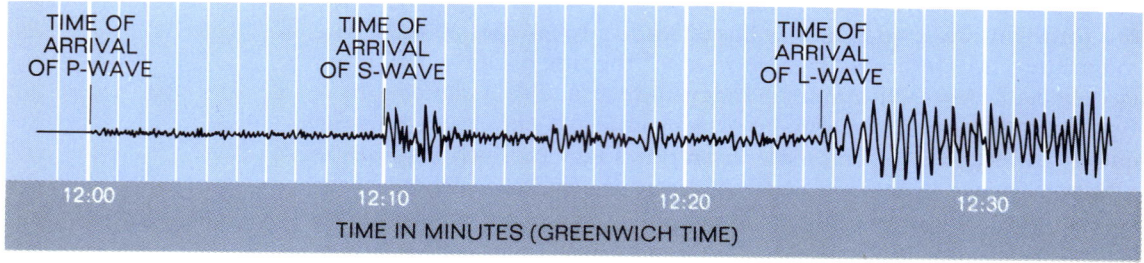

A record showing the different arrival times of P-waves, S-waves, and surface or L-waves.

Circles or arcs are drawn from major seismograph centers in order to locate a quake's epicenter. The epicenter will lie somewhere within the triangle of error.

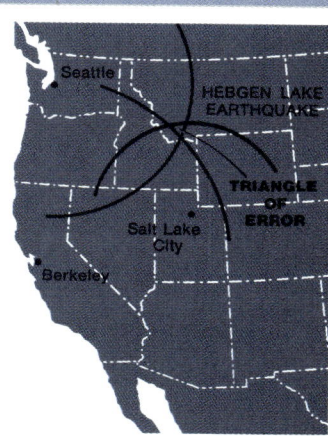

some famous earthquakes

Probably no earthquake in recorded time has been felt over such a large area as the one that occurred in Lisbon, Portugal, early in the morning of November 1, 1755. The shock was felt over an area five times as large as the United States. Tens of thousands of people lost their lives, some as far as 650 km away from where the earthquake originated.

A series of shocks completely devastated the city of Lisbon and killed about one-fourth of its population. Some people were crushed by falling buildings, others were burned by the fires that swept the rocking city, and many were drowned by the huge wave of water that surged in from the bay.

The water 1,900 km away in Loch Lomond, in Scotland, was

the earth's interior and earthquakes

tossed into waves several feet high by this earthquake. In Lisbon the shock drove the water out of the harbor, exposing the sea bottom. A wave was then thrown back as a wall of rushing water more than 15 meters high, which swept into the ruined city. A smaller wave raced across the Atlantic Ocean at several hundred miles an hour. This was an earthquake-induced **tsunami** or **seismic sea wave.**

Many other earthquakes, if less horrifying in their toll of human life, have been dramatic enough to satisfy any thrill-hunter. On

A massive seismic sea wave sweeps ashore in the Hawaiian Islands. The height of the man at left (arrow) provides a gauge of the height of the wave.

The tsunami that accompanied the Alaskan earthquake of 1964 dropped this fishing boat in the devastated streets of Seward.

the earth's interior and earthquakes

September 10, 1899, a severe earthquake shook the southeastern part of Alaska. Because it occurred in a sparsely populated area, it caused little damage. Investigators found that in Yakutat Bay the land had risen nearly 16 meters, the greatest earth movement ever measured in association with an earthquake. It so disturbed the glaciers in the region that their rate of advance was accelerated during the next few years.

A 1968 earthquake in the Midwest was probably caused by a very slight shifting of rocks along the New Madrid fault, in an isostatic rebound connected with postglacial adjustment. The New Madrid fault extends northeast from Missouri and then follows the course of the St. Lawrence River toward the sea. Severe earthquakes—the most severe ever felt within the continental United States up to that time—occurred in the same general region over a century and a half earlier, in December, 1811, and early January, 1812.

can earthquakes be predicted?

Considering the damage that earthquakes can do, the desirability of learning to predict them seems self-evident. One new technique that looks promising is based on radon measurement. **Radon** is a radioactive, inert gas formed by the decay of the trace amounts of uranium found in most rocks. Research projects in China, the Soviet Union, and the United States have shown that when rocks expand just before an earthquake in response to mounting strain, greater amounts of radon are released and dissolved in groundwater. A correlation between changes in the radon content of well-water and ongoing seismic activity has been established.

The upper atmosphere may give advance warnings of seismic sea waves. The connection was discovered by two University of Hawaii scientists. They found that earth tremors that can produce tsunamis also generate vertically polarized waves in the earth called **Rayleigh waves.** These waves pass into the air as vibrations, moving upward through the ionosphere. Because the ionosphere reflects radio waves, changes produced by the vibrations can be detected by their effect on radio waves.

All over the world groups of concerned scientists and laymen, such as California's Earthquake Prediction Evaluation Council, are searching for better prediction technologies. They are also planning programs of preparedness and programs for relief after an earthquake strikes. But dealing with earthquakes clearly calls for a massive effort of education, and perhaps some unavoidable social dislocation. To illustrate, let us look briefly at three examples of how people have reacted after an earthquake.

the Chinese example

The north-central region of mainland China is one of the most earthquake-prone areas in the world. The People's Republic of China has at its command the full resources of a highly motivated and disciplined population. Its citizens are fully aware of the importance of

EXPERIENCING AN EARTHQUAKE

Most people who have had the good fortune to live through a violent earthquake look back on the event with horror, and perhaps a prayer of thanksgiving. During a severe quake in a populated area buildings may be reduced to rubble, the ground may part, fires break out, and the cries of the injured and dying can be heard everywhere. But American naturalist John Muir, who experienced a serious earthquake in Yosemite Valley in 1872, paints a different picture of the cataclysm. The passage below is from his book, *Our National Parks.*

> In Yosemite Valley, one morning about two o'clock, I was aroused by an earthquake; and though I had never before enjoyed a storm of this sort, the strange, wild thrilling motion and rumbling could not be mistaken, and I ran out of my cabin, near the Sentinal Rock, both glad and frightened, shouting, "A noble earthquake!" . . . The shocks were so violent and varied, and succeeded one another so closely, one had to balance in walking as if on the deck of a ship among the waves . . . Then suddenly . . . there came a tremendous roar. The Eagle Rock, a short distance up the valley had given way, and I saw it falling in thousands of great boulders . . . pouring to the valley floor in a free curve luminous from friction, making a terribly sublime and beautiful spectacle—an arc of fire fifteen hundred feet span, as true in form and steady as a rainbow, in the midst of the stupendous rock-roaring storm. The sound was inconceivably deep and broad and earnest, as if the whole earth, like a living creature, had at last found a voice and were calling to her sister planets. It seemed to me that if all the thunder I had ever heard were condensed into one roar it would not equal this rock roar at the birth of a mountain talus . . . Storms of every sort, torrents, earthquakes, cataclysms, "convulsions of nature," etc., however mysterious and lawless at first sight they may seem, are only harmonious notes in the song of creation, varied expressions of God's love.

From *The Wilderness World of John Muir* by Edwin Way Teale. Boston: Houghton Mifflin Company, 1954, pp. 166–169.

IN FOCUS:
can the next California quake be predicted?

There were strange goings-on at the zoo in Tientsin, China, one warm summer day in 1969. The usually playful panda sat dejectedly in its cage, holding its head and moaning. The swans suddenly abandoned the water, and a tiger stopped its relentless pacing. Around Tientsin, rats ran out of buildings, cows and horses refused to enter their stalls, and snakes slithered from their holes. The animals, say Chinese scientists, were trying to tell humans that an earthquake was on the way.

The scientists, meanwhile, were detecting more subtle signs of the imminent tremor. Certain kinds of seismic waves were mysteriously changing velocity, electrical conductivity and the local magnetic field were changing, and the water level in wells was fluctuating. The amount of radon gas in the well water was on the rise, and the landscape near certain faults began to tilt upward. Shortly after these warnings, a quake convulsed Tientsin.

All of these phenomena can be warning signals of an impending earthquake that might take thousands of lives and cause billions of dollars in property damage. An increasing number of scientists in the United States, Japan, China, and Russia are paying more and more attention to such signals, in the hope of developing an ironclad system of earthquake prediction.

The science of earthquake prediction, however, is still in its infancy. Much that we know about the mechanics of earthquakes has come from advances in plate tectonics theory which have been made only in the past decade or so. Indeed, by demonstrating that most quakes occur on the boundaries of the earth's lithospheric plates, this theory has enabled scientists to designate the general areas that are most likely to experience an earthquake.

One of the first contributions to quake forecasting came from the Russians in 1971. Soviet geologists found that the velocity of P-waves generated by quakes, mine blasts, or other such events decreased for certain periods of time before a quake. Then, just before the quake struck, the waves returned to their normal speeds. The Russians also found that the longer the time in which the waves deviated from their normal velocities, the higher was the magnitude of the ensuing quake. Here, then, was a powerful tool for forecasting the approximate time and magnitude of a quake.

Wave velocity variation and the other physical phenomena that warn of an impending quake have been linked in an explanation of earthquake dynamics known as the dilatancy theory. According to this hypothesis, as rock at shallow depths in the earth is squeezed by tectonic forces, it swells. At the same time, thousands of tiny cracks appear in the rock. The P-waves passing through the rock slow down because they cannot travel through the newly opened spaces as rapidly as they can through solid rock. Next, groundwater seeps into the rock's microcracks and the P-waves climb to their normal speeds. But the water also weakens the rock to the point where it yields to tectonic stress and moves. The result is an earthquake.

The dilatancy theory also explains other quake indications. Electrical resistance in the rock rises because the air in the microcracks does not conduct electricity well. Changes in the local magnetic field are probably connected with this shift in electrical resistance. The swelling of the rock exposes more of its surface area to the water, which absorbs radon, thus raising the level of this radioactive gas in the groundwater. Finally, the swollen rock causes the overlying earth to bulge.

Have scientists actually used this theory for accurate quake prediction? By successfully forecasting the huge 1975 tremor in Liaoning Province, the Chinese were able to evacuate one million people in advance of the quake, saving an estimated 10,000 lives. Thus far, about ten other quakes have been predicted, including one in New York's Adirondack Mountains and one near Riverside, California, both in 1973.

Currently, scientists are keeping a nervous watch on a 4500-square-mile bulge in the earth at Palmdale, California, some 35 miles north of Los Angeles. The Palmdale bulge lies on the San Andreas fault and has risen as much as ten inches in some places during the past seventeen years. Telltale anomalies in seismic wave velocity have been detected, and the tectonic plates in the area are long overdue for movement. In fact, they have been locked in place since 1857, when a cataclysmic quake rocked the region. As a result, some scientists believe that a major quake can be expected sometime within the next ten years. Experts estimate that such a quake would kill 12,000 people. Others, though, believe that the bulge is merely a "false pregnancy," and point out that there have been cases of land uplift without accompanying quakes.

For the future, some scientists foresee that exact earthquake prediction will become a reality within the next decade, provided adequate funds are available. However, other people doubt the wisdom of earthquake prediction because the possible economic losses due to such a forecast might outweigh the losses in life and property. The answer to earthquake problems, say these researchers, is to build more quakeproof buildings, bridges, and dams, and to enforce existing building codes in earthquake-prone areas. Obviously, those involved in this new field have their work cut out for them.

the earth's interior and earthquakes

adequate advance warnings, and are practiced in precautionary measures.

In 1975 the Haicheng earthquake registered 7.3 on the Richter scale, and in 1976 the great Tangshan earthquake struck. For a long time reports of damage and casualties caused by the Tangshan quake were unofficial and very sketchy. Now, however, Chinese authorities have acknowledged that this quake was the worst to strike China in more than four hundred years. In addition to the terrific amount of structural damage caused by the Tangshan earthquake, it appears that as many as half a million people may have died in this major disaster, which also shattered the industrial output of the city.

This theater was shaken down by the earthquake that struck Long Beach, California, in 1933.

The populace of Tangshan quickly set about repairing the damage. But those who have been through an earthquake disaster often suffer psychological damage also, and such damage is not as easy to repair as a collapsed building or a broken bridge. For the Chinese, this problem was severe. During the period in which the earthquakes occurred, there were also several meteor showers, and Chinese tradition has always associated both events with the impending death of an emperor. Many Chinese did not consider it a coincidence that the two greatest political figures in the People's Republic—Premier Chou En-Lai and Chairman Mao Tse-Tung—died shortly after the earthquake.

the Guatemalan example

A major earthquake destroyed Guatemala's then-capital of Antigua in 1773. The capital was moved to the present site of Guatemala City, on what was thought to be safer ground. Since then, three major earthquakes have leveled the city—in 1917, 1942, and 1976. The 1976 quake registered 7.5 on the Richter scale and killed more than 25,000 people. Many more were left injured and homeless.

The Motagua fault zone is northeast of Guatemala City, and

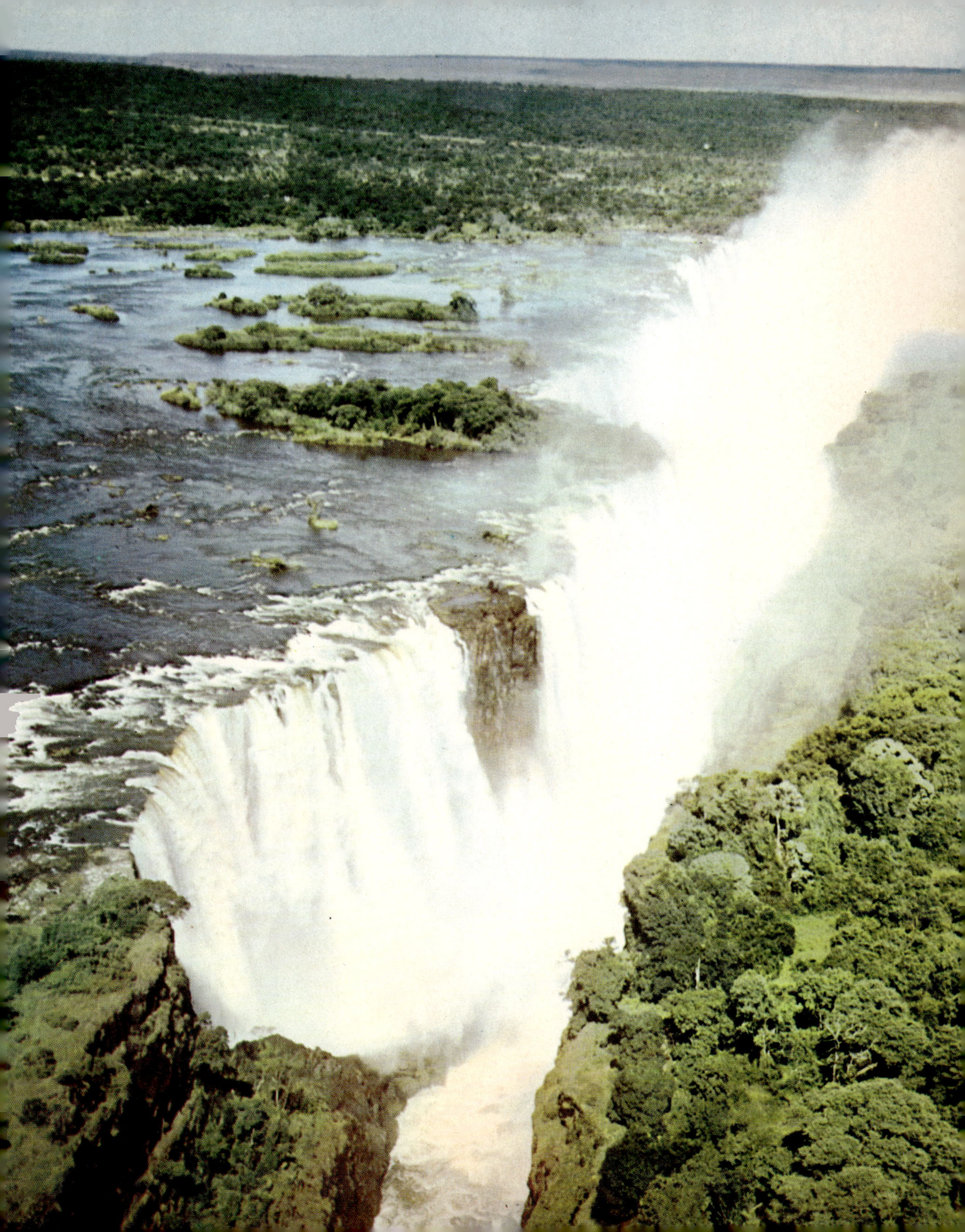

GEOLOGY

As seen from the depths of space, Mars is the red planet, and the planet Venus appears green. The earth looks blue — the Big Blue Marble, as one of the United States astronauts called it. At the surface of the earth, however, we see a panorama of color — a sky ranging through an infinity of blues; soil, rocks, and minerals in every hue from blackest basalt to brilliant ruby; the greys, greens, and turquoises of the ocean; the blinding white of cumulus clouds. In the following pages, we can see the infinitely varied display of color offered by our planet, alone of all the members of the solar system.

(Above) The Devil's Postpile National Monument, near Yosemite National Park in California, is a fine example of columnar jointing. This popular tourist attraction was formed when basalt shrank and split as it cooled.

(Preceding page) Victoria Falls, spanning the boundary between Zambia and Rhodesia, Africa, is one of the earth's great scenic wonders. Zambia is presently constructing two hydroelectric power plants which will harness part of the energy of Victoria Falls.

Many primitive societies built cliff dwellings such as this one in Arizona. By taking advantage of the natural structure of the cliff, the builders of this dwelling provided themselves with a safe fortress almost inaccessible to enemies.

(Above) Wizard Island in Crater Lake, Oregon, is the remains of a volcanic cone formed in a caldera.
(Right) Mount Hood, in Oregon's Cascade Range, is a composite cone. This volcano may have erupted as recently as the nineteenth century.

(Below left) Sierra Grande, in northeastern New Mexico, is a shield volcano. *(Below right)* Mount Lassen, in California, erupted in 1917. This volcano is a cinder cone.

When a meandering stream floods, the stream may form a new channel, cutting off a loop which then becomes an oxbow lake such as the one above. This stream, in Canada, shows a great amount of movement.

The White Mountains of New Hampshire are a part of the Appalachian Range. These comparatively low mountains are geologically mature. The White Mountain National Forest provides a variety of facilities for many winter sports.

Mountain glaciers have often been called "rivers of ice." The resemblance to a river can be clearly seen in the patterns of flow of this Alpine glacier, which has carved out a wide trough. The rugged peaks in the background and surrounding the glacier are perpetually covered with snow.

The brittle top ice of a glacier develops great crevasses as it adjusts to the slower flow of the ice near the ground. The top ice may also crack as the glacier flows over uneven ground. Some crevasses may be as much as 30 meters deep.

(Left) A U-shaped valley carved by a glacier. Compare with the picture at far right. *(Middle)* Canada has many impressive glaciers, such as this one at Mount Robson. *(Right)* Rivers, unlike glaciers, tend to carve deep, V-shaped valleys.

Shiprock, in New Mexico, is the eroded remains of a plug of basalt which once filled the throat of a volcano. This unusual structure, a popular tourist attraction, has long been considered sacred by the Navajo Indians of New Mexico.

These fossils date from the Ordovician Period. Brachiopods similar to scallop shells and fragments of trilobites can be seen embedded in limestone, along with the remains of other marine creatures from the same period.

Water freezing in the joints of a rock results in irregular breakage, caused by expanding ice crystals. Many loose chunks have broken away from the face of this granite cliff.

Tiny cracks, or microcracks, are common in many minerals. Microcracks can be observed in this sample of pink calcite. The sample was discovered in Franklin, New Jersey.

Some kinds of limestone are formed from pieces of shell. This type of limestone, known as coquina, is common in Florida, where it is frequently found around beaches.

the earth's interior and earthquakes

Frame houses suffered the smallest amount of damage in the great San Francisco earthquake, but were often tipped sideways.

the movements of two lithospheric plates there are similar to those along the San Andreas and related faults in California. Future earthquakes in this fault zone are inevitable, so it would seem sensible to move the capital city once again. Yet survivors of the 1976 earthquake immediately sought to re-establish control of their former home sites along the fault. This points up the major problem in educating the public about earthquakes: it is extremely difficult to tear people away from their homes and familiar surroundings on the basis of a disaster that will not occur until some vague, unknown time in the future.

the earth's interior and earthquakes

the California example

The earthquake which devastated San Francisco on the morning of April 19, 1906, was only one in a long series of shocks and tremors which indicate extensive and continuing seismic activity west of the Rocky Moutains. A number of "moderate" earthquakes on the order of 4.0 to 5.5 on the Richter scale have been reported on the west coast since the 1906 quake. The earthquake that occurred north of Los Angeles on February 9, 1971 was caused by movement on an east–west fault, not the San Andreas fault. The Richter scale intensity of this quake was 6.5, and aftershocks lasted more than a week.

The coastal area of California is frequently shaken by tiny earthquakes which are felt as minor jolts lasting only a couple of seconds. Dwellers in this region tend to ignore these little shocks; their

Free soup kitchens were set up to feed the many persons left homeless by the Long Beach earthquake of 1933.

the earth's interior and earthquakes

complacency is a major obstacle to efforts at public earthquake education. Home builders and developers, for example, have had no difficulty at all in selling buildings located right on top of the San Andreas fault, even though an occasional severe shock may split a wall or crumple a sidewalk.

After the severe Long Beach earthquake of 1933, California building codes were rewritten to include construction standards for supposedly earthquake-proof public buildings. During the San Fernando earthquake of 1971, however, many structures that conformed to these standards collapsed or were badly damaged. They included highways, overpasses, hospitals, and a large dam. Clearly, the building code for both public and privately-owned structures needs further revision, based on a more realistic appraisal of the earthquake hazard.

Geophysicists have publicly stated for many years that the question is not *whether* a major earthquake disaster will strike California in the future; the question is *when* it will strike—tomorrow, next week, ten years from now. Earthquake education is more important now than ever before, because one thing is perfectly certain. Sooner or later, the San Andreas fault will move again.

An aerial view of a huge landslide that occurred in the Madison County, Montana, earthquake area.

the earth's interior and earthquakes

SUMMARY

On March 27, 1964, Anchorage, Alaska was struck by the strongest earthquake ever recorded on the North American continent. It was completely unexpected, as most earthquakes are. But a better understanding of the earth, together with modern techniques, should soon make it possible to predict earthquakes.

The earth has a density of 5.52 g/cm^3. Surface rocks are less dense, with densities between 2 and 3 g/cm^3. The earth's interior, therefore, must be very dense.

The interior of the earth is hot from the energy released by the slow decay of radioactive elements. The heat flows through the earth and escapes into space at the surface. This prevents the internal temperature from increasing without limit.

Seismology is the study of earthquakes and the waves they produce. Seismographs record these waves. There are three types of waves. Primary (P) and secondary (S) waves travel through the various layers of the earth's interior on their way to us from the earthquake's focus. A careful analysis of these waves tells us various things about the internal structure of the earth. Surface waves travel only around the outside of the earth through its crust, and are responsible for most earthquake damage.

At the earth's center is the iron and nickel core, with a radius of 3,470 km. Its density is 9.5 g/cm^3. The outer part is liquid and the inner part is solid.

The mantle extends from a depth of 40 km down to 2,900 km. It contains 67 percent of the earth's mass. It is made up of three layers: the upper mantle, the transition zone, and the lower mantle.

The asthenosphere is the region between 75 km and 175 km in depth. It is a plastic solid. On it float the lithospheric plates, consisting of the crust and part of the upper mantle material.

The earth is a large magnet. The magnetism is believed to arise from fluid motions in the core. (Another theory proposes that the magnetism comes from convection cells in the mantle.) The magnetic field has reversed itself several times since the earth was formed.

Earthquakes are vibrations of the crust caused by events in the interior. According to the elastic rebound theory, the lithospheric plates push against each other. Regions in contact become stretched. Then something gives way and the stretched regions snap back. This rebound results in an earthquake.

The older Mercalli scale grades earthquakes according to the effects they produce. The modern Richter scale is based on the amount of energy released by the earthquake.

Earthquakes can be located by studying the difference in time between the arrival of the P-waves and the S-waves. The destructiveness of earthquakes is often multiplied by the seismic sea waves that they cause.

It is becoming possible to predict impending earthquakes. But the problem of educating the public is more difficult. Despite the several earthquake disasters in this century, people still build their homes on top of major faults.

the earth's interior and earthquakes

REVIEW QUESTIONS

1. The average density of the earth is 5.52 g/cm³. Why do we think the core is much denser?

2. What is seismology? Describe a seismograph and tell what it records.

3. How do earthquakes help us to understand the internal structure of the earth?

4. How do P-waves differ from S-waves?

5. Describe the observations that indicate that the outer core is liquid.

6. Why is the interior of the earth hot?

7. Describe the mantle. How is the upper mantle different from the lower mantle?

8. What is the asthenosphere? Why do we think that it is not entirely solid?

9. What is the source of the earth's magnetism? How do we know that the earth's magnetic field has reversed itself several times in the past?

10. How does the elastic rebound theory explain earthquakes?

11. About how many earthquakes occur every year? How many of these are sensible earthquakes?

12. Where do most earthquakes occur? Why?

13. Describe the Richter scale. Earthquake A is of magnitude 2 on the Richter scale and earthquake B is of magnitude 4. How much more energy is released by B compared to A?

14. How can earthquakes be predicted?

15. What steps would you take to reduce widespread damage due to earthquakes?

16. Describe an earthquake disaster in which seismic sea waves added to the destruction caused by the earth tremor.

17. What did the second earthquake say to the first earthquake that had just destroyed San Francisco?

(It was not my fault!)

CHAPTER SIX

VOLCANOES AND IGNEOUS ACTIVITY

Few of the ancient myths are as powerfully compelling as those about the lost civilization of Atlantis. Fabled as a cosmopolitan center of art and technology, one day it sank beneath the ocean waves and was gone forever. Although earlier generations of scholars regarded the stories of Atlantis as products of the imagination, there is a growing body of evidence that the myths are based on reality—the reality of a volcanic eruption 3,500 years ago.

Around 1600 B.C. the island of Atlantis was a dominant cultural influence in the Aegean, and its major port was a thriving center of commerce. Then within twenty years, multiple disasters struck. First its major city was buried by lava and pyroclastic debris from repeated volcanic eruptions. Then a mighty volcanic explosion—the third most violent known to history—blew away an estimated 24 square kilometers of the island. The explosion was probably caused by the rapid quenching of super-hot lava in sea water. It has been demonstrated that more than 2,000 foot-pounds of energy are released when a single cubic centimeter of lava is cooled from 1,100°C to 100°C.

Hard on the heels of the explosion came giant seismic sea waves. They completed the destruction of Atlantis, and engulfed coastal settlements all over the Aegean world. In a final aftermath, earthquakes triggered by the blast rumbled over the island of Crete 125 kilometers away, shaking down the palace of Knossos and devastating the countryside. Within a few hours, the entire Minoan empire lay in ruins. Atlantis itself was so utterly destroyed that for centuries it was unrecognizable. Only a thin crescent of land around the rim of the island was left above water—the present-day island of Thera. No one thought to identify this small barren crescent with the glories that were Atlantis—until recent finds by archeologists, combined with studies by geologists, began to reveal the past.

volcanoes and igneous activity

magma formation

The origin and formation of magma are still subjects of controversy among geologists. Of the many hypotheses advanced, only two are widely supported. According to one hypothesis, all magma originates in the asthenosphere of the earth's mantle. This **primary magma** later becomes differentiated into basaltic and granitic forms. The second hypothesis of magma origin presumes that there are at least two varieties of primary magma. Basaltic magma originates in the upper mantle, according to this theory, but granitic magma is formed when some of the rocks in the crust are melted.

why do rocks melt?

In the last chapter we discussed temperature distribution and melting points within the depths of the earth. We saw that an increase in pressure resulted in a corresponding rise in the melting point. In addition to the internal heat produced, frictional forces that occur in the lithosphere as the result of crustal movement cause local "hot spots." These forces also open up cracks and gateways to the surface. Such openings cause sudden drops in pressure, and this in turn causes lowering of melting points.

Since magma is lighter than the surrounding rock, it tends to rise. But remember that pressure decreases as the magma moves up, and so melting points drop. Thus the drop in temperature can be canceled out by the dip in melting point; if the magma moves up quickly enough, it can stay liquid until it reaches the surface.

Melting points are also lowered by the presence of water, and this effect is amplified by increased pressure. Rocks that contain little water will thus have a higher melting point than comparable rocks with a greater content of water.

composition of magma

The silica content of magma provides a handy means of classifying igneous rock. We must remember that igneous or metamorphic rock that melts to produce magma is already in place; it does not melt down entirely or completely. The individual minerals it contains melt at different temperatures, resulting in **partial melting.** Partial melting is thought to be responsible for almost all magmas. Basaltic rock melts at temperatures in the 1,200°C range, and granitic rock liquifies at temperatures around 800°C.

Basaltic magmas which produce mafic rock contain about 50 percent silica. Granitic or rhyolitic magmas give rise to rocks containing about 70 percent silica. Intermediate rock types, such as andesite and quartz diorite, contain about 60 percent silica.

Liquids vary widely in the rate at which they flow. For exam-

TAPPING GEOTHERMAL ENERGY

Deep in the earth lie extensive reservoirs of energy in the form of hot water and steam. These geothermal resources were once used by the Romans, Japanese, and others for cooking, bathing, and medicinal and recreational purposes. Even today, it is not unusual for a physician to recommend that a wealthy patient take a "water cure" in Europe, Japan, or Africa. Now, pressured by the world-wide energy crunch, experts are eyeing these underground reservoirs as possible sources of relatively pollution-free energy.

The heat source for most geothermal reservoirs is deep-lying magma, which is capped by a thick bed of crystalline rock. This impermeable rock transfers the heat of the magma to the overlying layer of water-containing porous rocks. The water becomes heated; if it boils, it may be converted into steam. In some places, the water or steam rushes to the surface as a fumarole or geyser.

The earliest large-scale use of geothermal energy began in 1904 at Larderello, Italy. Today, geothermal power is employed at various sites in Iceland, Japan, Mexico, New Zealand, Russia, and the United States. For example, most of the homes in Reykjavik, Iceland, are heated by hot water from underground springs. The world's largest geothermal facility, located in northern California, produces enough electrical energy to provide power for a city of 500,000 people. All told, the capacity of the world's major geothermal plants is equivalent to the output of a large nuclear power plant.

If this figure seems small, it is because the exploitation of geothermal resources is hamstrung by a number of problems. For one thing, exploration techniques are costly and far from developed. Because of the high temperatures involved, drilling is twice as expensive as drilling for oil or gas. Finally, geothermal resources do not attract profit-oriented firms because the product, unlike oil or gas, cannot be packaged, shipped, and sold worldwide. Thus, the future of geothermal power as a major energy source is quite cloudy. In fact, some scientists are willing to bet that this form of energy will never really come into its own.

volcanoes and igneous activity

ple, it is much easier to pour water than it is to pour honey. The property of resistance to flow is called **viscosity**. A magma's viscosity is directly related to its silica content. The more silica molecules present, the slower the flow of the magma.

magmatic differentiation

In chapter 2 we discussed Bowen's reaction series, which results in the fractional crystallization of minerals. This is believed to be the main process that accounts for magmas of different composition. Localization of this process, or the localized aggregation of crystals, can produce a magma containing dissimilar elements. The extremely slow rate of cooling in the deeper parts of the crust may permit significant settling of crystals. This would give rise to a differentiated material, with those crystals having a low silica content being at the bottom and those with a high silica content above.

Lava flows often take fantastic shapes. The expanse of lava shown here looks like a turbulent sea that was suddenly frozen.

The first mineral to crystallize in the discontinuous reaction series is olivine. This is followed by pyroxene, hornblende, and biotite. Thus there is evidence to support the view that after partial melting occurs, low-silica minerals tend to crystallize and settle out first, leaving a magma richer in silica content than the original material.

intrusive rock masses

We are usually more aware of the dramatic forms of igneous activity, such as volcanic eruptions. But the movement of magma under the earth's crust, called **intrusive** igneous activity, is not always so highly visible.

Sometimes magma being forced upward through a fissure finds a weak place between strata of sedimentary rocks. There it flows horizontally between the strata, forming an intrusion known

volcanoes and igneous activity

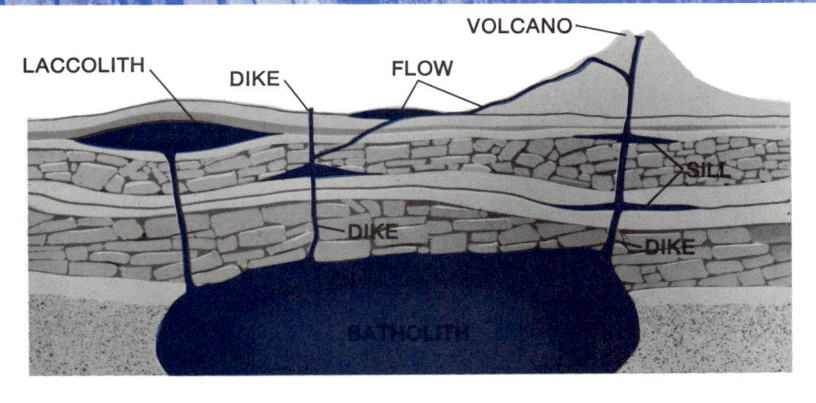

Several types of intrusive rock masses are shown here. Where the laccolith has forced sedimentary strata upward, a rounded dome has formed.

as a **sill.** One of the best-known sills is the Palisades in New Jersey, which extends along the banks of the Hudson River for 64 km. In some places, this sill is more than 300 meters thick. The Palisades sill is characterized by vertical **columnar jointing,** which occurs when basaltic magma cools slowly.

When an intrusion of magma between strata causes the overlying rock to arch upward, a characteristic dome is formed. The solidified magma is then known as a **laccolith.** Where many masses of magma have intruded under a column of sedimentary or other rocks over a large area—more than 100 km²—the intrusion is called a **batholith.**

dikes

Another kind of intrusive rock mass can be seen in the strange formation called Shiprock, in New Mexico. Shiprock is a 1,400-meter plug of basalt that once filled the throat of a volcanic cone. Erosion has removed the softer rock covering that once hid the solid basalt. It has also exposed long vertical walls of intruded rock called **dikes.** At the time of volcanic activity, magma flowed into great cracks that radiated from the vent of the volcano. This solidified into rock that was harder than the surrounding rocks.

what is a volcano?

The more familiar kind of volcanic activity is the erupting volcano. A volcano is an opening in the earth's crust through which lava, gas, and rock fragments are thrown out. The accumulation of this material as it falls back to the ground around the opening, or simply flows a short distance out of the opening, is called the **cone.** When magma solidifies on the earth's surface, it is called **extrusive.**

Cinder cones are the result of explosive eruptions that throw out rock fragments, with little or no lava flow. These cones have uniform, steepsided slopes. The cones of **shield volcanoes** have more

volcanoes and igneous activity

rounded and gently sloping sides, and are usually formed by a less violent eruption. The third type of volcanic cone is called **composite** or **stratovolcanic.** Composite volcanoes are made up of alternate layers of lava flows and **pyroclastic** debris (the word means "fire-shattered"), the material formed during explosive eruptions when rock is broken into small fragments and thrown out.

The **crater** is the cup-shaped depression on top of a volcano. Sometimes the crater and the whole top section of a volcano are missing. In their place we find a **caldera,** which is many times wider than the original crater. Calderas are formed when the top section of the cone collapses into the hollow area which was formerly filled by magma. In some instances—Wizard Island in Crater Lake, Oregon is an example—a new cone may be formed within the caldera by later eruptions.

Magma forms in the asthenosphere. As it moves up through the lithosphere it may reach the surface in the form of **lava flows** and volcanoes. If the eruption is "quiet," the discharge is mainly lava containing little gas and very small crystals. This type of lava contains comparatively little silica. In an explosive eruption, more siliceous materials are discharged. Explosive eruptions are noted for the large amount of gaseous materials they expel. These may include steam, carbon dioxide, hydrogen, hydrochloric acid, hydrogen sulfide, and sulfur dioxide.

The Devil's Tower, in Wyoming, is an excellent example of columnar jointing. Columnar jointing occurs when basaltic magma cools slowly.

volcanic eruptions

In 1815, Tamboro, a volcano on the Indonesian island of Sumbawa, erupted, discharging some 58 cubic kilometers of material—the greatest known volume of matter discharged. A crater 11.3 kilometers in diameter was formed, and Tamboro lost some 1,250 meters of its height. It has been estimated that the Tamboro eruption was caused by an internal pressure of 46,500,000 pounds per square inch.

volcanoes and igneous activity

On August 27, 1883, the Indonesian island of Krakatoa established a record as the greatest verified volcanic explosion when it blew up with the estimated force of a 5,000-megaton hydrogen bomb. The explosion was heard over an eighth of the earth's surface. Approximately 36,380 people died, and 163 villages were destroyed. Much of the 6.8 cubic kilometers of volcanic rock that exploded was so finely fragmented that it reached the stratosphere and was scattered around the globe by wind currents, causing spectacular red sunsets for several years. Larger rock fragments were thrown to a height of 54 kilometers, while dust continued to fall for 10 days at distances up to 5,000 kilometers away. The Krakatoa explosion also set the record for the highest recorded seismic sea wave or tsunami of 41 meters. Other tsunamis have been recorded racing across oceans at speeds of up to 784 km/hr. Much of Krakatoa's 29 square kilometers vanished, leaving a hole some 330 meters deep and several kilometers across.

What exactly happens during an eruption? An analogy may help clarify events. Imagine the scene at a jam-packed stadium. The game or performance has just ended. The crowd's advance to the exits is slowed by aisles, ramps, walkways, and steps. Exit points are few and far between, and they are being held closed from the outside. Thousands of people surge forward, temporarily forcing the exit open, and many escape before the exit gate swings shut again. The movement of the crowd inside is checked. People waiting to get out start to back up in lines and bunches scattered throughout the stadium's structure of tiers and levels. The crowd pushes hard on the closed exit gate. If they can't develop enough force, they may have to move on, seeking another exit gate—or they may even burst through a weak point in the wall. This is a rough analogy to the way magma moves through the earth's interior seeking a way to the surface.

Every volcano has its own unique characteristics, but most follow a general pattern. As magma moves toward the surface its progress may be slowed or halted as conduits become sealed or channels shift. There may be long periods of inactivity lasting hundreds or even thousands of years. During this time, however, magma continues to accumulate in chambers near the surface. Breakthrough may occur when pressure below continues to build. The overlying rocks may weaken through stress, or the vaporization of seeping groundwater can trigger explosive breakthrough. Magma accumulation sometimes causes the earth's surface above to bulge and buckle. Local earth tremors, subterranean noises, changes in the flow and temperature of nearby springs, odd behavior of horses and livestock, swirls of gases drifting near the cone—these are all early warning signs of volcanic eruption.

Steam and vapor emission are the universal hallmarks of volcanic eruption. A mushroom cloud may form, or the vapor clouds may take the shape of an immense tree, with its thick trunk billowing

Above, the 1977 eruption of Mt. Usu on the Japanese island of Hokkaido. Below, the relationship between a magma pool and a fissure eruption.

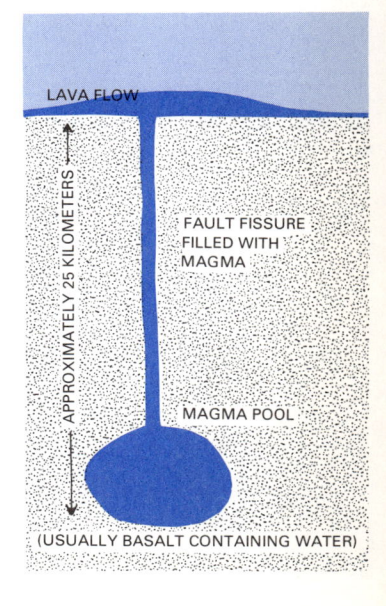

volcanoes and igneous activity

up out of the crater and its branches intertwining with the clouds overhead. Whatever its shape, vapor is among the first signs of impending eruption, and it may linger on in wisps long after the eruption is over.

Huge steam clouds may form over the area of an eruption and condense into heavy rain. Fierce hot currents rip up into the cooler air, causing violent updrafts and ionization of the atmospheric gases. Violent thunderstorms can occur, with intense discharges of lightning. The rains can be deadly, both for their suffocating sulfurous content and for the torrential volcanic mud flows they may set off.

The most bizarre and death-dealing variation on the emission

This photo was taken shortly after a *nuée ardente* destroyed the city of St. Pierre, Martinique. Observers viewed the event from several ships that were anchored in the city's harbor.

of gases during an eruption is called a fiery cloud or *nuée ardente*. This cloud is made up of incandescent volcanic debris mixed with toxic gases, and can be blasted through the side of a composite volcano if its crater is clogged. It can roar down volcanic slopes like an avalanche, at speeds of more than 160 kilometers per hour. To add to the danger, rock fragments may be carried along in the lower portion of the fiery cloud, which is much denser than air. A *nuée ardente* erupted on the West Indian island of Martinique on May 8, 1902, quickly sweeping down on the town of St. Pierre. The temperature of the cloud was more than 1,500°F. The burning that accompanied it probably used up most of the oxygen in the air. This combination of factors killed all but two of the 30,000 people in the city within minutes.

Gases, lava, and pyroclastic debris are expelled by eruptions in varying amounts. Mud flows sometimes occur near the end of a major eruption. But most curious aspect of volcanic eruption is the transformation of magma into lava as it rushes up the last few tens or hundreds of meters to the surface. The condition of lava depends

volcanoes and igneous activity

upon its viscosity. This, in turn, is related to its temperature, silica content, and the amount of water vapor and other volatile chemicals it contains. The higher the temperature of extruded lava, the less viscous it is. Basalt flows may solidify into a smooth, ropy lava known as **pahoehoe,** or into a rough, cinder-like texture called **aa.**

Rhyolitic and andesitic magmas are viscous. Their dissolved gases cannot easily escape and may build up enough internal pressure to cause a violent eruption. The superheated gases of rhyolitic and andesitic magmas expand a thousandfold as they are released, forcing lava and pulverized rock from the vent. The pyroclastic materials blown out by escaping gas may range from enormous chunks to superfine dust. The larger chunks are usually globular or streamlined in shape from having hardened as they flew through the air. They are called **bombs.**

A coating of dust or fine ash may cover the countryside for miles around a volcano after an eruption, settling like a gray snow several inches deep. This mixture invades households, finding its way into every crevice even if doors and windows are shut. It may fall in such quantities that the sheer weight alone does much structural damage.

Cinder cones generally have a steep slope of some 30 to 35 degrees; they tend to be rather low (under 300 meters). Cinder cone symmetry is frequently disturbed by local conditions. For example, a strong wind may blow loose materials over to one side. Cones of this type can build up very rapidly. Monte Nuovo in Naples, according to contemporary reports, formed a hill 122 meters high and 2.4 kilometers in circumference in only a few days.

If lava is very silica-poor and therefore flows easily, it can

Each member of the Hawaiian Island chain is the peak of a huge shield volcano. It is possible that volcanic activity in the future may add new islands to the chain.

IN FOCUS:
volcano watch: predicting the next blow-up

On July 6, 1975, a part of the earth near Tolbachik volcano on Russia's Kamchatka Peninsula erupted violently. Four fissures opened, hurling ashes seven kilometers into the air and blanketing 200 square kilometers of the surrounding area with volcanic ejecta. A month after the eruption, a cone 250 meters high had accumulated. It was a spectacular performance, as volcanic eruptions go. But more important, it had been predicted four years in advance By P. I. Tokarev, a Russian vulcanologist.

Tokarev's forecast was no mean feat in a field where scientists can watch a volcano stew away for years and not have the slightest inkling of when or how violently it will erupt.

For one thing, a unified theory of prediction is difficult to come up with because each volcano is a case unto itself, different from all others. In some instances, vulcanologists have no idea of the size and location of the volcano's magma reservoir, the rate at which new magma is entering it, the amount of gas being released as the magma cools, and whether pressure is being built up in the interior of the structure—all important elements in predicting an eruption. When and if a volcano does blow, the lava may trickle quietly down its flanks, or it may be rocketed out in a spectacular display of pyrotechnics.

How are scientists coping with these problems? For starters, they have established a number of observation stations in regions where volcanoes are notoriously spirited. Along with the Hawaiian observatory, there are others in the United States (in California and Washington), as well as in Japan, Chile, New Zealand, and Indonesia. Vulcanologists at these stations keep an eye on certain volcanoes that seem ominous. They study the cone's history, geology, seismicity, geochemistry, any deformation of its surface, and any variations in its electric, magnetic, and gravitational fields.

In Hawaii, for example, scientists have a fairly complete eruption history of Kilauea since 1750 and of Mauna Loa since 1832. They know when these volcanoes blew, how long each event lasted, the length of their dormant periods, the area covered by the lava flow, and the volume of lava produced.

For short-term analysis, vulcanologists have an arsenal of devices at their disposal. Seismometers, for instance, are particularly valuable in detecting the shallow earthquakes that usually foreshadow an eruption. The quakes begin as the magma starts its rumbling ascent, increasing in number and magnitude just before an explosion. At Kilauea, swarms of quakes—sometimes as many as three or four thousand—can be detected before an eruption from the flank. However, seismicity can fool volcano watchers, as it did in Coseguina, Nicaragua, in 1912, when the volcano there blew without sensible quakes. Conversely, Vesuvius has been known to produce tremors without erupting at all.

Another important tool for the vulcanologists is the tiltmeter, a device which detects swelling or deformation of the surface of the cone as the rising magma begins to push outward. But again, scientists must compile long records of such swelling before they can make a prediction, since some volcanoes bulge before erupting, and others don't.

Since a sharp rise in temperature often presages an explosion, vulcanologists plant special thermometers around a suspect volcano. More recently, infrared images taken by aircraft and satellites have also pinpointed hot spots in and around an active volcano. A related technique is the analysis of gas emitted from a volcano. In most cases, temperature increases cause a change in the composition of volcanic gases. Thus, at Mount Mihara, in Japan, the sulfur, fluorine, and chlorine content of the gas increased just before the volcano exploded in August, 1957.

Scientists have recently set up a volcano watch on Mount Baker, in the state of Washington. This peak, more than 10,000 feet high, has lain quiet since the late 1860s. But in March, 1975, it began to behave strangely. The volume of noxious sulfur fumes emanating from the stewing magma has more than tripled, and steam pluming from the volcano's maw is visible in Seattle, 90 miles away. The temperature at the steam vents has soared to 260 degrees Fahrenheit—the hottest detected in any volcano in the Cascade range during the past 90 years or so. Moreover, microscopic filaments of a glassy substance, produced by gas passing through molten rock, have been detected in the area. Scientists are predicting that a major eruption is in the offing. If Baker does blow its top, 40 million cubic meters of rock and hot cinders barreling down the flank of the cone at 320 kilometers an hour would devastate the tiny town of Concrete at the foot of the mountain. Not surprisingly, geologists have ringed Baker with a network of seismometers and tiltmeters, and are keeping a careful eye on these recorders.

So far, vulcanologists know a great deal about volcanoes, but relatively little about predicting their outbursts. Hard work is needed before forecasting techniques can be routinely employed.

spread out over a great area before solidifying, forming a nearly flat cone. If the lava is perfectly liquid, it will spread out to form a sheet, with no perceptible slope at all. In fact, some lavas are so fluid that they can flow down an incline at 1°, and shield volcanoes of basalt have been observed that have a slope of less than 3°.

Eruptions of basaltic magma can produce shield volcanoes of enormous proportions. The largest volcano in the world is a shield volcano called Mauna Loa, which together with four other shield volcanoes forms the island of Hawaii. These volcanoes were created by eruptions of mostly basaltic materials over the course of almost a million years.

fissure eruptions

Fissure eruptions appear to have shaped a far greater share of the earth's geological development than have cone-and-crater type eruptions. Most fissure eruptions occur on the ocean floor. Scientists have only recently begun to investigate the dynamics of "live" fissure eruptions along mid-ocean ridges where new sea floor is being created. We will discuss this subject in more detail in chapters 7 and 8. Actual pictures of these extrusions show them as looping coils, rather like toothpaste being squeezed from a tube, or heaps of **pillow lava,** strewn about like clusters of bowling balls.

On land, basaltic lava may be explosively erupted from a fissure, forming a platform called a flood basalt. Forests take root and grow quickly in these areas. The vast area known as the Columbia River Plateau in the northwestern United States is an imposing example of a flood basalt.

studying volcanoes

The states of Hawaii and Alaska contain many active volcanoes. Each of the Hawaiian Islands is the peak of one or more gigantic shield volcanoes that have their bases some 4,250 meters below sea level. At the summit of Kilauea, a volcano on the island of Hawaii, is the Hawaiian Volcano Laboratory, operated by the United States Geological Survey.

One of the many eruptions of Kilauea occurred in 1959. For over a year, J. Eaton and K. J. Murata had been observing and measuring the seismic waves being produced under the volcano. They had also set up tiltmeters around the flanks of Kilauea to measure any changes in the shape of the mountain. They discovered that the intensity of the seismic waves and the bulging of the sides of the volcano were rapidly increasing. From such data, they were able to predict that an eruption would soon take place. It did!

When it finally happened, a pit crater of Kilauea became filled with a lake of lava 120 meters deep. As the lava poured out of the

volcanoes and igneous activity

interior of the volcano, the sides began to contract and resume their normal shape. The eruption lasted until April 1960. For the first time, scientists had predicted a volcanic eruption.

There are currently about 530 active volcanoes around the world. Roughly 15 percent of these are on the ocean floor. In chapter 8 we will discuss the theory of plate tectonics, which presumes that the earth's crust is composed of several gigantic plates. At the points where these plates butt against each other, new crustal material is being formed. When we compare regions of volcanic activity with the earthquake belts, and then with the plate boundaries, we see that earthquakes and volcanoes are associated with plate boundaries.

According to this theory, three things can happen at plate boundaries. First, there can be rifting; that is, two plates can be moving apart. Second, transform faulting can take place; two plates can be sliding past each other. Third, the ocean floor can be destroyed as two plates move together.

Along the fringe of South and Central America, two plates are moving together; one is in the process of destroying the other. As the South American plate moves westward, it is overriding the oceanic plate to the west. Behind this zone of plate consumption there is violent volcanic activity in the Andes. Some of the worst earthquakes in recent times have taken place in Chile and Peru. All of this disturbance is caused by the downgoing oceanic plate.

Farther north, other things are happening. Los Angeles will be as far to the north of San Francisco as it is to the south of it now—in about twenty million years. The largest plate in the Pacific Ocean—the Pacific plate—is moving 6.3 centimeters northwestward per year, taking a piece of North America with it. The North American plate is moving more directly west, so transform faulting is taking place.

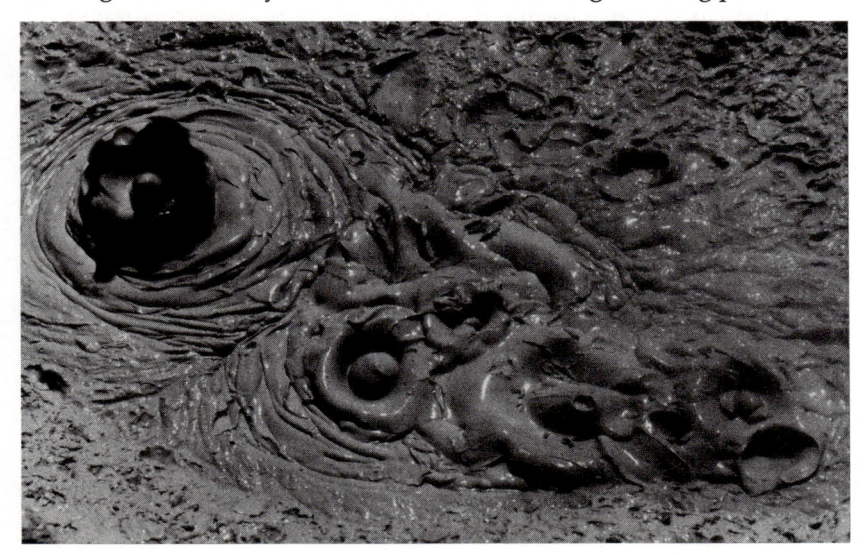

volcanoes and earthquake activity

Volcanic activity often takes bizarre forms. This region of Auckland, New Zealand, appropriately known as Hell's Gate, displays pools of boiling mud.

volcanoes and igneous activity

About forty million years ago there were many more volcanoes in the western United States, which marked the subsidence of an oceanic plate now gone. The string of volcanoes still active in the Pacific Northwest are the last vestige of another plate slowly being destroyed. It appears that the North American and Pacific plates tend to pull apart and press together as they slide past each other. A great deal of the ocean floor has been destroyed by the westward motion of the American plates.

Across the Pacific, intense volcanic activity calls our attention to another area where plate material is going down into the mantle and being destroyed. The island arc of Indonesia indicates that the Pacific plate is going down at a 45-degree angle about a hundred miles away. Indonesia's seventy-seven volcanoes are the result of processes generated by that plate's descent.

other igneous activity

Water vapor and gases contained in magma and then released when there is a drop in surrounding pressure can often be seen rising as ghostly plumes from the ground. In addition, groundwater may sometimes percolate down to great depths. Here it is heated up by the surrounding temperatures, like water in a teakettle on a hot stove. Later, it may find a new way to the surface through a fracture in the rock, bursting out as hot springs or steam.

There are two types of so-called "mud volcanoes." One variety are actual volcanic eruptions which eject a mixture of fine volcanic material and water—in other words, mud. More common are the cases in which water (or melted snow) from the volcano mixes with soil from the slope of the volcano on the way down.

Volcanic mud was one of the advance warnings of the violent eruption of Mount Pelée in Martinique, in 1902. Volcanic mud in great quantities swept down from Vesuvius in 79 A.D. to bury the city of Herculaneum (right next to Pompeii), sealing it under a hard crust of mud which was then covered by lava from later eruptions. Volcanic mud can even be formed when hot ashes mingle with stream or lake water. In Iceland, melted snow has been known to produce volcanic mud as hot ash falls on it.

Geysers are a type of hot spring whose intermittent eruptions occur in cycles or at set intervals. **Fumaroles** are vents in the ground on or near volcanoes, from which issue vapors and fumes. The "Giant" geyser in Yellowstone Park shoots up a spire as high as 61 meters during its eruptions, at intervals from 7 days to 3 months. It discharges 2,649,500 liters per hour, and currently ranks as the tallest active geyser. In 1909, New Zealand's Waimangu geyser erupted to a

volcanoes and igneous activity

height of some 305 meters, but it has not been active since its last violent eruption in 1917.

Upward-moving magmatic gases may vaporize ground water on contact, or a geyser's water conduit may be heated by conduction through the surrounding rock. As the temperature rises, water temperature in the bottom of the conduit reaches the boiling point, but the water is prevented from boiling by the pressure of the water above it. Vapor pressure in the lower portion eventually becomes greater than the pressure of water on top. The suppressed portion suddenly turns to steam, which pushes out water at the top by expansion. Pressure drops, causing more water to flash into steam. This, in turn, blows out the whole column of water and causes an eruption. In some geysers, this show repeats itself with clockwork regularity.

Geysir is the Icelandic word for "gusher." In everyday usage it is the name given to the Great Geyser. The Great Geyser is located about 48 kilometers northwest of the Icelandic town of Hekla, in a broad valley containing over 100 hot springs which differ greatly in character and dimension. The Great Geyser is a circular pool about 1.3 meters deep and 18 meters in diameter. Its conduit is 3.5 meters in diameter and 21.4 meters deep. Eruptions vary from columns of 30 meters in height occurring at 30-hour intervals, to columns reaching altitudes of 24 to 46 meters and occurring at 6-hour intervals.

Another view of thermal activity in Auckland. This area features a variety of hot springs, fumaroles, and boiling mud holes.

volcanoes and igneous activity

SUMMARY

The greatest known volcanic explosion occurred in 1883, when the Indonesian island of Krakatoa blew up, killing some 36,000 people. A similar explosion apparently destroyed the island nation of Atlantis around 1600 B.C., leaving only the crescent-shaped island of Thera. Such explosions are probably caused by the rapid quenching of superhot lava in sea water, releasing immense amounts of energy.

Magma probably forms in the low velocity zone of the earth's mantle layer. When it solidifies at the surface it is called extrusive. When it solidifies in the crust below it is called intrusive. Fractional crystallization is considered to be the basic reason for the difference between types of magma.

A volcano is an opening in the earth through which molten rock or magma is discharged. Volcanic action may form cinder cones, shield volcanoes, or composite cones. Magma emerges from the crater at the top of the cone, through vents in the sides, or from a caldera formed by the inward collapse of the crater's walls.

Most volcanic eruptions follow a general pattern. Earth tremors, changes in the temperature and flow rate of springs, strange behavior by animals, and gases and vapors emerging from the cone are all warning signs of a coming eruption. Heavy rains and thunderstorms are often associated with the eruption. Perhaps the most deadly manifestation is the *nuée ardente*, a superheated cloud of volcanic debris and gases. Such a cloud destroyed the town of St. Pierre, Martinique, in 1902.

Cone-and-crater eruptions are responsible for the formation of islands such as the Hawaiians. But fissure eruptions have had more influence in the shaping of the earth. Most of these occur on the ocean bottom where new sea floor is being created along mid-ocean ranges.

Full-time monitoring of volcanoes is now being performed in Hawaii, and some eruptions can be predicted. Monitoring is also taking place in the Pacific Northwest, where volcanoes such as Mount Rainier and Lassen Peak show signs of reawakening. Volcanic activity along the western coast of the Americas is connected with earthquake activity, and both are related to the fact that gigantic plates of the earth's crust are forming and subsiding in this area.

Mud volcanoes eject a muddy mixture of fine volcanic material and water. Volcanic mud can also form when hot ashes mix with stream or lake water. Mud volcanoes are closely related to geysers, hot springs which erupt intermittently. Geysers are often found in the same volcanic area with fumaroles, vents in the ground that issue vapors and gases.

volcanoes and igneous activity

REVIEW QUESTIONS

1. Describe the three types of volcanic cones.
2. How does a caldera probably form?
3. How are volcanoes related to earthquake activity? In what particular areas do both occur?
4. Where does magma form?
5. List some of the common advance warnings of an impending volcanic eruption.
6. Name at least two sites where frequent volcanic activity takes place.
7. Distinguish between extrusive and intrusive magma.
8. What happens to the melting point of rock when a sudden drop in pressure occurs?
9. Describe the events that cause a geyser to erupt.
10. What is a batholith?
11. What is fractional crystallization?
12. Describe two ways in which volcanic mud is formed.
13. What is partial melting?
14. Why do heavy rains often accompany an eruption?
15. What is a *nuée ardente*? Describe an example.
16. Describe the events that probably caused the destruction of Atlantis.
17. Where is the world's largest volcano?
18. Where and when did the greatest verified volcanic explosion occur?

CHAPTER SEVEN

YOUR BOOKS MAY BE WORTH UP TO 50% WHEN BROUGHT BACK DURING THE WEEK OF FINALS!

!!!!!!!!!!!!!!WARNING!!!!!!!READ THIS!!!!!

DON BOOKSTORE — SANTA ANA COLLEGE

REFUND POLICY!!!!!!!!!!!!!!!!!!!!!!

DO NOT WRITE OR MARK in your books until you are certain you will actually use them !!!
REFUNDS WILL BE GRANTED IN FULL provided that the following conditions are met:

1. Bring your items to the Refund Counter which is located at entrance of the store. Please do not bring books or supplies into the selling area of the store at any time.
2. **YOU MUST PRESENT YOUR CASH REGISTER RECEIPT** & Completed Drop Slip.
3. **NEW** texts must be in brand new condition. If they contain any markings or names, the refund allowance will be the used-book purchase price. Obviously, this requirement does not pertain to the refund of a **USED** text.
4. **THE LAST DAY FOR A FULL REFUND IS THE END OF THE SECOND WEEK OF THE SEMESTER FOR WHICH THE BOOK WAS PURCHASED! SUMMER REFUND DATES WILL BE POSTED.**

!!!!!KEEP!!!!!THIS!!!!!RECEIPT!!!!!

THE CASE FOR CONTINENTAL DRIFT

Far to the north of most of the United States, the Alaskan peninsula juts out westward between the Arctic Ocean and the Bering Sea. Facing it, scarcely more than a hundred kilometers away across the Bering Strait, is Siberia's Chukchi Peninsula, the easternmost projection of Asia. Once the two continents were connected by a bridge of land, over which many species of animals migrated in search of better grazing, more abundant prey, or escape from the cold climates of the great ice age. Among them were the first humans to enter the New World.

The old land bridge existed because of the glaciers, which locked up so much water in mountainous sheets of ice that the level of the oceans fell and shallow areas were exposed. But what if, some day, Alaska and Siberia should again be joined—not by a drop in water level but by a massive intercontinental crash?

There is every reason to think it will happen, though no one now living will be around to witness the event. North America and Asia are moving toward each other at the rate of a few centimeters per year, and the Bering Strait is where they will eventually meet. When that happens, there are likely to be some rather spectacular earthquakes, as the two continents press and slip against each other. Very likely a new range of mountains will be pushed up, running more or less north and south across the joined land, as the Alps and the Himalayas have been pushed up by similar collisions.

the case for continental drift

The closing of the water passage between the Arctic and the Bering Sea could have important effects on the climate. At present about 20 percent of the Arctic's water comes from the Pacific via the Bering Sea. This water brings both warmth and a rich supply of nutrients. When it is cut off, there will be less food for fish and sea mammals in that part of the Arctic, and the climate will become even chillier than it is now. Probably whatever glaciers are in the area will grow and advance southward, cooling the climate and shortening the growing season in southern Alaska and Canada.

Furthermore, the Aleutian Islands, the southern boundary of the Bering Sea, are involved in the same westward movement as Alaska. As they are jammed more tightly against Asia, will they close up into a nearly continuous land barrier? Then the seals could no longer get through to breed on the tiny, barren Pribilof Islands, and the Pacific Ocean waters to the south would no longer be cooled by chill subarctic currents. Warm weather would spread farther north all around the ocean—probably bringing tropical storms with it.

All this is not likely to happen for a few million years yet. But if human beings are still around when it does, we can be reasonably sure someone will find a way to take advantage of it. Some enterprising developer will soon be distributing prospectuses for Aleutian Marina Homes (the Aleutians are among the most earthquake-prone spots on earth, but earthquakes never kept anyone from building on the San Andreas fault). And prospective buyers will anxiously weigh the high cost of typhoon insurance against the joys of salt water sports from one's own private dock on the shores of beautiful Bering Lake.

For now, however, we may leave the future to take care of itself, while we turn our attention to the past. Over millions of years, movements of the earth have produced mountains, valleys, lakes, and oceans—sometimes in ways still not fully understood. The evidence for these movements lies in the earth itself. It can be interpreted by informed laymen and knowledgeable scientists as clearly as if it had been published in a book. The earth is a restless, dynamic planet, even though the changes take place very slowly as compared to our lifetimes or even the span of human existence.

Until recently, most scientists believed that the large landmasses—the continents—had always been relatively stable, anchored in the same places throughout geologic history. But startling evidence against this belief has accumulated rapidly in the last twenty years. It now appears that the continents were all once part of a single landmass, or **continental nucleus.** They may have separated and then moved together again, not once but several times.

Most earth scientists now favor this idea, which is known as the theory of **continental drift.** However, there is still some disagreement on details of how this migration of the continents took place. Much more investigation must be done, with the aid of new equipment and technology.

the earth's crust in motion

In our earlier discussion of the earth's structure, we saw that it is composed of three general parts: a central core made of dense metals in both molten and solid states; a compact rock layer called the mantle, about 3,000 km thick; and an outer crust, about 40 km thick, made of less dense rocks that "float" on the heavier rocks beneath. We know most about the crust, because its surface is exposed. But our firsthand knowledge of even this rather thin layer is limited; so far, we are able to drill only about ten kilometers into it. Most of our speculations about the structure of the earth are based on scientific studies of the planet's heat, magnetism, gravity, and motion.

We know that the crustal rocks fall into three basic classifications. Sedimentary rocks were formed by consolidation of layers of sediments. Igneous rocks result from the solidification of magma. And metamorphic rocks are those of either type that have undergone changes due to heat or pressure.

mountain formation

Many kinds of movement have taken place in the earth. Rocks have been lifted, depressed, tilted, and twisted. Collectively, these movements are called **diastrophism** (from Greek *diastrophe*, "twisting"). While the exact causes may not always be apparent, the rock record clearly demonstrates that movements such as folding and faulting have occurred.

Folding and faulting are largely responsible for the mountains of the world—mountains with distinctive characteristics. Folds in thick rock layers, for example, produce gentle, rounded mountains. Faulting produces conspicuous cliffs, or scarps, where one rock mass is lifted vertically. Overthrusting of one mass by another for a considerable distance often creates mountains with sharp faces at the edges of the layers.

But the world's most impressive mountain systems were formed in quite a different way. This fact was not recognized until recently, for these mountains are not located on the continents, where they could be easily studied. Instead, they are found in the oceans. The Mid-Atlantic Ridge, for example, extends north and south for about 16,000 kilometers, far longer than any continental mountain range. It contains peaks that rise three kilometers or more above the ocean floor. Similar ridges are found in the other oceans. The dis-

the case for continental drift

covery of these systems, and our growing understanding of how they originated, have provided support for the theory of continental drift.

a theory: continents are adrift

Over the centuries both scientists and laymen have been intrigued by the jigsaw-puzzle appearance of the earth's land areas. It had often been suggested that if the continents could somehow be pulled together, their edges might fit fairly well. Alexander von Humboldt, a German geographer–explorer of the mid-nineteenth century, suggested that the east coast of South America looks as though it once fitted into the west coast of Africa. At the time hardly anyone took this idea seriously. Almost all scientists were certain that the continents were firmly anchored in their places.

In 1912, Alfred Wegener, a German scientist, published a book that reopened the subject. Its English title is *The Origin of the Continents and Oceans*. Very few people, however, accepted Wegener's idea that the continents once were united in a supercontinent which he called **Pangaea** (Greek for "all land"), and have since drifted to their present positions.

Wegener conceived his idea of continental drift from some longitude measurements that he made while serving on two expeditions to Greenland, in 1906 and 1912. He found that the longitudes he calculated for the same place in each of the two years differed. He believed the differences were greater than the probable error of the measurements. If so, there was only one logical explanation — Greenland was moving.

The drift of Greenland that Wegener thought he had measured probably does exist, but — ironically — it may have nothing to do with true continental drift. All landmasses have a slight movement to and fro that is related to lunar rock tides, which we will discuss in chapter 15. This movement could account for the discrepancy in Wegener's calculations. There also may be a special movement of Greenland. As the glaciers on that island melt, the island may be drifting northward because of lessened centrifugal effect. This would cause a difference in latitude, as well as longitude.

the debate over continental drift

There were both strengths and weaknesses in Wegener's arguments. He and other scientists had recognized similarities between fossil plants and animals in the rocks of continents now separated by great distances. For instance, remains of *Mesosaurus*, a small freshwater reptile that lived about 270 million years ago, have been unearthed in both Brazil and South Africa — and nowhere else. Fossil leaves from a fern, *Glossopteris*, dating from 136 to 395 million years

the case for continental drift

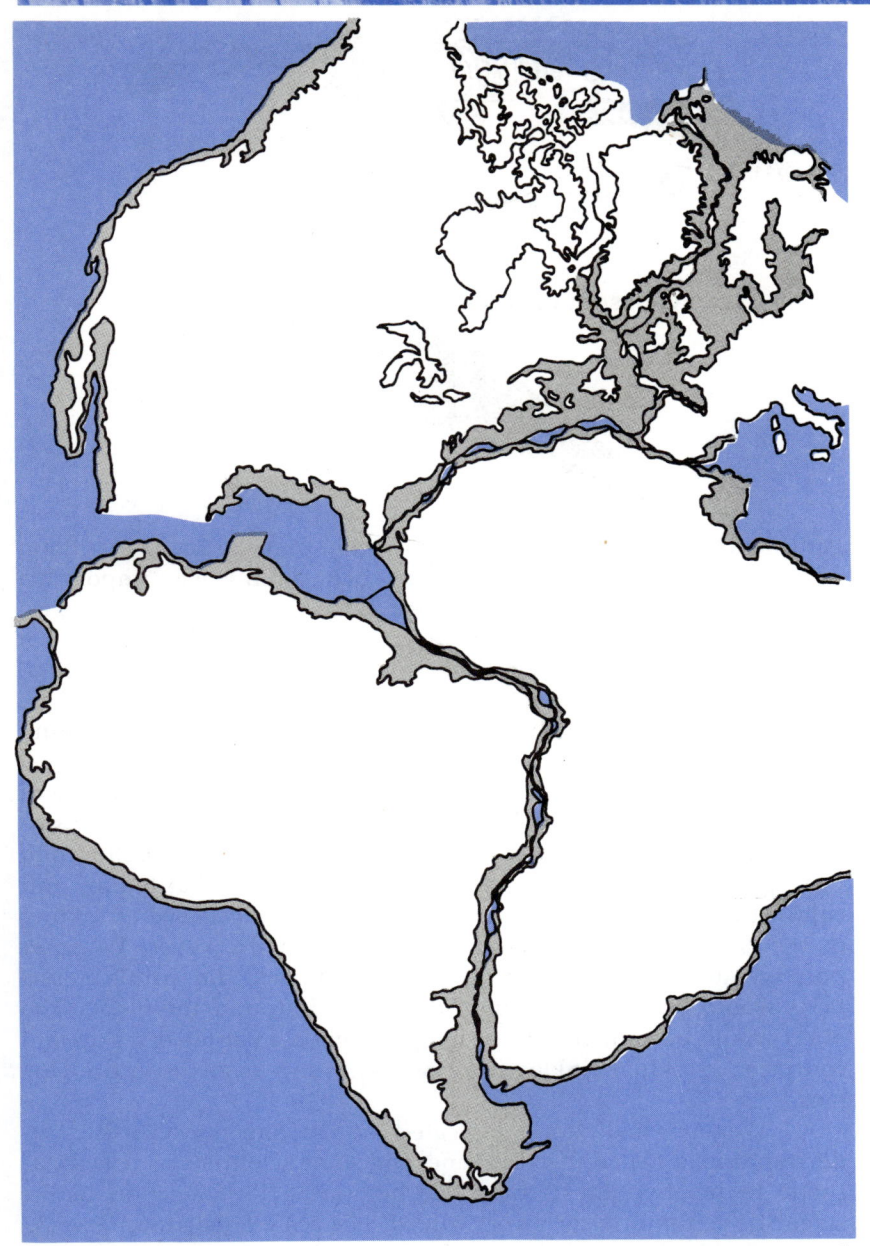

The supercontinent of Pangaea probably looked much like this before it began to separate. This illustration shows the remarkably precise fit of the continents. Dark areas indicate the continental shelves.

ago, have been found in Africa, India, Australia, Antarctica, and South America. (Austrian geologist Edward Suess proposed that all these areas were once part of a supercontinent he termed **Gondwanaland.**) It seemed unlikely that the appearance of these remains in such widely separated locations could be mere coincidence.

Wegener's opponents explained the data by proposing that land bridges or additional continents had once existed between the

the case for continental drift

Mesosaurus

Glossopteris *flora*

The skeleton of *Mesosaurus*, a reptile whose remains have been discovered in Brazil and South Africa, appears above. Below is the fern *Glossopteris*. Both species flourished on Pangaea before the continents separated.

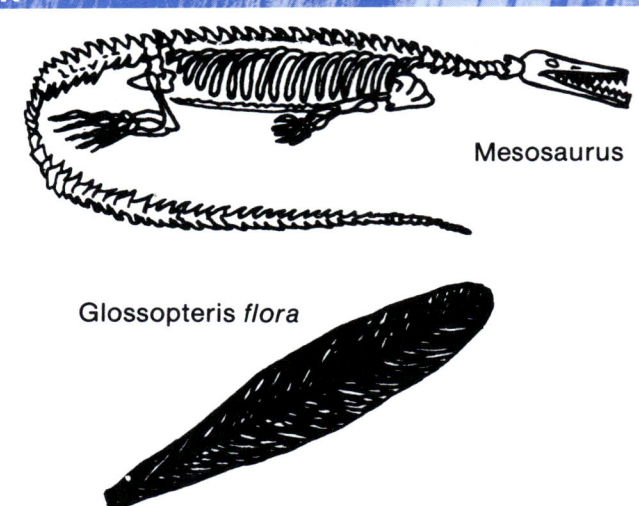

continents as they are today, but had later sunk to the ocean floor. Wegener, however, maintained that such sinking was impossible because the surface layers are lighter than those beneath. In this respect, his view has been upheld by more recent discoveries. For instance, though *Mesosaurus* lived almost 300 million years ago, no fossils older than about 150 million years have yet been found by drilling into the Atlantic Ocean floor, nor have any sunken land masses been discovered.

The accepted belief among geologists of Wegener's time was that continents had not drifted, but had risen and fallen, so that continents became ocean floors and vice versa. The earth was believed to be a sphere of molten material that was still cooling, hardening, and contracting. According to this theory, contractions caused wrinkling of the crust, and thus mountains were born. However, Wegener pointed out that if such wrinkling had taken place, the earth's mountains would be more uniformly distributed around the globe. The effect would be similar to what happens when an apple dries up and contracts; its skin wrinkles more or less evenly over the entire surface.

Wegener had clearly indicated the flaws in the scientific understanding of the earth at the time, and some scientists were sympathetic to his hypothesis from the beginning. But reaction in the scientific community was very mixed. His antagonists found weaknesses in his thinking. One of them, Harold Jeffreys of Cambridge University, attacked Wegener's explanation that plates of the earth's crust were moved by the gravitational pull of the moon and sun. Jeffreys was correct on this point. The moon and sun cannot exert enough gravitational force to move large portions of the earth's crust. In the 1920s, Jeffreys further criticized Wegener's concept in a treatise of his own.

the case for continental drift

The controversy over continental drift is a good example of an important part of the scientific method. No major new theory is ever accepted without question, just as it is originally proposed. There is always a long period of arguing, testing, and improving. The theory's opponents may in truth be its best friends, for they find the weaknesses that those who favor it may overlook. These weaknesses may involve lapses in logical reasoning, or failures to account for known data, or new data that cannot be fitted into the theory as it stands. Only if the theory can be modified to deal satisfactorily with these objections will it eventually become generally accepted.

Thus, recognizing the validity of some of their opponents' criticisms, several scientists who sided with Wegener "mended" his hypothesis by offering explanations of their own. But for practical purposes, the continental drift hypothesis was discredited. Wegener himself was too busy with new research—he was primarily a meteorologist—to spend much time defending his ideas. Then, several decades later, new discoveries about the earth's structure were made, and the hypothesis was revived and restudied.

The debate over continental drift was difficult to settle in Wegener's day because so little genuine evidence was available to support either side. Only lately have scientists developed equipment and techniques for collecting the relevant data. Today we are aware that Wegener was remarkably far ahead of his time, although he was not correct in every detail about continental displacement. He is now credited with the most thorough and thoughtful description of continental drift, and is recognized as an important original thinker in the field of earth science.

continental drift and geologic features

Let us examine some of Wegener's supporting evidence for continental drift in the light of our present scientific knowledge. First of all, he claimed that the continents had their origins in one supercontinent. Scientists today generally believe, like Wegener, that existing continents once fitted together rather precisely to form one landmass. Computer studies have demonstrated a good fit around the edges of the continents, especially if the continental shelves and slopes extending outward from the shorelines are considered. Wegener himself, in fact, drew up several maps illustrating the close fit of the South American and African coastlines.

Rock formations on the continents of South America and Africa appear to match up about as well as their coastal margins. If we draw up a geological map of these two continents, belts of the same rocks can be seen in both—belts that could have been continuous if the continents were once joined. This is a strong indication in favor of the theory. And, as we have seen, the fossil record in those rocks also supports Wegener's beliefs.

Other patterns of rock and mineral deposits around the world also seem to bolster the argument for an original supercontinent that

the case for continental drift

This figure shows the distribution pattern of similar fossil remains which have been found in South America and South Africa. The pattern provides additional support for the theory of continental drift.

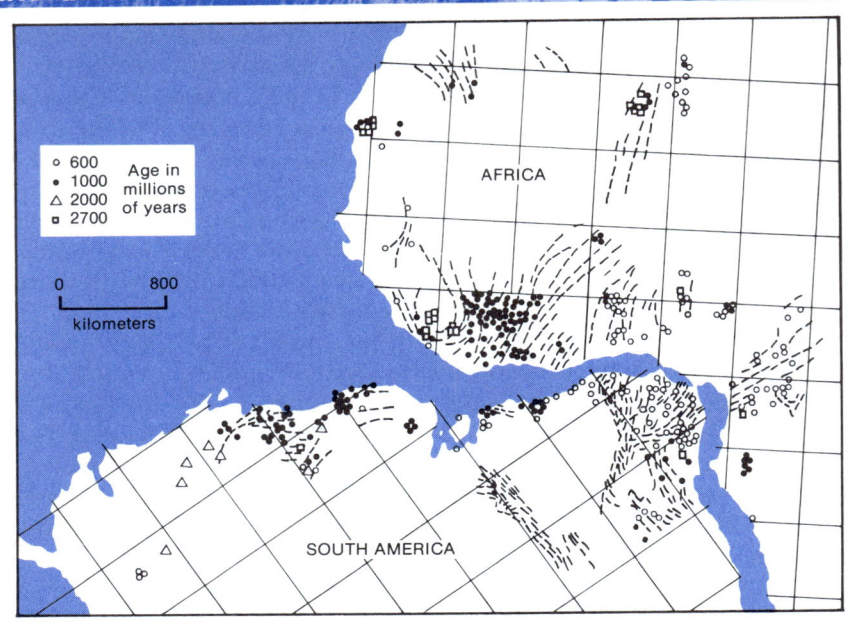

more drift evidence from ice action

later broke up. Wegener believed that in Pangaea coal was laid down along a tropical belt that was centered on the ancient equator during a period about 300 million years ago. With the drifting of continents these beds of coal, once rather close together, were dispersed to what is now the eastern United States, Great Britain, France, Germany, the Soviet Union, and China.

Wegener also traced deposits of other materials through the ages. In climatic zones beyond the old equator of Pangaea, toward 30° north latitude and 30° south latitude, he reasoned that dryness prevailed. Salt, desert sandstone, and gypsum would thus have been formed in these regions. Wegener believed that those areas became the western United States and North Africa of recent times, which have just such arid-climate mineral deposits in rocks of the correct age.

Glacial tillites—rocks formed from the clay, sand, gravel, and boulders once carried by glaciers—have been found among rocks 250 to 500 million years old. The youngest known examples are in South Africa, South America, and Australia, which are now warm regions. These relics of ancient glacial activity have been cited by scientists as evidence that the continents were once one great landmass, centered near the South Pole.

The direction of ice movement indicated by glacial striae in the rocks suggests a very curious state of affairs. It appears that if the continents were then as they are today, the glaciers must have invaded the land from the oceans! This, of course, would have been impossible, because glaciers could not have formed on the oceans nor could they be anchored there. Without being anchored, they could not have pushed onto and across the land.

There are two possible ways that we can explain these ancient glaciations in the Southern Hemisphere, assuming that the geological evidence has been correctly interpreted. First, during this period land might have extended seaward far beyond its present boundaries, allowing anchorage to the glaciers. However, what little evidence we have suggests that this is not a good explanation.

Second, perhaps the glaciated areas were part of a much greater landmass. A continental glacier may have existed over a region that has since split up and drifted apart. When the ice spread outward from its center, it left grooves on rocks that now show which way the ice moved. The directions of scratches on present-day continents, even those now dry and free of ice, point to the conclusion that the continents came from a large landmass. Those who accept the theory of continental drift are confident that this explanation accounts for the glaciated areas.

Wegener made another original discovery. He reasoned that if all continents and ocean bottoms rose and fell randomly, as his antagonists believed, the various altitudes of the earth's terrain would be distributed in a fairly regular and predictable way. Very small amounts of land would be found at the highest and lowest altitudes. The largest single amount would be about halfway between these altitudes, and gradually increased amounts would be found at the levels in between. On a graph, the distribution would form a single bell-shaped curve, with its low ends at the levels of the highest mountaintops and the deepest ocean trenches, and its peak at about 2.44 kilometers below sea level. But when Wegener analyzed the existing measurements, he found quite a different pattern. The largest amounts of land were at two separate levels: sea level, and about 5 km below sea level, with far smaller amounts in between. Such a distribution made no sense on the old hypothesis. However, it makes very good sense in relation to what we now know about the composition of the earth's crust. If the crust is of two fairly distinct kinds— relatively light continental crust and relatively heavy oceanic crust— we would expect to find them floating at two fairly distinct levels in the mantle.

Further strong evidence for the theory of continental drift comes from indications that the ocean floor is spreading. This is something that could scarcely have been learned before our day. Until very recently, the ocean depths were inaccessible to exploration. Present-day technology, however, has enabled us to expand our knowledge of this new frontier.

the crust: new levels of understanding

IN FOCUS
drifting continents, changing climates

What were ancient climates like and how did continental drift affect them? The earth itself offers some clues. For example, the distribution of coral reefs, which form in shallow, warm oceans, gives the researcher some idea of the climate in which the reefs were born. Similarly, the presence of evaporates—rock salt, potash salts, and gypsum—is a dead giveaway that the climate at the time these materials were deposited was hot and arid. The so-called "red beds"—sandstone and shale sediments which are colored red by the mineral grains and iron compounds they contain—tip off the investigator that the ancient climate was hot, with seasonal rainfall. Large coal deposits point to the presence of a moist climate, the result either of lots of rain in a warm region, or moderate rainfall in a cold one. Cold climates are revealed by traces of glaciers, either in the form of moraines or by grooves left on rocks by the moving ice as it passed over them.

Using these and other clues, earth scientists have reconstructed ancient climates, or paleoclimates as they are technically called, from the Precambrian period to the present. They have also plotted the drift of the continents over the millenia. For example, at the end of the Permian period some 225 million years ago, all the continents were welded into a single supercontinent known as Pangaea. The southern tip of this huge landmass was capped with glacial ice, whereas the northern hemisphere was ice-free. The equator passed through what is now Western Europe, the American Southwest, and the Canadian Maritime Provinces. Thick layers of evaporates and red beds are found in the Permian strata of these regions. In much of North America and Europe the climate was warm and rainfall was heavy, an indication that tropical or subtropical conditions must have existed.

During the following period, the Triassic of 200 million years ago, Pangaea broke up into two supercontinents of more or less equal size. They were separated, roughly at the modern equator, by the Tethys Sea. To the north was Laurasia, so called because Asia was joined with the Saint Lawrence region of North America by way of Europe. To the south was Gondwanaland, named after the region in India where fossil plants common to all of the continent were first found and studied. The southern supercontinent included Africa, India, Antarctica, South America, and Australia. The extreme southern regions of this continent lay under ice, while its northerly climes basked in tropical warmth. Red beds and evaporates found in Laurasia indicate that Europe and Asia were quite hot. So was the United States, where red beds were deposited to thicknesses of three thousand feet in the Rocky Mountains. Coal deposits in Siberia, Western Poland, and South Australia show that these regions were moist and warm, with lush vegetation.

In the mid-Triassic, Gondwanaland and Laurasia began drifting apart. By the late Jurassic, 135 million years ago, the continents had already assumed the shapes we know today. At the same time Africa and South America began to separate. The drift of the continents at this time was not great enough to produce any dramatic climatic changes. In Europe, coral reefs are in evidence, particularly in the Alpine regions. The temperature of the European seas stood about 10°C higher than at present. Evaporates in Asia point to a dry climate there. Lush plant life flourished in Greenland, Germany, Japan, and Sweden.

At the end of the Cretaceous period, some 65 million years ago, the Americas drifted westward and the Atlantic opened. Africa and India continued moving northward, while the Eurasian plate rotated toward the east. Australia began to split off from Antarctica. As the continents wandered, the climate began to cool off. Ocean temperatures around the world dropped an average of one degree. The arid desert belt migrated south, almost to its present position. The belt of coral reefs in Europe began to creep further south. So did the belt of evaporates that girdled the upper regions of the Northern Hemisphere. Warm climate evaporates all but disappeared in Europe, and the red beds in North America were similarly vanishing. Arid conditions existed in parts of Africa and South America, as shown by the presence of Cretaceous evaporates. Warmth-loving ferns and cycads grew in the Arctic, Alaska, and Greenland.

At the dawn of the Tertiary period, about 60 million years ago, the continents reached their present positions. Today's climate is influenced by their topography and distribution, as well as by interaction between solar radiation and the earth's atmosphere. As the continents continue their inexorable wanderings in the future, their movements are sure to create climates as different from today's as were those of the Permian.

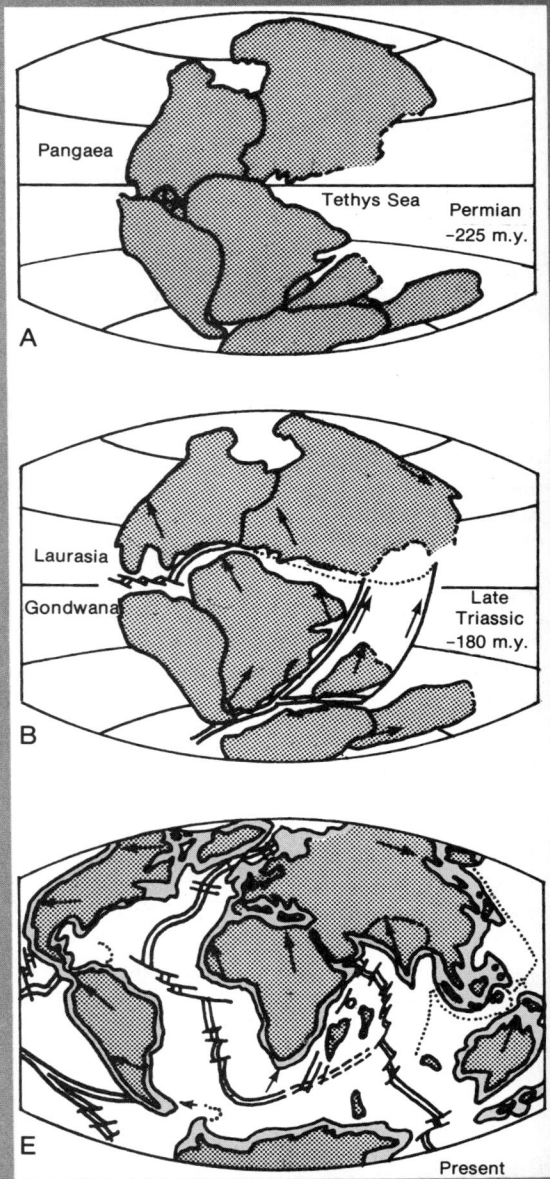

the case for continental drift

how the ocean floor grows

When extensive exploration of sea bottoms began, scientists made many puzzling discoveries. For instance, they learned that the Mid-Atlantic Ridge is not only quite high but is also split down the middle by a deep gash, or **rift.** Little sediment covers the mountains themselves, but the thickness of sediment increases with distance away from the rift. Moreover, the youngest rocks lie close to the rift and older ones are farther away. Broad bands paralleling the range differ from one another in fossil magnetism, and temperatures under the ridges are very high.

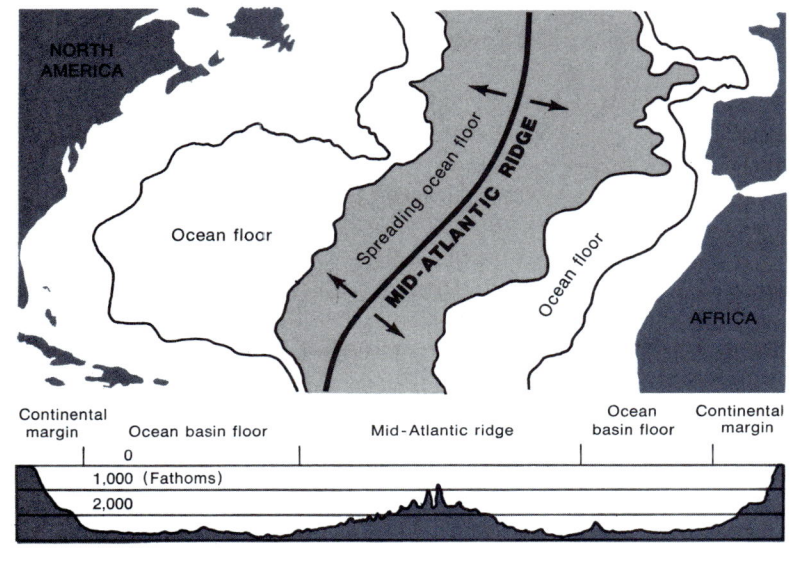

It is now believed that the ocean floor is spreading in both directions away from the Mid-Atlantic Ridge. Depth measurements are given in fathoms (1 fathom = 6 feet, or about 2 meters).

As more and more data were collected, scientists gradually developed a hypothesis to explain what had happened and is still happening in the Atlantic Ocean. Strange as it may have sounded at first, their explanation was that the rift in the mountain range is one long slash in the earth's crust. Hot magma from the mantle beneath the crust is oozing out to form mountains. Furthermore—and most surprising of all—the experts came to believe that the ocean floor is spreading eastward and westward, away from the Mid-Atlantic Ridge. Moving with it—carried as parts of the same huge sections, or plates, of the lithosphere—are most of the earth's continents: North and South America, Africa, and the combined mass of Europe and Asia.

Further investigations have shown that the ocean bottom is indeed moving, at an average rate of 1.5 to 5 centimeters per year in each direction. Most research findings have tended to confirm the basic hypothesis, although much more research is needed to fill in all the gaps. Today, most geologists agree that the ocean floor is spreading from underwater mountain-range rifts, not only in the Atlantic but in all the oceans. And because so much evidence has come to

the case for continental drift

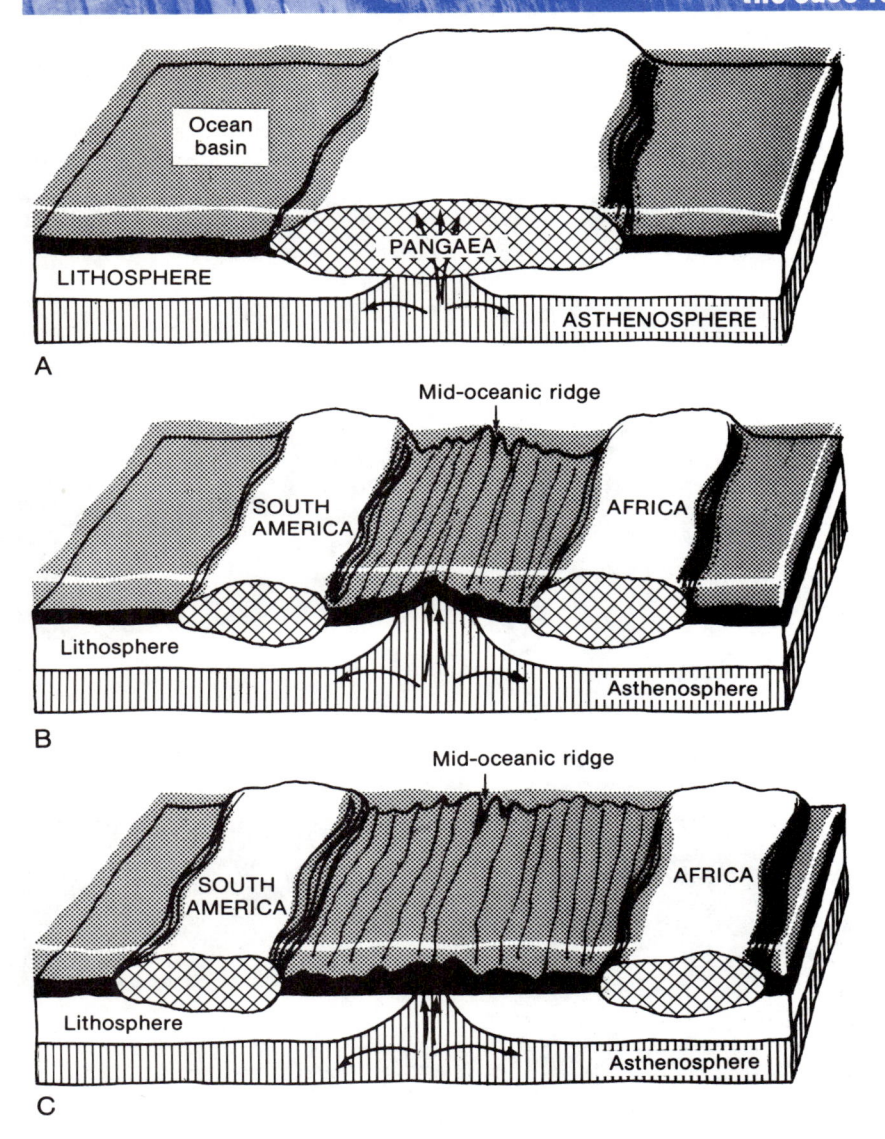

A model illustrating the probable stages in the breakup of Pangaea and the formation of the mid-oceanic ridge.

light, the explanation can now be considered a theory rather than a hypothesis.

Much valuable information was collected in the 1960s and in 1970 by the *Glomar Challenger*, a ship specially rigged with a drilling platform on its deck. This oceanographic research vessel, operated by scientists and technicians from several nations, brought up a number of rock cores from holes drilled in the Atlantic Ocean floor. Analyses of these samples indicate the age of the rocks below and the extent of their magnetism. They also show that basaltic rock has been accumu-

underwater data: key to new concepts

the case for continental drift

lating on the ocean floor for millions of years. It has arrived there by means of eruptions through the ocean bottom at intervals of 200 to 10,000 years.

Since then, even more sophisticated techniques have been used by scientific explorers. For example, experts from two nations

Alvin, one of the three research submersibles that took part in Project FAMOUS. This tiny vessel has penetrated to depths of more than 3,000 meters.

Alvin lies alongside her mother ship while preparing to dive in search of a hydrogen bomb, lost off the coast of Spain by a U.S. Air Force plane. *Alvin* located the bomb after 34 dives.

the case for continental drift

with advanced oceanographic equipment collaborated to explore an area of the Mid-Atlantic Rift in a venture given the name FAMOUS (French-American Mid-Ocean Undersea Study). Perhaps the most exciting research was performed by three submersible research vessels, the *Alvin*, *Archimède*, and *Cyana*. In these tiny vessels scientists were able to descend to depths of more than 3,000 meters, where they could see and photograph the wonders of the seascape at close range.

The investigators employed various methods of collecting information. Sound waves were transmitted from a surface ship and reflected from the bottom to show where heights, depths, and special features were located. Drills were used to take cores of material from the bottom. Dredges brought up samples of rock that had been thrust out of the rift as molten lava. Intense spotlights were able to illuminate small areas around the vessels, so that television and still pictures could be taken. The temperatures of both water and sediment

Archimède, another of the Project FAMOUS submersibles. Here *Archimède* is undergoing a battery recharge before resuming her exploration of the mid-ocean ridge southwest of the Azores.

the case for continental drift

were measured. And beacons emitting sound were used for precise navigation and mapping. In addition, magnetometers measured the magnetism of various rocks along the rift and demonstrated that broad bands of rock with certain magnetic characteristics were moving away from the mountains.

A basalt lava flow, photographed by *Archimède* at a depth of more than 2,500 meters. This is pillow lava, so called because of its odd shapes.

the ocean floor dips into trenches

Elsewhere, other investigators have been at work. They have discovered not only mountain-and-rift systems but also deep trenches, often associated with arcs of surface islands such as the Aleutians in the north Pacific. In the western Pacific, there are many trenches that penetrate nearly 11,000 meters below sea level.

Here portions of the earth's crust are believed to be pushing together. Where they meet, the crust on one side plunges downward, forming the trench. At the same time, this downthrusting material tends to cause a buildup of rock on the other side where the crust is relatively stationary. If the buildup is great enough, it extends above sea level to form islands. The islands of Japan and the Philippine Islands are large groups resulting from volcanic activity associated with this action, though many smaller, more widely dispersed islands are located alongside Pacific trenches as well.

the convection cell hypothesis

If the crust is being born at rifts and disappearing at trenches, then we must speculate further. Where does the new material come from, and where does the old material go—and how?

In chapter 5 we considered the possibility that convection cells in the mantle might be the cause of the earth's magnetic field. Scientists have now suggested that these same convection cells could also provide an explanation of the mid-ocean ridges and the movements of the ocean floor.

the case for continental drift

A convection cell involves a patterned flow of material, caused by the movement of heat along a thermal gradient. The lower part of the earth's mantle, which is in contact with the core, is believed to be considerably hotter than the upper part, which is in contact with the crust. Therefore, if the mantle — or part of it — were fluid, there would be convection currents carrying heated material up from the lower mantle or the core to the cooler lithospheric region. The latter would absorb the heat and eventually radiate it out into space. The cooled mantle material would then sink back downward, replacing other heated material that was rising. A circular pattern of flow would be set up — hot material constantly rising, flowing along near the surface as it cooled, then sinking back down, to pick up more heat and rise again.

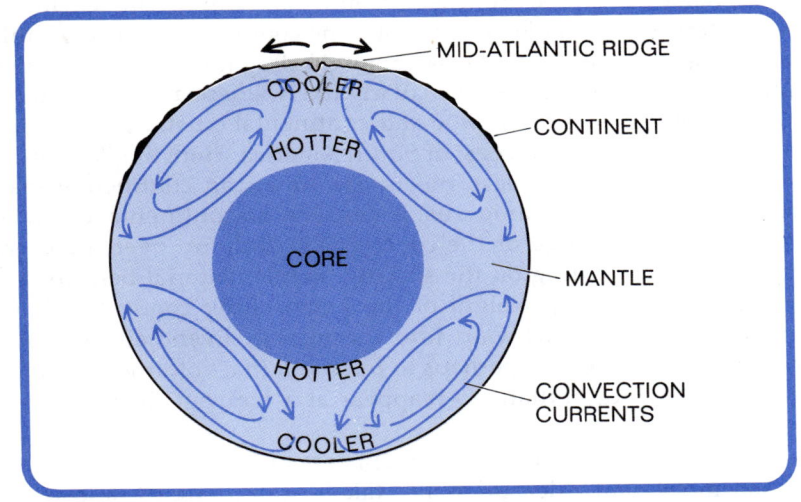

It is possible that very slow convection currents in the earth's plastic mantle may be responsible for the creation of mid-ocean ridges and ocean trenches.

The mantle, however, is believed to be solid. Can such activity take place in a solid? We do not know, for we have no way to observe it. But some scientists believe that at least some parts of the mantle, though solid, are plastic enough to flow very slowly. As we noted in chapter 5, substances which flow in the solid state — such as tar — are known. At the heat and pressure presumed to exist in the mantle, other substances might do the same. If so, convection currents in the mantle are possible.

How does the convection-cell hypothesis help to explain sea-floor spreading and the oceanic ridges and trenches? Remember that the crust is a very thin skin, spread over a very large area. If it were being pushed at one edge, we would expect it to buckle into folds, like a thin sheet of pizza dough being pushed across a pastry board. If it were being pulled by some force tugging on the opposite edge, we would expect the pizza dough to tear. But no one has found fractures of the expected sort.

convection and sea-floor movements

the case for continental drift

However, if the mantle material is flowing in a convection current, it might carry the crust, or the whole lithospheric plate, along with it—as we might safely move the pizza dough by picking up the pastry board and carrying it on that. This is what many scientists believe is happening. The upper parts of the convection cells, nearest the bottom of the lithosphere, may exert a drag on the lithosphere, causing it to move.

Convection-cell movement can also explain the buildup of the mid-ocean ridge. As the hot magma wells upward from the depths of the mantle, some of it would naturally press up through the rift between the plates. Since the pressure here would be less than below, it would liquify, spill out to either side, then cool and harden, forming the ridge.

What about the trenches? Here, too, convection-cell movement offers a reasonable explanation. The cooling magma, as it sinks back toward the center of the earth, might drag the edge of the lithosphere with it. This seems particularly likely where a trench is found at the meeting place of oceanic and continental crust. Such a trench exists along the western coast of South America. Here the floor of the Pacific Ocean, moving east, meets the American continental mass moving west. Since oceanic crust is denser than continental crust, we would expect the edge of the continent to slide up over the ocean floor. The latter, riding on the sinking mantle material beneath and pressed down by the weight of the encroaching continent above, would be carried downward into the mantle. Eventually it would melt back into magma—perhaps to burst out in a volcanic eruption a few miles inland, perhaps to reappear at a faraway rift millions of years later.

Most of what data we have appears to support this picture of crustal movement. Measurements taken around trench-and-island systems show that the crust is, indeed, in motion. Earthquakes are another indicator. If trenches are places where crustal material is disappearing downward, we would expect to find considerable earthquake activity here, because the edges of huge lithospheric plates would be moving with respect to one another. And in fact, the locations of earthquake centers as detected by seismographic instruments often correspond remarkably well with the locations of trenches. Earthquake activity is also common along the mid-ocean ridges, as we would likewise expect if plates are being pulled apart here. In these areas systems of numerous parallel fracture zones cut across the ridges. Fault movement in these fracture zones is mostly of the strike-slip sort, which would be natural if some sections of the plates are moving apart faster than other sections.

does the crust move to or from ocean ridges?

There is one problem with the theory that the sea floor is spreading outward from the mid-ocean ridges. If the ridge is being formed by the upwelling of magma from the mantle below, its material ought to resemble that of the mantle. But studies of the mid-At-

ANTARCTIC COAL: CAN WE GET AT IT?

Antarctica is truly the edge of the earth, the bottom of the world. Yet its frozen wastes attract scientific interest because of its unique meteorological and biological conditions. Even more important to an increasingly energy-conscious world, there is coal in Antarctica—lots of it. In fact, this bleak continent possesses one of the largest coal fields on earth. The deposits, consisting of semi-anthracite, anthracite, and natural coke, are generally poor in quality. They were laid down millions of years ago when Antarctica, covered by steaming tropical jungles, was part of Gondwanaland.

Is there a way to get at this resource? Unfortunately, there isn't—at least, not for the moment. The problem lies not in a lack of technology—we have that—but in economic feasibility. For one thing, most of the Antarctic continent is covered by a cap of ice nearly a mile thick, so the cost of mining the coal would far exceed its market value. Bringing in supplies and shipping the coal out would require icebreakers, a factor which further drives up expenses. The cost of operating a mining camp would be similarly high: living quarters, entertainment, and special medical care would be needed.

The cost of labor would be prohibitive, too. What's more, the frigid conditions are nearly impossible to work in. Temperatures in Antarctica have been known to reach 126 degrees below zero (Fahrenheit), and the unceasing winds raise the wind chill factor considerably. Under such conditions, even the simplest task becomes a major undertaking. On one occasion, for example, Soviet workers using flashlights in the winter darkness took nearly three weeks to repair a roof that had been blown off a shack.

The high incidence of snowblindness, frostbite, and the intense psychological isolation would further tax human effort. Because of these pressures, it would be necessary to rotate both mining personnel and maintenance staff periodically for "rest and recuperation" leaves. The hostile terrain requires that all personnel and equipment be transported by air in special ski-equipped planes, and such aircraft cost up to five times as much as commercial air transports. In short, most scientists now believe that the Antarctic coal will have to remain where it is until some future generation discovers a way to extract it economically.

the case for continental drift

lantic Ridge suggest that it does not. Differences have been detected both in its chemical composition and in its response to seismic waves. It is a basaltic rock, less dense than the mantle, and appears to be most similar to the thin basaltic crust under the rest of the ocean floor.

What does this mean? Can the ridge have been built out of crustal materials that were somehow pushed up? Perhaps—if the sea floor at the ridge is not spreading outward but, on the contrary, coming together. This forces us, then, to reconsider our theory. Which way is the ocean bottom really moving?

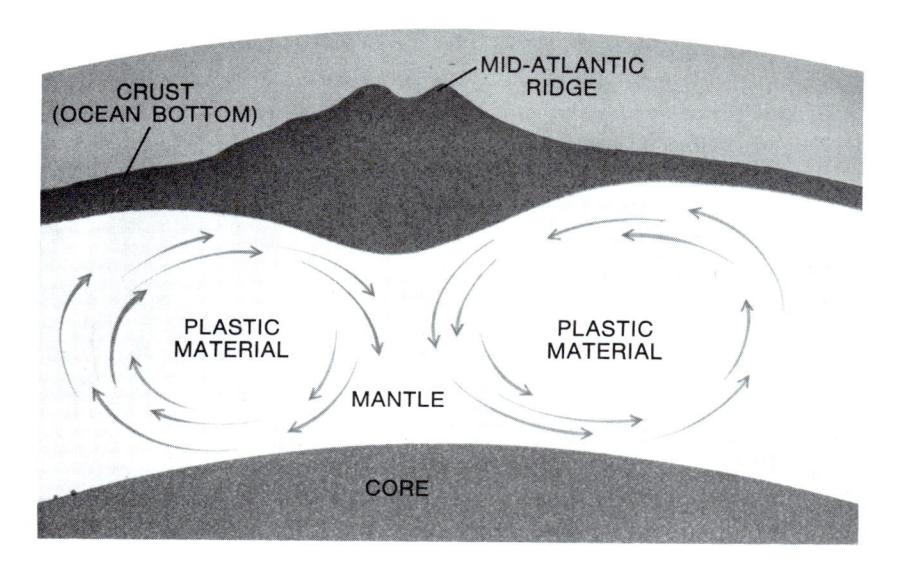

It was once suggested that flow patterns in the mantle might look something like this. Why has this pattern now been largely disproved?

The presence of the ridge is no help toward an answer. Laboratory experiments, using apparatus designed to imitate convection cells in the mantle, have shown that a ridge can be produced no matter which direction the material below is moving. But could a coming-together movement produce the type of ridge that exists? Not by simply jamming the plates together, for in this case the ridge would resemble the top, not the bottom, of the crust. Perhaps two convection cells are dragging together material "scraped" from the bottom of the crust, moving it toward the rift rather than away. The ridge would then be pushed up where these two masses of material come together.

Such movement, continued over millions of years, would slowly move the entire crust, and thus narrow the ocean basin. However, the evidence from cores—magnetism, the ages of the rocks, and so on—suggests that just the opposite has happened. Also, the trenches would be much harder to explain on this hypothesis than they are under a hypothesis of sea-floor spreading.

the case for continental drift

Considering the evidence overall, then, it seems more probable that the ocean basin is really spreading, not shrinking, and that the ridge is composed of material from the mantle. If so, we will eventually find out why the ridge, contrary to expectation, differs as it does from the mantle. Possibly the small chemical difference is insignificant. Possibly, too, the difference in seismic response is caused by the difference in pressure between the mantle below and the ridge above. These matters will surely be investigated in the future. Today we simply do not know enough about them to formulate any sort of final answer.

the earth as a giant magnet

Magnetism provides us with some further important evidence to support the theory of continental drift. The magnetic nature of the earth has fascinated people for centuries. Early travelers and explorers found that bits of lodestone or pieces of iron stroked by lodestone always oriented themselves in a north–south direction when allowed to swing freely. Christopher Columbus navigated with the aid of a crude compass made of a magnetized needle floating in water. The needle of any compass aligns itself with the invisible lines of magnetic force that surround the earth. In the laboratory, iron particles line up around a bar magnet, demonstrating where the lines of force lie. The earth is similar to a giant bar magnet, with North and South magnetic poles.

We mentioned in chapter 5 that the magnetic poles do not coincide with the geographic poles, around which the earth rotates. At

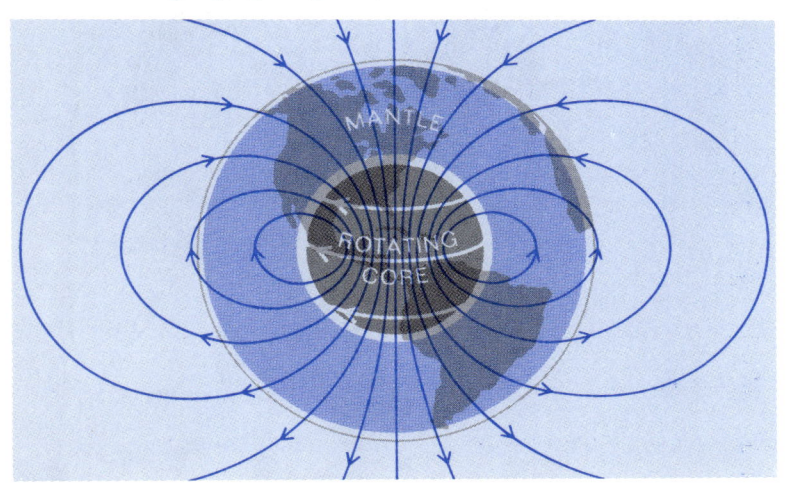

One hypothesis suggests that the earth's magnetic field may be caused by a convection cell of flowing ions in the fluid outer core.

the case for continental drift

most places on earth, therefore, a compass needle does not point precisely toward the geographic north. But the magnetic and geographic poles are close enough together to make a compass useful. The relatively small error, called **magnetic declination,** can be calculated and then allowed for at any place on the planet. In some places, a compass points east of true north; in others, west of true north. For example, in northern Maine, the declination is 20° west, so a compass needle points 20° west of true north. In central Indiana, the compass needle

A compass needle points toward magnetic north rather than geographic north. This figure shows the relative positions of the magnetic and geographic poles.

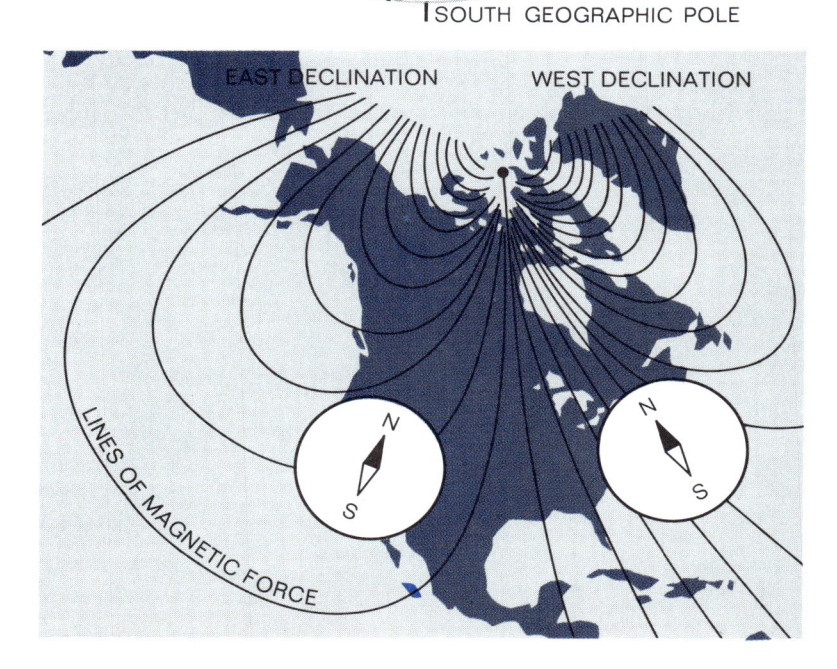

The lines of magnetic force illustrate east and west declination from true magnetic north.

the case for continental drift

shows no error at all, while in southern California, magnetic declination is 15° east.

the wandering poles

As we noted earlier, the earth's magnetic field has reversed itself from time to time. But it has also changed in other ways. Since scientists began to measure the strength of the field more than a century ago, for instance, that strength has waned about five percent. Even more interestingly, the magnetic poles are not securely anchored in one spot—they "wander" a bit. We know this because magnetic declination changes slowly from year to year in a given location. As a result, in most places, a compass needle does not point today in exactly the same direction as it pointed ten years ago.

It is possible to study the shifting of the magnetic poles over a much longer period, back through the long span of geologic time. As we saw in chapter 5, many minerals respond weakly or strongly to magnetic lines of force. When they were formed from cooling magma, their crystals aligned themselves with the existing lines of force. So today investigators can examine minerals of a certain age and determine what the earth's magnetism was like then. In a sense, the magnetic field was "frozen" in the rocks for all time.

Hundreds of measurements have been made on magnetic minerals to learn how magnetism has varied from the present back through more than 500 million years. When the data are plotted on a single map for rocks of all ages, there is a surprising result: the magnetic poles appear to have shifted their positions by thousands of

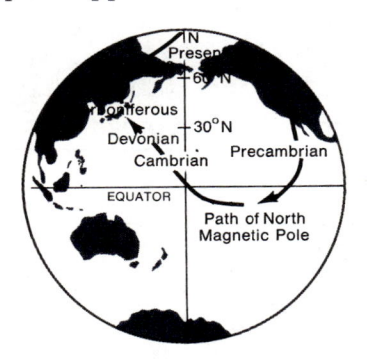

The magnetic poles have shifted position during the past 500 million years. For a key to the date of each geological period shown, see the table between chapters 9 and 10.

kilometers during the period. If this is true, magnetic "north" must once have been south of the present-day equator.

However, the rocks of each continent indicate a different track for the wandering poles. Since, obviously, a pole cannot have been in two or three places at the same time, some other explanation is needed. The most satisfactory explanation is that it was not the poles that wandered—at least not very much—but rather the continents themselves. So these findings, like those about sea-floor spreading, suggest that the continental drift Alfred Wegener described actually took place.

the case for continental drift

magnetic records on the ocean floor

Frederick J. Vine of Princeton University compared sea-bottom magnetic findings with the continental rock record. He found close agreement between them in regard to normal or reversed polarity. Successive bands of volcanic rock parallel to the mid-ocean rifts showed different polarities—a northward orientation in one band, a southward orientation in the next. The ages of the rocks in the bands matched the ages of continental rocks that showed similar polarities.

The rocks' polarity was determined, of course, when they were formed—when the magma that composed them hardened after being spewed out of the rift. When the earth's magnetic field later reversed itself, these existing rocks were not affected. Newly forming rocks, though, were marked by the new polarity. Vine devised a useful analogy; he compared the bands of rock to wide bands of recording tape, gradually unreeling away from the rift. As the "tapes" move past the

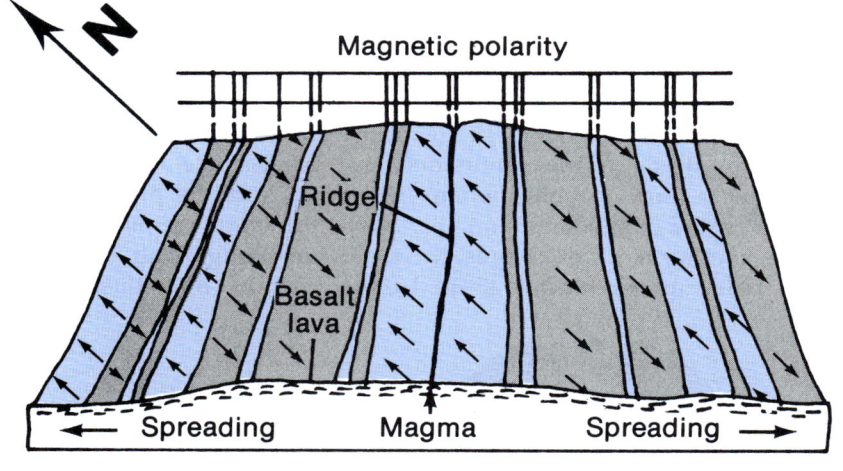

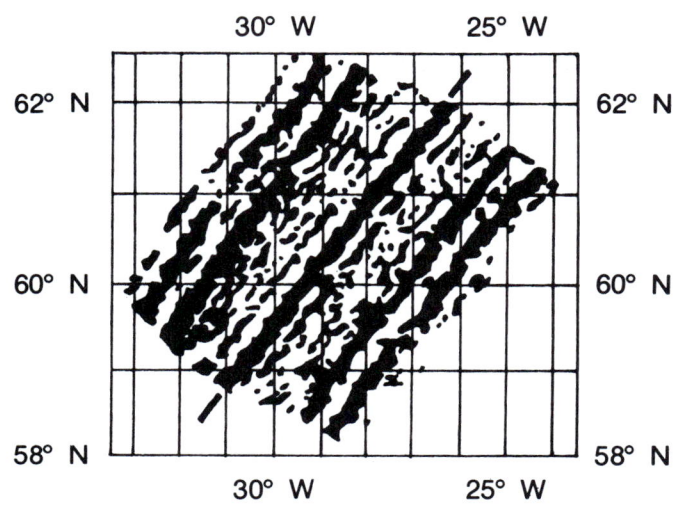

(a) Rocks around a mid-ocean rift exhibit bands of alternating magnetic polarity. (b) A record of magnetic striping around a rift.

the case for continental drift

"recording head" of the rift, they register whatever magnetic polarity is currently being "played" there. Then they hold it for later "replay" by scientific detectors, just as the familiar recording tape in a cassette registers magnetic changes representing talk or music, and stores them to be played back later when we throw the proper switch.

The presence of these magnetic stripes alongside rifts is another convincing argument for the theory of continental drift. When we add this evidence to our knowledge of rock formations and fossils, the physical fit of continents, "wandering" magnetic poles, rift-and-ridge and trench-and-island formations, earthquake activity, and movements of crustal plates, the drift theory is compelling. Whereas twenty years ago most earth scientists were still highly skeptical, today the majority of them agree that continental drift offers a useful and quite possible explanation for many otherwise puzzling phenomena.

Preparing for a series of Project FAMOUS dives, *Alvin* and her mother ship *Lulu* are positioned near the Bahamas. *Alvin's* first deep dive was made in 1965.

the case for continental drift

SUMMARY

Continents move. They have separated from the continental nucleus and come together again several times since the earth was formed. Until recently, however, scientists did not know this.

Movements that lift, depress, tilt, or twist the layers of the earth are collectively called diastrophism. Folding and faulting produce mountain ranges on the continents. But the most impressive mountain systems are found in the oceans. They are formed by a process related to continental drift.

In 1912, Wegener proposed that all the continents once formed a supercontinent he called Pangaea. Africa, for example, was in contact with South America. Remains of *Mesosaurus*, unearthed both in South Africa and Brazil, support Wegener's theory.

The continental drift theory was not accepted for a long time. Wegener's opponents argued that the evidence could be explained by the former existence of land bridges or now-sunken continents. Others attacked his suggestion that continents were moved by the gravitational attraction of the sun and the moon. This is a valid criticism, since the forces exerted by the sun and moon are not strong enough to do this.

Wegener's theory fell into disrepute until recently, when researchers found strong evidence to support it. For example, computer studies have shown how well the continents would fit if brought together. Patterns of mineral deposits support the idea of a supercontinent. In addition, glacial tillites have been discovered in South Africa, South America, and Australia, all of which are now warm in climate. This indicates that they were once polar continents.

Support for the continental drift theory also comes from the fact that the ocean floor is spreading. The rift centered in the Mid-Atlantic Ridge is a slash in the earth's crust. Magma from the upper mantle oozes out of this rift, forming the underwater mountains. The floor on either side of the right slowly moves away at a rate of a few centimeters per year. Under ocean trenches the material in the crust is sinking into the mantle. These phenomena are believed to be caused by convective motions in the mantle. However, this hypothesis has not been fully worked out.

The earth is a giant magnet. Its magnetic poles are close to the geographic poles, but they are not fixed. There is evidence that their positions have changed by as much as thousands of kilometers during the last 500 million years. Furthermore, the earth's magnetic field has reversed itself several times during the same period. The gradually spreading ocean floors have recorded the changing magnetic fields. These records provide additional support for Wegener's theory of continental drift.

the case for continental drift

REVIEW QUESTIONS

1. What is diastrophism?
2. What are the physical differences between mountains formed by folding and those formed by faulting?
3. Describe the observations on which Wegener based his theory of continental drift.
4. How do fossil remains support Wegener's theory?
5. Why was Wegener's theory not accepted at first? What were some of the objections made?
6. What was Wegener's objection to the theory that continents rose and fell?
7. How do modern observations of rock formations and mineral deposits support the theory of continental drift?
8. How does glacial evidence support Wegener's theory?
9. What is the Mid-Atlantic Ridge? How do geologists think it was formed?
10. Describe some of the modern research techniques used to study the ocean bottom.
11. What are ocean trenches? How were they formed?
12. How can convection in the earth's mantle explain ocean floor spreading?
13. Why is it not possible to find portions of the crust that date back to the formation of the earth?
14. Why can we describe the earth as a great magnet?
15. How has the earth's magnetic field changed during the last 500 million years?
16. Where do we find a record of the changing magnetic field?
17. How does the magnetic record support Wegener's theory?

CHAPTER EIGHT

PLATE TECTONICS

There have been plenty of arguments between environmentalists and commercial interests in the last few years. One of the hottest disputes has concerned the filling in of coastal wetlands to provide land for housing and industrial development. Environmentalists point out the ecological importance of the wetlands, for humans as well as for fish. Developers cite the pressing need for more usable land on a crowded planet.

If we wait a few hundred million years, nature may settle the argument for us. It is quite possible that the amount of land surface on the earth is slowly growing, so that eventually we may end up with more land than ocean. Of course, whether we will like it or not when we get it is another question. Loss of our present vast ocean surfaces could do unpleasant things to the climate.

But why should the land be growing in the first place? Chiefly because continental crust is lighter than ocean crust. This means that once it is formed, continental crust can never be pulled back down into the mantle and destroyed. At the same time, new continental crust is continually being formed, by volcanic action and other igneous activity. Logically, therefore, it seems as if sooner or later the whole earth will be covered with it. Some water would seep underground; the rest would be widely dispersed in streams and small lakes, from which much of it would evaporate.

On the other hand, a different scenario is possible. Consider what may happen when America, drifting west, finally meets up with Asia and Australia. As the great masses of continental crust smash into each other, neither will be able to sink down. One will have to ride up and over the other; or else one or both will have to buckle and fold back like a closing

plate tectonics

accordion. Either way, the result will be a chain of mountains that will stretch from Alaska to the southern tip of Chile, and will probably make the Himalayas look like foothills. Maybe there will be a gap around Panama—after all, America hasn't much crust to contribute at that point. We can imagine the Panama Pass becoming the key link in east—west trade routes, great cities growing up on either side of it, wars being fought for its possession. . . .

At any rate, the creation of such a mountain chain could use up a lot of continental crust—enough to put off the final disaster for a hundred million years more. Of course, the Pacific Ocean would get lost in the process, but the Atlantic would have grown big enough to make up for it. And think of the skiing in those new mountains!

a new view of the world

The first comprehensive concept of continental drift, which we discussed in the last chapter, originated with Alfred Wegener earlier in this century. Because he had little concrete evidence to support his hypothesis, many scientists naturally disagreed with him. So Wegener's belief that our present-day continents were originally one large supercontinent remained scarcely more than a curiosity until a decade or two ago.

During the 1960s, new exploration of the earth's crust, especially on the ocean bottom, revealed ridges and rifts from which the ocean floor was spreading. Furthermore, it appeared that sea bottom was disappearing into enormously deep trenches of the seas. These exciting developments led to a new examination of Wegener's hypothesis. All the evidence seemed to demonstrate that, for the most part, the German meteorologist's ideas had been correct. As new evidence accumulated, through the use of more efficient equipment and techniques, it became clear that the crust of the earth was indeed moving. But since crust was being born along certain boundaries and consumed along others, some revisions of Wegener's explanation were needed to account for these movements.

moving plates at the surface

Because of recent discoveries, scientists now have an entirely new theory of the earth's composition. Textbooks published in the late 1950s and early 1960s explain the anatomy of the earth in only a very general way. For instance, they define the lithosphere as the stony outer layer of the earth, extending down approximately 80 km from the surface. It encompasses the crust and upper mantle and is the rather rigid skin of our planet. Just below the lithosphere lies the asthenosphere. It was believed to be a fairly resilient region, a few hundred kilometers thick, in the upper mantle. Below the astheno-

plate tectonics

sphere is the mesosphere, which comprises the remainder of the mantle and probably has little to do with the changes taking place continually in the lithosphere and asthenosphere.

In chapters 4, 5, and 7, we described some of the movements of material in the outer layers of the earth. Over millions of years, such movements have altered the outer portion of the planet that we can see and explore. These changes are part of a system of **tectonics,** a term that refers to the structures of the earth and the forces that cause the changes.

When scientists began serious study of the world's rock formations and bodies of water, they found that, even now, much evidence remains of the activities that shaped the earth's surface. Sedimentary rock layers, igneous and metamorphic masses, folds, faults, fossil remains, changes made by wind, water, and ice—all contribute to a record that can be interpreted by researchers. New tools and techniques have made the interpretations more accurate. Furthermore, scientists have developed new and challenging concepts to account for the evidence they have turned up. The widely accepted consensus today is that the theory of **plate tectonics** explains many of the changes that the earth has experienced over the ages.

birth and death of the lithosphere

Earlier in this century most geologists agreed, on the basis of what they knew then, that the continents had always been fixed in their present positions. To explain the similar rock formations and fossils in South America and Africa, they theorized that land bridges had once spanned the gap and had later disappeared. Some scientists even argued that ancient continents, now submerged, had filled the spaces between the landmasses. But research showed no evidence of destroyed land bridges or lost continents on the ocean floor, so another interpretation became necessary.

Today many geologists are convinced that the lithosphere actually consists of large plates, something like gigantic ice floes, that move about on the plastic asthenosphere beneath. According to the accepted theory, the earth's crust is made up of about seven large plates and twenty small ones. Admittedly, not all researchers are entirely in agreement about the number of plates or the location of their boundaries. Nonetheless, the basic theory of plate structure and movement is well established, and will probably rank as one of the most important scientific theories of the twentieth century.

The boundaries of the crustal plates have not been arbitrarily assigned. They are well defined by the tectonic activity that takes place along plate edges. At some edges, plates are being created by molten magma which rises through ocean-bottom rifts, cools, and then hardens into rock. The material forms ridges and slowly slides away to both sides of the rifts, toward continental landmasses on the far sides of the plates. The sites at which the plates are being born are called **divergent boundaries.**

plate tectonics

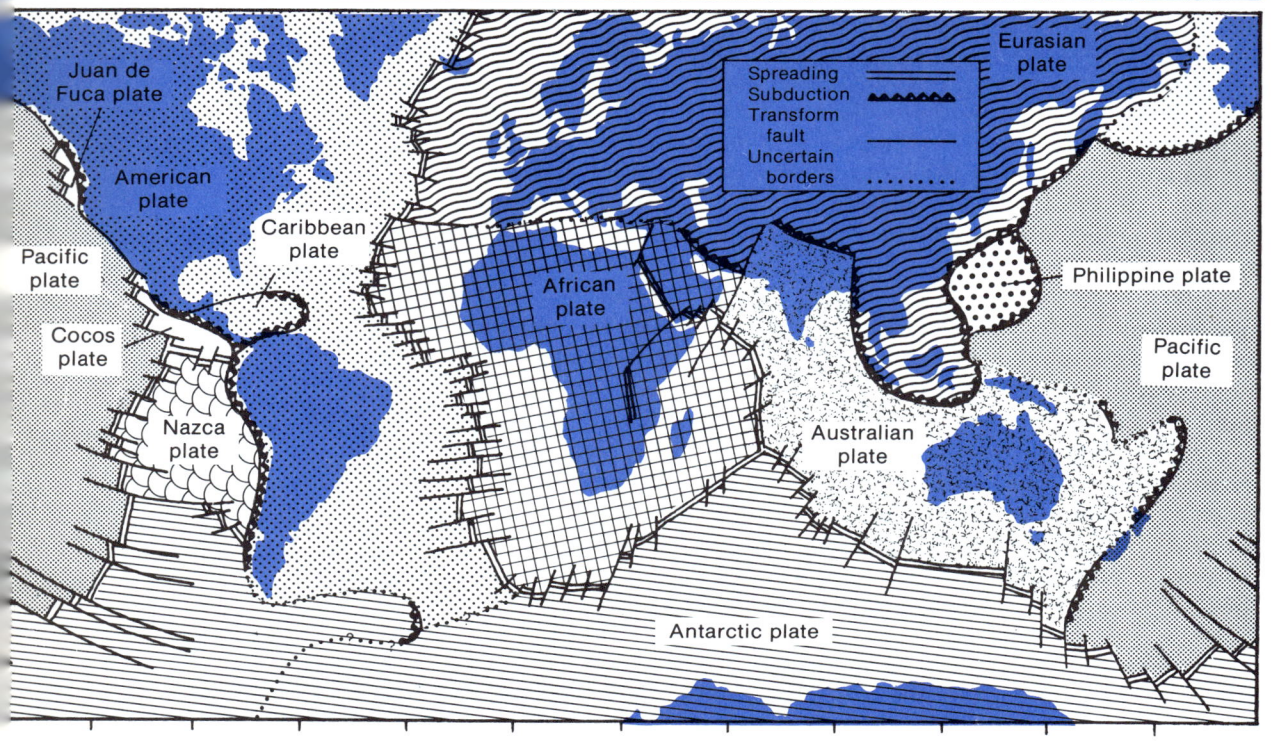

Scientists are generally in agreement about the boundaries of the earth's major lithospheric plates, as shown above. Boundaries of some of the smaller plates are still in doubt.

By contrast, plates slowly disappear along other well-defined lines. At the coastlines of certain continents—the west coast of South America, where the Andes rise, is an example—the earth's lithosphere is being subducted. That means that the ocean-floor lithosphere is sliding downward into the asthenosphere under another plate, along a line called a **convergent boundary.** The Andes are believed to have risen to their present elevations because the ocean-floor plate has exerted strong horizontal force against the plate that forms the South American continent.

At the divergent boundary (left) two plates are being created. An ocean-floor plate is being subducted at the convergent boundary (right).

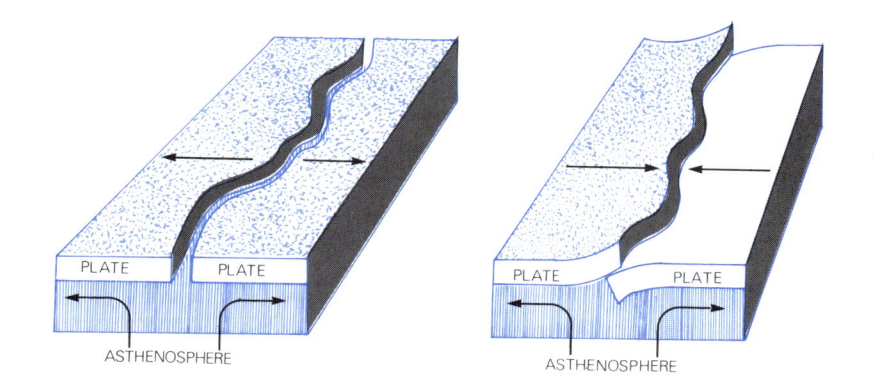

plate tectonics

Usually the trenches formed by such underthrusting are V-shaped, but some have flatter bottoms because sediments have been deposited in them. One of the most imposing of the trench systems alongside a continental landmass is made up of the Peru–Chile Trench and the Middle America Trench, the latter lying just seaward of the western mountains of Central America. These trenches stretch for thousands of kilometers north and south, varying in their depth from a thousand meters or so to about 8,000 meters below sea level. One low point at a depth of about 6,500 meters lies almost opposite the highest elevation in the Andes—Cerro Aconcagua, about 6,500 meters high, near Santiago, Chile. This might indicate that the greater the force exerted by the ocean plate in creating a deep trench, the greater is the mountain-building force on the opposite side of the trench.

Plates are also disappearing into the earth's asthenosphere in some of the deepest mid-ocean trenches ever sounded. Most of these areas of subduction are found in the Pacific Ocean—many of them beside some of the principal Pacific island chains. Plate motion here resembles that along the Peru–Chile Trench: one plate is slipping downward beneath another plate, creating a convergent boundary.

plate features in the Pacific Ocean

One of the most striking of all the trench systems runs north and south just a short distance from the Tonga Islands, in the southwestern Pacific Ocean. Parallel to it are a series of coral islands built on volcanic foundations. On the eastern side of the islands the sea bottom is about 10,000 meters deep at its deepest point. To the westward a line of volcanic islands lies more or less parallel to the coral islands.

It is significant that earthquakes are especially common in a slender band extending through the islands. The band appears to represent a line of lithospheric material that slopes downward into the trench and underneath the islands. Analysis of seismic waves and other physical evidence seems to indicate that the earthquakes originate in the Pacific Plate that moves westward and slides under the Australian Plate along this line. Shock waves along the band might be due to occasional large movements between the plates, rather like seismic waves caused by mechanical slippage at a fault. Both strings of islands probably originated when crustal rock carried downward was subjected to high pressure and heat. Under such conditions it tended to disintegrate and melt, producing gases and magma which made their way upward through the edge of the Australian Plate to the surface.

The Pacific Plate, which here underthrusts the Australian Plate, is created at a ridge far to the east, known as the East Pacific Rise. It has always moved as one mass at the same rate of speed. The edge that has been disappearing has remained solid till it reached a certain depth, and then decomposed. About 80 km thick, the Pacific

plate tectonics

Plate floats along on the earth's mantle, much as an ice floe is carried by water, to the trench where it slowly disappears.

In subduction zones, researchers have found that gravitational force fluctuates; in the depths of the trench it is less than expected. This indicates that the rocks beneath the trench contain less mass than is needed to maintain isostatic balance, as discussed in chapter 4. This suggests an additional reason for the frequency of earthquakes and earth movement in these areas.

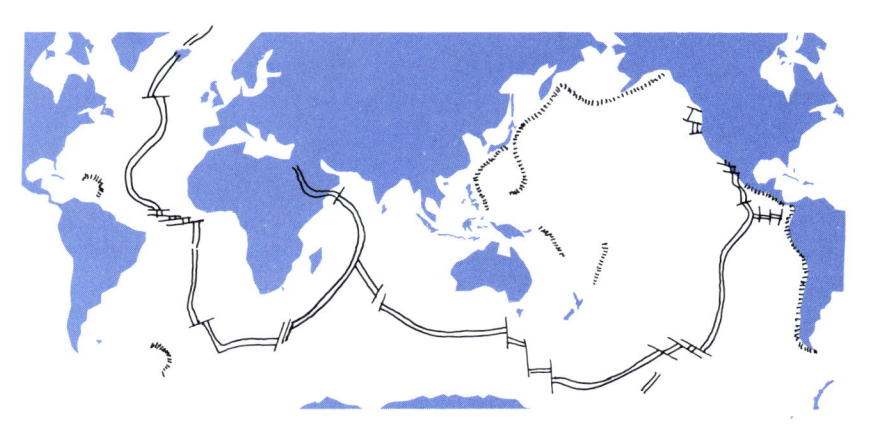

The major earthquake belts of the earth are closely linked to the points at which subduction of plates is taking place. Volcanic activity is also common at these boundaries, shown in the figure at right.

Investigators have made another discovery that leads to fascinating speculation about the movement of the Pacific Plate over millions of years. A chain of islands and underwater mountains, extending from the Kamchatka Peninsula of Siberia through the Midway group to Hawaii, is believed by some researchers to be the result of a stationary "hot spot" in the earth's mantle. According to this theory, as the Pacific Plate moved toward the northwest, the magma sent up by the hot spot built, one by one, a trail of mountains that marked its path. Some of these mountains grew high enough to break through the ocean surface and become islands. For instance, dating of rocks on Kauai, the most northerly Hawaiian island, shows that it is 3 to 5.6 million years old. The islands to the southeast are progressively younger; Hawaii, the last, is also the youngest. Recent volcanic activity on this island might indicate that it is now above the hot spot. If this theory is correct, it seems likely that more islands will eventually be added to the Hawaiian chain as the Pacific Plate continues its northwesterly drift.

Surveys of sea floors have made it relatively easy for scientists to determine where rifts and trenches occur. Moreover, seismologists have recorded the focuses of earthquakes by tracing the shock waves back to their sources, as we saw in chapter 5. When these sites are plotted on maps, they form a pattern of earthquake activity around the globe that coincides nicely with the location of subduction zones.

plate tectonics
transform faults

Still another kind of motion is taking place within plates and along their edges. This is a sliding movement along faults. Above sea level, masses of rock may move in various directions along fractured surfaces. But on the ocean floor, the chief movement is horizontal, producing what in chapter 4 were identified as strike-slip or transform faults. It has only recently been understood just how these masses of ocean-bottom material are able to move in this way.

Transform faults are found principally in the oceans. They generally occur in parallel series along the underwater ridges and rifts, with each fault cutting the ridge more or less at a right angle. Discovered in the 1950s, these faults at first puzzled scientists, who called them **fracture zones.** Because some of the fractures continued for thousands of kilometers across the Pacific Ocean basin, it was assumed that displacement of the rock masses between the fault lines occurs over the entire surface.

These transform faults were fully explained only when more data became available. It turns out that different zones of the diverging plates move apart at different rates of speed. This builds up horizontal stresses between slow-moving and fast-moving sections of each plate. Eventually the stress becomes severe enough to cause fractures and strike-slip faulting between the segments. But this happens only close to the original rift. Further away, the movement somehow becomes adjusted, so that all the parallel segments are moving at the same speed, and therefore no more fault movement takes place. So earthquakes originate in the fault zones near the rifts and ridges, but the shock waves dissipate rapidly as they travel. They are called **shallow-focus earthquakes,** and although they occur frequently, they cause almost no damage in populated areas. As we noted above, seismologists have used them to locate plate boundaries.

Convection in the earth's mantle probably follows a pattern similar to the one shown below left. At right, arrows indicate the direction of plate movement and the formation of transform faults.

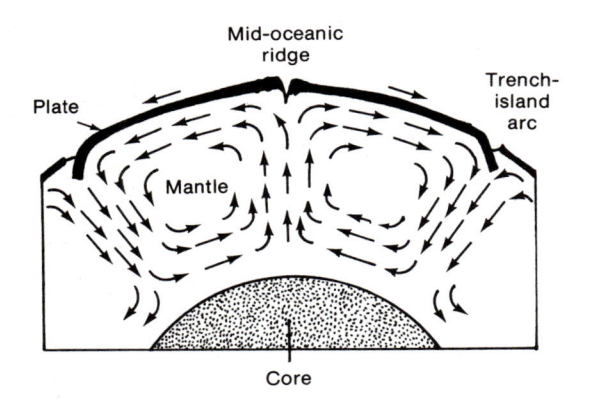

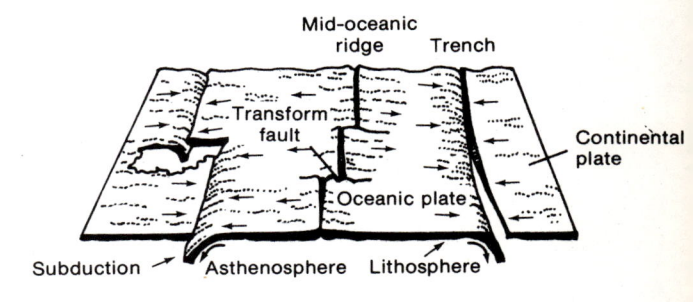

plate tectonics

Sudden cooling of magma on the ocean floor as it comes into contact with sea water produces basalt formations such as the one shown at right. This photo was taken by *Archimède* at a depth of more than two thousand meters.

where plates come into existence

The movements of the crustal plates are rather more complex than we have indicated thus far. Since plates move about on the earth's surface like tightly packed ice floes, what happens where their edges contact each other? No gaps exist between them. For practical purposes, the earth's crust is continuous.

It is easy enough to imagine a single point at which magma rises from the asthenosphere and is spewed over the landscape. Even today many volcanoes — in Hawaii, Indonesia, Japan, Zaire, and Martinique, to name a few locations — are active. Now imagine magma spilling from north–south cracks in the ocean floor for thousands of miles, and being abruptly cooled by sea waters. This gives you a picture of the process that brings new material to the ocean bottom and builds substantial ridges, as the huge lithospheric platforms we call plates are pushed apart. All this action on the sea floor takes place very, very slowly. Earth scientists estimate that spreading of plates at a rift amounts to only a few centimeters a year. That is enough, though, for the continents to have been separated and rejoined possibly ten times during the nearly five billion years of our planet's existence. The slowness of the process is one of the reasons that scientists were late in discovering these movements. Another reason, of course, is the fact that the physical evidence has been buried at the bottom of the sea, out of reach of human observers until very recently.

Most of the rifts from which the magma oozes run roughly north and south near the midlines of the oceans. It is probable that the rifts first formed on the continental landmasses and then shoved the split-off portions aside at a roughly uniform rate to form new, separate landmasses. For example, a fairly recent split occurred between the present-day continent of Africa and the Arabian peninsula, creating the Red Sea and the Gulf of Suez. If the spreading continues, the gap may grow as wide as the Atlantic Ocean itself within a few tens of million years.

In every case, the rifts are located at the tops of sea-floor ridges. Produced partly by the accumulation of solidified lava, the ridge probably is also built up by expansion of rocks due to heating.

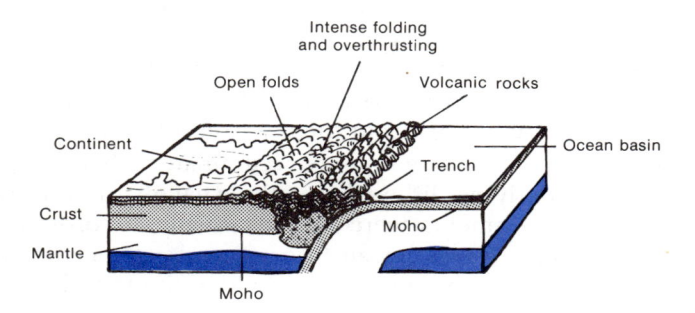

Subduction of an ocean plate where it meets a continental plate produces a characteristic pattern of folding and overthrusting. Some of the earth's major mountain chains were formed as the result of such tectonic activity.

However, the lava cools as it moves to the flanks of the ridge and thus shrinks somewhat, so the ocean floor away from the ridge tends to be lower. Ridge tops usually extend at least to within 3,000 meters of the water's surface, and may come much higher. In some cases, the ridge may actually rise above the water. The island of Iceland, for example, is a part of the Mid-Atlantic Ridge that appears above the ocean surface.

Ocean sediments are rather thin on a ridge and its flanks, because the ridge is made up of the youngest rocks on the ocean floor. But as the material moves away from the ridge, sediment drifting down through the water accumulates in a thicker and thicker layer. The thickest deposits of sediment are usually found along continental margins, where the rocks are oldest—especially where one plate slides under another.

island arcs

The Pacific Ocean is bordered by several chains of volcanic islands flanked by trenches on the ocean floor. Each of these chains, known as an **island arc,** has a deep trench on its ocean side and a shallower depression between the arc and the continent. The islands of Japan and the Aleutian Islands are almost perfect examples of island arcs.

Geologists use a working hypothesis to explain the formation of island arcs. It is believed that they originated when billions of tons of magma poured from a fracture in the ocean floor. The removal of this material weakened the lithosphere on either side of the rift, causing the ocean floor to sink slightly. Eventually, two troughs were formed, with a long line of volcanic islands between them at the site of the original rift.

plate tectonics

As we have noted, most island arcs are found around the perimeter of the Pacific Ocean, although a few others exist elsewhere in the world. You will recall that in chapter 4 we stated that some researchers believe that portions of the Appalachian Mountains were once an island arc. If we accept this theory, we can account for the difference between the eastern and western parts of the Appalachian range.

Let us suppose that while the Appalachians were forming, sediments accumulated alongside the volcanic islands. The debris that settled in the outer, deeper trench was almost entirely of volcanic origin. The sediments that settled in the shallower trough on the landward side were washed into the sea from the continent.

As more and more sediments accumulated, their weight depressed the underlying lithosphere; the shallow trough became deeper, and the sediments were buried. Finally, according to this hypothesis, tectonic forces began to exert pressure on the sides of both troughs. The trench on the ocean side of the island arc underwent the greatest amount of compression. As the walls of the trench were forced closer together, some of the volcanic material within the trench was converted to metamorphic rock and was forced upward. The sediments in the shallower trench to the landward side of the arc, however, were uplifted much more slowly, and did not undergo metamorphism. Although this hypothesis has not yet been proven, it does, at least, provide a plausible explanation for the difference in composition of the eastern and western Applachian Mountains.

raw material for plates

Where does magma flowing out of a rift come from? It originates in mantle material that has been melted. Some of this material is so hot that it melts the solid rocks lying above it and makes its way up to the surface of the crust through slits or holes, creating volcanoes. Even today volcanoes burst through into the earth's oceans. An example is the island of Surtsey, near Iceland, which was formed by a volcano that grew upward from the bottom of the ocean in 1963 and expelled smoke, gases, and lava.

A large part of the magma escaping through the crust appears at underwater rifts. An exception is found in Iceland—that section of the Mid-Atlantic Ridge which, as we noted, has bulged up above the ocean floor. This accessible portion of the ridge serves as a kind of field laboratory where researchers can study igneous processes and other phenomena of sea-floor spreading.

Why do some portions of the crust rise so high, while other portions, such as ocean floors, are relatively low? What supports rocks at higher elevations? Scientists now believe that the theory of isostasy accounts for irregularities in the crust, as we discussed in chapter 4. As we saw, continental crust tends to be less dense than ocean-bottom crust, and as a result is pulled upon less forcefully by gravity.

SURTSEY: THE BIRTH OF AN ISLAND

In November of 1963, signs of an impending volcanic eruption were detected off the southern coast of Iceland. The surface of the ocean became violently agitated. Thick, noxious clouds of gas and steam floated over the area. Scores of dead fish, along with foamy fragments of pumice and cinder, swirled around a point in the water which marked the center of the submarine eruption as it reached for the surface.

By November 24, a huge mound of volcanic debris had emerged from below the surface. It was nearly a kilometer long, with a summit more than 100 meters above sea level. For months afterward, awesome clouds from the new-born volcanic island continued to billow over the area. Almost continuous barrages of thunder and lightning added to the spectacular display. Appropriately enough, the new island was named Surtsey, after the fire god of Icelandic myth.

Awesome as the emergence of Surtsey was in its own right, the sudden appearance of such volcanic islands is a fairly frequent event. Iceland is located in one of the most active volcanic regions in the world. Whereas Surtsey is a volcanic peak built up from the ocean floor by successive lava flows, Iceland is actually the summit of a huge bulge in the Mid-Atlantic ridge.

Iceland sits astride the diverging American and Eurasian lithospheric plates, which are moving apart at the yearly rate of two centimeters. The frequent earthquakes and submarine volcanic activity in this part of the world are caused by the upwelling of magma at this point. Surtsey appears to be a more or less permanent addition to the scene. Many such volcanic islands, however, emerge for only a short time and then are once more swallowed up by the sea. Two smaller islands that appeared near Surtsey have long since vanished as the result of erosion by waves and winds.

At the moment, Surtsey seems to be inactive. As soon as the new island became cool enough to visit, scientists from many countries flocked to study it. Several long-term research projects are now based on Surtsey, examining it from the standpoint of such disciplines as geology, biology, and ecology. If Surtsey reawakens, of course, these projects will be halted very suddenly. With this in mind, the scientists on Surtsey are working feverishly to amass all the information they can — occasionally glancing over their shoulders to reassure themselves that the sleeping fire god of Iceland is still sleeping!

plate tectonics

Most of the earth's mountains are no higher than about five and a half kilometers. One notable exception is the Himalayas, located where the plate bearing present-day India collided millions of years ago with another plate carrying Eurasia. The collision caused — and continues to cause — a massive uplift that created the world's highest mountains, up to nearly 8 km in elevation. Some scientists believe that these mountains stand so tall because there is a thicker-than-average layer of low-density material under them. Refinements in the theory of isostasy are still being worked out, so it may be years

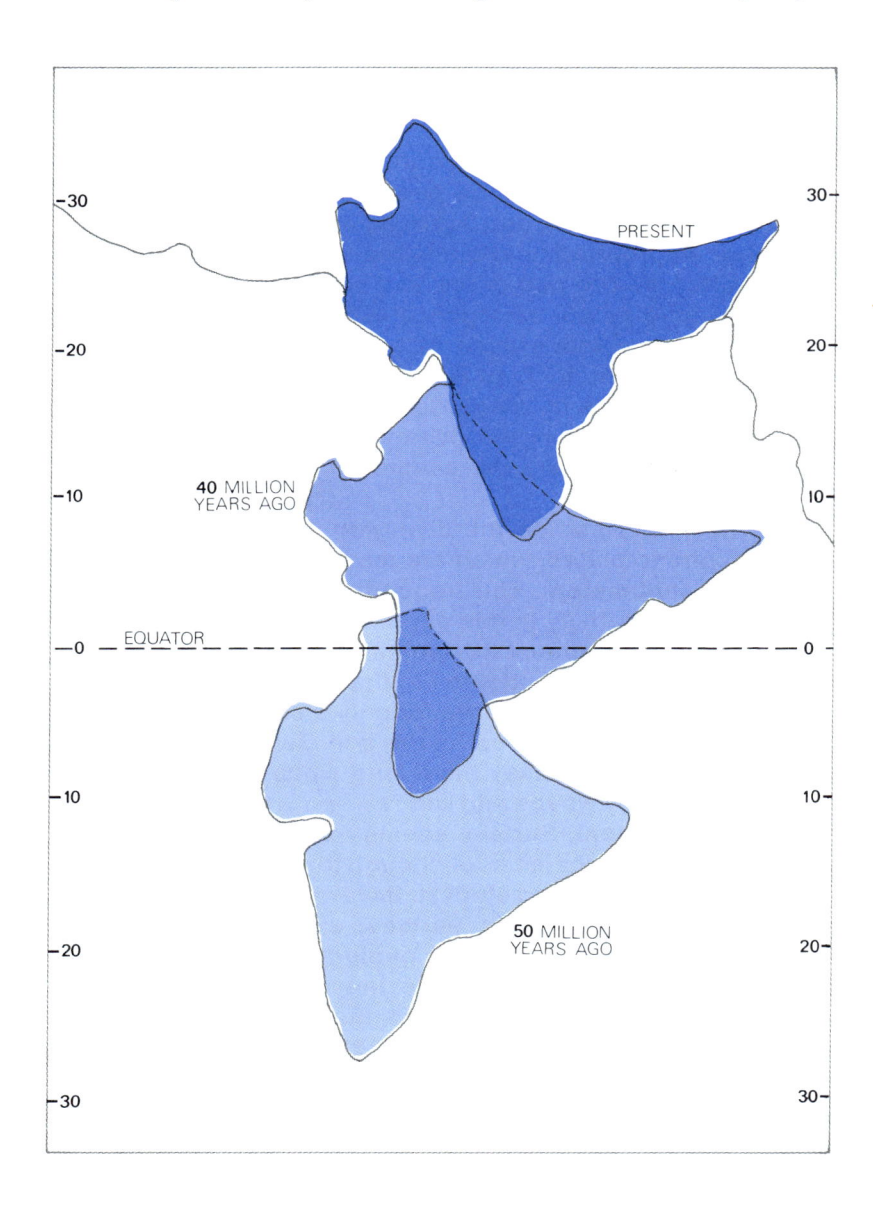

The great Himalayan mountain range was created when India drifted northward and collided with the plate carrying Eurasia. India continues to move; its former and present positions are shown in the figure at left.

plate tectonics

before all earth conditions are accounted for. And because technology still limits the distance we can penetrate below the surface of the earth, some questions about the nature of our planet may always be matters of speculation.

The collision that created the Himalayas occurred about 45 million years ago, after India had drifted several thousand km northward from its old position in the supercontinent Pangaea. Since the collision, India has moved even further north—some 2,000 km into Eurasia. This movement has rolled up a mass of relatively light rock material to form the Himalayas. Explaining the displacement of such a vast area of the crust is no easy task. But data compiled from satellite photographs, fossil finds, rock analysis, and earthquake-site determinations have been combined with careful reasoning to produce a satisfactory hypothesis. The so-called **suture area** today has been found to be laced with faults, of both strike-slip and thrust varieties (see chapter 4). The squeezing that produced these faults has caused much of the crust to shorten and thicken, in a kind of accordion effect. Furthermore, much land area has been displaced eastward, in the direction of China.

Not only has this continuing collision made sweeping changes in the shape of the land, but the raising of the crust and creation of mountains and high plateaus has profoundly affected the climate.

a collision that added to a continent

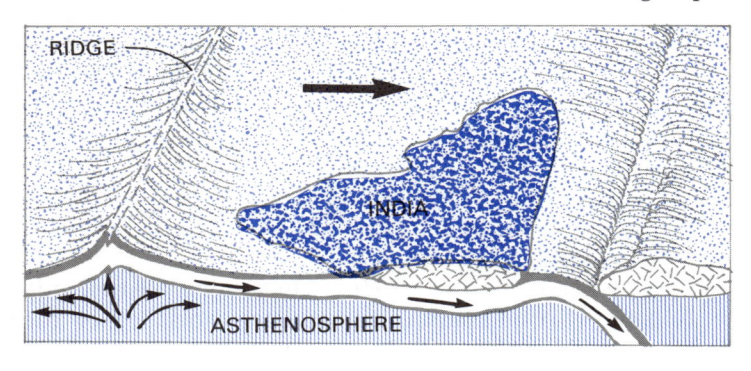

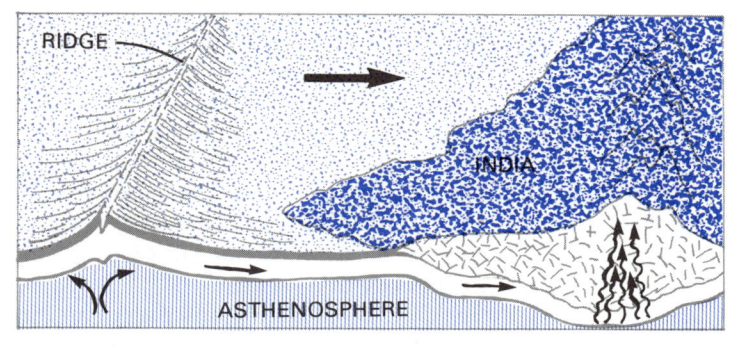

The steps that led to the formation of the Himalayas are shown here. The relatively light rock material of the Eurasian plate was virtually accordion-pleated by the pressure of the northward-moving Indian plate.

plate tectonics

The temperatures, rainfall, drainage patterns, wind directions and velocities, and many other climatic factors in this region of the world are the direct result of this clash between two continental masses. Another important byproduct of the collision was the creation of one of the world's most active earthquake zones. Because India exerts continuing horizontal force against Eurasia, slippage along fault lines sends shock waves that occasionally cause havoc in China, Siberia, Tibet, Afghanistan, and Nepal.

evidence from the ocean floor

Alfred Wegener, who first conceived the hypothesis of continental drift, was forced to depend largely on his intuition to explain how the continents were once part of a supercontinent that broke up. If he had possessed the evidence that has now been made available by the use of sophisticated equipment and techniques, he would have been acclaimed for his remarkable insight into the question of how the physical world evolved. The conclusions he arrived at despite lack of supporting evidence at the time are now largely supported by information collected in recent years.

One of the most important efforts to accumulate new information was made by technicians and scientists who sailed on the *Glomar Challenger* during the first phases of the Deep Sea Drilling Project, set up jointly by the United States Government and American institutions. This oceanographic ship was no ordinary vessel. It was equipped with a derrick looking much like those used in oil fields, to allow drilling bits and sections to be assembled and then rotated to bore into rock layers. The *Glomar Challenger* was seeking core samples of sea floor rocks. The samples it brought up from the bottom showed what kinds of rocks and minerals were located below, where they were located, how the basaltic rocks had been magnetized, and where fossils existed. All this information was gathered to help scientists learn more about the history of the earth.

The work done by the *Challenger* is only one part of an international effort to learn more about the earth's crust. Several nations have furnished people and money to promote further studies. The *Challenger's* crew has run drill bits down through more than 3 km of water into as much as a kilometer of solid rock, repeatedly pulling a string of drilling sections out of a hole and replacing them in the hole by remote control. This technique has made it possible to drill a series of holes to new depths at varying distances from a rift. Thus scientists have learned a great deal about the chronology of events in formation of the crust. The samples brought up to the surface show, for example, that the material closest to a rift is the youngest of all,

plate tectonics

213

Piston corers such as the one illustrated are used to bring up cores of sediments from the ocean floor. Studies indicate that different kinds of sediments are found at different locations, and that the rate of sediment deposition is very slow.

and that which lies farther off is proportionately older. Evidence of this kind, showing changes in development of the earth, will be discussed a little later.

While many nations have been involved in ocean research in recent years, two of the most experienced are the United States and France. The latter has produced a number of outstanding sea scientists and explorers and built several useful underwater craft as well. During the 1970s the two nations cooperated in a joint venture called the French–American Mid-Ocean Undersea Study, or FAMOUS. The goal was to explore the Atlantic Ocean bottom with manned submersible craft, which would allow closer examination than before, and to apply other modern techniques of exploration. Most previous observations had been made from the surface or at modest depths, employing sonar, drilling holes, or dredging. Collaborators from the two nations chose an area of the Mid-Atlantic Rift near the Azores, where depths run to more than 3,000 meters, to hunt for clues to the mystery of continental drift and plate tectonics.

The first human excursions underwater were made inside artificial atmospheres — trapped air within some sort of rigid vessel. The first practical submarine appeared in the seventeenth century, and the first submersible intended for war was designed during the American Revolution. By the nineteenth century, submarines had become larger and were made of iron, and by World War I undersea craft could travel long distances carrying cannons and torpedoes for sinking surface vessels. While research continued to create machines of war, diving devices of increasing sophistication came into use. In the 1930s bathyspheres, suspended by cables from the surface, were developed for exploration. Later came bathyscaphs — mobile, manned, and steerable. One of these, the *Trieste*, navigated to a rec-

submersibles supply firsthand evidence

plate tectonics

ord depth of 10,830 meters in the Mariana Trench, where water pressure is approximately eight tons per square inch.

Since then, France has developed submersibles with differing capacities. One of these was the *Archimède*, which was limited to a depth of about 3,000 meters, was not highly maneuverable, and had limited visibility. For the FAMOUS project, French designers produced a new, more versatile diving machine, the *Cyana*. At the same time, Americans improved their submersible, the *Alvin*, by adding a titanium hull, so that it could descend to 3,000 meters or even deeper. Fitted with three portholes, the *Alvin* was especially well suited for observation and photography, as well as for a host of other tasks.

sounding the depths

The FAMOUS project had many tools of scientific research at its disposal, and its findings reflected the new state of oceanographic investigation. Some of those findings gave considerable impetus to the development of more detailed explanations of the earth's evolution and of the changes its crust is likely to undergo in the future.

However, significant data contributing to our knowledge of plate tectonics had already been accumulated during World War II, when surface and underwater vessels could be detected with considerable accuracy by sensitive, long-range sound-echo equipment. Sonar not only pinpointed large man-made objects but also helped to chart the world's oceans. For the first time, seamen could probe and map ocean bottoms with relative easy and accuracy, allowing ships and submarines to travel where they had never been before. Eventually, nuclear submarines penetrated even to the North Pole.

Sonar equipment sends out sound impulses through water, and those impulses travel in a straight line until they encounter large, solid objects. Some of the sound energy is absorbed or scattered, but a substantial amount is reflected to the sonar set, which then gives off audible signals or presents a picture of the objects on a screen. Because sound travels at a uniform speed through water, the equipment can measure the time required for impulses to reach an object and return. The distance to the object can then be calculated.

Earth scientists in the FAMOUS project took advantage of sound devices to detect earthquake activity. Three buoys were set afloat in a triangular pattern. Each of them contained a **hydrophone** or underwater microphone for picking up sound transmitted by disturbances in the crust, and a radio transmitter. Earth shocks picked up by the hydrophone were radioed to the surface vessel and their origins plotted, in much the same way that earthquakes are located by triangulation from seismic stations on land. The survey data aided project personnel in determining what movements were occurring around the rift where the American Plate and the African plate were being born.

Sound energy was utilized in still another way. It was important to keep track of the submersible *Alvin* in relation to its surface

plate tectonics

support ship and the bottom features, shown on contour maps of the bottom that had been prepared. **Transponders** are devices that send out sound waves of their own when triggered by sound signals from another source. Three transponders were anchored in a triangular pattern close to the bottom where they could not be disturbed by surface currents. The surface ship could establish its position by sending impulses to the transponders and measuring the time required to get signals back from each of the three. Then the *Alvin* sent out impulses to the ship and the transponders; by comparing the times when all four signals were received, operators on the ship could determine the exact position and depth of the *Alvin*. This made it possible to establish precisely where the *Alvin's* crew was taking photographs, picking up samples, and making visual observations.

    Sonar mapping of the ocean bottom was accomplished by three different techniques. To start with, one vessel sent sound impulses to the bottom in a broad beam for fast coverage of large areas. Then another acoustical instrument sent waves in a much narrower beam for more precise measurements of features. Finally, a cooperating British vessel towed a mechanical device which beamed sound energy sideways close to the bottom, to highlight hills and walls. When all the resulting data were integrated, scientists had an accurate three-dimensional picture of the bottom they wished to explore.

As a crowd of interested onlookers watches, *Alvin* is being prepared for a dive. *Alvin* is particularly well suited for undersea photography, and can descend to depths of at least 3,000 meters.

plate tectonics
bringing the bottom to the top

A traditional way of gathering information about the sea floor has been collecting samples. When oceanographic research vessels first obtained samples of glassy, volcanic rock near the Canary Islands, scientists were at a loss to explain why such material would appear on the bottom. But later expeditions brought up more volcanic rock, and researchers concluded that the Mid-Atlantic Rift was a long slit through which new material was upwelling to the surface of the earth's crust. Similar finds were later made along ridges in the South Atlantic, Indian, and Pacific Oceans.

The *Glomar Challenger* added to the body of knowledge when it brought up, along the United States coastline, drill cores that proved to be about 150 million years old—the oldest bottom material found up to that time. Sediments, too, were older and thicker near the coast than in mid-ocean. Scientists concluded that the coastal rock

Scuba divers have made important contributions to our understanding of the ocean floor at shallower depths. This diver has acquired several "friends," curious grouper fish who are not at all afraid of this strange visitor to their domain.

must have emerged from the Mid-Atlantic Rift shortly after North American and Africa split away from each other and the sea floor began to spread. Because the earth is between four and five billion years old, it is probable that the continents have separated and rejoined a number of times, as we indicated earlier. All evidence confirms that our planet has been anything but static over the eons, and that its surface continues to change.

During the FAMOUS project, many enormous chunks of lava were found on the sea bottom. These volcanic rocks took many shapes, each reflecting the way in which the lava had been squeezed out through the rift. Some looked like solid or hollow pillows, others like tubes, buds, waterfalls, or toothpaste ribbons. Some volcanic rock was found to be built up into slopes and walls, with valleys or clefts between.

more tools for probing the sea floor

Thermocouples are sensors that measure temperatures. Mounted on the *Alvin* or towed near the bottom, they were used to take the temperature of sea water, especially where volcanic material might be expected to heat it. Surprisingly, however, there was no indication that the water in the rift area was appreciably hotter than any sea water at that depth. One rock fracture, however, showed evidence of heat flow, and certain manganese deposits were thought to have been created by hot water percolating up through the sea floor. The nature of the features around the rift suggested that the area had gone through alternating periods of activity and quiet, and that vulcanism might be starting up again in two piles of rock that the explorers called Mount Venus and Mount Pluto. It is possible that the rate of plate spreading is so slow that magma below cannot flow constantly through the opening and thus is not always in direct contact with sea water.

Nonetheless, project members found traces of heat flow in the sediment if not in the water. Some sites were considerably warmer than others, especially along the flanks of the rift valleys. High heat-flow rates were calculated for a transform-fault section of one fracture zone. Veins within rocks and cementing of rocks, both of which often require heat, were also found. No active lava flows could be observed, but expedition explorers speculated that the lava they saw and photographed might be only a few hundred years old, quite young by geological standards.

Oceanographers in previous decades had measured above-average amounts of heat flow from within mid-oceanic ridge crests. Rising magma can reach temperatures as high as 1300°C, so that the edge of an emerging tectonic plate is considerably hotter than the rest of the plate.

Special photographic techniques were also developed to record conditions on the sea floor during the FAMOUS undertaking. First, from a surface vessel, large-area photographs were made of the

plate tectonics

Cyana, one of the submersibles of Project FAMOUS, is gently lowered into the sea from her mother ship. Studies performed by *Cyana* have added greatly to our understanding of how the American and African plates move.

bottom, with the aid of a stroboscopic-flash lamp positioned above the camera. Each picture covered an area about 100 times larger than conventional methods would have allowed. These photographs allowed scientists who were to descend in the *Alvin* and *Cyana* to become familiar with the terrain in advance.

In recent years, earth scientists have been challenged by the discovery of magnetic variations in the crust at the bottom of the ocean. When aircraft towed magnetometers over certain areas, it was found that the earth's magnetism was slightly different than expected. Magnetic surveys on the ocean's surface substantiated the findings. Instruments were towed close to the sea floor, with the result that the magnetic irregularities were confirmed and described.

A number of scientific studies have shown that molten rock solidifying alongside ocean-bottom rifts contains iron-oxide particles oriented according to the magnetic lines of force existing at the time (see chapter 7). A band of rock close to a rift and paralleling it would show a certain magnetic polarity, while its neighbor a bit farther away from the rift would demonstrate opposite polarity. This pattern of alternating polarities repeated itself many times.

Scientists now believe that the earth's polarity has rapidly reversed itself many times, and have a good deal of evidence to support their theory. As molten rock emerged from a rift in easterly and westerly directions, the lava "froze" and its particles lined themselves up with the pattern of earth's magnetism then in effect. Later,

plate tectonics

when the magnetic polarity was reversed, particles in two new bands of lava, one on each side of the rift, oriented themselves in the opposite direction. Over a long period of time, one pair of bands succeeded another to create **magnetic striping** across the ocean bottom, parallel to a rift. Today it is possible to measure the locations and widths of the stripes, determine when they emerged from rifts, and thus measure the rate of plate movement over millions of years. At the mid-Atlantic site examined by FAMOUS explorers, it was learned that the American Plate had moved westward from the rift at the rate of 0.74 centimeter a year, and the African Plate had moved eastward almost twice as fast, 1.3 centimeters yearly, over millions of years.

Investigators detected some discrepancies in magnetic readings between the bottom and the ocean's surface. The deep rock layers below the lava may reveal different patterns of magnetism in future studies. But there is little doubt that all magnetic evidence supports the conclusion that mid-ocean rifts give rise to the plates, which move away gradually with the support of the mantle beneath.

Many strange forms of life exist on the ocean floor. Here the bottom is blanketed with dozens of fragile-appearing sea stars, a variety of deep-water starfish. This scene was photographed by *Trieste*.

what is ahead in plate tectonics

While researchers are studying the forces that shaped the earth's crust, they are also looking ahead to further changes that seem inevitable. If the continental drift–plate tectonics theory is valid—and there is every indication that it is—then crustal movements will continue at about the same rates. Modification of crustal plates, slow-

IN FOCUS:
the ring of fire

Around the fringes of the Pacific Ocean runs a distinct line of volcanic activity, popularly known as the Ring of Fire. This belt is marked by island arcs and active volcanoes on the western edge of the Pacific, and by evidence of recent volcanic activity along the western coastlines of North and South America.

Are these signs of volcanic activity isolated phenomena, or are they somehow related? The answer to this question had to wait until the early 1960s, when scientists arrived at a unified hypothesis of plate tectonics. This hypothesis provides us with a comprehensive picture of the earth's history and development. It accounts for most of the world's major surface features, including continents, mountain ranges, and ocean basins. Furthermore, it provides an explanation for the location and occurrence of earthquakes and volcanoes, and for the distribution of the various types of rock on or near the surface of the earth.

The tectonic movements that shape the earth are constantly taking place. New oceanic crust is created at mid-ocean ridges by magma moving up from the mantle; the ocean floor spreads, and continents drift; oceanic crust is subducted. As the oceanic crust pushes down toward the mantle, partial melting produces new igneous rocks, and new granitic material is added at continental margins. Eventually, as oceanic crust is subducted, no more ocean is left, and neighboring continents collide on their lithospheric skate boards. This, in turn, leads to the creation of great mountain ranges such as the Himalayas.

Most of us are familiar with ecological recycling centers where old, worn-out, or used-up materials can be taken to be sorted out for reprocessing into new materials. The earth's recycling program—lithospheric subduction—has been in operation for at least four billion years.

The Ring of Fire displays several different aspects of this recycling process at work. When two plates meet, the oceanic plate, with its denser material, bends and is pushed under the less dense continental plate. At first the angle of descent is shallow, but it becomes steeper and steeper with increasing depth.

Usually the descending plate remains cooler than the mantle it is passing through, until it reaches a depth of about 600 kilometers. Beyond this point, it heats up more rapidly, until finally, at about the 700-kilometer mark, the descending plate can no longer be distinguished thermally as a separate unit.

Generally speaking, the more rapidly a plate descends, the greater is the maximum depth of earthquakes associated with it. For example, along the west coast of South America, the small Nazca plate is being subducted by the westward-moving South American plate, in a zone some 6,700 kilometers in length. The rate of subduction at the Peru—Chile subduction zone, where the Nazca plate is descending, is 9.3 centimeters per year, the second highest for all major zones. Earthquakes here have been recorded to a depth of 700 kilometers.

In the Middle American subduction zone, the Cocos plate is descending under the North American plate at the highest known rate of 9.5 centimeters per year. We know that this is a relatively new process, because earthquakes in this region have been detected to a maximum depth of only 270 kilometers.

The intense earthquakes that occur where two lithospheric plates are pressing on each other result from stresses in the subducting plate. Others are caused by deformation of the lithosphere and by isostatic adjustments. Below a depth of 700 kilometers no earthquakes occur, because the now-hot slab of lithosphere is no longer a rigid elastic medium prone to fractures or faulting. Thus the small stresses that exist below this level are not relieved by the sudden rebound associated with earthquakes, but rather by slow plastic deformation.

A major feature of subduction zones is vulcanism that gives rise to andesite, which often takes the form of a fine-grained grey rock. Andesite is formed from deposits of sediment that accumulate in the deep ocean trenches over the course of time. When two lithospheric plates press together, these sediments may be trapped and subjected to deformation, heating, and pressure. Sometimes the sediment can be pulled down to great depths by tectonic forces. There it melts, and eventually may be returned to the surface by volcanic eruption.

Is the Ring of Fire a permanent feature of the earth's topography? No, because as the Pacific plate moves westward and is subducted by the Eurasian plate, the Pacific Ocean itself will cease to exist. The American and Eurasian plates will collide, throwing up a mighty mountain chain. Although volcanic activity will halt, compressional forces are sure to create a new zone of violent earthquake activity running roughly from pole to pole.

But what of the newly-expanded Atlantic Ocean? Where seafloor spreading takes place at the great Mid-Atlantic ridge, volcanic activity is likely to form, not a Ring of Fire, but a Band of Fire, extending from the South Pole to the North Pole. And when the Band of Fire and the belt of earthquake activity connected with the new mountain chain meet at the poles, some interesting things are sure to happen!

plate tectonics

Archimède rests on the surface after a dive, waiting for her batteries to be recharged. The submersibles used in Project FAMOUS are highly complex, but we can expect that future ocean-floor explorations will use even more sophisticated craft.

moving when considered over a human lifetime, will radically alter the planet's geography and other aspects of its environment. The world several million years from now will appear considerably different from the way it does today, just as the world of previous eras bore little resemblance to the one we see now.

what new knowledge will yield

From the short-range viewpoint, though, there are sure to be changes that affect generations of people now living, especially their notions of how the earth has evolved over the ages. As we said earlier, earth scientists are largely in agreement about the basic theory: the planet's crust is made up of plates that are actively moved about, constantly being created and destroyed along their edges; continents on the plates have drifted apart or collided; and most deformations of the crust are associated with basic internal processes. Of course, the theory may be modified as new data, new analyses, new comparisons, new trials, and new thinking provide more detailed information about the earth. Here are some of the areas in which we can expect to see substantial progress within a few decades:

plate divergence and convergence

Further deep-water exploration with highly instrumented surface vessels and manned submersibles should supply a wealth of information about the crust lying under the seas. This is particularly true of ridges and rifts, because they lie closer to sea level than trenches and are thus more accessible.

earthquakes

Shocks transmitted through the earth originate in movements along fault planes in the continental masses of the world, or at diverging and converging boundaries of plates. Although shallow-focus earthquakes in the oceans cause few problems, shocks created at convergent plate boundaries may be extremely harmful to human settlements. Now that it is easier to locate their sources, measure shock intensities, and draw on previous experience, scientists will develop increasingly accurate methods of earthquake prediction.

earth's composition

Detailed analysis of the earth's materials is critical to understanding our planet. More samples must be obtained on the surfaces of the continents, on the ocean floor, and from the depths of the crust. Because it is still difficult to penetrate the crust, new, indirect techniques—utilizing gravitational force, heat, and magnetism, for instance—may have to be developed.

magnetism

Much remains to be learned about how the magnetism of the earth has changed and, in particular, why the earth's magnetic field has reversed polarity so many times. In the matter of reversal, scientists have so far been reluctant even to make educated guesses. But more sensitive magnetometers should provide more information about the earth's changing magnetic field.

heat

Studies of the planet's heat content and heat flow will tell us more about the earth's creation and development. It is even possible that heat emitted by rising magma may eventually be used as a source of energy, just as geothermal steam geysers generate electricity at some places today.

international research

Because scientific advances today depend so much on large-scale collaboration, there will be greater emphasis on research that cuts across political and national boundaries, such as the work performed by the Apollo–Soyuz astronauts. Expeditions by the *Glomar Challenger* and Project FAMOUS have also set a pattern for future joint efforts.

computer processing

As scientists amass greater quantities of information, it becomes increasingly difficult to store, analyze, abstract, and distribute the data that contribute to a better understanding of the earth. But computer technology has grown so sophisticated that this task can be managed far more efficiently than ever before, with work performed in days that would once have taken years. Theories can now be readily tested and amended.

surveying and mapping

Since many regions of the earth are still little known and little explored, photographs made from aircraft and earth satellites represent a major advance in our study of the world's widely varying surfaces. Sonar and other navigation tools have simplified the chore of mapping ocean floors, and computers are handling the data more efficiently. These trends will continue and accelerate.

plate tectonics

SUMMARY

During the 1960s, explorations of the ocean bottom provided evidence to support Wegner's hypothesis of continental drift. It became clear that parts of the earth's lithosphere were indeed moving. These changes in the lithosphere are part of a system of tectonics.

The lithosphere is made up of about seven large plates and about twenty small ones. Their boundaries are defined by the tectonic activity at their edges. At some edges, called divergent boundaries, plates are being created by magma rising through ocean-floor rifts. At other places, plates are sliding downward and disappearing under other plates. These points are called convergent boundaries. Where plates push against each other mountain chains, such as the Andes, may be formed.

Trenches are formed where one plate slips beneath another. Earthquakes occur around these points, probably because of occasional large movements between the plates. Gravitational force in the depths of the trenches is less, indicating that the rocks beneath them are not very dense.

Movement along undersea transform fault boundaries is horizontal. It produces shallow-focus earthquakes. These quakes usually cause little damage, but they can be used to locate plate boundaries.

Most of the rifts at which magma rises run north and south. They are always located at the top of a sea-floor ridge. The magma itself consists of mantle material which has been melted.

During recent years personnel from many nations have cooperated in exploration of the ocean bottom. Submersibles have reached depths of as much as 9.5 kilometers. Sonar and other electronic equipment has been used to map the contours of the bottom. Special techniques have been developed for underwater photography, and cores of the ocean-bottom sediments have been brought to the surface for detailed laboratory examination.

Continuing movement of the lithospheric plates in the future will alter the earth's geographical features. Another abrupt reversal of the earth's polarity may occur as soon as 10,000 years from now. As the study of plate tectonics continues, a wide variety of careers will be opened to persons trained in the various fields of earth science.

plate tectonics

REVIEW QUESTIONS

1. Explain why Wegener's hypothesis needed revision even though his essential premise was correct.

2. How did early geologists explain the similar rock formations and fossils in widely-separated areas?

3. Describe the general structure of the lithosphere.

4. How can the boundaries of the lithospheric plates be determined?

5. Explain the relationship between ocean trenches and the formation of mountain chains on land.

6. Where does the magma originate?

7. Describe in as much detail as possible how crustal plates are formed and destroyed.

8. Distinguish between a convergent boundary and a divergent boundary.

9. What is the greatest depth so far reached by a submersible? Where did this occur?

10. What is a transform fault boundary?

11. Explain how the Himalayas were formed.

12. Describe the investigations carried out by Project FAMOUS.

13. What is magnetic striping?

14. Distinguish between a hydrophone and a transponder.

CHAPTER NINE

THE ROCK RECORD

Suppose you try a simple experiment. Take a sheet of paper and, relying solely on your own memory and impressions, draw a map of the largest community adjacent to where you live — preferably a city or big metropolitan area. Include as much detail as you can: major streets, expressways, airports, railroad terminals, parks, and other points of interest. Try to be very accurate, noting street and avenue names, and identifying specific areas such as business districts, theater districts, shopping centers, even individual buildings and shops. Let your memories work for you. Include on your map individual neighborhoods and communities. If you can think of nothing else, use orientation as reference points — downtown, or uptown, or crosstown. If orientation fails, then use site settings based on happenings in your own life — "Somewhere in here I was stuck in a traffic jam," for example.

When you have really exhausted your total memory storehouse in filling in the details, compare the results with your classmates. There will probably be a lot of similarities, and maybe even some identical spots, on the different memory maps. But there are also sure to be many gaps, distortions, and wide disparities. Some people will have visited one part of town and not another. Some will have special interests that alert them to some things and let them ignore others. "A bike shop on that corner? I walk past there at least once a week and I've never seen it." "You can't see this little garden from the street — I only know about it because I used to have some cousins who lived in the building." "There's a whole bunch of these old pre-Civil War houses in that neighborhood, but they're so run down that you can't recognize them unless you know what you're looking for." And after all the comparisons are made, there will undoubtedly remain some large blank spots on the master map — places no one in the group knows anything about.

the rock record

This was the situation faced at the beginning of the nineteenth century by researchers in the infant science of geology. They were certain that a record of the earth's history lay in the rock strata that make up the crust—but interpreting that record was like trying to read a book in which some pages are duplicated, some are out of sequence, and some are missing entirely.

the problem of chronology

One of the most challenging aspects of all scientific research is the task of establishing guidelines and developing working categories for the data under investigation. If basic agreement cannot be reached on these matters, it is almost impossible to establish rules for research. The early pioneers in geology responded to this problem in a variety of ways.

Plato taught us to search for the underlying truth below the surface appearance of things. Aristotle took us a step further. He demonstrated that unless we classify and categorize concepts so that we can see how they are related to one another, we will not be able to understand their entire significance. Francis Bacon made induction an important means of scientific inquiry, and thus set the scene for the amazing profusion of discoveries and breakthroughs which occurred during the seventeenth and eighteenth centuries.

the rock controversy

The "raw materials" of geology—the rocks themselves—had been literally lying around for millions of years. Throughout human history, there had been endless confusion and speculation about the weird and often fascinating things that could be found all over the face of the earth.

Scientific investigation often involves thinking backward—from a premise or hypothesis that has already been agreed upon. In most cases, the investigator searches for evidence to support the hypothesis, working backward from the original premise and gathering the data necessary to back it up. Thinking backward is a perfectly legitimate way of working on a problem; but it can be dangerous as well.

The origin of igneous rocks became the focal point for a lengthy and bitter controversy during the late eighteenth and early nineteenth centuries. From our discussion of igneous processes in chapter 2 and chapter 6, you will remember that it can be a very tricky proposition to explain which came first, lava or rocks. During the period between 1775 and 1825, many instances were discovered in which rocks such as granite seemed to have invaded newer formations while still in a molten condition. Rocks found in the oldest for-

the rock record

mations strongly resembled some of the materials that had been produced by recent volcanoes. Many different theories were put forward to explain geological origins and types of rocks, each theory with its own categories and labels. The picture became progressively more confusing. Obviously, a lot of people were thinking backward.

William Smith, an English surveyor, was thinking backward, too. But unlike his more scholarly contemporaries, Smith came from humble beginnings and was mostly self-taught. His inquiries and investigations were based on direct observation and logical reasoning, and were not strongly influenced by the current wrangling within the scientific community. In his work as a surveyor, he had opportunities to study the layers of rock that were exposed by the digging of an important industrial canal.

This model of the Grand Canyon clearly shows the many sedimentary strata through which the Colorado River has carved its path over millions of years. Study of these strata has provided a detailed picture of the region's geological history.

Wherever William Smith worked, he made diagrams of the sequence of the rocks and of the fossils he found in them. From his records, he made a kind of calendar. This was not a calendar of days, months, and years, but a calendar of older and younger fossils. Smith combined information gathered from many places — stone quarries, mines, and places where canals and roads cut through sedimentary rocks. When he finished, he had a long record of the sequence of sedimentary rocks and the fossils found in them.

Smith's calendar was arranged so that at the top of the list were the youngest rocks and fossils, and at the bottom, the oldest. He called the list a **geologic column.** Smith sent copies of his geologic column to many other geologists. They saw how useful it was, and built similar geologic columns for the regions they were studying. Slowly, a geologic column containing information from all over the world was assembled.

the geological column

the rock record

building a geologic column

In the accompanying figure you can see how a geologic column is constructed. At the left are cross sections of three hills, showing the various strata of rocks in each hill. Beside each stratum is a diagram of the principal kind of fossil found there. Hill A contains the youngest rocks. Hill C, at the bottom, contains the oldest rocks. The bottom stratum in Hill C is the oldest rock type found in any of the three hills.

The shale at the top of Hill A contains a large, curled fossil shell. Below it is a sandstone in which no fossils were found. Beneath the sandstone is a stratum of conglomerate, and below that is a fossil-bearing limestone. The bottom rock stratum in this hill is a sandstone.

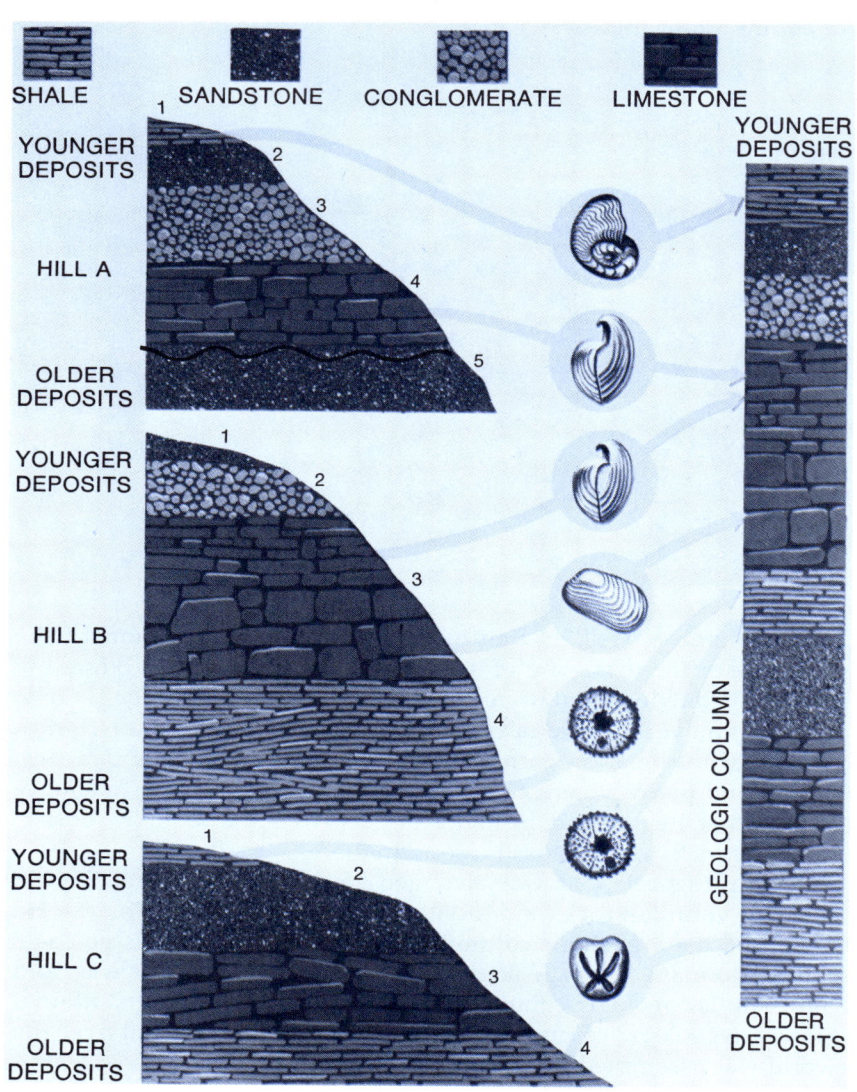

Notice that fossils are not found in every stratum. Limestones and shales usually contain many fossils. Few fossils are found in conglomerates, and these are often fragmented.

Now look at the diagram of Hill B. Notice that the top three strata repeat the same sequence of rock found in strata A-2, A-3, and A-4, in Hill A. Notice also that the limestone strata A-4, in Hill A, and B-3, in Hill B, contain the same kind of fossil shell, which can be used as an index fossil. Because of this, we correlate these two strata. This means that we consider them to be the same age.

In a similar way, the various strata in Hill B are correlated with those in Hill C. When all the information from the three hills is put together, we have the geologic column seen at the right side of the figure. From this we can develop the geologic history of the region represented by the three hills. Other geologists can use the same column to find the relative ages of strata in other parts of the world.

There is an unsolved problem in this example. Where do we place the sandstone stratum A-5? Since limestones A-4 and B-3 appear to be the same age, sandstone A-5, which is older than limestone A-4, would have to be older and thus below limestone B-3. But there are no sandstones deep down in Hill B. There is one near the top of Hill C. Perhaps sandstones A-5 and C-2 are about the same age. However, we cannot be certain that this is true. If there were fossils in these sandstones, that would help solve the problem.

Notice that the geologist who studied Hill A used an irregular line between A-4 and A-5. This means that he thought the limestone had been laid down on a sandstone that had previously been partly worn away. An unknown amount of rock from the top of the sequence below A-4 was removed before the limestone layer was formed. Let us see what may have happened.

sedimentation interrupted

How could sandstone A-5 have been eroded before limestone A-4 was deposited? The best explanation for such a condition is that after the sandstone had formed, it was uplifted out of the sea and worn away by surface and ground water. Then the seas covered the land again, and the material that formed the limestone was deposited on the irregular surface. Thus, the sediments that formed the rocks in Hill A were not laid down continuously, one upon the other. The period of sedimentation was interrupted. There may have been other layers between A-4 and A-5 that also were removed.

Geologists call such a situation an **unconformity**. It means, in this case, that an unknown period of time elapsed between the times the sediments for sandstone A-5 and limestone A-4 were deposited. The best we can say is that A-5 and C-2 may be the same age, but such a statement is only a guess.

naming the rocks

Smith's geologic column attracted the interest of two other Englishmen. One of them was Adam Sedgwick, a minister who taught at a college in Cambridge. The other was Roderick Murchison, a retired army officer. Their common interest in the study of rocks made them friends.

One summer, in 1835, Sedgwick and Murchison decided to

the rock record

map part of Wales. Sedgwick went to the northern part of Wales and Murchison to the southern part. Murchison found that many of the fossils in Wales were different from those in central England.

Murchison had found his rocks in the area that had once been inhabited by an ancient British tribe called the Silures. So he called them *Silurian* rocks. He urged Sedgwick to call his rocks *Cambrian*, from the ancient name for Wales. We use these names today for rocks all over the world that are the same age as the two sets found in Wales.

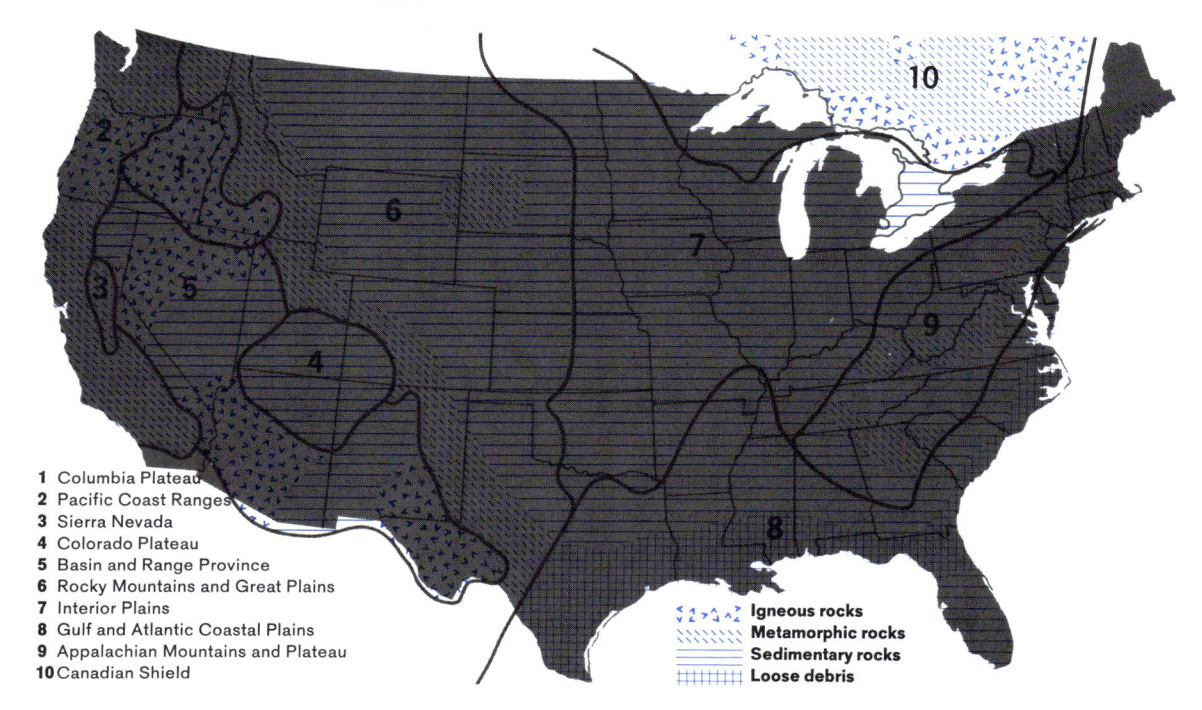

1 Columbia Plateau
2 Pacific Coast Ranges
3 Sierra Nevada
4 Colorado Plateau
5 Basin and Range Province
6 Rocky Mountains and Great Plains
7 Interior Plains
8 Gulf and Atlantic Coastal Plains
9 Appalachian Mountains and Plateau
10 Canadian Shield

Igneous rocks
Metamorphic rocks
Sedimentary rocks
Loose debris

The map above shows the general distribution of surface rocks in the continental United States.

Studies of these two sequences of rocks and fossils revealed that the fossils at the top of the Cambrian sequence were just about the same as those at the bottom of the Silurian sequence. This suggested that the Cambrian rocks were the older of the two. At about the same time, other sequences rich in fossils were being found in England and other parts of Europe. None of these fossil-rich sequences contained fossils older than the ones found in Cambrian rocks.

The man most responsible for the geologic column we use today was a Scottish lawyer named Charles Lyell. Lyell was an avid reader and an acute observer, and he traveled widely during his long life. His keen scientific curiosity, international perspective, and geological bent enabled him to see the need for a standard system of identifying rocks.

Lyell spent most of his life organizing what other geologists

the rock record

had discovered. In 1830 he published the first modern textbook on geology, a book which had great impact on Charles Darwin's thinking about evolution. By 1872, Lyell had developed a geologic column much like the one we use today, applying one name to rocks of the same age from all parts of the world. These names separated rock layers into major time periods according to their age.

fossil evidence

The fern fronds above were preserved as carbon prints, in which every detail of the plants can be seen.

The problem of classifying and categorizing fossils is particularly challenging. It has become increasingly evident that the history of life on earth extends very far back in time. In 1953, the 65,000,000-year-old fossil of a flowering plant was discovered in Colorado, and many much older fossils have also been found.

how fossils are formed

How are fossils formed? There is no single, simple answer to this question. In all cases, the remains of a plant or an animal are buried in sediments. Then any of several things may happen. In the case of recent fossil shells, after the soft tissue of the animal has decayed and sediments have filled all the cavities, the actual shells are often preserved practically unchanged.

Sometimes, after the soft parts of the animal have decayed, and the sediments have hardened into rock around the hard parts, the hard parts either decay or are removed through chemical action. This removal leaves a **mold** of the original hard parts. Water containing dissolved silica, calcium carbonate, or some other mineral may seep into the mold, and the minerals may be deposited there when

the rock record

the water evaporates. Such an event will produce a **cast** of the original parts.

Petrified wood is a common kind of plant fossil. When the original piece of wood is buried, mineral-bearing water fills the open spaces inside the cell walls of the wood. Gradually the cells become filled with silica or some other mineral. Slowly the woody cell walls decay, and the spaces that are left also become filled with a mineral deposit. When either animal or plant material is wholly replaced by a mineral, the fossil is a stony cast of the original.

Plant material that is buried in mud containing little or no oxygen decays only partially, and much of the carbon from the cellulose of the plant remains. This kind of fossil is a **carbon print** of leaves or stems. Under similar conditions, carbon prints of soft animal tissues may be formed. Some plant saps, when buried in sediments, harden and become a substance we call **amber.** Insects that were trapped in the sticky plant sap are sometimes found in amber.

Perhaps the most interesting fossils are **microfossils**—fossils so tiny that you rarely notice them in rocks. These are the remains of microscopic organisms that had hard parts. Among this kind of fossils are the tests, or shells, of single-celled organisms. The ooze that forms at the bottom of the deepest parts of the sea is often composed largely of tests.

Microfossils are difficult to study. Thus far, we have discovered ways of extracting them from shales and certain limestones. For other kinds of rocks it is necessary to make very thin, polished chips of the rocks and study these under the microscope.

Trapped millions of years ago in sticky plant sap which later turned to amber, this ant is perfectly preserved today.

where fossils are found

Fossils are found only in sedimentary rocks. The conditions under which igneous and metamorphic rocks are formed prevent the preservation of fossils, since both heat and pressure will quickly destroy the fragile remains of plants and animals.

Deep-sea sediments are composed mainly of microfossils. The deep parts of the oceans contain little life as compared with the shallow parts, because there is little food and light available. Thus, few recognizable remains of large animals are preserved in deep-sea sediments.

Sedimentary rocks that are formed from deposits made in lake beds sometimes contain the fossils of land animals and plants. Throughout the past 350 million years, a great variety of animals have roamed the continents. What we know about these we have learned from fossils.

When an animal dies on the land, scavengers usually eat it. However, the body of an animal that has drowned in a river sometimes becomes buried in mud or sand. Then there is a good chance that the bones will become fossils. For this to happen, the sediments in which the bones are buried must not be disturbed by violent earth

OUR DWINDLING FOSSIL FUELS

The coal, petroleum, and natural gas that feed the voracious industries of the developed countries were formed by natural processes over hundreds of millions of years. Yet nearly all of these fuels have been gobbled up in the space of one or two human lifetimes. Indeed, fully fifty percent of the coal ever mined from the earth was devoured between 1940 and 1969. Similarly, half of the world's known supply of petroleum has been used up since 1959.

Coal, petroleum, and natural gas are considered non-renewable resources, not only because they are destroyed when first used but because their formation takes millions of years. They are also called fossil fuels because they have been formed from the remains of plants and animals that lived eons ago. Coal is formed from the remains of land plants, petroleum and natural gas from the remains of aquatic plants and animals. After these organisms died, they were buried under layer upon layer of sediments. As millenia passed, more and more sediments accumulated. The organic materials were transformed into coal, oil, and gas by pressure and heat in the earth's interior.

Even with the knowledge that these fuels are quickly being eaten up, industrialized countries continue to be profligate in their use of them. Energy experts predict that if the world's current energy consumption continues to double every ten years, as it is now doing, ninety percent of the earth's petroleum and natural gas will be exhausted by the year 2030, and ninety percent of the coal will be gone by 2400. Sobered by these figures, scientists and government officials are looking for other sources of energy. At present, nuclear fuels—radioactive isotopes of uranium, for example—appear to be the best bet for the future, but even these are associated with serious environmental and political problems.

the rock record

movements or stream action. Microorganisms that decay bone must also be absent. Such conditions are not common; that is why we find land fossils so infrequently.

dating the earth

As the nineteenth century opened, a lot of theories seemed to be falling into place. Hutton's idea of uniformitarianism represented a whole new way of thinking and looking at things. It meant predictability. Smith's geologic column demonstrated orderly time sequence. Lyell contributed nothing new of his own, but he filled the role which was to become the trademark of the new age: he was the master organizer. The rush of scientific discovery was pushing out a brand-new leading edge of development called technology. Anything seemed possible, if only the right formula could be found.

This horseshoe crab was caught and buried when a mud bank collapsed. Eventually, sediments accumulated and formed into rocks, in which the fossil of the crab was preserved. The horseshoe crab still exists today in essentially unchanged form, a living relic of the past.

calling cards from the past

Geologists in the eighteenth and nineteenth centuries began finding fossils in strange places. Many widely separated formations were found to contain similar fossils. These formations, then, must have been created at approximately the same time. The fossil "graffiti" had much to tell them about their own British Isles. They discovered Cambrian rocks in Wales, the Western Midlands, and also

the rock record

in the extreme northwest of Scotland and the Outer Hebrides. At the latter locations they observed major unconformities. In general, the British Isles are like a stack of dominoes which has fallen over from north and west to south and east. The oldest rocks are found to the north and west; the newest to the south and east.

Fossils tell the geologist, "Hey—I was here!" and allow him to chart an area's history in stone. For example, Hillside Avenue in Jamaica, Long Island, parallels the deepest southward advance of the last glacier about 25,000 years ago. The shrunken remains of this glacier now cover Greenland and higher elevations of Baffin Island. At the intersection of 169th Street and Hillside the terrain is some 17 to 23 meters higher along the north side. To the south, the ground smooths out in its familiar flat, sandy Long Island topography and runs off to the south shore beaches and the Atlantic Ocean.

How can we date a formation such as this one with such a high degree of accuracy? An estimate can be made from the rate of erosion and deposition of sedimentary rocks, but this is not really precise. However, there is a better way. As a glacier shrinks, its melting waters form a temporary lake at the retreating edge of the ice. Sediment carried by the flowing meltwater is deposited on the lake bottom in layers—fine, dark clay in the winter when the water is quietest, and lighter-colored, somewhat heavier silt in the spring and summer. This double-layered accumulation of sediment is known as a **varve,** and each varve represents one year. Since the lake moves northward with the retreating glacier, each year's varve only partly covers the previous one, so the total effect is rather like slices of ham arranged on a platter. A count of the total number of these overlapping varves between a glacier's present location and its ancient furthest limits tell us the number of years since it began to retreat. Between Hillside Avenue and the edge of the Greenland glacier, scientists have counted 25,000 varves; hence we can assume that the glacier was in Jamaica 25,000 years ago.

nuclear dating

An atom of the common isotope of uranium, U-238, changes to an isotope of lead, Pb-206. However, the change does not occur in all atoms at the same time. For a large number of uranium atoms, such as might be found in a large piece of rock, the change from uranium to lead appears to take place slowly.

The change, or **decay,** to lead occurs because the characteristics of the uranium atom cause it to lose both nuclear particles and outer electrons. The loss of these particles is called **natural radioactive decay.** The atom decays until its new properties cause it to become stable. At this time, decay stops. So the nucleus of a U-238 atom undergoes natural decay until it becomes a stable Pb-206 nucleus, as shown in the accompanying figure.

Nuclear fission, which we discussed in chapter 1, is a form of artificial radioactive decay. In this case, all the material undergoes

the rock record

ROCK CONTAINING U-238

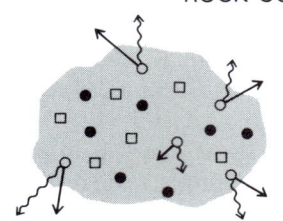

KEY
- URANIUM ATOMS
- URANIUM ATOMS IN PROCESS OF DECAY
- LEAD ATOMS

The lead content of a uranium-bearing rock increases over time, as U-238 atoms decay to stable Pb.

decay in fractions of seconds. The rapid release of the energy that held the nucleus together is what causes the destructive explosion. The first atomic bomb explosion was caused by the fission of a uranium isotope, U-235.

The rate of radioactive decay depends on chance. In about every 4.5 billion (4.5×10) years, there is a fifty-fifty chance that a particular U-238 atom will decay. Once the decay begins, it proceeds rapidly. In a relatively short time, the U-238 nucleus is reduced to a mass of 206, the mass of the lead isotope.

Because there is a fifty-fifty chance that decay will occur, scientists can measure the age of rocks by the amount of decay that has already taken place. Imagine a mineral that contains uranium as part of its chemical makeup. At any time, some of the atoms are changing to lead, as shown in the accompanying figure. In the case of U-238, about one half of the original atoms will have decayed to Pb-206 in approximately 4.5×10^9 years. This length of time is called the **half-life** of U-238.

Scientists can use this knowledge to date mineral specimens that contain uranium. When the minerals were first formed, they contained no Pb-206 — only U-238. As the crystals aged, some of the uranium changed to lead. By carefully measuring the amount of Pb-206 and U-238 in a specimen, it is possible to calculate how long ago the mineral contained no Pb-206 — that is, when it was formed. For instance, if about 50 percent of the uranium has changed to lead, the mineral is about 4.5 billion years old.

Certain Precambrian granites in the Rocky Mountains have been dated this way and found to be about 1.2 billion years old. This means that much less than one-half of the original U-238 has decayed to lead. Among the oldest rocks dated with the aid of uranium are some igneous rocks from Rhodesia, in southern Africa. These are about 3.8 billion years old.

other radioactive materials

Minerals containing uranium in measurable quantities are found in few rocks. This means that the uranium–lead method cannot be used very often. There is, however, another method that is useful for dating rocks. Radioactive potassium is found in many igneous and metamorphic rocks. The chemist's symbol for potassium is K. The important radioactive potassium isotope is K-40. K-40 slowly changes to argon-40 (Ar-40). The half-life of K-40 is 1.35×10^9

the rock record

years, only one-third as long as the half-life of U-238. Therefore, the potassium–argon method can be used on younger rocks more accurately than the uranium–lead method can. Many crystalline rocks have been dated using the potassium–argon method. Some are as young as a million years old. But none of the radioactive dating techniques are easy to perform. All require special laboratory equipment and very skillful chemists and physicists.

Several other radioactive elements are used for measuring the ages of rocks. It is often possible to use two radioactive-decay methods with a single sample. When this is done, and the two ages thus determined are reasonably close, we are more confident about the accuracy of the date.

In any case, the results of radioactive dating are much more accurate than any of the guesses previously made. The basic assumption of this method is that the rate of radioactive decay of an isotope is constant. The method does not give us an age for the earth, but it does give us a good estimate of the age of the earth's thin crust, which we can see and measure.

In addition to the chemical and physical problems involved in measuring the age of a rock, the geologic situation must be understood. For example, metamorphism affects many age determinations. The determination of the age of the gneiss found in the vicinity of Baltimore, Maryland, is a good example. The uranium–lead method gives an estimate of about 1.1 billion years. The potassium–argon method gives us an estimate of only 300 million years.

The reason for this great difference is that the minerals used for the uranium–lead method are affected little, or not at all, by metamorphism. Those used for the potassium–argon method are strongly affected by metamorphism. Geologists believe the Baltimore gneiss was formed from igneous rock that crystallized about 1.1 billion years ago. Then, about 300 million years ago, the rock was changed into gneiss by the metamorphism that occurred when the Appalachian Mountains developed. This seems to explain the difference.

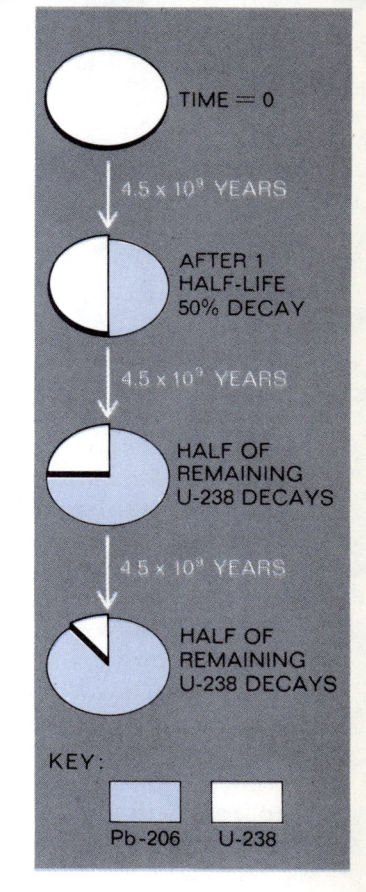

TIME = 0

4.5×10^9 YEARS

AFTER 1 HALF-LIFE 50% DECAY

4.5×10^9 YEARS

HALF OF REMAINING U-238 DECAYS

4.5×10^9 YEARS

HALF OF REMAINING U-238 DECAYS

KEY:

Pb-206   U-238

In this explanation of half-life, the time interval between each decay is 4.5 billion years.

radiocarbon dating

All organisms contain some radioactive K-40. In addition, one of the substances of which living things are composed is carbon. Most carbon consists of the common isotope, C-12, but a tiny fraction of the carbon in living tissue is radioactive. It is called **radiocarbon**, or **carbon-14** (C-14). The half-life of carbon-14 is only about 5,800 years.

Living things acquire both C-12 and C-14 through the food they make or eat. They gain amounts of radiocarbon in proportion to their weight. The ratio of C-14 to C-12 remains the same in an organism's body as it is in the earth's atmosphere.

When plants and animals die, they no longer gain new radiocarbon, and the radiocarbon they already contain continues to de-

IN FOCUS:
reconstructing the past from human coprolites

Scientists are a resourceful lot. They often resort to techniques that members of other disciplines often consider highly unusual, if not downright bizarre. For example, the study of seismic waves generated by earthquakes has enabled earth scientists to analyze the structure of the earth. Photographs taken from artificial satellites have helped pinpoint the location of valuable resources. And analyses of earth materials have aided in crime detection. Now, from archaeology—a sister science of geology—comes yet another extraordinary technique. This involves analysis of fossilized human feces, or coprolites (from the Greek *copros*, "dung"; and *lithos*, "stone").

Vaughn Bryant and Glenna Williams-Dean have unearthed human coprolites that date back as far as 300,000 years. Examination of their contents has provided highly detailed information about the habits of late prehistoric farmers and food gatherers in both the Old and New Worlds. Not only can coprolites tell us what the people ate, but they can also shed light on the ways they prepared food, the environment in which they lived, and how they made their living from it. Most surprisingly, these strange fossils can also give us clues to the health of the ancient people.

An investigator coming upon coprolites at an archaeological site cannot tell at a glance whether their origin is human or non-human. But when they are placed in a trisodium phosphate solution for about three days, the specimens betray their source by the ways they alter the fluid. A human coprolite, for example, renders the solution opaque and turns it dark or black, whereas specimens from other animals affect the fluid differently.

Another test involves passing the specimen through screens of successively smaller mesh, using distilled water to facilitate the process. The solid residue is removed and the liquid portion drained to gather any solid materials it may include. Both residues are then examined under a microscope. If the residue contains a large variety of materials—charcoal from cooking, plant fibers; pollen, seeds, hairs, pieces of bone, fish and reptile scales, nut and egg shells, and feathers, for example—the investigator can be almost certain that the specimen is from a human. Very few other animals are as omnivorous as we are.

The ways the prehistoric people prepared their food can be determined by examining the seeds of plants, whose tough outer jackets pass through the digestive system virtually unchanged. Scientists can identify the plant family and sometimes even the species of such seeds. Analysis of the millet found in some Mexican samples showed that the jackets of the grain had been crushed, which meant that it had been pounded. Other millet samples had been split, indicating that the grain had been prepared by rolling it back and forth on a hard surface, probably a flat stone. This method is still used today in some non-industrialized societies. Sometimes the plant seeds had not been processed at all, and were apparently eaten whole.

Analysis of seeds found in coprolites also reveals that people in many parts of the New World ate foods such as tomatoes, grapes, chili peppers, and squashes. Similarly, by examining the feather fragments in the coprolites at one Nevada site, investigators were able to identify the waterfowl species that were included in the diet of the occupants. Among the birds consumed were herons, geese, and mud hens. The inhabitants of a site in Texas, by contrast, were particularly fond of grasshoppers, rodents, and lizards.

Occasionally, parasites are detected in coprolites, enabling investigators to infer the health of the people. For example, some Peruvian specimens, between 3,000 and 5,000 years old, were found to contain tapeworm eggs. Some 10,000-year-old coprolites from a site in Utah included the eggs of the parasitic thorny-headed worm.

Bryant and Williams-Dean's own study of the pollen in coprolites at a Texas shelter has given us a picture of the ancient environment of the site. The shelter consists of eight strata, occupied at various times between 500 B.C. and 800 A.D. Windborne pollen found in the specimens indicates that hackberry, oak, and pine, as well as several types of flowering plants and seed-bearing grasses, grew in the area. Analysis of the pollen in specimens from all eight strata showed that the shelter was occupied only during the late spring and early summer by groups of hunter-gatherers. Undoubtedly, the same groups stayed at other locations at other times of the year, an indication that they had worked out some kind of seasonal round which depended on the availability of certain plants and animals. This evidence is important because it suggests that the inhabitants did not simply wander from area to area and depend solely on chance to find food. Rather, they were ingenious amateur botanists who knew when and where certain plants would be ripe for the picking. In short, our ancient ancestors were experts at wresting their livelihood from a bleak, arid environment.

Coprolite analysis is a growing discipline. It may eventually solve some of the mysteries of our prehistoric past, including when and why humans made the transition from nomadic food gathering to growing their own food. This development, perhaps more than any other, played a key role in the rise of civilization.

the rock record

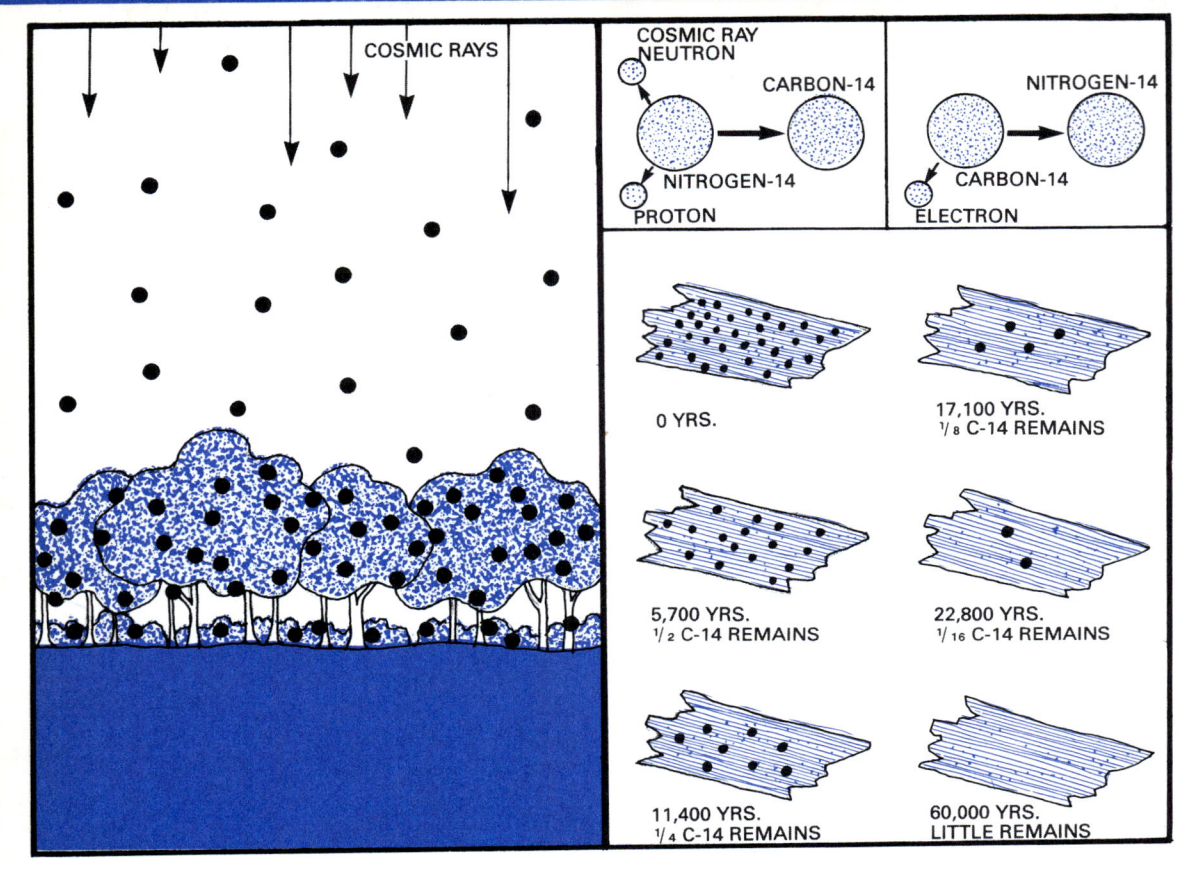

Cosmic rays enter the atmosphere and change nitrogen-14 to carbon-14. Since carbon-14 has a comparatively short half-life, it can only be used for dating of events within the last 70,000 years.

cay. By comparing the amounts of C-12 and C-14 in the actual remains of living things, we can estimate their ages. Because of the short half-life of C-14, radiocarbon dating is useful only for geologically recent events. It is now possible to estimate ages as great as 70,000 years by this method.

One of the events geologists have studied closely is the movement of the great ice sheets that covered much of the Northern Hemisphere in the recent past. In North America the land from the Ohio and Missouri rivers northward was at one time buried under thousands of feet of ice. The southward movement of this mass of ice buried parts of trees under great piles of rock debris deposited by the ice. When the remains of these trees are studied for the radiocarbon they contain, we get an estimate of when they were buried. On the basis of such evidence, it appears that the ice was advancing in the northern part of our Midwest as recently as 11,000 years ago.

absolute and relative dating

The dating of the beginning of the Cenozoic Era is an example of how geologists estimate the age of a particular era or period. In Gilpin County, Colorado, there is a deep mass of igneous rock that contains radioactive minerals. This mass lies below a layer of Creta-

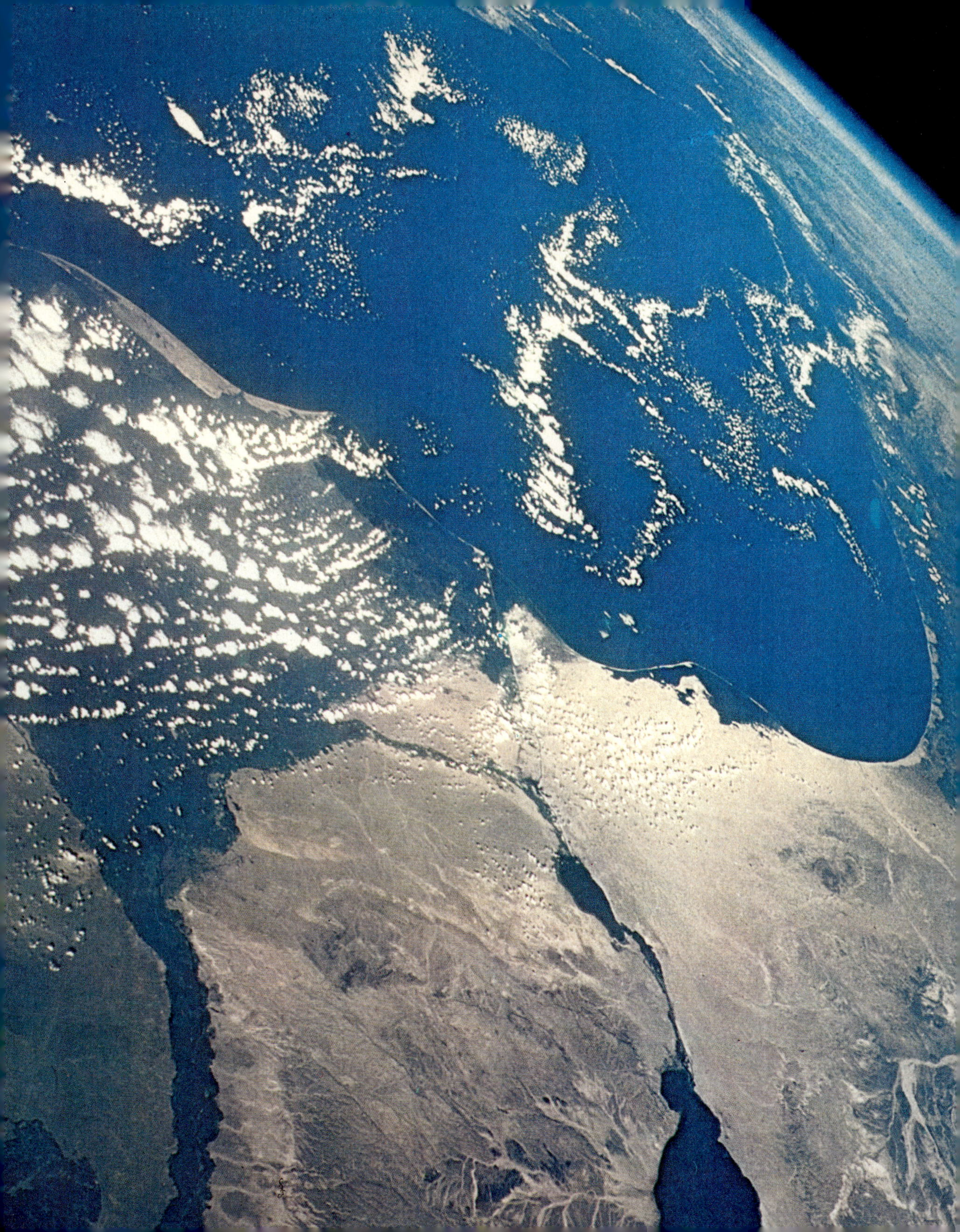

METEOROLOGY

Without the thin envelope of gases that surrounds our planet, life could not exist on the surface of the earth. At the same time, our atmosphere seems deliberately designed to test our endurance. The highest temperature ever recorded on the earth's surface was 135°F, in Libya. Temperatures in Antarctica drop so low that, for all practical purposes, the continent is uninhabitable. Even in more moderate climates, we are subject to teeth-rattling thunderstorms, pellets of ice falling from the sky, swirling snowstorms, and a variety of winds ranging from light, cooling breezes to the fury of hurricanes. But how dull our planet would be without its weather! And with all its faults, our atmosphere is still responsible for many of the beauties of the earth.

Our great oceans greatly affect the earth's climate. Water evaporates from the oceans; it is returned to them in the form of rain or snow. In this photograph we see a fog, frequently generated where land meets cold ocean currents.
(Preceding page) During the past two decades photographs from orbiting satellites have proved particularly useful for weather prediction. This spectacular view from space shows the Nile River and its prominent delta.

A young hurricane can be seen in this satellite photograph. The swirling pattern of the clouds clearly shows the direction of winds spiraling in toward the central eye of the storm, where winds drop to a dead calm and the barometric pressure may fall to several inches below normal. Identification and prediction of coming hurricanes is one of the most valuable achievements of meteorology.

A wedge-shaped cold front is advancing under a layer of warm, moist air. As the warm air is forced upward, a layer of nimbus clouds forms just above the cold front. The meeting of these two air masses will soon bring rain.

The energy of the sun is the basic force that moves the atmosphere. Solar energy evaporates water, causes winds to blow, heats the air and the surface of the earth, and enables plants to carry out the process of photosynthesis. Life on earth could never have arisen if no solar energy had been available.

In this aerial photograph, the green "rain belt" of the Sierra Nevada Mountains in California is contrasted sharply with the brown and comparatively barren Mojave Desert in the background. The Mojave is one of the emptiest and most inhospitable regions in the United States. Because of the intense heat and very low humidity, only plants adapted to store water, such as cactuses, can survive here.

The region around Stone Mountain, Georgia, displays temperature and humidity variations typical of a temperate climate. Here rainfall is plentiful, and the landscape is green and inviting. Stone Mountain is a monadnock; it protrudes from an almost level peneplain, the result of erosion over immense periods of time. Some day in the future the land around Stone Mountain may once again be uplifted to form a new mountain range.

With rare exceptions, the aurora borealis can be seen only in the cold northern regions above the Arctic Circle. Light from the aurora is often reflected off the snow-covered ground to make the Arctic night an unforgettable sight. Auroral displays seem to coincide with periods of intense sunspot activity, and are associated with problems in radio transmission.

Thunderstorms often occur in clusters or lines, but sometimes a single isolated storm may form within a few minutes, release a heavy downpour, and then dissipate as quickly as it formed. Such isolated storms can often be observed over oceans, plains, or desert regions. Here we see the rain falling from a late afternoon thunderstorm in a semi-tropical region of the Atlantic Ocean.

Death Valley, nearly 100 meters below sea level, is the hottest, driest spot in the United States. Even wells dug in this region often bring up water so heavily contaminated with various toxic salts that it can be dangerous or even fatal to drink.

This formation of sedimentary rocks in Arizona contains layered deposits of coal and sandstone, which allow us to trace the climatic history of the region. At the time the coal strata were laid down, the climate must have been semi-tropical and the ground swampy. The red color of the sandstone strata tells us that they originated in a hot, arid climate.

the rock record

ceous sedimentary rock that lies, in turn, below a layer of Tertiary sedimentary rock. In this particular area, the igneous rock has pushed upward through a fracture in the Cretaceous rock, but not through the Tertiary rock above it. The age of the intruding igneous rock is therefore between that of the Tertiary and the Cretaceous layers.

Using the principles of superposition and uniformitarianism, geologists have established three dates for these rocks. The first is a relative date: the Tertiary Period occurred after the Cretaceous. The second date is also relative: the crystalline rock was formed after the Cretaceous sediments but before the Tertiary. The third date is an absolute age: by means of radioactive dating techniques, the minerals in the crystalline rock were found to be about 60 million years old.

Geologists concluded that the Mesozoic Era ended and the Cenozoic era began at least 60 million years ago. But since we know that this particular Tertiary sedimentary rock is not the oldest Tertiary rock that has been found, we have added 10 million years to the age of the Tertiary Period. Thus, the Cenozoic Era is thought to have begun approximately 70 million years ago.

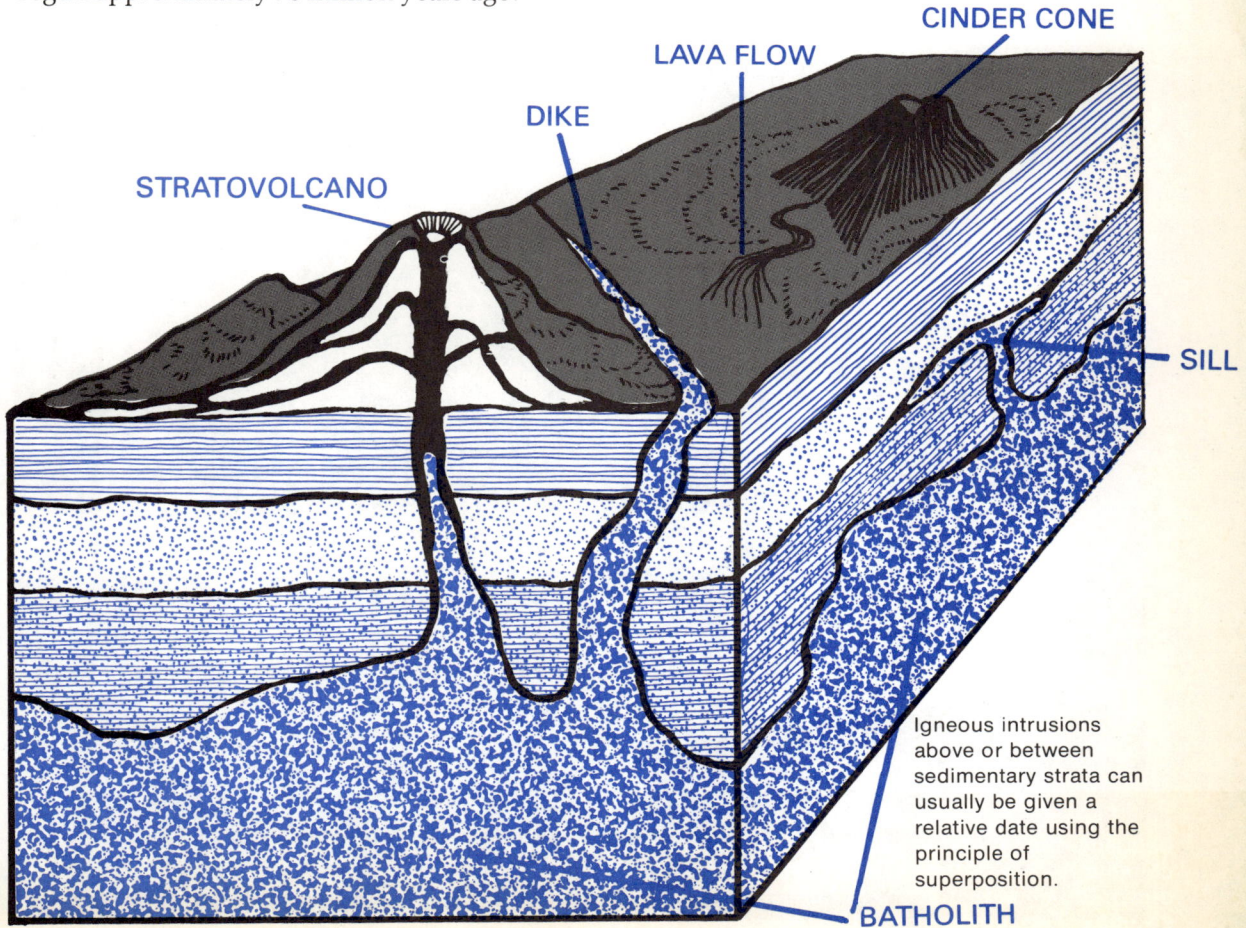

Igneous intrusions above or between sedimentary strata can usually be given a relative date using the principle of superposition.

the rock record

the age of the earth

In chapter 1 we discussed the method used by Archbishop Ussher to arrive at the age of the earth. As we saw, he concluded that the earth was created in 4004 B.C. Today his pronouncement seems misinformed.

The first really scientific attempt to measure the age of the earth was made around 1770 by the French naturalist Georges Buffon. Buffon came up with a figure of 75,000 years. Modern scientists who compare the ages of rock formations from all over the world have done much to establish a more or less complete geologic time scale. But there are still problems. The earliest period for which we have detailed dates is the Cambrian, which began only about 600 million years ago — less than 15 percent of the current best estimates of the earth's age.

This fossil cichlid was found in Wyoming. It bears a remarkable resemblance to present-day cichlids, which are commonly known as sticklebacks.

the rock record

There is no detailed classification of Precambrian rocks. Radioactive dating techniques are not of much help, and the age of a mineral in a sedimentary rock can only tell us that the rock is no older than the mineral. According to one radio-dating method that measures the rate of decay of rubidium into strontium, the oldest reliably dated rocks are some gneisses from Greenland, which appear to be about 3.8 billion years old.

Recently, an English scientist, Arthur Holmes, carefully studied information from all over the world about the ages of rocks determined by radioactive dating. From this evidence he calculated that the thin crust of rocks on the surface of the earth cannot be older than 4.5 billion (4.5×10^9) years.

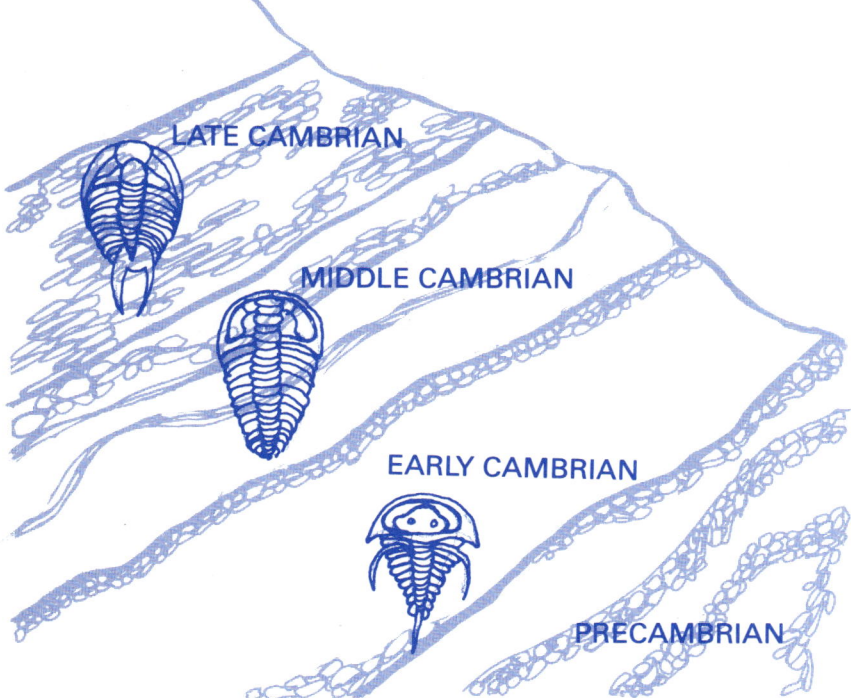

Trilobites were an important and widespread form of Cambrian marine life. Here we can see the evolutionary changes that took place over a period of approximately 100 million years.

This is close to the estimate of most astronomers that the earth is about 4.6 billion years old. Some astronomers arrived at this estimate by observing the speed at which distant parts of the observable universe are traveling away from us. This method is based on the theory that in the beginning all the matter in the universe was concentrated at one place, and has since scattered. Most astronomers now believe it has taken about 20 billion years for this expansion to occur. We will discuss this theory in detail in chapter 16.

The astronomers used one method, and Holmes used an entirely different method. But since both methods give about the same results, it appears that we can reasonably assume that the earth is about 4.6 billion years old.

the rock record

SUMMARY

Early geologists were faced with a rock record that was full of seeming discrepancies. Investigators formulated a variety of hypotheses about the sequence of rock formations. But the overall picture became very confusing because each researcher was using different categories and labels.

William Smith was among the first to sort out the chronological sequence of rock formations. He constructed a geologic column showing the sequence of sedimentary rocks and fossils. Adam Sedgwick and Roderick Murchison made similar studies of Cambrian and Silurian rocks. Finally, Charles Lyell organized and classified the information uncovered by geologists. He developed a geologic column much like the one we use today.

Fossils are formed when the remains of organisms are buried and preserved in sediments. They may exist in nearly unchanged form, or in the form of molds or casts. Plant fossils which have decayed only partially may exist as carbon prints. Fossils probably occur in all kinds of sedimentary rocks, but are most common in those which were deposited in comparatively shallow water. Deep-sea sediments are composed mainly of microfossils.

Nuclear dating of rocks or fossils is based on the rate of radioactive decay of one element into another. The length of time required for half the atoms of an element to decay is known as the element's half-life. Radioactive dating may be based on uranium–lead decay, potassium–argon decay, or the decay of rubidium to strontium.

Dating of organic materials can also be based on the amount of C-14 or radiocarbon present. This form of dating is useful only for comparatively recent geological events. Radiocarbon dating of trees preserved within glacial ice indicates that our most recent ice age was about 11,000 years ago.

Buffon made the first scientific attempt to determine the age of the earth, arriving at a figure of 75,000 years. Rubidium–strontium dating, however, indicates that some rocks in Greenland are about 3.8 million years old. Recent studies by geologists seem to agree with those of astronomers, who estimate that the earth is about 4.6 billion years old.

the rock record

REVIEW QUESTIONS

1. Explain why so many scientific breakthroughs were made during the seventeenth and eighteenth centuries.

2. What were Charles Lyell's contributions to modern geology?

3. Describe at least three ways in which fossils are formed.

4. How is radiocarbon dating performed?

5. How did William Smith's investigations differ from those of most scientists of the time?

6. Explain why fossils are found only in sedimentary rocks.

7. What is uniformitarianism?

8. Explain the principles behind nuclear dating.

9. What is a geologic column? What does it tell us?

10. Explain how geologists and astronomers have both arrived at the same estimate of the earth's age.

11. What is an unconformity?

12. Describe the investigations made by Sedgwick and Murchison.

13. What do we call the oldest fossil-containing rocks?

14. Distinguish between absolute and relative dating.

15. What is the main component of deep-sea sediments?

16. Describe the conditions necessary for the formation of a carbon print.

17. What is an isotope?

18. Explain why Precambrian rocks cannot be classified in detail.

19. What is a varve?

20. Who made the first scientific attempt to estimate the age of the earth? What was his conclusion?

HISTORY OF THE EARTH

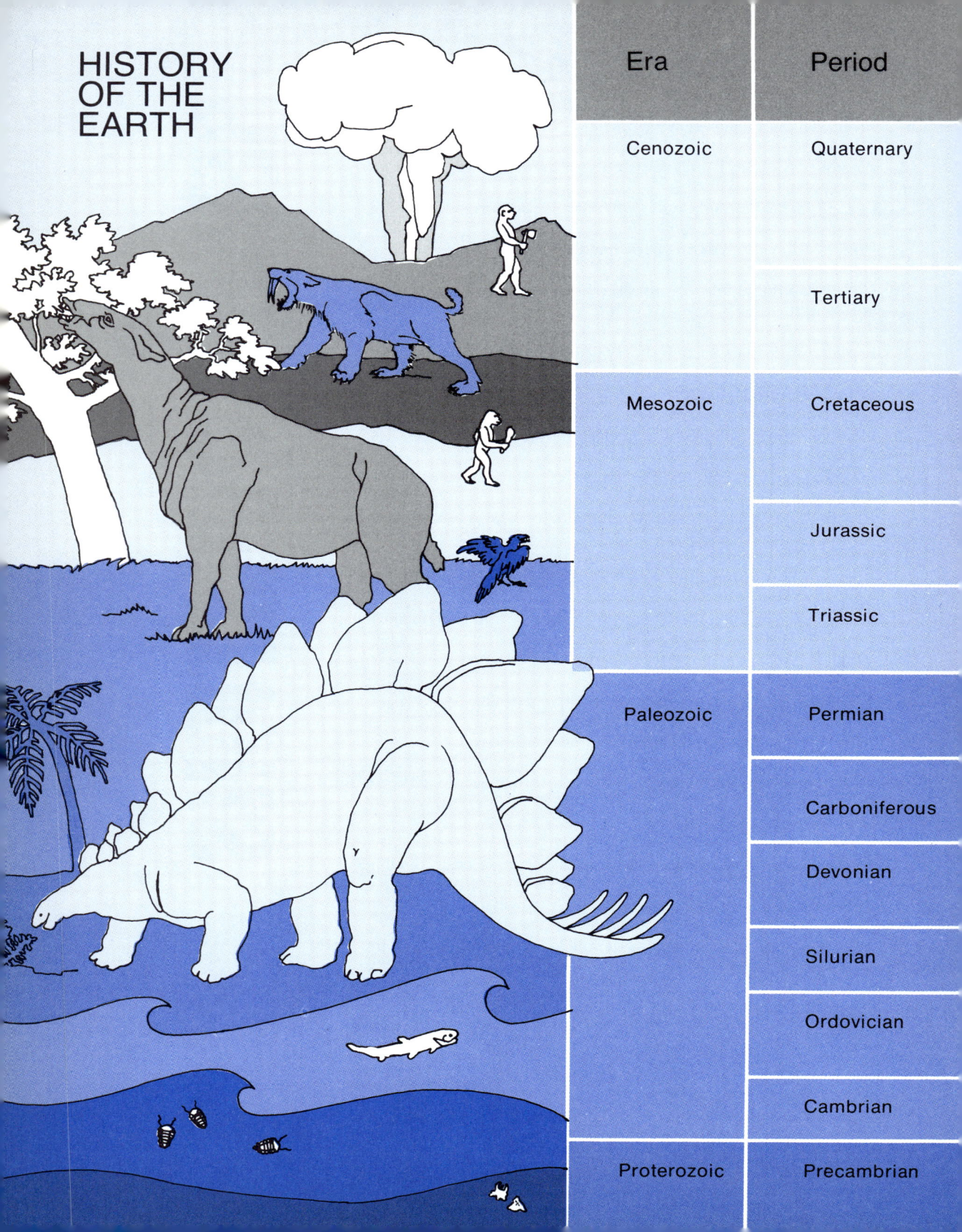

Era	Period
Cenozoic	Quaternary
	Tertiary
Mesozoic	Cretaceous
	Jurassic
	Triassic
Paleozoic	Permian
	Carboniferous
	Devonian
	Silurian
	Ordovician
	Cambrian
Proterozoic	Precambrian

Major Geological Events	Major Biological Events	Millions of Years Ago
Continued series of ice ages; climate stabilizes toward end of period	First proto-humans; modern birds and mammals; Ramapithecus, Australopithecines, Neanderthals; Homo sapiens toward end of period	7
Climates fluctuate; general cooling of earth toward end of period	Modern birds and mammals; first primates toward end of period	70
Alternating periods of cold and drought; North American volcanic activity; Appalachians and Rockies uplifted	Birds and mammals flourish; dinosaurs become extinct near end of period; angiosperms dominant	135
Sierra Nevadas and coastal ranges uplifted	First true bird; diatoms appear	190
Pangaea separating	Reptiles dominant; first true mammals appear	225
Appalachians uplifted; widespread mountain building	Primitive seed plants; extinction of some marine life	280
Seas retreat; Pangaea forming	Insects widespread; reptiles appear	345
	Fishes dominant; amphibians appear	395
	First land invertebrates	440
North America covered by inland seas	First vertebrates; marine vertebrates dominant	500
Land flat and low; warm climate	First shelled invertebrates	600
Glaciation on most continents	Soft-bodied marine invertebrates; very few fossils	3.5 billion (?)

CHAPTER TEN

THE EARTH'S HISTORY

It was supposed to be a large family reunion and outing. But the weather has turned uncooperative. Heavy rain is falling, so here we all are, stuck in the house together with a lot of time to kill until dinner is ready. What shall we do?

Someone gets down a big jigsaw puzzle in a box marked "The Earth's History." We all sit down around the dining room table and start sorting out the pieces. We aren't certain that we have them all. Maybe the dog ran off with a few key parts, and some of the pieces might have been lost over the years. We ask Uncle Jim, half-dozing in his favorite easy chair, but his only answer is a faint smile. We'll just have to put together what we have and find out.

As we start to fit the pieces together, we can see a pattern beginning to emerge. It begins at the center of the puzzle and appears to move outward in a spiral. But until we add more pieces, the picture is still far from complete. What will we see as we put it together?

Someone at the table has found the first puzzle piece which marks the inner core of the spiral. It says only "Something Happened Here 4.6 Billion Years Ago." We accept this, and place the puzzle piece in the center of the table where it belongs. As the spiral grows, its first billion years are hazy. We catch glimpses of the earth forming from dust and interstellar gas; we see the sun burst into a nuclear-powered source of light and warmth for its family of planets. The earth's crust begins to cool and harden, but as yet it possesses neither sedimentary rocks nor life.

the earth's history

before the beginning

The earth is estimated to be about 4.6 billion years old. However, the geological time frame used by scientists who study the earth's history covers only the last 600 million years of this period. Enough data have been accumulated to fill in the sequence of events during those 600 million years. They turn out to be nicely spaced and paced, like a well-integrated three-act play. Act I is called The Paleozoic Era; Act II is titled The Mesozoic Era; and Act III, which takes the action into recent times, is called The Cenozoic Era.

But what about the other four billion years of the earth's history which took place before the curtain went up on our tidy little three-acter? Until very recently, earth scientists simply didn't have the tools or the perspective to light up this dark and mysterious backstage area. Now, however, an enormous flood of new information is coming in from the field of astrophysics, from superbly equipped space probes, and from the research being done on meteorites. A story line and a cast of major and minor characters are beginning to appear.

The oldest known rocks, as we saw in the last chapter, are a series of metamorphosed sedimentary and igneous rocks which have been radio-dated at 3.8 million years. These rocks are different even from those formed later during the Archaean time interval. **Archaean** rocks are those known to be older than 2.2 to 2.8 billion years (this time variation allows for the fact that rock structure and stratification are not uniform around the world).

No large masses of granite date from the Archaean, but shales and sandstones are present, formed by weathering and reworking of volcanic rocks. The dominant Archaean series consists of basalts and andesites, volcanic rocks which are silica-poor but rich in iron and magnesium.

the beginnings of life

We know that liquid water has been present on the earth for the past 3.5 billion years. Our knowledge is based on the fact that the oldest fossil cells to be discovered thus far are approximately that age, and life on earth must have liquid water to exist. But how did the life begin?

The Archaean atmosphere is believed to have been composed chiefly of water, methane, and ammonia. There was no free oxygen (O_2) because most free oxygen is a product of organic activity. Because there was no O_2, there could be no layer of ozone (O_3) to block out the sun's intense ultraviolet radiation. It has been theorized that this extreme radiation could have triggered the synthesis of several

the earth's history

organic compounds, including the amino acids which are an essential component of life.

During Archaean times, great mountain ranges were being formed and there was widespread volcanic activity. All over the earth the older schists and gneisses were invaded by newer igneous rocks. As we fit more pieces into our jigsaw puzzle, we see strange-looking plains, mountains, and plateaus; we watch the work of waves, winds, and streams. Even though they are now highly metamorphosed, folded, and intruded, we know that sedimentary rocks were being deposited during this period, to great thicknesses in some regions. An enormous heat zone probably existed near the earth's surface, causing the extensive metamorphism and igneous intrusion. The crust of the earth was still very thin at this formative stage. Internally, crust and mantle material had not yet stabilized and settled into their present-day configuration. The turbulent Archaean Era ended on a very dramatic note as a period of uplifting formed the great mountains of the Adirondacks.

an open-call audition

In the world of the theater, an open-call audition creates a lot of excitement among the hordes of hopeful actors who appear for it. You don't need an agent or a special introduction to attend an open call; you just show up, do your thing, and if they like it, you get the job.

This was what was happening during the **Proterozoic Era,** which began about two billion years ago. A lot of hopeful life forms showed up for an audition. We can see a lot of milling around backstage—after all, opening night for our little three-act play is only 1 billion, 400 million years away!

As new puzzle pieces are fitted into place, we see bacteria, algae, and other primitive single-celled organisms starting to appear on the land as well as in the sea. For example, paleobotanists, who study prehistoric forms of plant life, have found fossilized algae in the Bitter Spring formation of central Australia, a formation which is about one billion years old. These fossils show many of the characteristics of present-day blue-green algae. Like all photosynthetic plants, they produced oxygen as a waste product. As the number of photosynthetic plants increased, the oxygen level in the atmosphere built up. By about one billion years ago, it had reached as much as 10 percent of its present level. Today, oxygen makes up about 20 percent of the atmosphere.

Toward the end of the Proterozoic Era, the oxygen supply had increased sufficiently to allow a new class of life to move out from backstage for their audition. These were the **metazoans,** many-celled, oxygen-dependent animal organisms. The cells of each organism were no longer identical to each other, but were beginning to specialize for particular tasks.

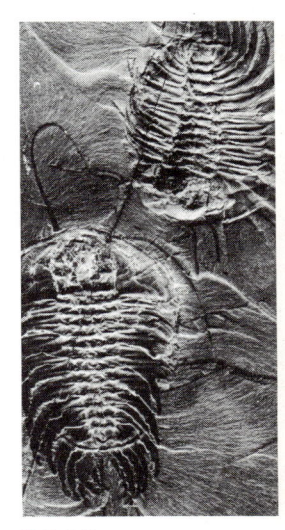

Trilobites were among the most successful marine creatures of the Cambrian Period. Their hard, segmented shells protected them against predators.

the earth's history

the play's the thing

People sitting around the dining room table are getting more interested in the jigsaw puzzle. A mosaic of events can now be seen which holds great promise for the future. Some pieces of the puzzle join to form a picture of the opening night program of a play. It is interesting to note that although the principal players and supporting cast were already assembled, auditions would be held while the play was in progress. The single performance of the play would last about 600 million years, with different scenes and bits of action taking place in different locales around the world.

act I: the Paleozoic Era

The program tells us that Act I was performed over a period of 375 million years, ending about 225 million years ago. Like many first acts, it was rather long. But it was broken up into six separate scenes or periods.

scene i. - Cambrian Period

As the curtain rose, much of the present North American continent was submerged. No animals were on the land, no plants clung to the bare rocks. The climate was mild, even warm. The land was flat and low.

There are several theories as to why organisms became so inventive at the dawn of the Paleozoic Era, changing and adapting in so many ways. Was it a happy coincidence resulting from the presence of just the right kind of environment? Or did the photosynthesizing algae inevitably pave the way for oxygen-dependent forms of life? The most widely accepted view today is that this explosion of biological inventiveness was triggered by the rising oxygen level in the changing atmosphere.

One of the most practical inventions was the hard shell developed by many primitive sea animals. The shell was like an external skeleton, serving as a base for the attachment of muscles and thus giving the organism a greater degree of mobility. It also functioned as very effective armor against predators.

Among the most successful of the shelled sea creatures were the **trilobites** (now extinct), whose big scene occurred during the Cambrian Period. They burst on stage in great numbers and many varieties. The earlier models of this series seem to have been blind. Later, improved models had many eyes, but since those eyes could not be moved, a set was provided to look in each direction. Eventually, a trilobite species appeared which could look in ninety-eight different directions at the same time; determined to get it right, nature had come up with an early Edsel.

the earth's history

During the Cambrian Period, the sea floor probably looked very much like this artist's conception. Note that the jellyfish shown still exist in essentially unchanged form.

scene ii.- Ordovician Period

During this period the greatest flood in history occurred. About 60 percent of North America was covered, and most of the rocks dating from this period are limestones. The first vertebrates appeared, in the form of fresh-water fishes. Some of the trilobites were still around, but now they had company; active, predaceous **cephalopods** had appeared on the scene. Now the trilobites needed those many sets of eyes. Their arch-enemy had arrived, and soon the trilobites were pushed entirely off the stage. By late in the Ordovician, the ocean was teeming with simple forms of marine life. No life was yet present on the land. The Ordovician Period ended with a great upheaval which folded the rocks in the Taconic and Green Mountain regions of New England.

Cephalopods, predatory sea animals related to our modern squid and octopus, were largely responsible for the disappearance of the trilobites around the end of the Ordovician Period.

the earth's history

scene iii. - Silurian Period

As the third scene opened, most of the central part of the North American continent was submerged. Fish-like vertebrates became more widespread, resembling present-day lampreys and hagfishes. Sponges and other similar marine forms were extremely common.

But now life was spreading to the land. Small fragments of fossilized terrestrial plants have been found in several Silurian rock formations. Other Silurian rocks display fossils of ancient animals that appear to be the direct ancestors of scorpions and millipedes. It is not absolutely certain whether these creatures were air-breathers or water-breathers, but some scientists believe that they were the first land-dwellers.

scene iv. - Devonian Period

The first jawed fish — the **placoderms** — made their appearance in this scene, which began about 395 million years ago. They were such a success that the Devonian is often called "The Age of Fish." The placoderms were able to move from shallow coastal areas to deeper parts of the sea, since they were no longer dependent on sucking their food from the bottom.

Nature was feverishly filling in all the gaps up and down the new line of life. It is not surprising that the placoderms' nemesis soon arrived on the scene. These were the ocean-dwelling **Chondrichthyes,** now represented on the earth by the sharks and rays. The placoderms are the only vertebrate class to have become extinct, largely because they were unable to escape these voracious new predators.

Now the bony fresh-water fishes began to move out to live in the open sea. This group of creatures had an improved jaw suspension which permitted a wider range of feeding habits. They possessed body coverings of overlapping scales, and protective gill covers like those of modern fish. If there is such a thing as the Loch Ness Monster, it might possibly be a living relic of this group which wandered into Loch Ness long ago and was trapped there when uplifting of the land closed off access to the open sea.

Such a living relic has been discovered before. In 1938, a living coelacanth was caught off the east coast of Africa. This lobe-finned

Bony fishes spread throughout the oceans during the Devonian Period. They resembled modern fish in several ways, including their possession of scales and gill covers.

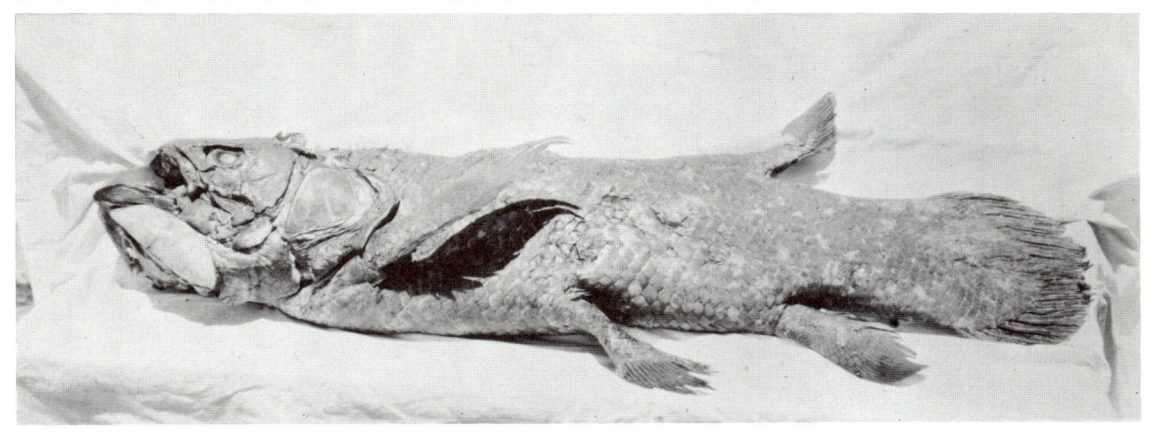

the earth's history

Until a coelacanth was caught near Madagascar in 1938, it was believed that this Devonian lobe-finned fish had been extinct for at least 70 million years.

fish of the Devonian Period had been believed to be long extinct, but since the first one was found many others have been caught in the same area of the ocean.

Another interesting development of the Devonian was the **lungfish,** of which several kinds survive today in Africa and South America. Lungfish had a limited ability to store air. Like the "walking catfish" of today, they possessed stout fins that allowed them to support their weight on land and move from one body of water to another.

The next step was the arrival of the **amphibians,** which breathed water during their immature stage and air when they became adult. The amphibians had well-developed lungs, an improved mechanism for hearing, and stronger skeletons which enabled them to support their weight on land. The loss of scales and gills was more than compensated for by the development of eyelids and tear glands to protect the eyes. But the amphibians were still tied to an aquatic life. Their soft skins needed to be kept perpetually moist, and their eggs, unprotected by shells, had to be laid in water.

scene V. - Carboniferous Period

The fifth scene began about 345 million years ago. As it opened, much of the land around the earth was low-lying, warm, and moist. The Carboniferous is sometimes called "The Age of Cockroaches" because some 800 kinds of these insects have been found in fossil form—including some specimens up to ten centimeters long.

There were vast, swampy forests of simple trees, and many varieties of ferns, but no flowering plants. In Devonian times, land plants were represented by the spore-bearing **Psilopsida,** which reached a height of one meter and had rudimentary leaves. But during the Carboniferous Period, these were superseded by the **Lycopsida,** straight-trunked, narrow-leaved "scale trees" which sometimes attained a height of 30 meters.

A lot of the land was slowly sinking. When trees and other plants died they were covered with water and preserved. As the con-

the earth's history

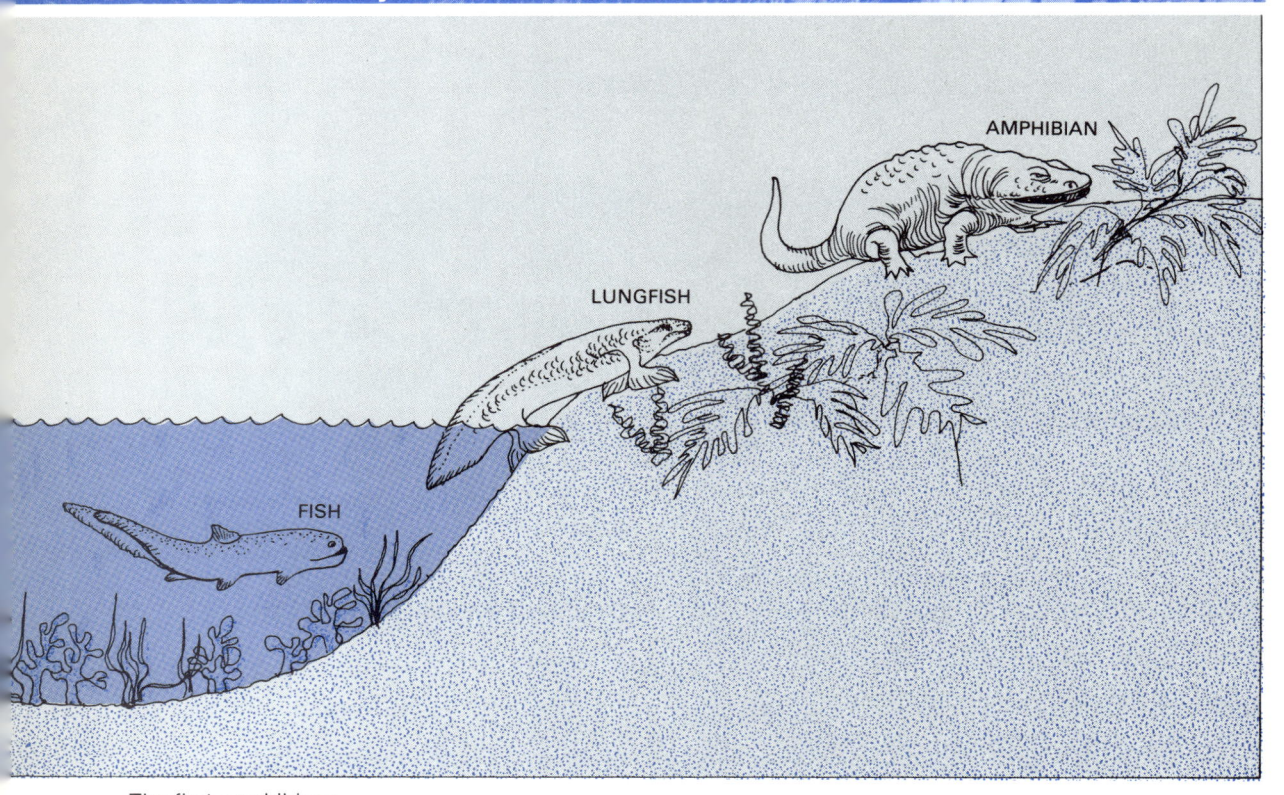

The first amphibians were probably direct descendants of the lungfish. Although adult amphibians are air-breathers, the immature forms possess gills and are water-breathing.

tinents began to move into position to form the supercontinent of Pangaea, the accumulation of dead plant material was gradually covered by sediments which piled up to depths reaching hundreds and thousands of feet. The plant material, under the pressure of the sediments, became coal. The sediment above it became shale, sandstone, or limestone.

The first true reptiles moved out on stage during Carboniferous times. They had two distinct advantages over the amphibians. First, their eggs were covered with hard protective shells, allowing them to be laid on land. And instead of soft, moist skins, the reptiles had scaly hides that protected them against both predators and sudden changes in their environment.

scene vi.- Permian Period

The final scene of Act I began about 280 million years ago. As the curtain rose, the supercontinent of Pangaea was forming. Parts of what are now Africa, Australia, and South America were covered by glaciers. Continents collided, destroying much of the shallow shelf surrounding each continent. The land rose and the great swamps disappeared, wiping out whole categories of plant life. Many marine invertebrates also became extinct as the oceans grew smaller and colder.

the earth's history

Life in three Paleozoic periods is reconstructed here. The left side of the figure shows a Devonian scene; the center represents the Carboniferous Period. At right are plant and animal forms from the Permian Period.

But new forms of life were waiting to come on stage. Primitive seed plants, apparently developed earlier on the higher and drier parts of the continents, began to become dominant over the early ferns. Terrestrial animals suffered few extinctions as compared to marine species; among the reptiles many new and adaptable forms developed and became abundant.

As the Permian Period came to a close, Act I reached a dramatic climax with an extensive period of world-wide mountain building. The Appalachian Mountains were uplifted. The seas retreated from the eastern United States, and so far have not returned.

act II: the Mesozoic Era

Following the pattern of most good plays, Act I reached its conclusion on a note of crisis and suspense. Continents came together; mountain ranges were thrown up; and hundreds of species of invertebrates became extinct. Act II, however, was considerably shorter, and things seem to have tightened up all around. The program indicated that wholesale changes in the cast had taken place. Some roles were completely written out of the script, and many new characters were introduced. The action took place in three fast-paced scenes.

Act II began with a bang about 225 million years ago. The supercontinent of Pangaea began to come apart at the seams. The Atlantic Ocean opened up, and continents began to drift toward their present positions.

scene i. - Triassic Period

the earth's history

Brontosaurus, shown in foreground, was the largest—and proportionately the stupidest—of the land-dwelling dinosaurs. It is believed that this giant reptile relied on the shallow water of swampy areas to help support its great weight.

The entire Mesozoic Era has often been called the "Age of Reptiles," since highly successful reptile groups now came to dominate land, sea, and air environments. The forerunners of these dominant reptiles were the **thecodonts,** agile animals up to a meter high which could use their large, heavy tails to balance their bodies while they ran on their hind legs alone. Some thecodonts gave rise to land reptiles or dinosaurs. Others evolved into forms that resembled crocodiles.

Dinosaurs were the most spectacular products of the Triassic. There was the lumbering *Brontosaurus,* 20 meters long, supporting its enormous weight on four pillar-like legs while its long, snaky neck probed for the plant food it lived on. With a brain weighing only a few ounces, *Brontosaurus* was probably not equipped to be alert for the approach of its most dangerous enemy, the carnivorous *Tyrannosaurus rex.* Waddling along on its hind legs, *Tyrannosaurus* stood six meters tall, and the herbivorous dinosaurs it preyed on were no match for its huge fangs and insatiable appetite. Only a few of the dinosaurs, including the armored, three-horned *Triceratops,* could hope to protect themselves against *Tyrannosaurus.*

In the meantime, while the dinosaurs continued to dominate the scene, the first true mammals were developing. Their precursors were a group of mammal-like reptiles called the **therapsids.** Eventually, mammals would become the dominant life forms on earth, and the rapidly-changing conditions during the Triassic may have been at least partially responsible for the new biological features they developed. The most important of these was the ability to give birth to

the earth's history

live offspring. The dinosaurs, like most reptiles, probably abandoned their eggs as soon as they were laid; when the young hatched, they had to fend for themselves or die. Infant mammals, however, were born in a semi-helpless state, and required further nourishment with milk. A young mammal, therefore, had a ready source of food and at least one parent to protect it; its chances of surviving to adulthood were much greater than those of a young reptile.

Other mammalian adaptations were also highly important. Mammals needed a great deal of food in proportion to their size, and their bodies used it up rapidly. But this high metabolism also enabled them to sustain higher levels of activity for longer periods of time than reptiles could. An insulating covering of fur or hair allowed mammals to maintain a constant level of body temperature. Unlike the cold-blooded reptiles, mammals did not become sluggish or overheated when external temperatures dropped below or rose above their optimum internal temperatures.

As Pangaea continued to break apart and the continents drifted away from each other, North America assumed its present appearance—minus the Atlantic and Gulf Coastal plains. The red color of the sedimentary rocks in areas such as Arizona's Painted Desert and Zion National Park, Utah, are indications that the land in these regions was high and arid, in contrast to the moist, low-lying lands so typical of the Paleozoic Era (Act I). Salt and gypsum deposits in these rocks provide more evidence of desert conditions.

The constantly blowing desert winds may have had a part in another development of the Triassic, the appearance of the first flying reptiles. Judging from the structure of their fossilized remains, these creatures needed all the help they could get to lift off the ground. Recently, an example of their problem with flight dynamics came to light when a highly detailed fossil specimen was discovered in the Texas high country. These animals, including such varieties as *Pterodactylus* and *Rhamphorhynchus,* possessed bodies that were much too heavy and wing muscles much too weak for them to have achieved a "power" take-off. The question is, how did they ever manage to get airborne?

Before the Texas find, speculation on this point had led to no firm conclusion. But working with data and measurements from the Texas specimen, engineering students were able to render a faithful reproduction of a flying reptile. It was complete with an enlarged leathery membrane that stretched between the hindquarters and the elongated fourth digit to form a primitive "wing." Hang glider enthusiasts, it seems, may have the best explanation for how these awkward creatures managed to get airborne. Instinctively, they found the right air current, and lumbering down an incline, they probably floated off into the desert breeze. (The hang glider theorists, however, have not explained how the reptiles managed to land again without smashing every bone in their fragile skeletons.)

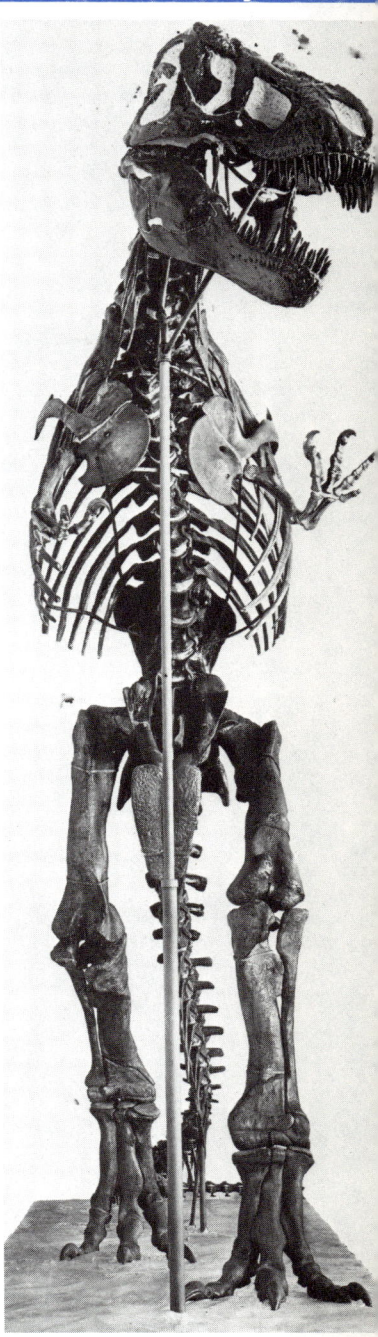

Tyrannosaurus rex was the most ferocious of the predatory dinosaurs.

IN FOCUS:
the hot-blooded dinosaurs

Dinosaurs are alive and well and living among us today. Yes, that's correct. Those pea-brained, graceless beasts that supposedly died out at the end of the Cretaceous period are flourishing today in the form of their descendants, the birds. What's more, they were warm-blooded, like mammals and birds, rather than cold-blooded, like reptiles. This is the latest word from scientists who have studied and reconstructed dinosaur fossils and their ecological systems.

We usually think of dinosaurs as hulking behemoths that lumbered across the tropical Mesozoic landscape, striking terror into the hearts of all living things. But in fact, dinosaurs ranged in size from the tiny *Compsognathus* to the huge *Brontosaurus*, the largest creature ever to walk the earth. Moreover, throughout their 100-million-year reign, dinosaurs adapted very successfully to a variety of environments.

Was the dinosaur really warm-blooded? To begin with, the terms "warm-blooded" and "cold-blooded" are misleading. The so-called cold-blooded animals, for example, are anything but cold; the body temperature of some lizards may reach 106 degrees Fahrenheit. The internal temperature of a cold-blooded animal, or ectotherm, depends upon the temperature of its surroundings. A reptile keeps its body temperature at optimum level by basking in the sun when its temperature is low and by moving to the shade when its body heat climbs to a certain point. Below this optimum, respiration and heartbeat slacken, and the animal becomes sluggish and inactive. It is easy to see that ectotherms cannot survive in environments with great temperature extremes.

The warm-blooded or endothermic birds and mammals, by contrast, can sustain an almost constant high body temperature by internal regulation. High body temperature guarantees a sustained high energy output from the muscles, heart, and lungs. Endotherms are thus able to perform more efficiently than ectotherms, and can exploit a wider variety of environments.

The idea that dinosaurs were endothermic rather than ectothermic has been advanced by a number of investigators, one of whom is Robert Bakker. Bakker bases his case on three intriguing threads of evidence. First, an examination of dinosaur bone reveals that its structure is much like that of the endotherm. It is extremely rich in blood vessels and contains many Haversian canals—the sites at which calcium and phosphates are exchanged between the bone and the blood. In the highly active endotherms, numerous blood vessels are required to move nutrients to and from muscles and other active sites. Ectotherms, however, do not need as many blood vessels or Haversian canals in their bones because their bodies require less energy.

The second piece of evidence is the fact that dinosaur fossils have been found where the Arctic Circle was located during the Cretaceous Period. This suggests that dinosaurs had adapted to extremely cold environments. If the great beasts had been reptiles, they would have been unable to survive such low temperatures because of their ectothermic metabolism.

Finally, Bakker analyzed animal remains in several fossil deposits. He found that the ratio of dinosaur predators to the animals that they preyed on closely matched the predator—prey ratio of mammals and birds. Compared to ectotherms, endothermic predators require a larger number of prey in order to satisfy their high energy needs. A stable natural population of animals, then, can support far more ectotherms than endotherms, because ectotherms make far fewer kills. Bakker has shown that the predator—prey ratio for the dinosaurs is much closer to that of warmblooded mammals than to that of reptiles.

As for the thesis that dinosaurs are related to birds, Bakker points to the extraordinary similarity between *Archaeopteryx*, usually regarded as the first bird, and a small carnivorous dinosaur known as *Deinonychus*. The anatomies of the two are similar enough to suggest that a small dinosaur of this type may have been the immediate ancestor of the first bird. If further study supports Bakker's thesis, his idea that dinosaurs live on today in the form of birds will revolutionize the present system of biological classification.

The dinosaur dynasties fell during the twilight of the Cretaceous. The reason for their decline is still being debated. Some suggest that a shift to a colder climate killed them off. But others, including Bakker, lean toward a geological explanation for their demise. At the end of the Cretaceous Period, the shallow seas were drained and mountain-building came to a virtual stop over much of the earth. As a result, the variety of habitats for land animals decreased, and delicate, complex ecosystems were destroyed. Not surprisingly, competition for the available ecological niches became fierce. The dinosaurs, largely because of their size, fell victims to this struggle. Smaller animals, including the early mammals, survived because they found it easier to eke out a livelihood from the new environment.

Obviously, this theory fails to explain why the smaller dinosaurs were wiped out also. Perhaps future research will reveal the entire story of the dinosaurs' decline. But whatever theories are advanced to explain this mysterious mass extinction must now deal with the possibility that the dinosaurs were warmblooded.

the earth's history

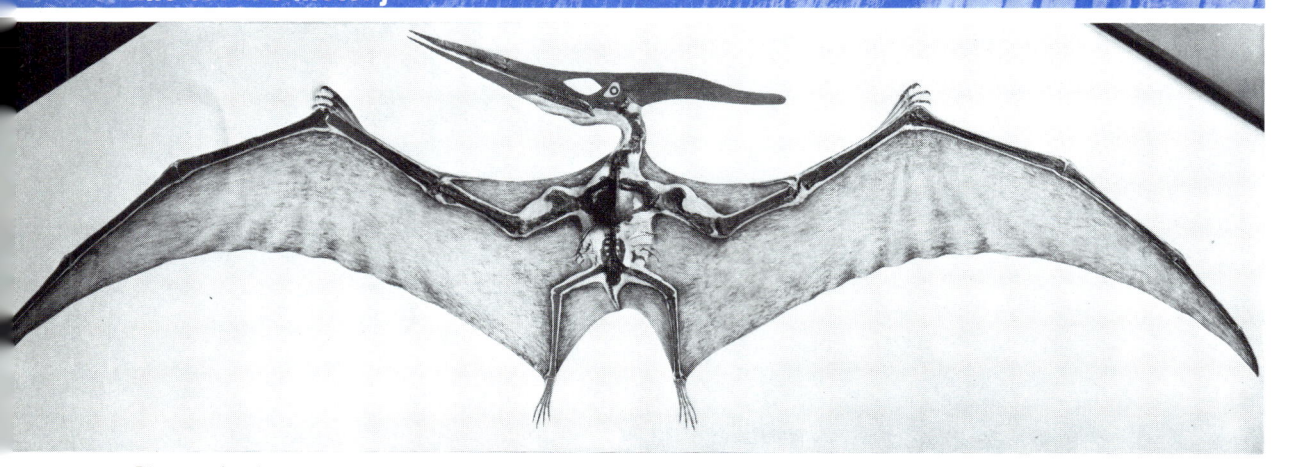

Pteranodon is representative of the many flying reptiles that appeared during the Triassic Period. It is hard to imagine how this unwieldy creature managed take-offs and landings!

scene ii.- Jurassic Period

The horseshoe crab was contemporary with the trilobites. This "crab"— actually an arthropod— survives today in almost identical form, a living relic of the past.

The end of the Triassic found almost the whole North American continent above water. Its close was signaled by the Palisade Disturbance, a moderate crustal movement probably associated with the breakup of Pangaea. The name "Palisades" is taken from a formation along the western bank of the Hudson river, where basalt has intruded into sedimentary Triassic layers. The resistant igneous rock has appeared at the surface where the overlying softer sedimentary rocks have been worn away in the course of time.

In this scene the first true bird was introduced. *Archaeopteryx* was about the size of a crow, and had the teeth and bone structure of a reptile. Unlike reptiles, though, it had a covering of feathers, a coat of insulation which indicates that it was warm-blooded.

This scene began about 190 million years ago. Reptiles continued to be masters of the seas, some of them reaching lengths of more than 12 meters. These animals were streamlined like fish and like the seagoing mammals of today.

The **ammonites** played out their ill-starred role. As the Jurassic opened, this large group of mollusks seemed to be approaching extinction. Then, in a dramatic reversal, they evolved into at least 23 specialized and highly successful families, and became dominant among the marine invertebrates of the Mesozoic. Some pieces of the puzzle are missing in this area. When next we hear from the ammonites, we find that they became extinct at the end of the Mesozoic Era; cause—unknown.

Diatoms also made an appearance at this time—single-celled creatures which secreted thin protective shells of silica. The diatoms were responsible for much of the primary photosynthetic production of organic matter in the sea. They formed an important part of a food chain which resembles the one that exists today.

The end of the Jurassic Period was marked by the Nevadan Disturbance. The western United States took on something close to

the earth's history

its present configuration as the coastal ranges and the Sierra Nevada mountains were lifted up. Gold-bearing solutions were being injected into the California rocks along with new igneous masses. In the eastern part of the country, the Appalachians were being worn down to a peneplain.

The final scene of Act II began some 135 million years ago. After a modest entrance, mammals began edging toward the center of the stage. Seven distinct groups were represented, several of which are now extinct. The mammals' less restrictive birth and feeding habits, and their greater flexibility in coping with sudden climatic changes, were beginning to have their effect on the competing land reptiles.

There are some indications that birds also flourished during the Cretaceous. However, they were probably a favorite snack for many of the predators, and very few made it to the fossil stage. Those that weren't eaten must have had a varied menu of their own to choose from; ants, bees, dragonflies, beetles, grasshoppers, cockroaches, moths, and flies were all plentiful at this time.

By the end of the Cretaceous Period about 90 percent of all land plants were **angiosperms;** their seeds were covered by a protective coating. Covered seeds were well adapted to seasons of cold and drought—which was just what they had to deal with during this time. The fact that the angiosperms could survive such conditions was extremely important to the continued survival of the struggling mammals who fed on them.

Act II closed with the presentation of a cue card, telling the audience to say farewell to the dinosaurs. People around the dining room table who have been working hard on this jigsaw puzzle are getting angry. So many pieces are missing— and at such a critical

scene iii.- Cretaceous Period

At left, a reconstruction of *Archaeopteryx*, believed to be the first true bird although it may not have been able to fly. Its feathers probably served only for insulation.

At right, the skeleton of *Archaeopteryx* from which the reconstruction was made. The structure of the skeleton is very similar to skeletons of running dinosaurs from this period.

the earth's history

point! We look at Uncle Jim with annoyance, but he snores on serenely. All we can find, now, is a piece of cardboard roughly cut to fit in the vacant section of the puzzle. Uncle Jim's writing has labeled it: "The Time of the Great Dying." This doesn't tell us much; but a little further along we see that not a single dinosaur survived this crisis, whatever it was. Many of the marine invertebrates have also disappeared. Volcanic activity breaks out violently, from Alaska to Mexico. The worn-down Appalachians and the Rockies receive a boost. Sedimentation and uplift have continued into the present time.

act III: the Cenozoic Era

As the final act of the play begins, we are caught up in the fast action. Act III is very short, only lasting about 70 million years. A program note tells us that during much of this act, the earth will cool and a series of ice ages will occur as a result.

Almost as an afterthought, the program also tips us off to be on the alert for a peculiar creature who will appear briefly near the very end of the play. In the credits, this character is listed as *Homo sapiens*. We know it as well as we know our own name — in fact, it *is* our own name.

scene i. - Tertiary Period

The Cenozoic Era has been called "The Age of Mammals." But it could just as easily be called the age of insects or of angiosperms, because all three groups widened their domain and influence during this era.

When all those dinosaurs received their eviction notices, a lot of good habitats were left vacant. Rat-sized mammals lost no time in moving right in. It had been a long, hard fight, but mammals were finally on the way toward a position of dominance in the world order. Free-lancers — the unspecialized mammalian orders — began to be weeded out. Specialization, in fact, became the name of the game. The first mammalian carnivores appeared: small, otter-like creatures with teeth and brains specialized for predation. Mammals resembling bears, dogs, cats, and horses came on the stage. And very quietly, almost like extras in a crowd scene, the first primates arrived. At first glance, they didn't look very important.

Many of the land animals were noted for their immense size. The extinct *Baluchitherium*, for example, was about 8 meters long and nearly 6 meters high at the shoulder. It looked something like a huge, hairy rhinoceros. Pigs two meters high rooted in the woods, beside the elephant-like mastodons and wooly mammoths. In the oceans, whales, porpoises, and dolphins evolved from terrestrial carnivores which had returned to the sea. The cast included the largest animal

THE UNREPEATABLE CARBONIFEROUS PERIOD

There was a time, eons ago, when the sites of present-day London, Paris, and New York were steaming jungles, green with luxuriant vegetation. During this time, known as the Early Carboniferous Period, the entire earth basked in tropical or subtropical warmth. This was an exceptional moment in the history of the planet, one that is not likely to be repeated.

The Carboniferous Period, named for the carbon-bearing sediments formed at the time, comprised 65 million years of the Paleozoic Era. It is divided into the Early Carboniferous (345 million to 315 million years before the present) and the Late Carboniferous (315 million to 280 million years ago). During the earlier part of the period, the earth's climate was uniformly warm. In the following part of the period, however, the Southern Hemisphere grew cooler, and much of this region was capped with thick glaciers.

From a geological standpoint, the entire Carboniferous Period is very important because it was during this time, particularly during the Late Carboniferous, that immense coal fields were laid down in North America and Europe. Shallow inland seas covered extensive areas of the continents, and tropical conditions gave birth to huge swamp forests and bogs. When the lush vegetation in these regions died it accumulated, layer upon layer, until tremendous pressures and high temperatures in the earth's interior transformed it into coal.

There are a number of reasons why such unusual climatic conditions existed during the Carboniferous. For one thing, the South Pole was located somewhat north and east of its present position, and the North Pole lay in what is now the northern Pacific Ocean. The modern-day southern continents were situated near the South Pole, and Europe and North America were at or near the equator. Then, too, all of the continents were welded together into the supercontinent of Pangaea. This configuration affected ocean currents and temperature, making the oceans somewhat warmer than they are today. Finally, the uplift and subsidence of continental landmasses also influenced the climate.

It is highly improbable that such a unique combination of factors will ever occur again. The coal laid down during the Carboniferous Period is, therefore, a nonrenewable resource. And according to many scientists, if we go on using it as we are, the entire supply will be virtually exhausted within a few hundred years.

the earth's history

scene ii. - Quaternary Period

ever to have lived, the great blue whale (currently appearing in cans of cat food at your local supermarket).

Now we have come to the final scene in our play, beginning only a few million years ago. Glaciers poised at the poles advance and retreat, back and forth, like giant sheets of sandpaper. The glaciers destroy whole species, wipe out environments, erase comfortable habitats. Animals and plant life adapt as best they can. Only in the narrow belt of warmer climate near the equator does life continue relatively unchanged. And there, something new is happening.

It is dawn on the barren African plain. Some shadowy figures move hesitantly across the open spaces, occasionally darting back to the protective shelter of the trees which border the area. These creatures are an anachronism. In this day of specialization, they are throwbacks to earlier, less specialized mammals. Now, as the sun rises, they are scavenging for the remains of animals killed during the night by the deadly predators. Look at them closely. They are neither apes nor humans. They are our proto-human ancestors, and slowly but surely they are evolving human consciousness.

These hominids we call *Ramapithecus* are freaks in the prevailing scheme of things. Mutations have happened all along the evolutionary road, allowing better adaptation to the existing environment. Until now, this has called for specialization; but these creatures seem to be evolving away from specialization. In fact, *Ramapithecus* doesn't seem to fit in anywhere.

But perhaps specialization isn't the only answer to survival. Unlike their fellow primates, the apes of tropical Africa and Asia, *Ramapithecus* did not have the right kind of teeth to strip leaves from branches. Somewhere during their history, one of these hominids found a solution to this lack of specialization: he started using his

The evolutionary history of the horse has been particularly easy to trace on the basis of fossil remains. This is *Eohippus* ("dawn horse"), the earliest known form. It was about the size of a collie.

the earth's history

An artist's reconstruction of a Neanderthal family scene. Current research suggests that the Neanderthals may have communicated with a form of primitive speech.

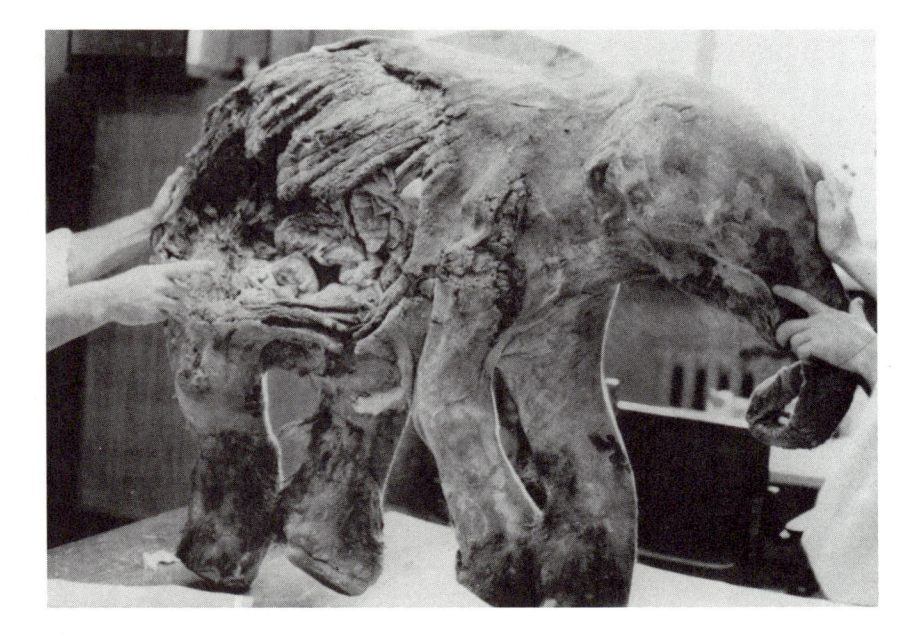

hands. This was a risky move, since hands as well as feet could be used for locomotion and quick escape from danger. But *Ramapithecus* had started something.

Look again, about a million years later. *Ramapithecus* is now gone, but his tendency to use his hands for food-gathering has been passed along to a new group, the *Australopithecines*. These bipedal proto-humans resembled apes in many ways, but their teeth, pelvises, and limbs were much more like those of present-day humans. They used their hands to make crude stone implements and weapons, and were able to compete with specialized predators in bringing

The frozen body of this baby mammoth was discovered in Siberia in 1977. It apparently died around 10,000 years ago, when it was about nine years old.

the earth's history

Our ancestors of about 15,000 years ago were skilled artists. Their paintings, which may have some religious significance, have been found in caves throughout the southern part of Europe.

epilogue: a review of the play

down prey.

Still later, we look again and see that something new has been added. Our more recent ancestors, *Homo neandertalensis*, may have had the capacity for primitive speech. Communication and group organization became much more efficient. Nonspecialization, once such a big handicap, was now becoming a big advantage. Through stealth, cunning, and cooperation, the Neanderthals could roam anywhere and adapt to any number of habitats and environments. Our ancestors became opportunists.

The secret of their success was really very simple. It lay in their possession of a brain unspecialized for any particular way of living, of getting food, or of surviving. The human brain, in fact, was specialized for only two things: verbal communication and gathering information. This gift was handed down to us through a long line of ancestors. We received it when we came out on the stage about 40,000 years ago.

Now that the play is over, we can reach a tentative conclusion: there seems to have been some sort of ordering principle behind the development of life on earth and its evolution from single-celled organisms into highly complex forms. The sequence of events appears to have been aimed toward some definite achievement: when specialization, which once seemed so necessary, started to block the sequence, it was abandoned. We are *Homo sapiens*—"wise man"— and so far, we are the end of the sequence. Look again a million years from now.

As we get up from the dining room table and start putting the puzzle

the earth's history

back in its box, we all want to know what time it is. How long have we been working on this thing, anyway? To some of us it has seemed like minutes. To others, it's been long hours, and some of us feel we have been at it for days.

Uncle Jim rouses himself and comes over to get the puzzle and put it away. People complain to him that two-thirds of the pieces are missing. But Uncle Jim just smiles. He explains that the next time we take the puzzle down it will have more pieces. In fact, it will continue to grow until all the missing pieces finally appear. And then he settles back into his easy chair, to resume a dream that mingles the past, the present, and the future.

the earth's history

SUMMARY

The earth is approximately 4.6 billion years old. Only about the last 600 million years are well understood. This period is divided into the Paleozoic, Mesozoic, and Cenozoic Eras.

Liquid water has been present on the earth for about 3.5 billion years, since the earliest known forms of life date back to this time. It is not certain just how life began, but ultraviolet radiation from the sun may have triggered the synthesis of organic compounds.

During the Archaean Era the crust of the earth had not yet stabilized. Uplifting and volcanic activity were widespread. During the Proterozoic Era, which began about two billion years ago, primitive organisms spread to the land. As photosynthetic plants increased, atmospheric oxygen rose rapidly, allowing oxygen-dependent organisms such as the metazoans to evolve.

The Paleozoic Era began about 600 million years ago. It is divided into the Cambrian, Ordovician, Silurian, Devonian, Carboniferous, and Permian Periods. This era saw the development of shelled sea creatures, cephalopods, marine vertebrates, insects, amphibians and early reptiles. The Paleozoic Era closed as continents came together to form Pangaea, mountains were uplifted, and many invertebrate species became extinct.

The Mesozoic Era, beginning about 225 million years ago, was divided into the Triassic, Jurassic, and Cretaceous Periods. This era is often called the age of reptiles, although the first mammals appeared at this time also. Dinosaurs were dominant for most of the era. The first true birds appeared during the Jurassic, and were warm-blooded. During the Cretaceous birds and mammals became more widespread, and most land plants were angiosperms. Toward the end of the Mesozoic Era the dinosaurs became completely extinct. The reason for their disappearance is not known.

The Cenozoic Era began about 70 million years ago. It is divided into the Tertiary and Quaternary Periods. Mammals became the dominant form, some of them closely resembling present-day species. The first primates and sea-dwelling mammals appeared. The first hominid form, *Ramapithecus,* appeared during the Quaternary. They were followed by the Australopithecines, who made crude tools and weapons. Primitive speech probably appeared later with the advent of the Neanderthals.

The success of humans as the dominant group on earth is due to their lack of specialization. The human brain was specialized only for communication and gathering information. Our own species, *Homo sapiens*, appeared about 40,000 years ago.

the earth's history

REVIEW QUESTIONS

1. Approximately how old is the earth?
2. Describe the Archaean atmosphere.
3. What physical characteristics made the trilobites so successful? Why did they finally disappear?
4. Name the periods of the Paleozoic Era and describe what happened during each one.
5. When did the Proterozoic Era begin?
6. Describe some of the life forms of the Devonian period.
7. What important group first appeared during the Triassic?
8. How long has liquid water been present on earth? How do we know this?
9. Describe the three periods of the Mesozoic Era.
10. Why did atmospheric oxygen increase during the Proterozoic?
11. Describe the development of life during the Cretaceous.
12. What was so unusual about the metazoans?
13. When did the first vertebrates appear? What were they?
14. Explain why mammals finally became dominant. What were some of their advantages?
15. What is unusual about the human brain?
16. When did the first true reptiles appear? What was their advantage over the amphibians?
17. What geological events occurred during the Archaean?
18. What is significant about *Archaeopteryx*? When did it appear?
19. When did the Cenozoic begin?
20. What were some of the advantages and disadvantages of specialization?
21. What were the thecodonts?
22. When did Pangaea begin to form? What effect did this have on the land?
23. What was the first known hominid species?

PART 3

METEOROLOGY

CHAPTER ELEVEN

THE ATMOSPHERE

Odysseus, that crafty hero of Homer's tales, was on his way home from the siege of Troy. Sailing with a favorable wind sent by a friendly magician, he was actually within sight of his longed-for island of Ithaca, in the Ionian Sea just west of Greece. The other winds—those that might blow him off course—had been imprisoned in a sack by the magician, and now the sack lay close to his hand on shipboard. All through the voyage he had kept awake, guarding the sack lest some foolish seaman open it and release the winds. But now, exhausted, and relieved by the assurance that he would soon be at home, he fell asleep. The sailors turned their eyes to the mysterious sack—surely it held some great treasure! What right had the captain to keep it for himself, when they had all shared in the fighting and danger? So they opened it. The winds leaped out in all directions, a furious storm arose, the ship was blown clear across the sea, and Odysseus awoke to disaster and despair. It was to be years before he finally made it home.

Homer knew what he was talking about when he introduced wind and storm into his story. The Mediterranean region boasts an incredible variety of winds, hot and cold, wet and moist, fierce and not so fierce, blowing from all directions. The ones that hit Odysseus' ship probably included the hot, humid sirocco from Africa, the cold, dry, gusty bora from the north, and perhaps the dusty ghibli from Libya; with these may have been mixed the trade winds from the west that are found in temperate zones throughout the world, and perhaps an eastward-moving circular storm system such as commonly develops in the fall and winter months. The bora can be particularly nasty; it is formed when cold air from Russia spills over the mountains of Yugoslavia and funnels down through a gap at the head of the Adriatic Sea. It commonly reaches speeds of 100 km/hr, and gusts up to twice that are known—

the atmosphere

obviously nothing we would care to meet in a small, open, unprotected sailing ship without even a life raft.

If Odysseus had been in the western Mediterranean, he might have met the bora's sister wind, the mistral, which pours down the Rhone valley of France and out through the gap between the Alps and the Pyrenees, or the levanter and vendaval, which rush through the Strait of Gibraltar. The westward-blowing levanter is so strong that, even today, it can almost bring an eastward-flying airplane to a stop in midair. The sirocco would have been there, too. When we come to think of it, it seems a wonder that the Mediterranean, of all places, was the home of so many ancient seagoing civilizations. How did the early sailors ever get up the courage to venture out from shore, knowing that a gentle breeze might turn in minutes into a raging storm?

Why do all these different winds happen? Why do they happen here rather than somewhere else? For that matter, what causes weather in general? The search for an answer to such questions leads us into a study of the earth's atmosphere.

the nature of the atmosphere

The **atmosphere,** or what we call the air, is made up of gases held to the earth by gravity. All life on earth is dependent on the atmosphere. Obviously, human beings and other land animals could not survive without it, but neither could marine creatures. If there were no atmospheric pressure, the oceans would boil away, and the earth would become as dry and barren as the moon.

Our study of the atmosphere will include an examination of its general structure as well as a view of the way it functions. As we will see a little later, an important part of this concerns the way in which the atmosphere interacts with the radiation received from the sun. This is so important, in fact, that many scientists like to think of the atmosphere as a heat engine that converts solar energy (heat) into mechanical energy. This energy is the basis of most weather, which we will discuss in the next chapter.

a capsule look at the atmosphere

The atmosphere extends from the surface of the earth to a height of thousands of kilometers. It is densest near the surface, and becomes less dense as the altitude increases. At a height of about 100 km the density is less than one percent of sea level density. But some traces of the atmosphere can still be detected as much as 50,000 km above the earth.

As this shows, the volume of the atmosphere is much larger

the atmosphere

than that of the earth. But its total mass is very small, far less than that of the oceans, because the atmosphere is not very dense even at sea level. One cubic centimeter of sea-level air weighs about a thousandth of a gram. In contrast, one cubic centimeter of water weighs 1 gram, and one cm³ of rock weighs about 3 grams.

The atmosphere divides naturally into several layers, and later in the chapter we will discuss these. First, however, we must examine the chemical composition of the atmosphere as a whole. For this purpose we can view it as being composed of only two layers, an upper and a lower.

This photograph, taken from space, shows a large African dust storm covering thousands of square kilometers. Photos and other studies performed by satellites have become extremely important to our understanding of weather, climate, and wind movements within the earth's atmosphere.

atmospheric gases

The lower atmosphere is not a chemical compound, but a mixture of gases. Five gases make up 99.997 percent of the lower atmosphere: nitrogen, oxygen, argon, carbon dioxide, and water vapor. The remaining .003 percent consists of traces of hydrogen, helium, neon, krypton, xenon, and radon. Except for water vapor, the composition of the air is quite uniform. However, the water content or **humidity** of the air varies widely from place to place and from day to day. In warm, humid regions such as the tropics, water vapor can make up as much as 4 percent of the air, by volume. As we shall see in this chapter and the next, humidity plays an important role in various kinds of weather.

Nitrogen makes up 78 percent of dry air. It exists entirely in the form of molecules consisting of two bonded nitrogen atoms (N_2). It is fortunate that it is so abundant in the atmosphere, since it is rela-

the atmosphere

tively rare in the solid earth and in the oceans. Nitrogen is vitally important to life, for it is the principal component of proteins; all living creatures are built at least partly out of nitrogen compounds. But although plants and animals need nitrogen, few of them can absorb it directly from the air.

As we saw in chapter 2, nitrogen is a rather inert gas. This is why it has remained in the atmosphere. Its unwillingness to react chemically makes it hard to "digest." Plants must therefore absorb nitrogen from the soil in the form of nitrates, which have been formed in two ways. Lightning produces nitric acid in the atmosphere from nitrogen, oxygen, and water vapor. When this acid is carried to earth by rain, it reacts with rocks to produce nitrates, which plants can use. In another process, some bacteria can produce, or "fix," nitrates from the nitrogen in the atmosphere.

Oxygen makes up 21 percent of the atmosphere, where it exists in the molecular form (O_2). Unlike nitrogen, oxygen is very reactive, which makes its presence in the atmosphere rather surprising. In theory, it should all have combined with the surface materials of the earth long ago. The fact that it still exists in the atmosphere shows that it is constantly being produced. We know that plants, under the action of sunlight, produce oxygen from carbon dioxide through the process of photosynthesis. This is not to say that completely "new" oxygen is being created. Rather, plants separate oxygen from carbon dioxide and perhaps other compounds, and release it into the air. Each plant, during its lifetime, produces a certain amount of oxygen. When it dies and decays in the open it uses up the same amount of oxygen. The oxygen in the atmosphere comes from plants that have not decayed in the open. Through the ages, some plants have been buried by geological activity. In time, they become oil, coal, and peat. If all these fossil fuels in the earth were burned, their combustion

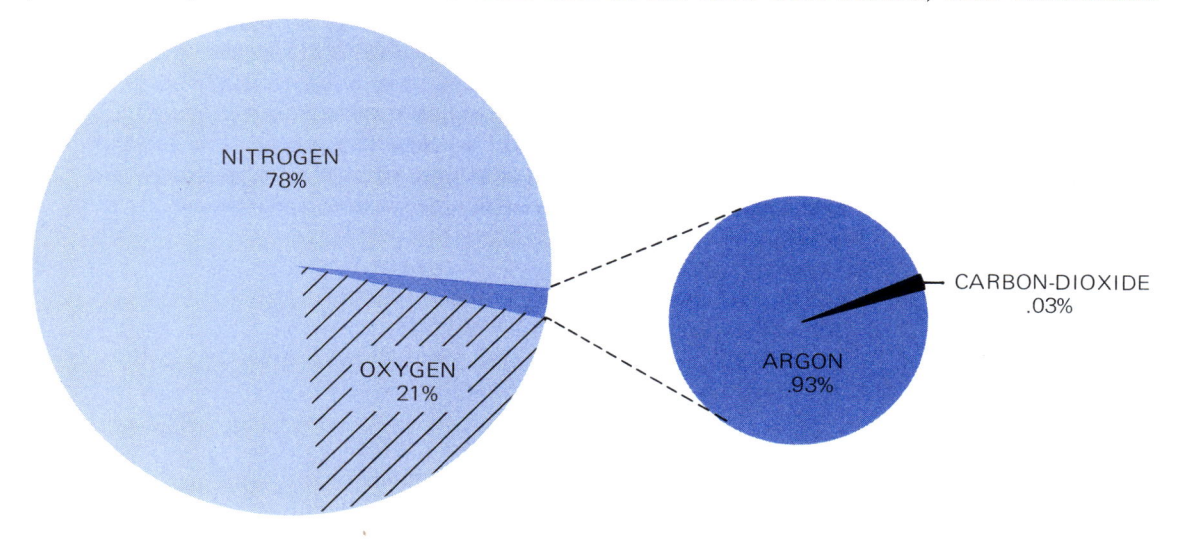

The graph shows the percentage of the most important elements in the atmosphere. Despite its very small percentage, carbon dioxide plays several major roles; without it, surface temperatures would be too low for most life to exist, and plants could not produce carbohydrates.

NITROGEN 78%

OXYGEN 21%

ARGON .93%

CARBON-DIOXIDE .03%

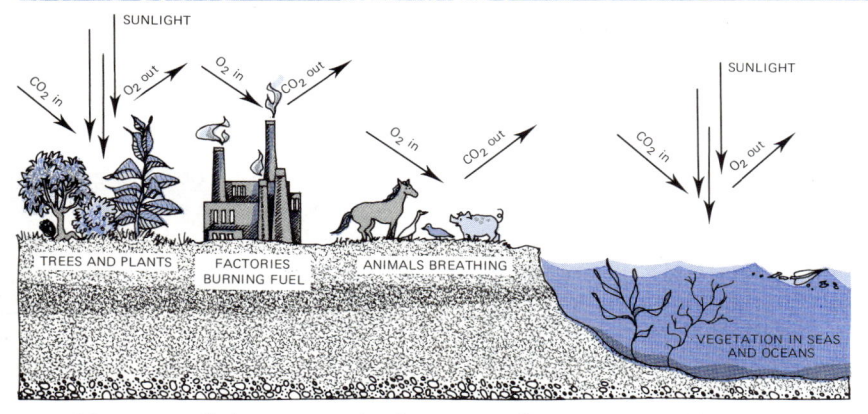

the atmosphere

Animal respiration and combustion processes convert atmospheric oxygen into carbon dioxide. The carbon dioxide is used by green plants for photosynthesis under the action of sunlight, and oxygen is thus released as a waste byproduct.

would use up all the oxygen in the atmosphere.

The importance of plants for earth's oxygen supply is evident when we look at past geological ages. About a billion years ago, only 1 percent of the atmosphere was oxygen, and there was far more carbon dioxide than at present. Then plant life evolved. As plants began to produce large amounts of oxygen, the percentage slowly climbed to its present level.

Nitrogen and oxygen make up 99 percent of dry air. Most of the remaining 1 percent is the inert gas argon, which does not form stable compounds with other elements and thus makes little difference to the atmosphere. But carbon dioxide (CO_2), which makes up only 0.03 percent of the atmosphere, has several important roles. For example, it traps solar radiation, making the atmosphere much warmer than it would otherwise be. It is also used directly by plants to produce carbohydrates, which nourish animals that feed on the plants. Carbon dioxide is then returned to the atmosphere by the animals' respiration.

The chemical composition of the upper atmosphere is very different. Above a level of about 120 km it becomes predominantly oxygen, mostly in the form of single atoms. Above this lies a layer of helium and hydrogen, which extends outward for hundreds of kilometers but contains very little matter and thus has a very low density. Beyond this is a layer of hydrogen, which is constantly escaping into space. In these outer regions we also find many charged particles or ions. Under the influence of the earth's magnetic field, the ions form two doughnut-shaped belts of radiation, known as the **Van Allen belts.**

how gases behave

In order to study the atmosphere, we must understand the basic pro-

temperature and pressure

cesses at work in it. Fortunately, the most important of these are approximately the same for all gases. We can approach them best by considering how a gas is affected by changes in temperature and pressure.

We can think of **temperature** as the ability of a body to give up heat energy. A hot body is more able to give up heat than a cold body, but this does not mean that the hottest body has the most heat energy to give up. A large tub of lukewarm water, for instance, can give up more energy than a red-hot poker, because the tub of water has a greater heat capacity.

To understand this more clearly, suppose we consider one gram of water and one gram of iron at the same temperature. By allowing the water to cool by 1°C, we can extract 1 calorie of energy. By allowing the iron to cool by the same 1°C, we can extract only ⅑ of a calorie. The gram of iron therefore has a smaller heat capacity than the gram of water.

The **specific heat** of a substance is the number of calories that must be added to or taken away from one gram of the substance to change its temperature by 1°C. The specific heat of water is 1.0, but that of most other substances is much less. The specific heat of an average soil, for example, is only about 0.2.

The large specific heat of water gives the ocean an important role in modifying the temperature of the lower atmosphere. To cool a 1-meter layer of ocean water by 0.1°C, a layer of air 30 meters thick will have to warm up by 10°C. In contrast, cooling the same amount of soil would warm that air by only about 2°C. In addition, evaporation of water requires a great deal of energy. To evaporate just 1 gram of water, without changing the water temperature, we need almost 600 calories—usually obtained by cooling the air. Thus the differences between the properties of land and ocean affect the temperatures of the air masses that lie above them, and cause many of the familiar complexities in the weather.

Pressure is the force acting on a unit of area. It is often measured in pounds per square inch, kilograms per square centimeter, or in some similar way. We are all aware that gases exert pressure. It is the pressure of air inside a soap bubble that keeps it from collapsing under the pressure of the air outside. The pressure of a gas depends on its other properties, especially its density and temperature.

Suppose you come down to your car in the morning and find that one of the tires is slightly flat. This means that the pressure in the tire is too low, and it cannot support the weight of the car. The obvious remedy is to put some more air in, thus raising the density of the air inside the tire. The pressure rises as a consequence, and the tire is no longer flat.

The pressure of a gas depends also on its temperature. Suppose you inflate the tires to the regulation pressure, and then drive

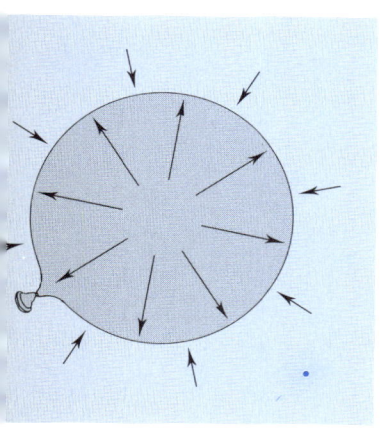

The gas pressure inside a balloon is greater than the air pressure outside it. This prevents the stretched skin of the balloon from collapsing.

the car all day. As a result of friction with the road, the tires become quite warm. If you measure the pressure at the end of the day, it will be well above the regulation pressure. (Tires are built to withstand this rise in pressure. Why doesn't the pressure have to be adjusted again in the evening after a day of hard driving?) Thus we see that the pressure of a gas is increased by an increase in either its density or its temperature. In the next section we shall see why.

By the same token, a change in pressure affects temperature. When a given amount of gas is compressed, without any energy being allowed to escape, the gas becomes warmer. This is the principle behind the diesel engine. In diesel engines, air is compressed much more than in an ordinary engine. It reaches a temperature high enough to ignite the fuel without the need for a spark plug.

Just as gas becomes warm when compressed, it becomes cooler when expanded. When a body of air flows up the side of a mountain its pressure decreases with increasing height. It therefore expands, and the expansion causes it to cool. When it flows over the top of a mountain and down the other side the reverse process takes place, and the air becomes warmer. Notice that no outside energy is added to the body of air, and no energy is taken away from it, to produce this warming and cooling. Such a process, in which temperature change takes place without any entrance or escape of energy, is called **adiabatic** warming or cooling.

kinetic theory of gases

The properties we have just discussed are not limited to any particular gas. Almost all gases behave similarly. The **kinetic theory of gases** provides an explanation for this similarity.

Toward the end of the last century, scientists found that the behavior of gases could be understood if they were thought of as collections of particles. We now know that these particles are molecules. The molecules are in constant motion, forever bouncing off each other and off the walls of the container holding gas. Because they are perfectly elastic, they keep on bouncing indefinitely — they do not slow down and stop, like a rubber ball that a child drops on the floor and forgets. It is this constant movement of molecules that explains the relationships in a gas between temperature, pressure, and density.

All moving objects are said to possess **kinetic energy.** Roughly speaking, kinetic energy can be defined as the ability of a moving object to do work — for instance, its ability to exert force on another object when it hits it. If you swing a golf club and hit a ball down the fairway, it is the kinetic energy of the golf club that has set the ball in motion.

The amount of kinetic energy possessed by an object is determined by two factors: its mass and its velocity. Suppose you hit the ball with a feather instead of a golf club. The feather has much less mass than the club, so even if it moves at the same speed it will not

the atmosphere

have enough kinetic energy to send the ball very far. Or suppose that, instead of swinging the club powerfully, you dangle it loosely from one hand and merely tap the ball. Again, the ball will not go far—this time because there was not enough velocity to give the club much kinetic energy. The same principle holds true when the moving objects are molecules rather than golf clubs. A heavy molecule has more kinetic energy than a light one moving at the same speed; and a fast-moving molecule has more kinetic energy than a slow-moving one with the same mass.

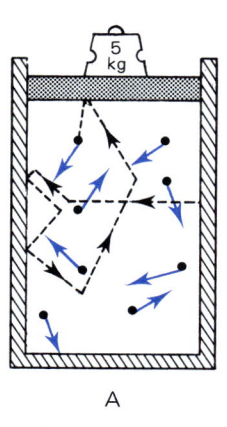

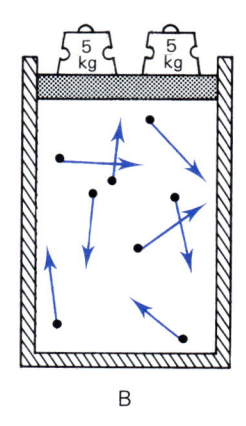

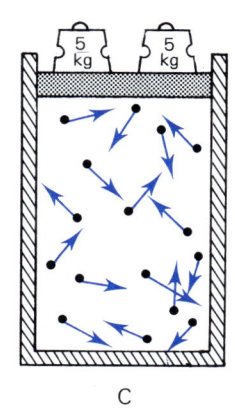

In A, a given quantity of gas at a given temperature is able to support 5 kg. At B, the temperature is doubled; the particles of gas thus travel faster, and the gas can support 10 kg. In C, the temperature is the same as in A, but the density of the gas is doubled, and it is thus able to support 10 kg.

A   B   C

The temperature of a substance is the average kinetic energy (or average energy, for short) of its particles. In a hot gas, molecules move fast and have a high average energy; the average energy of a cool gas is lower, because the molecules have slowed down. When molecules bounce off the walls of the container, they exert a pressure. Since they move faster at higher temperatures, they hit the walls harder and more often, and therefore exert more pressure. This explains the relationship between temperature and pressure. Furthermore, the more molecules there are in the container, the more will bounce off the walls every second, resulting in greater pressure. This explains the relationship between pressure and density.

The average energy of the molecules in a substance is the same for both heavy and light molecules, since both are at the same temperature. Take, for example, a mixture of hydrogen and oxygen. The oxygen molecules are 16 times more massive than the hydrogen molecules, but both types have the same average energy. Since the hydrogen molecules are lighter, they must be traveling much faster. Their high velocity suggests why the atmosphere contains so little hydrogen. Hydrogen molecules travel fast enough to escape the gravity of the earth. But the molecules of heavier gases do not travel as fast and cannot escape.

The kinetic theory also explains the adiabatic process. Consider a given amount of gas in a cylinder, being slowly compressed by a piston. As the piston moves inward, the molecules of the gas bounce off it. In doing so, they not only keep the energy they have, but also pick up additional energy from the energy of the moving piston. This is much the same as the way a tennis ball picks up energy when it hits a moving tennis racket. The increase in the energy of the molecules is also an increase in their temperature. This is why the temperature of a gas increases when it is compressed.

the pressure of air

Scientists use a variety of ways to express standard gas pressure. Several of these are shown on the next page. The National Weather Service originally expressed atmospheric pressure in terms of pounds per square inch, but in 1940 it began to measure and report atmospheric pressure in **millibars.** Since most weather reports and weather maps available to the public still give the pressure in inches, we will use this system throughout our discussion of the atmosphere and weather.

At sea level the atmosphere exerts a force of nearly 15 pounds on every exposed square inch of area. We can illustrate this very well with a simple experiment, using a can that can be tightly closed. The atmospheric pressure on the outside of the can is balanced by the pressure of the air inside, and so the can does not collapse. Next, we expel the air inside the can by keeping the can open and boiling a small amount of water in it. The water becomes a large volume of steam, which expels most of the air inside the can. Now we close the can tightly and cool it. The steam inside condenses back into a small amount of water. Air pressure within the can becomes virtually zero. The atmospheric pressure on the outside is now unopposed, and it crushes the can.

We do not usually pay much attention to atmospheric pressure, although the relatively high pressure associated with fine weather often gives us a sense of well-being, and the low pressure before a storm can make us feel dull and depressed. But if the pressure were removed entirely we would become painfully aware of it—for a few seconds, that is. The breath would expand out of our lungs, our blood would boil at normal body temperature, and the question of air pressure—or anything else in this life—would very soon cease to concern us.

the barometer

In the seventeenth century an Italian physicist, Evangelista Torricelli, invented a device that enabled him to measure accurately the pressure exerted by a column of air. Torricelli, educated in Rome as a mathematician, had written a book about the branch of physics

the atmosphere

AVERAGE ATMOSPHERIC PRESSURE

MILLIMETERS OF MERCURY

MILLIBARS

INCHES OF MERCURY

POUNDS PER SQUARE INCH

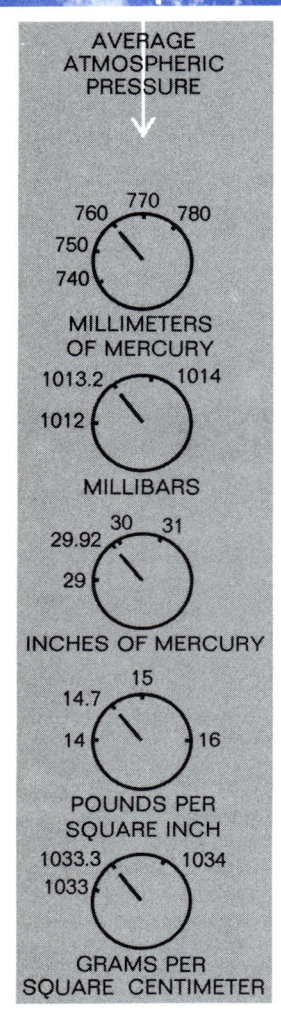

GRAMS PER SQUARE CENTIMETER

Scientists use several different systems of expressing standard gas pressure. Atmospheric pressure is usually described in inches or millibars of mercury.

aneroid barometers

known as mechanics. Impressed by the book, Galileo invited him to Florence.

Galileo suggested that Torricelli investigate the question of why a lift pump can raise water no more than 33 feet. To Galileo, this seemed contrary to Aristotle's statement, "Nature abhors a vacuum." If Aristotle were correct, there should be no limit to the height water can be raised by a lift pump.

Torricelli had performed many experiments having to do with the flow of fluids. It occurred to him that the behavior of the pump might be related to that subject. He had the idea that air pressing down on the surface of water pushed the water up into the tube beneath a pump.

To test this idea, Torricelli used mercury, which has a density of 13.6 g/cm^3 — far greater than the density of water. He sealed a long glass tube at one end and filled it with mercury. Then he stood the tube upright, with its open end in a bowl of mercury. Some of the mercury ran out of the tube, but most of it remained inside, standing high above the level in the bowl.

Next, Torricelli measured the column of mercury that remained in the tube, and found it to be 30 inches high. By further measurements, he determined that the weight of this column of mercury was the same as the weight of a column of water about 33 feet high in a tube of the same diameter. This led Torricelli to a daring conclusion. Contrary to established belief, contrary even to Aristotle, air must have weight. Torricelli reasoned that the column of mercury in the tube and the column of water drawn up by a lift pump were each supported by the weight of a column of the atmosphere. The column of air pressed down on the surface of the mercury in the bowl and on the surface of the pool of water supplying the pump.

Torricelli kept his partly-filled tube of mercury standing in the bowl of mercury. He observed that from day to day the height of the mercury in the tube changed slightly. Presumably these changes were caused by changes in the weight of the column of air pressing down on the mercury in the open bowl. Today we know Torricelli was correct. Atmospheric pressure does vary from day to day, for reasons which we shall see a little later.

Since that time, meteorologists and other scientists have used tubes of mercury like Torricelli's to measure atmospheric pressure. These instruments are called **barometers** (from Greek *baros*, "weight" + Old English *metan*, "to measure"). Many weather stations now measure atmospheric pressure at various altitudes. From these measurements, it is possible to construct atmospheric pressure maps for a given altitude. Pressure maps show lines called **isobars,** along which the atmospheric pressure is the same.

Torricelli's barometer was very cumbersome. The modern mercury barometer is more convenient, but it is still not very porta-

the atmosphere

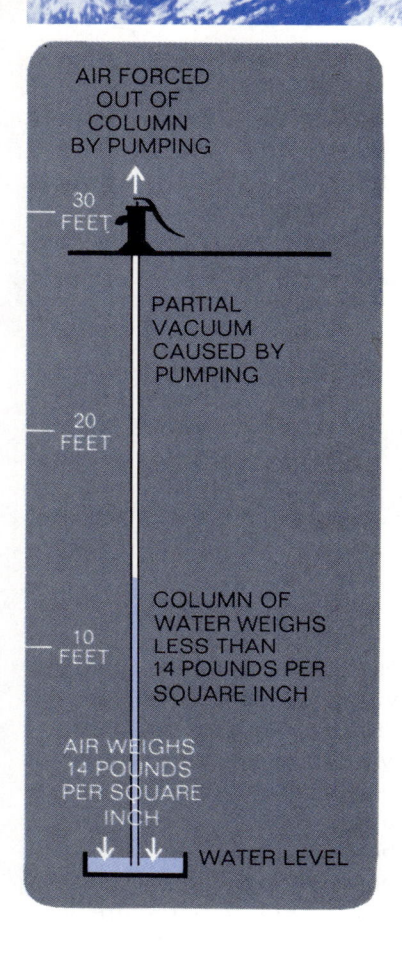

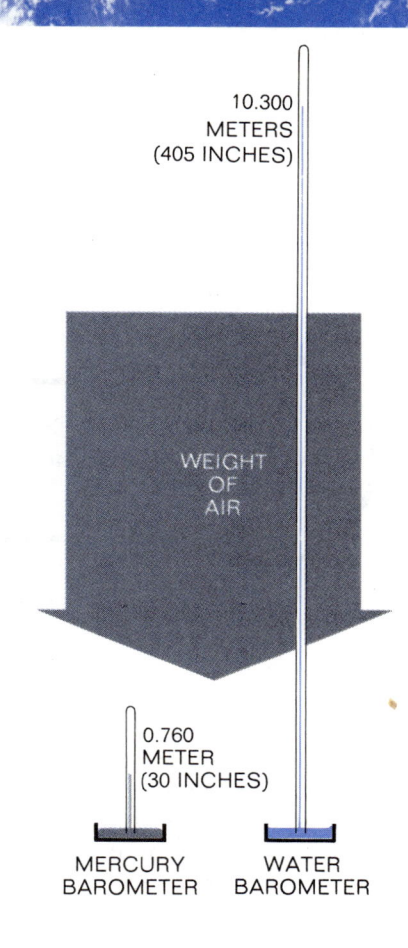

(Left) This diagram shows how air lifts water with the aid of a pump. No matter how perfect the vacuum created, the water cannot be lifted more than approximately 33 feet.

(Right) Atmospheric pressure on mercury and water is the same, and will support the same weight of both. The column of mercury is shorter because the mercury is much denser than the water.

ble. A more convenient but less accurate device is the **aneroid barometer,** which consists of a sealed chamber with thin, flexible metal walls. Most of the air in the chamber is removed, and the chamber is kept from collapsing by a strong metal spring. The force exerted by the spring resists the atmospheric pressure acting on the thin walls. This force depends on how much the spring is bent. If the atmospheric pressure increases, the walls of the chamber will tend to collapse a bit more. The spring will bend, exerting a greater force and limiting the extent of the collapse. The bending of the spring is therefore related to the pressure of the atmosphere. The amount of movement is small, but a series of levers are used to amplify it. The levers move a pointer around a dial, suitably marked to indicate the pressure in inches or millibars.

The aneroid barometer is small and easily carried. It is often used to measure the altitude of a given place, because the atmospheric pressure decreases as the altitude increases. We will examine the reasons for this in the next section.

the atmosphere

Diagram labels: GEAR ARRANGEMENT, FLEXIBLE DIAPHRAGM, SPRING, PARTIAL VACUUM, SCALE (HIGH 31, 30, MED 29, LOW 28)

The spring in an aneroid barometer bends in response to changes in atmospheric pressure. The variations are indicated by a pointer moving around a dial. Aneroid barometers are more convenient but less accurate than mercury barometers.

pressure and altitude

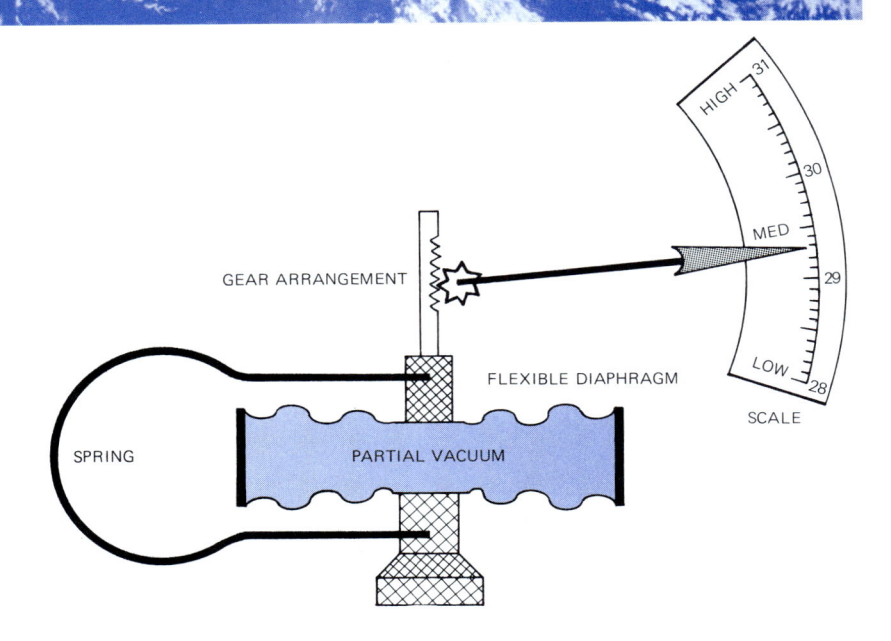

As we know, the atmosphere is held to the earth by gravity. If left to itself, the earthward pull of gravity would tend to collapse the atmosphere. But gravity is opposed by the atmospheric pressure. This is because the pressure of a body of air is exerted in all directions, upward and sideways as well as downward. The pressure at any level of the atmosphere is just enough to balance the air standing above it. The column of air standing above 1 square inch of the earth at sea level weighs approximately 15 pounds, and so sea level pressure is expressed as 15 pounds per square inch. If the pressure were not sufficient, the atmosphere would contract under the force of gravity. The contraction would raise the pressure and restore the balance. (If a local region has low pressure, air rushes in from high pressure regions. This is one factor affecting the strength and direction of winds.)

Shortly after Torricelli conducted his experiments, the French mathematician–philosopher Blaise Pascal became interested in them. Pascal reasoned that if Torricelli was correct in assuming that air pressure supported the mercury in a barometer, then the column of mercury should become shorter on a high mountain, where there would be less air above the barometer than near sea level.

To test this idea, Pascal's brother-in-law took a barometer to the summit of a mountain. He discovered that the level of mercury in the tube was about 80 millimeters lower at the top of the mountain than it was at the foot, confirming Pascal's idea. Thus, almost as soon as scientists learned that air had weight, they found that the pressure exerted by air varies from place to place.

Today, pressure variation due to height has been thoroughly

investigated. Measurements in the outermost reaches of the atmosphere require sophisticated techniques. But close to the earth Pascal's seventeenth-century method is still adequate.

After Pascal's discovery, scientists quickly realized that measurements made on gases at one place could not be compared with measurements made at other places unless air pressure was taken into consideration. This led them to adopt a standard pressure and a standard temperature for all information about gases. As a standard pressure they used the average atmospheric pressure at sea level — the pressure necessary to support a column of mercury 30 inches or 760 mm high. For convenience, 0°C was chosen as the standard temperature. As we have already seen, other units of measurement have been found useful for various special purposes.

solar energy

The sun is the source of enormous energy. In a millionth of a second it radiates more energy than human activity has used throughout all of recorded history. The earth, at an average distance of 150 million km from the sun, intercepts only 2 billionths of that energy. Even so, each square centimeter of area, at the outer edge of the atmosphere, receives about 2 calories of energy every minute. This is an immense amount of energy. To put it another way, think of a situation when the sun is directly overhead. If there were no atmosphere to interfere, each square meter of the earth's surface would receive 19,200 calories of energy from the sun each minute, enough to light twelve 100-watt electric bulbs.

The solar energy received at the surface of the earth is known as **insolation** (**in**coming **sol**ar radi**ation**). Insolation is measured in units called **langleys,** after Samuel P. Langley (1834–1906) of the Smithsonian Institution. A langley equals 1 calorie per square centimeter.

The energy output of the sun has remained fairly constant for a long time. Studies of fossils seem to indicate that its output cannot have changed by more than 25 percent during the last 3 billion years. During the last 50 years the sun has been closely observed, and the amount of energy being received from it appears to have been reasonably constant. Measurements from the earth's surface must make allowances for the effect of the atmosphere, but all things considered, we can say that the amount of radiation emitted by the sun has not varied by more than 2 percent over the last 50 years.

The fact that the sun shines quite steadily does not mean that the insolation at a given place is constant. Various factors can cause

the atmosphere

changes in it. For the moment we shall ignore the effect of the atmosphere, and speak as if the measurements were being made from points on an airless earth.

As we saw in chapter 1, the orbit of the earth is an ellipse, so the earth's distance from the sun varies. This has considerable effect on insolation. When the earth is closest to the sun, the insolation is 7 percent greater than when it is farthest. However, the times of greatest insolation are not necessarily the warmest times of year. For instance, we are nearest the sun on January 3, during the Northern Hemisphere winter, and farthest from it on July 4. (Another interesting point: because the earth is close to the sun during the Northern Hemisphere winters, we would expect them to be warmer than Southern Hemisphere winters, but they are not. The greater area of ocean in the Southern Hemisphere masks the effect of the greater insolation. Consequently, Southern Hemisphere winters are actually warmer than those in the Northern Hemisphere.)

Another factor that modifies insolation is the length of the day. The duration of daytime remains fairly constant at the equator, so the insolation there remains at around 900 langleys per day. In the temperate zones the length of the day varies greatly, causing variations in the insolation. Thus the insolation at the latitude of New Orleans ranges from 500 langleys per day in the winter to 1,000 langleys per day in the summer. Near the poles, this variation is even greater. Summer brings perpetual daylight, and the insolation goes up to 1,100 langleys per day. During the winter, and perpetual night, insolation drops to zero.

Insolation is also related to the position of the sun in the sky. During the summer the sun is high in the sky and its energy strikes the earth at an angle near the perpendicular, as we saw in chapter 1. Therefore the insolation increases. During the winter, when the sun

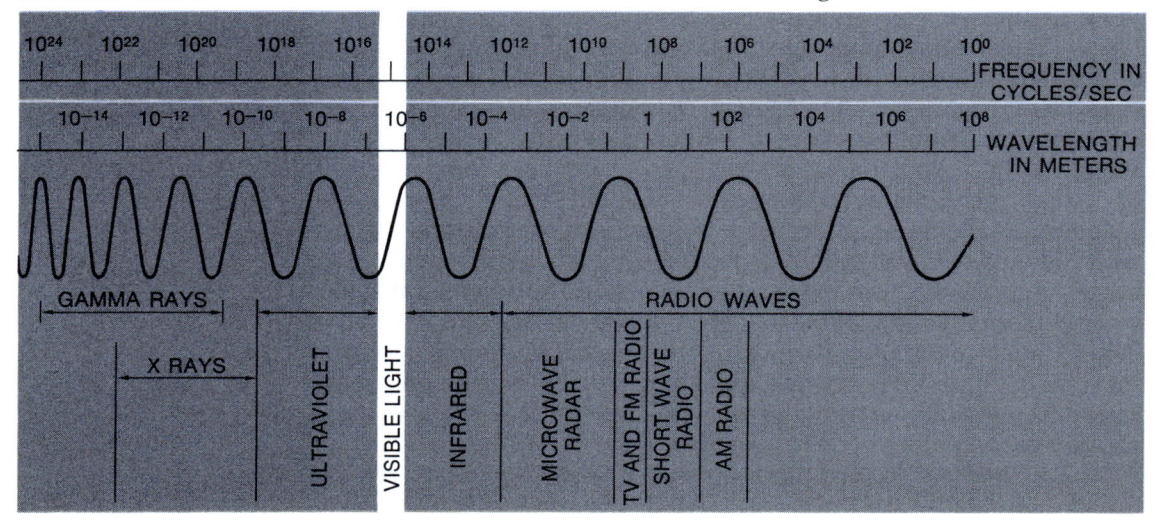

This diagram illustrates that only a small portion of light energy is visible to the human eye. Prolonged exposure to energy of the shortest wavelengths, such as X-rays and ultra-violet, can be dangerous to living things. Earth's atmosphere acts to screen out much of the damaging forms of shortwave radiation.

is low in the sky and its rays strike at a more oblique angle, insolation is less.

the nature of solar radiation

So far, we have been considering the rate at which solar energy would reach a place if there were no atmosphere. The presence of an atmosphere, of course, drastically modifies our estimates. It is important to understand both how solar energy interacts with the atmosphere and how the atmosphere is affected by this interaction. This, in turn, requires an examination of the nature of solar radiation.

The sun is a hot body, with a temperature around 6,000°C. It radiates electromagnetic energy at all wavelengths. However, the amount of radiation emitted is not the same at all wavelengths. Most of the sun's radiation has wavelengths around 0.45–0.5 microns, corresponding to the colors blue and green. Considerable amounts of radiation are also emitted at other visible wavelengths. In addition, some radiation is emitted at wavelengths too short (gamma rays, X-rays, and ultra-violet) or too long (infra-red and radio waves) to be seen by the human eye. Not all of this radiation reaches the earth's surface. The upper part of our atmosphere blocks out many of the extremely short rays, and carbon dioxide and water vapor in the lower atmosphere absorb much of the infra-red.

the energy budget of the atmosphere

Let us look more closely at what happens to solar radiation on its journey through the atmosphere. First, the gas molecules of the atmosphere reflect some of the energy back into space. More of this reflection takes place close to the surface of the earth than at the top of the atmosphere.

Second, some of the energy is absorbed by the molecules of the atmosphere. When matter absorbs energy, the temperature of the matter increases. This absorbed energy is later released as longwave radiation, which we feel as heat.

Third, incoming solar radiation converts some molecules of atmospheric gases to ions. The energy also changes the nucleus of some gas atoms and the molecular structure of other gases. As a result, at heights of more than 200 km above the earth's surface, gases exist only in ionic form.

Everything that happens to solar energy during its passage through the atmosphere takes place in less than one-thousandth of a second. Even before it can effectively enter the atmosphere, about 35 percent of the incoming sunlight is reflected back into space. (We can easily observe one effect of this. When the moon is a crescent, the rest of the moon is faintly lit by the light reflected from the earth. On the surface of the moon, this "earthlight" is 60 times brighter than moonlight is on earth. Before the Space Age, one way of measuring the reflectivity of the earth was to measure how brightly it lit up the moon.)

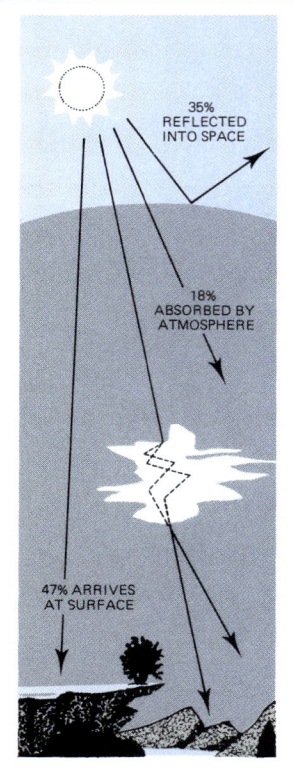

the atmosphere

After 35 percent of the sunlight has been reflected, the remaining 65 percent proceeds through the atmosphere. The atmosphere absorbs about 18 percent, and the surface of the earth receives the remainder—about 47 percent. The surface, in turn, reflects back about half of what it receives. This reflected light may be absorbed by the atmosphere or reflected back to earth again by clouds, or it may simply escape into space.

But what happens to solar energy that is captured by the earth and its atmosphere? If energy accumulated in the earth and the atmosphere indefinitely, the earth would be much too hot to support life. In fact, the oceans would long since have dried up, and the lower atmosphere, at least, would have become a steamy blanket.

Fortunately, the earth also emits radiation, though on a much lesser scale than the sun. In fact, the radiation absorbed by the earth is balanced by the radiation it emits. In the process, many interesting things happen. Solar radiation that enters the atmosphere is shortwave (high-energy) radiation, most of which passes through the atmosphere as visible light. Solar radiation that is absorbed by the atmosphere and the earth's lithosphere is eventually changed to longwave infra-red radiation, which is radiated back into space as heat.

At this point, though, a problem arises. We noted that some of the incoming infra-red radiation from the sun is absorbed by the molecules of carbon dioxide and water vapor in the atmosphere. The same thing happens to much of the outgoing infra-red radiation from the earth. It is trapped and held in the atmosphere for a time before it is finally re-radiated into space. This produces what is known as a **greenhouse effect**—a buildup of heat in the atmosphere.

The principle of a greenhouse is fairly simple. The glass walls and roof allow the fairly short waves of visible light to enter freely, but interfere with the escape of the longer infra-red heat waves. Hence the temperature inside the greenhouse becomes higher than the temperature outside, and plants can be grown all winter.

Not all of the sun's energy reaches the surface of the earth. Much of it is reflected back into space at the top of the atmosphere, and some is absorbed by the atmosphere before it can reach the surface.

Shortwave radiation passes freely through the glass of a greenhouse, but the glass traps the longer infra-red rays inside, causing the temperature to rise. In the same way, CO_2 traps infra-red radiation in the earth's atmosphere.

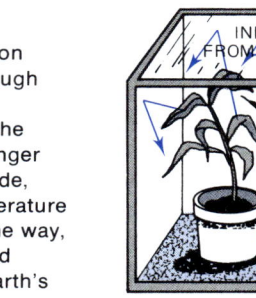

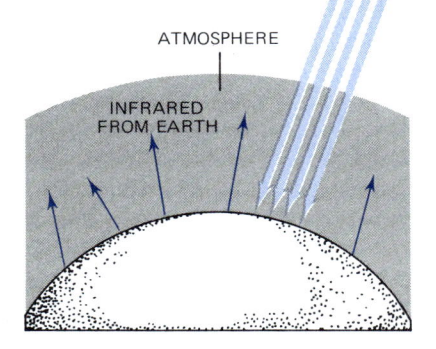

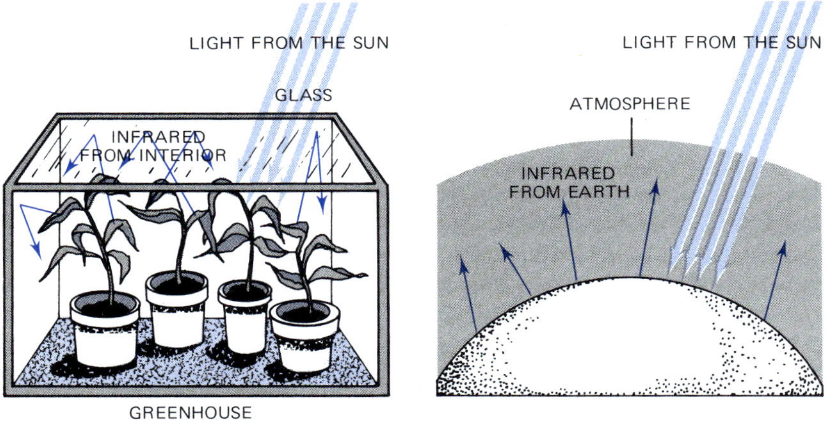

GREENHOUSE

the atmosphere

Carbon dioxide and water vapor in the atmosphere act rather like the glass in a greenhouse. This is useful, up to a point. The heat that is held in the atmosphere by the greenhouse effect is what keeps the earth from cooling off too much at night, or during the winter. Without it, our nights might be as cold as those on Mars or on the moon. But heat is like other good things—there can be too much of it. And if the amount of carbon dioxide in the atmosphere rises above its present level, the greenhouse effect is likely to heat up the earth even further. Already, since 1900, the CO_2 level has risen by at least 20 percent, partly as a result of the air pollution that is a byproduct of modern industrial technology. The long-term consequences of a large rise in the CO_2 level are hard to predict, but they might well be disastrous. We will discuss this problem in detail in chapter 17.

energy at work: the tropical zone

The energy of solar radiation is inversely proportional to its wavelength—that is, the shorter the wavelength, the greater the energy. When shortwave radiation is changed to longwave radiation, such as infra-red, some energy is given off. This energy heats the gases of the atmosphere and evaporates water, leading to the production of winds and the formation of clouds. To understand how radiation is converted into motion, we will examine the Tropical Zone in some detail.

The Tropical Zone is the part of the earth that receives the greatest amount of solar energy per unit of space. At every place in that zone, the sun is directly overhead at some time during the year. The Tropical Zone receives about 1.05×10^5 cal/cm²/yr (105 kilolangleys per year). Of this energy, about 11 percent heats the soil, rocks, water, and air. About 76 percent evaporates water from the oceans and the land. This accounts for 87 percent and leaves us with 13 percent.

A very small amount of this 13 percent is used for photosynthesis and not released by plants or animals through oxidation. If we add that to the small amount that oxidizes minerals at the earth's surface, we can account for about 1 percent. What about the remaining 12 percent?

We are all aware that the air moves. We feel it nearly every day as a breeze or a wind. This is what the remaining 12 percent of the energy does—it moves the air. It can do so because it is converted to heat at the earth's surface. This heat powers the atmosphere—a huge heat engine.

world use of solar energy

When we look at a globe we see immediately that much more of the Southern Hemisphere is covered by oceans than by land. In the Northern Hemisphere, about 40 percent of the surface is land, whereas south of the equator only 20 percent is land. This distribution of land and water affects the way solar energy is used in the Northern and Southern Hemispheres. Refer to Table 11-1 for a com-

the atmosphere

parison of the energy budget for the two hemispheres.

The average receipt of energy at the surface for the world as a whole is 72,000 cal/cm²/yr. In the Southern Hemisphere, a large frac-

	Evaporation	Heating	Energy Transferred
Northern Hemisphere	78%	21%	1%
Southern Hemisphere	86%	15%	−1%
Entire earth	82%	18%	0%

Table 11-1. How incoming solar energy is used by the two hemispheres of the earth.

tion of this is used to evaporate water—much more than in the Northern Hemisphere. In the process of evaporation, a great deal of energy is used without any change in temperature. Thus, less insolation is converted to heat in the Southern Hemisphere than in the Northern, and the temperature change between summer and winter is less in the Southern Hemisphere.

In general, the Southern Hemisphere is cooler than the Northern Hemisphere. We might expect the Northern Hemisphere to get warmer and warmer and the Southern Hemisphere to become cooler and cooler, but this does not happen. The excess heat that is generated in the north is spread by winds and ocean currents In this way, the temperature pattern over the whole world stays approximately the same from year to year. In chapter 13 we will discuss the way in which oceans help to distribute heat.

the structure of the atmosphere

The earth's atmosphere can be divided into distinct layers, largely on the basis of temperature. There are three warm layers, separated by two rather cold layers. The accompanying figure shows the overall structure of these layers.

As we saw earlier, we can also make a distinction between the lower and the upper atmosphere. The lower atmosphere is constantly stirred by turbulent air motions. As a result, its chemical makeup is quite uniform. Samples of the lower atmosphere have been taken all over the world, from sea level to a height of about 100 km. The oxygen content in each of these samples is very nearly the same. The lower atmosphere extends upward to about 120 km above sea level. Its layers (from the ground up) are the troposphere, the stratosphere, and the mesosphere.

the troposphere

The region closest to the earth's surface, where most of the weather occurs, is called the **troposphere.** In this layer of the atmosphere there is constant change, which we can observe from day to day. The temperature of the troposphere gradually falls with increase in altitude. The rate of temperature decrease is influenced by many factors, but in general it is about 6°C per 1,000 m (3.3°F per 1,000 ft).

The upper limit of the troposphere, at about 12 km above sea level, is called the **tropopause.** The temperature at the tropopause averages about −65°C (−85°F). But there the direction of temperature change reverses. Above this region, the temperature becomes warmer with increasing altitude.

Above the tropopause lies the **stratosphere,** extending upward to about 60 km above sea level. It temperature increases from about −65°C to about freezing at the upper limit or **stratopause.**

Clouds form so rarely in the stratosphere that we can say there are no clouds there. Molecules of water vapor in this region are too far apart to be attracted to one another, and there seem to be no dust particles on which ice crystals might form. (We will discuss cloud formation in greater detail in the next chapter).

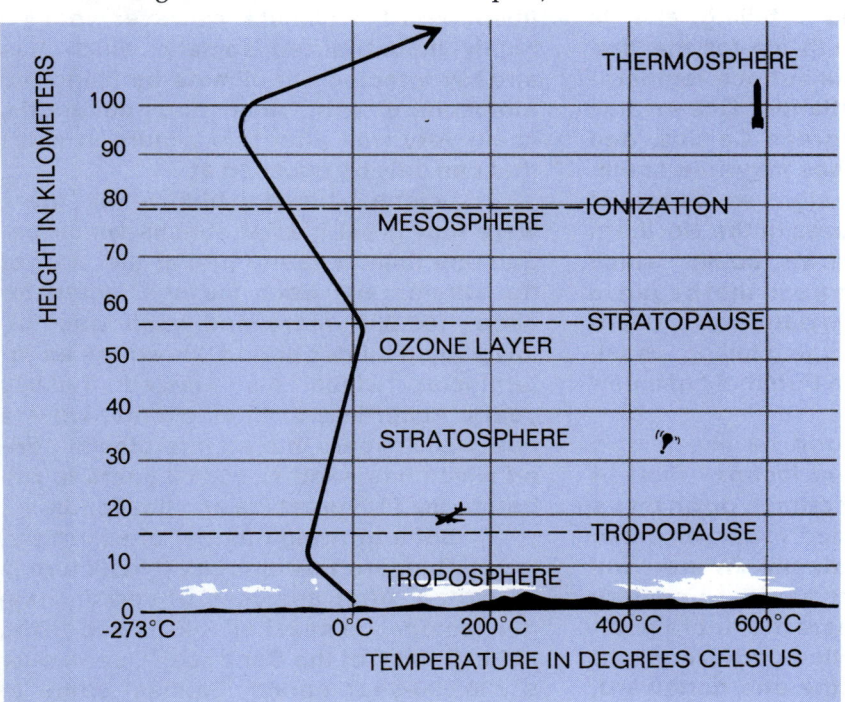

the stratosphere

The chart shows that temperature in the atmosphere undergoes a series of pronounced changes with increase in altitude. At the upper limit of the atmosphere, it may reach as high as 2,000°C.

At an altitude of about 50 km, solar radiation causes some of the oxygen molecules (O_2) to break up into atoms of oxygen (O). These combine with O_2 molecules to form molecules of O_3, a gas called **ozone.** The presence of this ozone-rich layer of atmosphere is very important to life on earth. Ozone absorbs the shorter-wavelength ultra-violet radiation from the sun and changes it to longer-

IN FOCUS:
is a new ice age on the way?

Armadillos are deserting their northern haunts in the American midwest for warmer southern climes. Some American farmers are complaining of mid-summer frosts and resulting damage to crops. The coldest winter on record chilled our continent in 1976–77, even bringing snow to sub-tropical Miami Beach. The average annual temperature of Iceland is falling, and its ports are socked in with ice for the first time in this century. The surface temperature of the North Atlantic Ocean has dropped by seven degrees Celsius, and the tepid Gulf Stream has migrated south. These are some of the signs which reflect the fact that temperatures in the Northern Hemisphere have grown colder since 1940, reversing a warm trend that began in the 1890s. In light of this data, some scientists have suggested the chilling possibility that we are on the threshold of a new ice age.

What will the world be like if it is transformed into a giant icebox? First of all, the glaciers will not crunch down on us suddenly and unexpectedly, but will make their appearance gradually. Winter will grow longer, year after year and century after century, until it lasts the entire year. The ice caps of Antarctica and Greenland will grow and spread, the one northward, the other southward. Glacial movement will halt when all the atmospheric moisture has been transformed into ice or snow. Mountains of ice thousands of feet thick will enshroud North America as far south as Long island, and much of Europe as well. Eventually, a full quarter of the earth will be clamped in their frigid grip. Meanwhile, the level of the oceans will drop by more than 300 feet. Equatorial regions will become much cooler than they are now, but life could survive there.

This scenario is based on conditions during past ice ages. It assumes that human technology will have little or no influence on the climate. However, today's highly industrialized society, which has already affected our climate by filling the atmosphere with dust and pollutants, could very well alter this picture in ways that can only be guessed at.

Scientists used to believe that there were four great glacial epochs which began one million years ago, at the start of the Quaternary. Each ice age lasted for about 100,000 years and each was followed by a milder period, known as an interglacial, which also lasted 100,000 years. According to this schema, we are now 10,000 years into an interglacial period which has another 90,000 years to run before the fifth great ice age descends.

But a growing number of scientists, led by Caesare Emiliani, say this picture is all wrong. After analyses of certain oxygen isotopes in seashell fossils deposited at the bottom of the Caribbean Ocean over a 700,000-year period, Emiliani came to the revolutionary conclusion that the ice ages were much more frequent and lasted for shorter periods of time than the geology textbooks stated. He found that in the past 300,000 years alone the world has seen seven ice ages, each followed by a warm period. None of the glacials or interglacials even came close to lasting for 100,000 years, as previously proposed. At its worst, for example, the third glacial epoch was around for only about 2,000 years, whereas the first interglacial dragged on for some 7,000 years. In addition, Emiliani found that these principal periods were themselves interspersed with less extreme cycles of warm and cold episodes.

If Emiliani is right, it is conceivable that an ice age might interrupt our current geological summer, and it could do so rather abruptly. Emiliani himself believes that a new ice age is just around the corner, possibly as near as two or three centuries. Other investigators hold different views. Some think it is less than a hundred years away, others that our planet will not be locked up in cold storage for another three or four thousand years.

What causes an ice age? No one is absolutely certain, but there is no shortage of opinions on the subject. The advent of ice ages has variously been attributed to changes in solar energy output, in the amount of volcanic dust in the atmosphere, in the earth's magnetic field, and in the amount of carbon dioxide in the ocean and atmosphere, as well as to the growth and melting of the ice caps and the circulation of deep ocean currents.

A theory that has been receiving a good deal of attention recently blames our deep freezes on slight but cyclic changes in the earth's orbit around the sun. This theory, advanced by J. D. Hays, John Imbrie, and N. J. Shackleton, is based on the observation that there are three cycles of orbital variation that overlap and combine to alter global climate. During each of these cycles—which last 23,000 years, 41,000 years, and 93,000 years respectively—the surface of the earth receives varying amounts of solar radiation. The amount received depends upon the earth's distance from the sun as it progresses in its elliptical orbit, and the degree to which its axis tilts as it rotates. By correlating this astronomical data with geological evidence—the examination of sediments from the floor of the Indian Ocean which are nearly a half million years old—these scientists predict that the earth will enter a new ice age within the next few thousand years.

While those who foresee an impending freeze currently hold the floor in this debate, it is probably safe to guess that neither the present generation nor its children will ever live in the shadow of giant glaciers like those that towered over the world millenia ago.

wavelength energy. Radiant energy in the ultra-violet part of the spectrum causes serious skin burns, and too much exposure to it may cause skin cancer.

Some scientists fear that nitrogen oxide produced by high-flying jet aircraft of the near future may interfere with the ozone layer. They estimate that in some places the present concentration of ozone may be reduced by as much as 50 percent. Other scientists are not so fearful, since the main ozone layer is close to the top of the stratosphere and not likely to be penetrated by jets. The problem of human activities that threaten the ozone layer will be discussed further in chapter 17.

the mesosphere

Above the stratosphere the temperature again drops, until at about 80 km above sea level it is approximately −100°C. The region from about 60 to 80 km above sea level is called the **mesosphere.**

The mesosphere is sometimes called the **ionosphere,** because in this region solar radiation changes the molecules of atmospheric gases to ions. Any water vapor present in the ionosphere decomposes to form oxygen and hydrogen. In this way, the earth is constantly drying out while the oxygen content of the atmosphere is increased. It is probable that the first oxygen in the earth's primitive atmosphere was produced this way, long before photosynthetic plants evolved.

the thermosphere

Above the mesosphere, in the **thermosphere** of the upper atmosphere, the temperature rises rapidly at first and then more slowly. At the limit of the atmosphere, it may be as high as 2,000°C. The high temperature in the thermosphere is the result of absorption of shortwave solar radiation. But since the particles in the thermosphere are more widely scattered than those in the lower atmosphere, not many calories of heat are produced here by the absorption of solar radiation.

The upper atmosphere lacks turbulence and is thin. The molecules of various gases in this region separate into layers, with the heaviest molecules lying closest to the earth and the lightest molecules in the most distant layer. As we saw earlier, at the very fringes of the upper atmosphere molecules of hydrogen are constantly escaping into space. For this reason, the outermost part of the upper atmosphere is sometimes known as the **exosphere.**

the spinning earth

The atmosphere can never be entirely steady. The heat it receives varies daily and seasonally, which results in alternate heating and cooling. Local conditions greatly modify the rate at which the temperature changes, and temperature may differ widely from region to region. Since pressure is related to temperature, regions of different pressure are formed. This, in turn, leads to atmospheric motion, as air from a region of high pressure rushes into a region of lower pressure. Furthermore, the motion occurs on a spinning globe, and this causes complications. We will look at one example of air motion to illustrate this. A more detailed study of air movements and the weather conditions they create will appear in the next chapter.

The uneven distribution of incoming solar radiation as it approaches the surface of the earth causes uneven heating of air, land, and ocean. Here we are interested in what happens to the air. We have seen that when a gas is heated, its molecules move faster. If the gas is in a closed container with firm walls, heating causes the pressure exerted by the gas to increase. The pressure is exerted in all directions—to the sides, up, and down. If the gas is not in a container, the increased pressure allows it to push out in all directions, and its original volume increases.

The air in the tropical heat belt between 10°N Latitude and 10°S Latitude is not in a firm-walled container. Expanding in all directions, the heated air pushes against the solid earth and sea below it, which do not yield. It also meets with resistance from the cooler and therefore denser air to the north and south. Thus the only direction in which this great mass of heated air can readily expand is upward. It flows upward until it spills to the north and south over the top of the adjacent walls of cooler air. The belt of rising air usually has only weak breezes at the earth's surface and hence is called the **equatorial calms,** or the **doldrums.**

What happens next to this tropical air? If the earth were not rotating upon its axis, the rising air would flow directly north and south away from the heat belt. But the earth is spinning, and this affects all movable things. How much they are affected depends on friction. Solids, like ourselves, are held in place by gravity and the friction between them and the ground. Fluids—gases and liquids— are also held to the earth by gravity, but there is little internal friction in such substances. The lack of friction makes fluids sensitive to the earth's spin.

Suppose we have a wheel that is 36 inches in diameter. Its circumference is about 113 inches. If the wheel rotates very slowly at one rotation per minute, its **angular velocity** is 360 degrees per minute, since there are 360 degrees in a circle. This is the angular velocity of the entire wheel—rim, spokes, and hub. But not all parts of the wheel move the same distance in one revolution. A point on the rim moves 113 inches every minute. That is, its **linear velocity** is 113 inches per minute. The linear velocity of a point nearer the hub is

the tropical heat belt

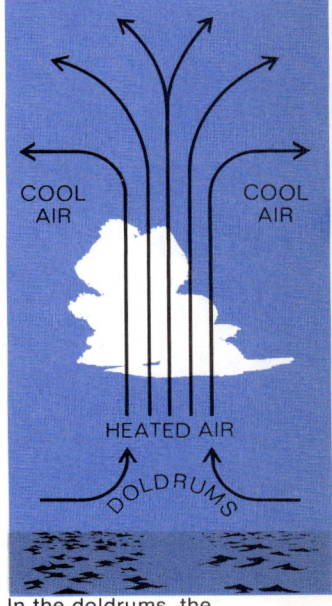

In the doldrums, the pressure of cooler air to the north and south causes the less dense, heated air to rise.

angular velocity and linear velocity

the atmosphere

much smaller.

Now consider the rotation of the earth on its axis. The earth turns 360 degrees in 24 hours. Therefore, in one hour it must turn 15 degrees. The earth and its atmosphere thus have an eastward angular velocity of 25 degrees per hour.

We can describe this eastward motion in another way. The distance around the earth at the equator is about 40,000 km. If you stand still at the equator, you and the ground are moving eastward at about 1,660 km/hr in relation to a point in space. But standing still in New Orleans (30°N), you are moving eastward at only about 1,440 km/hr. The reason for this is that at 30 degrees latitude you are nearer to the earth's axis, and the distance around the earth is only about 34,560 km instead of the 40,000 km it is at the equator.

At each place the angular velocity is the same—15 degrees per hour. But the linear velocity changes with latitude, from more than 1,600 km/hr at the equator to nothing at the poles. (At the poles you are standing right on the earth's axis, so you do not travel; you just turn around.)

the coriolis effect

Now let us get back to the air we left spilling northward and southward from the equatorial heat belt—more specifically, to the air spilling northward at 10°N Latitude. On that parallel, a degree of longitude is 109.686 km from east to west, so the air, as well as the earth, is moving eastward at 15 times this, or 1,645 km/hr—about the same as at the equator. A little further north, at 11°N Latitude, a degree of longitude is slightly less, or 109.333 km, so the eastward travel is only 1,639.8 km/hr.

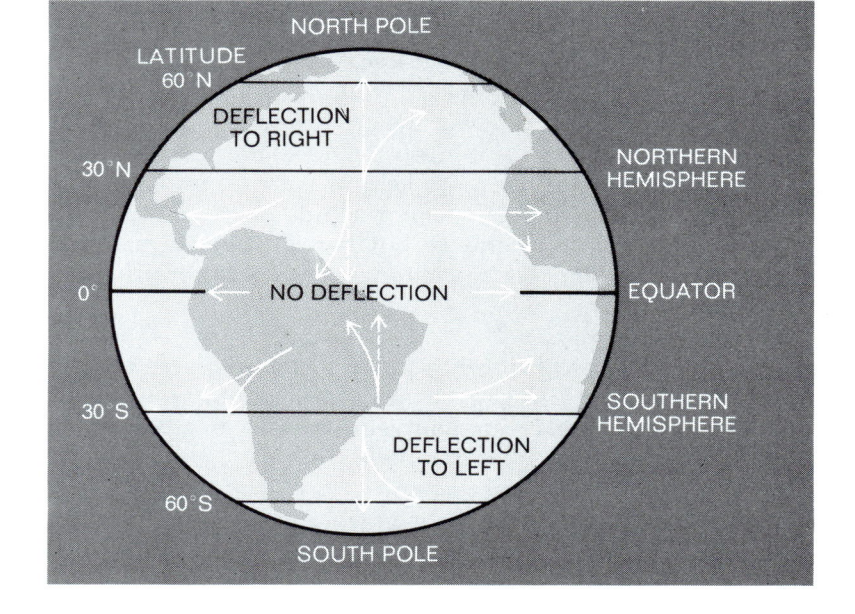

The Coriolis effect created by the earth's rotation causes moving objects to be deflected to the right in the Northern Hemisphere and to the left in the Southern Hemisphere. When ballistic missiles or satellites are launched, this effect must be considered in planning their trajectories.

ARE WE DESTROYING THE OZONE LAYER?

Our rapidly growing technology has often been compared to the Frankenstein monster: originally created to make life easier, it may instead be destroying us. For example, pollutants released by combustion engines and industrial plants are blamed for the rising rate of emphysema and heart disease. Almost daily, some new report tells us that the chemical additives which make our foods taste better or last longer can cause cancer in laboratory animals. Now comes word that technology may be blotting out the ozone layer, a development which, among other things, may produce a jump in the incidence of skin cancer.

The ozone layer in the stratosphere protects life on earth from the sun's deadly ultra-violet rays. Many scientists believe that this layer is being wiped out in a three-pronged attack inadvertently launched by humans. First, there are the fluorocarbons—synthetic chemicals composed of fluorine, chlorine, and carbon—which are sprayed into the atmosphere, chiefly by aerosol spray cans. These inert chemicals drift up to the stratosphere, where their chlorine atoms destroy ozone. A few experts estimate that the fluorocarbons have already reduced the ozone layer by one or two percent.

A second threat is the supersonic transport, or SST. Nitric oxide contained in the exhausts of these huge aircraft interferes with the natural formation of ozone. So do the oxides of nitrogen produced by the atmospheric testing of nuclear weapons.

Some scientists believe that those who predict impending doom are merely crying wolf. In the eyes of these researchers, even an 80 percent reduction of the ozone would not endanger life on earth. Still others say that we simply don't know what depletion of the ozone is doing to us. This debate might be settled by a complete testing program, including the use of air-sampling rockets and ground-based stations to monitor ultra-violet radiation. But such a program has yet to be proposed, let alone launched. In the meantime, we can only hope that those who foresee the worst are wrong.

the atmosphere

Now think of what this means for air that is moving northward and 1,645 km/hr eastward because of the earth's spin. As the air moves northward, it moves over a part of the earth's surface that is nearer the axis, and is therefore moving eastward more slowly than 1,645 km/hr. But the air itself tends to keep its original speed. The result is that from the surface at 11°N Latitude the air appears to be moving not only northward but also a little eastward, at a rate of about 5 km/hr (1,645−1,640). Look at the accompanying figure for an explanation of this.

The first person to explain the details of this phenomenon was a French mathematician, Gaspard de Coriolis (1792−1843). Today it is known as the **Coriolis effect.** It operates on any moving fluid, such as the winds of the atmosphere or the currents of the ocean. In the Northern Hemisphere, currents and winds veer to the right when you face in the direction of flow, and in the Southern Hemisphere they veer to the left. The Coriolis effect must be taken into account when long-range artillery or ballistic missiles are aimed. For example, a rocket fired from New Orleans and aimed directly at Chicago would land quite a bit to the east of its target.

The Coriolis effect has played a significant role in human history. It produces a steady easterly wind near the equator and steady westerly winds in the temperate zones. These winds have been very useful to sailors and seagoing traders. How are they produced?

Air rises high in the troposphere as a result of equatorial heat. It drifts away in a northeasterly direction north of the equator and in a southeasterly direction below the equator. Eventually this air from the heat belt cools and dries, and thus increases in density. When this happens, it sinks toward the surface of the earth at about 30°N Latitude and a little closer to the equator in the Southern Hemisphere. Such a circulation pattern is called an **air cell.** Since fluids do not pile up but must spread out, the sinking air must flow away in all directions once it reaches the surface.

The belts around the earth where the air sinks are called the **horse latitudes,** a name that dates back to the days of the Spanish exploration of the New World. These regions have very light and variable winds, and most of the air movement is downward. The Spanish galleons were often becalmed there, and their horses suffered from lack of water. When the horses died they were heaved overboard. The regions thus became known as the horse latitudes.

When air sinks to the earth's surface, as it does at the horse latitudes, it becomes compressed. The molecules it contains move faster and the air thus becomes warmer, an example of adiabatic heating. The horse latitudes are therefore a zone of sinking, hot, dry air, with only variable gentle winds.

Air that pushes northward along the surface from the horse latitudes does not travel directly north. The Coriolis effect causes it to

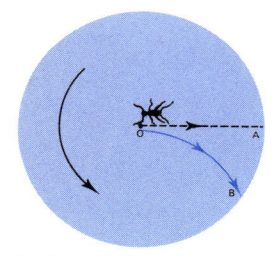

As the rotating disc turns counterclockwise, the bug trying to crawl from O to A actually travels the path from O to B.

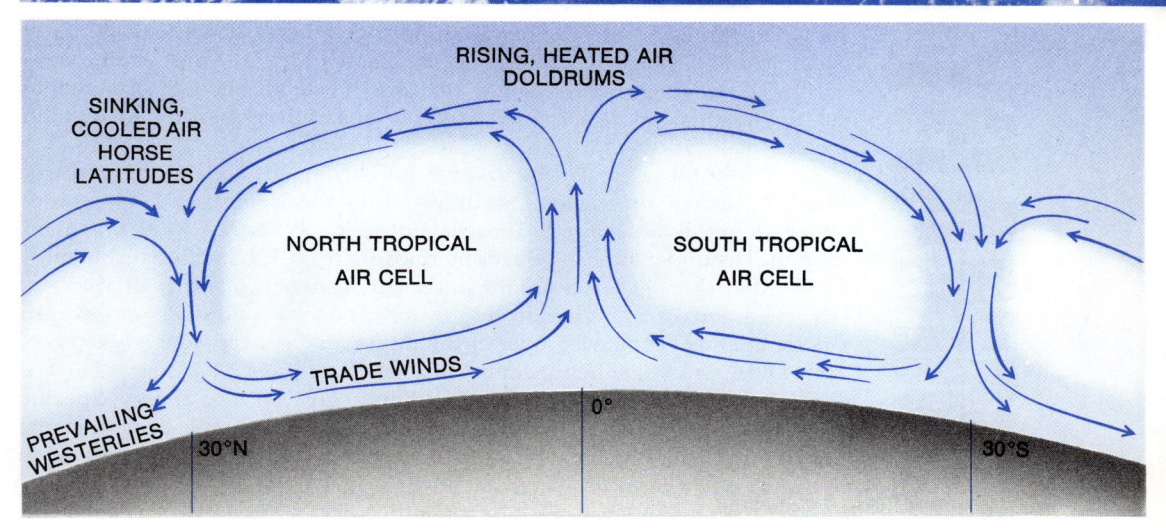

the atmosphere

The trade winds and the prevailing westerlies are created by convective motion of rising warm air and sinking cool air near the equator.

veer toward the right in the Northern Hemisphere. As a result, the winds north of the horse latitudes blow from the southwest to the northeast. These winds are known as the **prevailing westerlies.**

Air pushing southward from the horse latitudes, back toward the equator, is also influenced by the Coriolis effect, which causes it to drift westward. The combined southward and westward movements result in a wind that blows from the northeast to the southwest. This flow of air replaces the air that is swept upward by heating in the zone of the doldrums. These winds from the northeast pushed the sailing vessels engaged in trade with the West Indies from Europe toward the islands, and thus became known as the **trade winds.**

Once the air moved by the trade winds enters the doldrums, it expands by heating, rises, and again begins the cycle that starts with air rising in the heat belt. The part of the atmosphere that lies between the doldrums and the horse latitudes is called the **north tropical air cell.** A similar sequence of events south of the equator produces a south tropical air cell, and another band of prevailing westerlies further to the south.

the atmosphere

SUMMARY

All living things on earth depend on the atmosphere, which is made up of gases held to the earth by gravity. The atmosphere acts as a heat engine, converting solar energy into mechanical energy. It extends about 50,000 km into space. The atmosphere is densest at sea level and decreases in density with increased altitude.

The atmosphere can be divided into layers on the basis of temperature. The lower atmosphere is divided into the troposphere, the stratosphere, and the mesosphere. These layers are characterized by turbulent air motion. The upper atmosphere, or thermosphere, is thin and not turbulent.

More than 99 percent of the lower atmosphere is made up of nitrogen, oxygen, argon, carbon dioxide, and water vapor, with nitrogen and oxygen predominating. The lower layer of the upper atmosphere is largely oxygen; above this is a layer of helium and hydrogen.

Temperature is the ability of a body to give up heat energy. Specific heat is the number of calories that must be added to or taken away from one gram of a substance to change its temperature by 1°C. Pressure is defined as the amount of force acting on a unit area.

Gases become warmer when compressed, and cooler when expanded. A process in which energy neither enters nor escapes is called adiabatic. Air is often warmed or cooled adiabatically.

In the seventeenth century, Torricelli demonstrated with his primitive barometer that air has weight. Barometers are used to measure atmospheric pressure and thus can be used to measure altitude, since pressure decreases with increased altitude.

The sun is the earth's primary source of energy. Insolation is the amount of solar energy received by each square centimeter of earth's surface. It is measured in units called langleys. A langley equals 1 calorie per cm^2. The amount of insolation received varies with the time of year, the length of the day, and the latitude of the location.

The radiation absorbed by the earth is balanced by the radiation that it emits. Part of the solar energy received is reflected back into space. Part of the energy emitted by earth is trapped near the surface by carbon dioxide in the atmosphere, producing the greenhouse effect. Solar radiation also leads to the formation of atmospheric ozone, which prevents too much damaging ultra-violet radiation from reaching the earth's surface.

Because the earth rotates on its axis, moving air is deflected from a straight-line path. It veers to the side instead. This is known as the Coriolis effect. The Coriolis effect is responsible for the steady easterly winds near the equator and the steady westerly winds in the temperate zones.

REVIEW QUESTIONS

the atmosphere

1. Why can we describe the earth's atmosphere as a form of heat engine?
2. How does the density of the atmosphere vary?
3. Describe what happens to solar energy as it reaches and enters the atmosphere.
4. Name the three layers of the lower atmosphere, beginning with the highest.
5. What is specific heat?
6. Give a general description of the chemical composition of the atmosphere.
7. What is pressure?
8. Where does the oxygen in the atmosphere come from?
9. Why is ozone important?
10. What is the thermosphere?
11. Describe what happens to a gas when it is compressed or expanded.
12. What is an adiabatic process?
13. Give a general description of Torricelli's simple barometer. What important conclusion did it enable him to reach?
14. How does the ocean modify the temperature of the lower atmosphere?
15. In what form does nitrogen exist in the atmosphere?
16. What is insolation? How is it measured?
17. Explain the Coriolis effect, and give an example of how it works.
18. How does an aneroid barometer work?
19. What determines humidity?
20. Explain why a barometer can be used to determine the altitude of a given location.
21. What are some of the factors that alter insolation?
22. Explain the greenhouse effect. What would happen if it did not exist?
23. What does the average energy of a particle depend on?

CHAPTER TWELVE

WIND AND WEATHER

On Wednesday, September 4, 1900, the weather bureau in Galveston, Texas, received a brief notice from Washington. A small tropical disturbance had been reported from Cuba and seemed to be moving north. The next day there was another report: the storm was now near Key West, Florida. On Friday, the barometric pressure in Galveston began to fall.

Galveston is on the coast of the Gulf of Mexico, and in that pleasant early September it was crowded with vacationers who had fled the oppressive heat further inland. As the wind rose on Friday afternoon and Saturday morning, huge waves began rolling in from the Gulf, and people flocked to the beaches to watch. When a lone weatherman, who knew something about Gulf hurricanes, raced along the beach shouting at them to get away while they still could, they laughed at him. Go home, and miss all this excitement? Nonsense.

By noon on Saturday, though, they had found that it was no nonsense. The winds rose to well over a hundred miles an hour—how much over, it was impossible to tell, for the measuring instruments at the weather station had been blown away. Even worse was the water, which surged past the beaches and poured into the low-lying city, rising more than a foot every hour. People grabbed what they could and struggled inland, yesterday's laughter forgotten in the sudden shock of a grim fight for survival. Streets became rivers of churning water; the wind was a solid hammer that knocked people over, pinned them to walls, threw them into the water to struggle helplessly till they were swept under and drowned. One tiny child survived only because someone, in desperation, actually nailed him to a floating roof, where he was lucky enough to be found when the storm was over. At around eight o'clock Saturday evening, the storm tide finally crested at some twenty feet above normal high tide level. The highest land in Galveston was five feet lower.

wind and weather

By Sunday morning the worst of the hurricane was over. Dawn revealed a ruined city; stinking mud covered the wreckage of handsome buildings, and the dead, human and animal, lay in heaps. No one ever knew for sure how many people died, but it was at least six thousand. Conventional burial was impossible; most of the bodies were loaded onto barges, towed out into the Gulf, and dumped. When the city was rebuilt, the planners saw to it that one item was given very high priority—a seawall seventeen feet high, and strong enough to withstand the pounding of any hurricane-driven tide that might ever again thunder in from the blue waters of the Gulf.

water in the air

Water plays a major role in most aspects of weather. As we saw earlier, liquid water has a high heat capacity. One calorie of heat will raise the temperature of one gram of water by one degree Celsius. Water freezes at 0°C. When it melts, each gram absorbs 80 calories. (This is why ice cubes are so effective in keeping your cocktail cold.)

evaporation

Liquid water evaporates at all temperatures. When it evaporates in an enclosed space it fills the space with water vapor, which is only a little more than half as dense as air. Evaporation proceeds until the concentration of water vapor in the enclosed space reaches a certain maximum. The space is then **saturated;** that is, the amount of water vapor in it cannot increase any further.

When one gram of water evaporates it absorbs nearly 600 calories of heat, usually from the surroundings, or from the remaining liquid. Unless there is a continuous source of heat, the evaporation of a little water cools the rest, and thus slows down further evaporation.

The earth's atmosphere holds only a small amount of water. If all of it fell as rain, the average water level in the world would increase by only about an inch. But since rain falls somewhere every day of the year, it is clear that the atmosphere's supply of water vapor is continually replenished by evaporation. This process is part of what is called the **hydrologic cycle.**

humidity

Humidity is the water vapor content of air. People often associate humidity with wetness, but water vapor is quite dry. (The moist cloud we see hanging in the air over hot teakettles and in front of frosty noses is not water vapor. It is the condensed droplets that form when the invisible, gaseous water vapor cools and becomes liquid water again.) Water can exist as a solid, a liquid, or a gas, and is constantly changing from one state into another. The water vapor

wind and weather

Thunderstorms often occur as small, isolated storms. Here rain can be seen falling from a localized mountain thunderstorm.

content of the atmosphere therefore varies greatly from place to place and from time to time.

Humidity has a strong influence on the weather, much of which involves the condensation of water into rain, sleet, or snow. Water also affects the energy balance of the atmosphere. It warms the air by absorbing a significant amount of the solar energy that passes through it. It also acts as a reservoir of heat energy. For example, during the day water evaporates, absorbing heat and keeping the day cool. At night this evaporated water condenses and gives back the heat, warming the night.

Humidity is measured in several ways. **Absolute humidity** is the number of grams of water vapor in one cubic meter of air. Warm air can hold much more water per cubic meter than cold air. **Relative humidity,** usually expressed as a percentage, is the actual water vapor content in a sample of air compared to the content it would have if it were saturated. For instance, suppose we consider a cubic meter of air containing 8.65 grams of water vapor. If our sample were saturated it would contain 17.3 grams, twice as much as it actually does. The relative humidity of the sample is therefore 50 percent.

We all know that hot summer days are more uncomfortable when the relative humidity is high. This is because we regulate our body temperature by perspiring. As perspiration evaporates on our skin, it produces a strong cooling effect. But when the air is nearly saturated—that is, when the relative humidity is near 100 percent—perspiration evaporates slowly. Our body struggles to cool itself by producing more and more perspiration, making us wet and miserable. Human beings can survive in dry air at temperatures of 95°C (203°F) for hours. But humid air at even 60°C (128°F) often causes heat prostration and death.

wind and weather

Altocumulus clouds are the "mackerel sky" dreaded by early mariners because they often heralded an approaching storm system.

condensation

clouds and fog

What causes condensation? For condensation to occur, the air must be saturated, and this condition is easily produced when the air is cooled. **Radiative cooling** occurs when solar heat that has been absorbed is re-radiated upward from the earth's surface or from the upper surfaces of clouds. This type of cooling is most effective on clear nights. If overlying clouds are present much of the radiation is trapped under the clouds and cannot escape. **Advective cooling** takes place when a warm body of air comes into contact with cold land or sea.

Cooling and condensation also occur when two unsaturated air masses, one warm and one cold, meet and mix. As we will see a little later, the clouds and rain that are produced when cold air meets warm air are an important feature of weather in temperate climates. Adiabatic cooling, which we discussed in chapter 11, can also lead to condensation. Dry air cools 10°C for every kilometer that it rises. Moist air in which condensation is taking place cools more slowly, because condensation releases heat.

Air masses rise when they encounter obstacles, such as a mountain range. This results in **orographic cooling.** In other cases, upslope motion can occur when a warm air mass encounters a cool air mass. The cool air, being denser, flows under the warm air. The warm air then rises over the cool air, usually at a shallow angle.

Fogs and clouds consist of water droplets suspended in air. These tiny droplets are only about 1/1000 cm across, on the average, although they can sometimes reach ten times this size. Fog or cloud droplets fall very slowly through the atmosphere, at a rate of less than 2 km per hour. Larger droplets fall faster, but even the weakest updraft can keep them suspended in the atmosphere.

wind and weather

Clouds are usually formed by rising air that cools adiabatically. When streams of warm moist air run into mountain ranges, or elevated regions, the cooling produces **stratus** clouds with a layered appearance. Such clouds are often seen around the summits of mountains or high hills. Similar stratus clouds are formed when warm air is forced to climb over a mass of colder air.

Billowy **cumulus** clouds are produced by convective cooling – that is, adiabatic cooling in a convection cell. These are the fluffy white "fair weather" clouds that we see on a fine day. Cumulus clouds are a few kilometers across and a few hundred meters thick. When conditions are right, they may pile up on each other and reach heights of up to 20,000 meters. As we will see a little later, thunderstorms may be formed under these conditions.

Fair-weather cumulus clouds are often seen during fine spring and summer weather. These clouds are created by convective cooling.

classifying clouds

Meteorologists classify clouds in terms of their height and form. This classification scheme is quite complicated, so here we are using a simplified version. We have already described stratus and cumulus clouds. Other clouds are classified as **cirrus,** meaning feathery; **altus,** meaning of middle altitudes, and **nimbus,** meaning rainy.

Cirrus-type clouds are high clouds, occurring above about 6,000 meters, and are usually composed of ice crystals. Plain cirrus clouds are wispy, and are often called "mare's tails." **Cirrocumulus** clouds appear as feathery clumps, and **cirrostratus** clouds are feathery layers.

Between 2,000 meters and 7,000 meters, overlapping the high clouds, are the clouds of the middle altitudes. **Altocumulus** clouds occur in regularly-spaced clumps, resulting in what we call a "mackerel sky." **Altostratus** clouds form thin layers at middle altitude.

Low clouds occur below 2,000 meters. They may be simple stratus or cumulus clouds, but often the two types are combined to form **stratocumulus** clouds, which appear as clumps in neat layers. **Cumulonimbus** clouds are also considered low clouds, but they have an extensive vertical structure, and may tower up to great heights.

wind and weather

Cirrocumulus clouds, largely composed of ice crystals, frequently move in ahead of a winter cold front. They are often an indication that snow is approaching.

These are the familiar thunderclouds that often bring rain.

precipitation

Precipitation is the process in which water vapor condenses and falls to earth. We are not certain just how precipitation occurs. Cloud droplets are so tiny that even the slightest updraft keeps them afloat. They have to grow to at least $1/10$ cm before they can fall with ease. If a smaller drop began to fall, it would fall slowly and evaporate long before it reached the ground.

A raindrop or snowflake contains a million times as much water as a cloud droplet. Early investigators believed that they were simply cloud droplets that grew by collecting water molecules. But this process would take many hours, or even days. Since thunderclouds can form in minutes and produce heavy rain, another theory is needed to explain raindrop formation.

two theories of raindrop formation

According to one theory, cloud droplets are formed in different sizes. The larger droplets fall faster, colliding with the smaller droplets and sweeping them up. After a falling droplet has grown very large it splits up into many medium-sized droplets, which in turn sweep up more smaller ones. This process of **coalescence** is believed to produce the raindrops that fall in tropical rains.

According to another theory, rains in temperate zones are produced by a different process. In these regions thin stratus clouds seldom produce heavy rainfall; towering cumulonimbus clouds are needed for a substantial shower. It is theorized that ice particles form in the cold upper parts of cumulonimbus clouds. When they fall to lower altitudes, they capture supercooled water droplets and grow larger. Finally, they fall out of the cloud as frozen pellets, which melt into rain before they reach the ground.

other forms of precipitation

Snow, sleet, and hail are also common forms of precipitation. Snowflakes are ice crystals which usually form in low, cold clouds. They grow as they move about in the cloud, often taking interesting shapes. The classic shape is a complex hexagonal "star," and it is believed that no two snowflakes are identical.

Sleet is made up of small frozen raindrops, either soft and grainy or hard and transparent. If conditions are not quite suitable for sleet to form, freezing rain may fall, consisting of supercooled raindrops that freeze when they hit a surface. Freezing rain produces a clear glaze over trees, houses, and roads.

True **hailstones** are balls of ice, much larger than sleet pellets. Hail is frequently associated with thunderstorms. According to one theory, the strong updrafts in the center of the storm carry raindrops to high altitudes, where they freeze. Downdrafts bring them down into warmer air and they add a new layer of water . After several repetitions, this process produces large hailstones.

air masses

We have already discussed air masses in terms of whether they are warm or cold. The properties of a given air mass, such as temperature and humidity, are more or less the same over horizontal distances of hundreds of kilometers.

We use certain standard symbols to refer to air masses. There are two principal types, distinguished by the amount of heat energy they contain: the cold polar air masses (**P**) and the warm tropical air masses (**T**)

Air masses that form over the ocean are represented by the symbol **m,** which stands for maritime (oceanic). Those that form over the land are represented by the symbol **c** (continental). The two sets of symbols can be combined to designate a particular air mass. Thus, a continental polar air mass is designated **cP,** and a maritime tropical air mass is written as **mT.**

wind and weather

the polar air cell

Because polar air is cold, dry, and dense, it tends to sink and spread out over the surface of the earth. Air from the North Polar Zone flows southward, and because of its greater density it slips under warmer, less dense air that is moving northward.

During summer the southern margin of the cold air is in the vicinity of 50°N latitude across most of North America and the North Atlantic and Pacific Oceans. Warm air moving north from the horse latitudes slides up the sloping front of the polar air mass. This rising air divides near the tropopause. Some of it flows southward, completing a circulating **air cell** between the horse latitudes and the polar air mass. The other part of this rising air moves northward and sinks in the North Polar Zone to complete the **polar air cell.**

As more and more air is added to the polar air cell, the atmospheric pressure increases at its base. The cold air at the bottom can thus press southward and flow into the region of the prevailing westerly winds. This action, called a **polar outbreak,** makes more room for the air descending in the North Polar Zone.

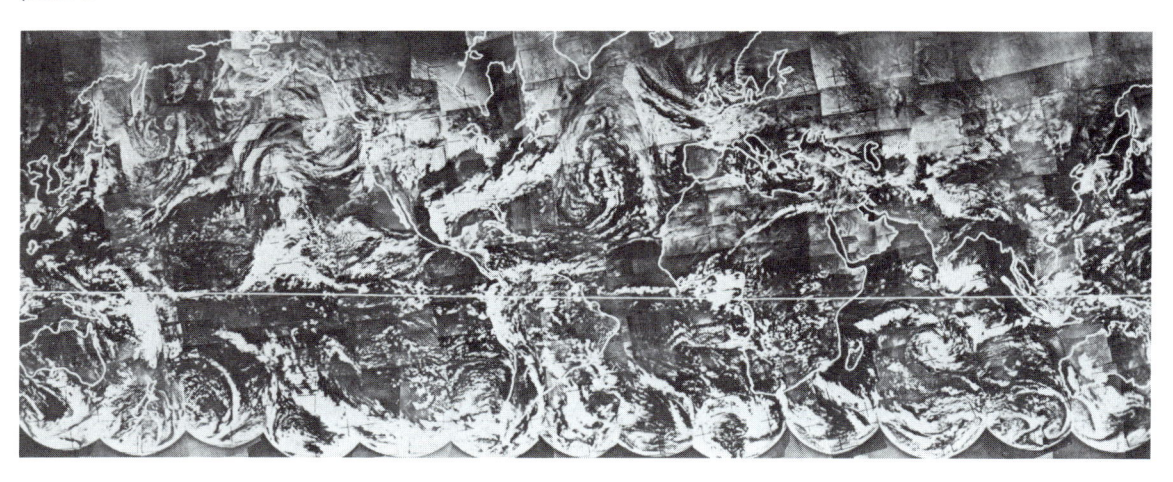

This photo, a composite assembled from a series of satellite photographs, is the first comprehensive picture ever made of the earth's entire weather pattern.

the battle between the air masses

The United States and southern Canada are wedged between the regions where a warm tropical air mass and a cold polar air mass form. Most of the United States, in fact, is a battleground for these masses of air. In the summer the tropical air masses dominate and push northward, bringing with them heat and moisture. Wherever the tropical air masses press against the polar air masses, storms are created. In the winter, the polar air masses are usually the more powerful.

Around 1915, meteorologists began to understand the importance of the border between air masses. Since the First World War was going on then, it is not surprising that they called the border a **front.**

Usually the line of a front is moving. If a cold air mass is advancing into a territory covered by warm air, we call it a **cold front.** If

wind and weather

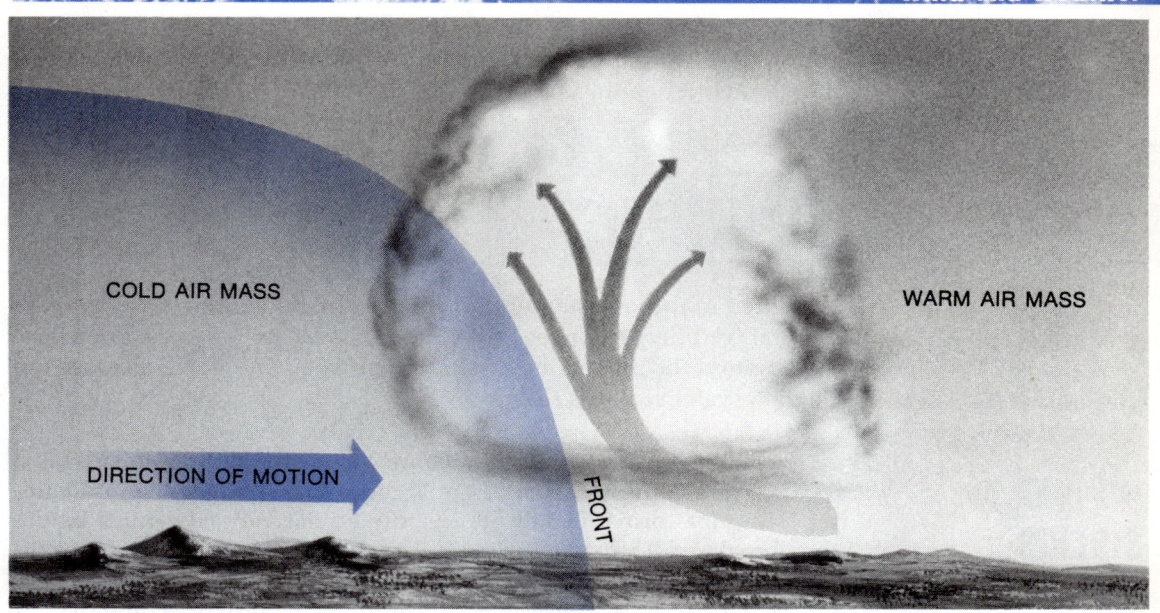

An advancing cold front forces a mass of warm air to rise by sliding under it. The approach of a cold front is usually accompanied by stormy weather.

the reverse is true, we call it a **warm front.** If neither air mass is invading the region of the other, the line is called a **stationary front.** The line of a front is usually smoothly curved or even straight. When this smooth line is disturbed, a storm may develop.

stability of air masses

An air mass may rise because of convection. When a "bubble" of air is heated by the warm surface of the earth it expands, becomes less dense, and begins to rise. The expansion cools it again, and soon the bubble is no warmer than the surrounding air. It then stops rising, and the situation becomes **convectively stable.**

On sunny days the ground can get very hot. The temperature of the atmosphere on such days often decreases very rapidly with height, so that a rising bubble of air will always be surrounded by cooler air and can therefore reach great heights. The situation is then **convectively unstable.** An unstable air mass can release its energy rapidly, and thus produce violent or stormy weather. Stable air, on the other hand, leads to fair weather.

temperature inversions

Often situations occur in which the surface of the earth is cooler than the air above it. This may happen, for example, when warm air moves in over cold land, or above a cold ocean stream. The layer of air in contact with the surface cools and becomes denser, while the air above it is still warm. This creates a **temperature inversion.**

Temperature inversions are very stable situations. The dense cold air has no tendency to rise, nor does the buoyant warm air above it have any tendency to descend, so the air layers do not mix. When a temperature inversion occurs over a city, air pollutants

wind and weather

emitted by industries slowly accumulate in the still air. In the course of a couple of days the air can become heavily polluted and dangerous to health.

some interesting weather

We noted in the last chapter that weather information available to the general public still uses the English system of measurements. Throughout the rest of our discussion of weather we will also use the English system for this reason.

thunderstorms

The **thunderstorm,** most common of the violent storms, is a local atmospheric disturbance rarely more than a few miles in diameter. Thunderstorms usually travel only a relatively few miles before they break up. They may be produced under several different conditions. For example, a line of thunderstorms often arises along a front, where two air masses with very different characteristics come in contact.

Conditions within a single air mass can often produce an isolated thunderstorm. First and foremost there must be unstable air. Second, the air must contain enough water vapor to produce the precipitation necessary for the storm.

Cumulonimbus clouds are the classic thunderheads. Note the distinctive anvil shape of the cloud tops. The anvil represents a region of highly turbulent air that pilots try to avoid at all costs.

Strong heating of the earth's surface is a major cause of thunderstorms. July and August, when the air is hot and humid, are the most common months for thunderstorms. Insolation is converted to heat and radiated back into the atmosphere. The air warms, expands, and rises. As a result, air moves in from the surrounding areas to replace the rising air. These are the strong, gusty winds that often precede a thunderstorm. The air flowing into the center of the forming storm is often saturated with water vapor. The column of hot, moist air rises thousands of feet with great speed. This rapid rise—a result of the highly unstable conditions—produces adiabatic cooling.

wind and weather

The water vapor in the expanding column of heated air condenses into **thunderheads,** huge cumulonimbus clouds. These clouds often grow in minutes as you observe them. More such clouds form as the rising air column climbs higher into the cool upper air. The flat, dark base of the clouds shows the altitude at which condensation is occurring.

Thunderstorms have always been accompanied by **lightning,** but during most of human history no one understood just what it was or why it occurred. Newton was the first to compare lightning with the sparks that are emitted by charged objects. Some objects become charged with electricity when they are rubbed against others. Anyone who has ever shuffled across a nylon carpet while wearing leather-soled shoes is familiar with this phenomenon, which we call **static electricity.** Lightning is also static electricity, although on a much grander scale.

In essence, lightning bolts are simply large sparks that leap between clouds, or between a cloud and the earth. The sequence of events that occurs when a lightning bolt strikes the earth has been captured in high-speed motion pictures. First, a negative charge rushes from the cloud to the earth at 50,000 kilometers per second. It appears in the photographs as a moving spot of light that travels downward in jerky steps. This **step leader** leaves a trail of ions behind it. When it reaches the ground a positive charge flows from the earth, following the trail of ions upward. We see this **return stroke** as a bright flash of lightning. It is the return stroke that is responsible for most lightning damage. A lightning bolt consists of repeated step leaders and return strokes, which continue until the cloud is completely discharged. Each stroke lasts only about 1/10,000 of a second.

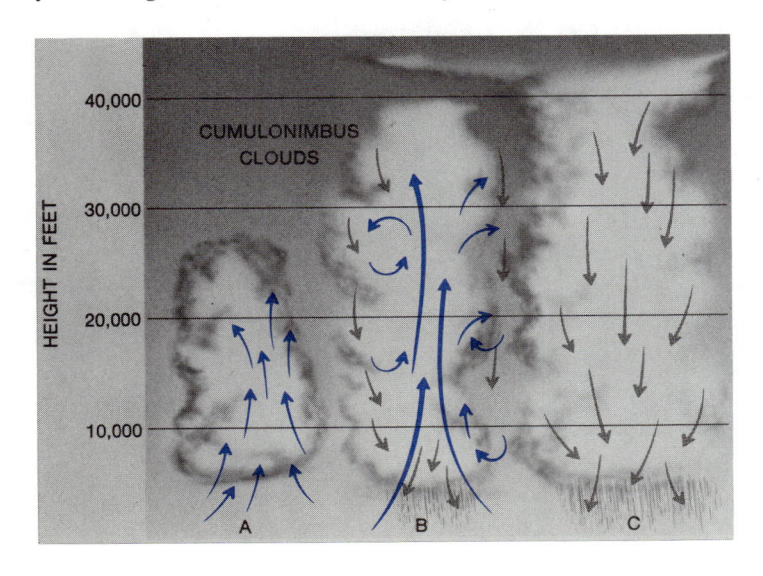

lightning

The pattern of updrafts and downdrafts during the life of a thunderstorm varies over three stages. A. Thunderstorm is forming. B. Storm is at its peak. C. Thunderstorm is dissipating.

Lightning usually accompanies storms associated with cumulonimbus clouds. It may leap from cloud to cloud or from a cloud to the earth.

wind and weather

the New England northeaster

Lightning bolts may involve nearly a billion volts of electricity. The air through which they pass is momentarily heated to as much as 15,000°C. This superheated air then expands with great violence, producing the explosive sound of thunder.

What happens when a tropical air mass and a polar air mass collide over the North Atlantic Ocean, where the irregularities of the continent do not interfere with the general winds?

During the winter the maritime polar (mP) air mass spreads southward to somewhere between 40° and 37°N latitude. At the same time, the northern front of the maritime tropical (mT) air mass is between 15° and 20°N latitude. Between these air masses is a broad region of mixed air. As the days become longer in the spring, the front of the mT air mass moves northward. Because the North Polar Zone is still largely in darkness and has not yet warmed, the southern front of the mP air mass remains stationary.

The mP air mass is cool and moist, and the mT air is warm and moist. The cool northern mass is therefore denser and contains less water vapor. At its southern edge it is a thin layer lying on the cool ocean water. Its upper surface slopes upward toward the north.

The warm tropical air expands as it rises and moves north-

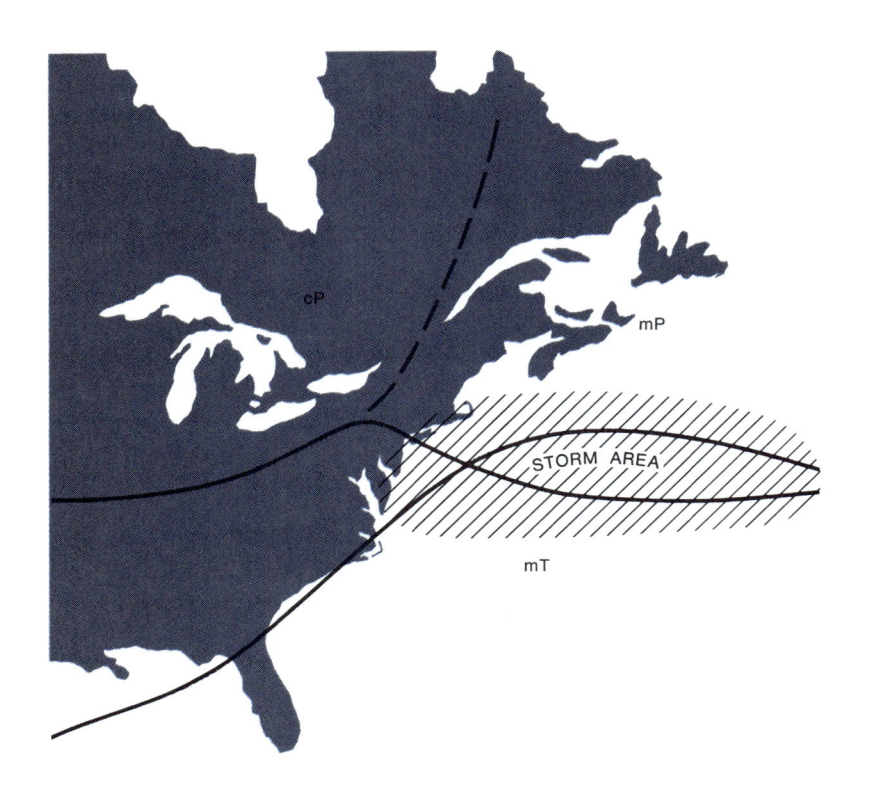

A characteristic pattern of air masses gives rise to a New England northeaster. This late winter storm often occurs when an mP and an mT air mass meet.

ward up the slope of the mP air mass. The expansion of the rising air causes it to cool adiabatically. Clouds form, releasing large amounts of energy. The increase in the energy content of the air, in turn, causes the forming storm to become intense.

Heavy rain and high winds blow from the northeast onto land still cold from the winter. This lowers the temperature of the humid air, increasing the rainfall. Low clouds and fog often accompany the rain. In short, when one of these late winter storms strikes New England, the weather is likely to be awful for several days. Soon, however, most of the excess energy has been drained out of the tropical air mass. It begins to weaken and retreats southward. Clear, cool weather sets in along the coast.

a Texas norther

Now let us see what happens when dissimilar air masses collide over the continent. In the broad and relatively smooth Mississippi basin, the opposing air giants are the cP air mass—cold, dry, and very dense—and the mT air mass—warm, moist, and less dense.

As the North Polar Zone goes into its long winter's night, the cP air mass increases in density and pushes southward. The mT air mass is the warmest and wettest of the air masses that invade the United States. In the winter it lies over the Gulf States as a warm, humid blanket. Because the winter sun is low, the land is colder than the air mass. This results in a temperature inversion, which at its northern boundary lies as high as about 1 mile above sea level.

As winter progresses, the cold cP air slides south under the mT air mass, and things begin to happen. The moisture of the warm air condenses and falls as snow. The heat energy released by this condensation causes strong updrafts, which in turn cause cold, dense air from the north to flow southward. The inflow of rapidly moving cold air drops the surface temperatures to well below freezing and often much lower. The wind sweeps snow before it, and a blizzard roars southward across the Texas panhandle. A **Texas norther** is in action.

When the pressure of the cP air mass has pushed the mT air mass far to the south, cold, clear weather follows. Slowly the tropical air mass regains power, and as the polar air retreats, the warm air slips northward over the land to replace it.

air pressures and winds

The first person to use meteorological information for weather prediction was Sir Francis Galton, an Englishman, who started weather mapping in 1863. Galton was a very able man, who contributed significantly to such varied disciplines as statistics, learn-

wind and weather

A typhoon, or Pacific Ocean hurricane, photographed from a satellite. If a typhoon occurs south of the equator the cyclonic flow is clockwise, in contrast to the counterclockwise flow of northern hemisphere cyclones.

ing theory, and biology. He also became an authority on weather and helped found the modern science of meteorology.

Galton discovered that there are several regions of the earth where the usual atmospheric pressure is slightly higher or lower than the earth's average. The belt of the doldrums, for example, has a continually lower-than-average pressure. As you will recall, this low pressure is caused by the equatorial belt of rising, heated air. Average sea-level pressure in the doldrums is about 29.7 inches (1,006 millibars), slightly lower than the 29.9 inches (1,013 mb) that is standard. In the region of the horse latitudes, where air is descending, the average sea-level pressure is about 30.1 inches (1,020 mb).

The difference in pressure between these two zones of vertically moving air is enough to develop a flow of air from the region of higher pressure to the region of lower pressure. These winds, blowing from the horse latitudes to the doldrums, are the trade winds we discussed earlier. The speed, or strength, of winds depends on the **pressure gradient** between the high and low. This is the difference in pressure between two points, divided by the distance between them. The greater the pressure gradient, the stronger the wind.

wind and weather

The best-developed tropical low pressure areas, or **lows,** are located over the land, and the best-developed tropical **highs** are over the oceans. This is an indirect result of the fact that, in the tropics, temperatures are usually higher over the land. Because of this, the air over the land can contain more water vapor than the air over the sea. But water vapor, as we noted earlier in the chapter, is less dense than air. So the more water vapor an air mass contains, the less dense it is, and the lower atmospheric pressure it exerts. The hot, humid air masses over tropical lands produce zones of relatively low pressure. The cooler, drier air masses over the oceans produce zones of relatively high pressure.

highs and lows of North America

Meteorologists use the term **cyclone** for any system of winds blowing in toward a low-pressure region. The wind pattern around a low is called a **cyclonic flow.** Conversely, an **anticyclone** is a high-pressure system, from which winds are circling outward.

Cyclonic winds flowing into the lows and anticyclonic winds flowing out of the highs disturb the pattern of the prevailing westerlies. In the United States the pressure gradient between highs and lows usually has a greater influence on the local surface winds than do the prevailing westerlies.

Winds aloft, the ones that fliers at high altitudes must consider, are usually produced by the rotational effect of the earth. These include the prevailing westerlies and the trade winds. Often, when the sky is patchy with clouds in several layers, you can see the uppermost layer of clouds traveling with the prevailing westerlies. At the same time, the lower clouds are traveling in a different direction, because they are influenced by local pressure differences.

storms

Thus far our discussion of the physics of the atmosphere has been very general. Now let us look at some practical applications of this knowledge. Prediction of our daily weather is probably the most useful. Hundreds of millions of dollars are spent every year for weather forecasting, but the money is well spent. Since many industries and private individuals rely on weather forecasts to plan their various activities, the economic return from the study of weather is at least ten times the outlay.

We do not know precisely how a cyclonic storm starts, but we do know that storms are associated with lows. Lows that become storms, in turn, are often associated with fronts. Very frequently they are associated with an **occluded front,** which is formed as the result of a complex set of conditions.

An occluded front frequently starts as a stationary front, which develops into both a cold and a warm front. For some reason, the cold air mass starts to move—usually southward—and forms a cold front. Adjacent to this, the warm air mass begins to move northward as a warm front. This produces a wave in the originally smooth line of the stationary front. As the advancing cold front speeds up, it pivots at the point where it meets the warm front. We could compare the two fronts to two arms of a pinwheel, with the rear arm (the cold front) gradually catching up to the forward arm (the warm front).

In about 24 hours the cold front turns so far around that it slips under the warm front and lifts a large volume of the warm, moist air. The pivoting cold front thus joins the cold air mass that has been ahead of the warm front. The occluded front is the boundary between the cold air next to the ground and the isolated warm air above it. Unlike other fronts, an occluded front does not come into contact with the ground.

As the occluded front is forming, additional warm, moist air blows into the deep dent in the original stationary front. The sides of the dent are not parallel; they converge toward the pivot around which the cold air is turning. The advancing air from a warm, high-pressure area is trapped between the converging walls of cold air, and cannot penetrate the cold air mass. The only thing it can do as it advances into the dent is to rise. This rising, warm air produces a low-pressure area.

The physics of this situation is complex. You can understand, of course, that a deep column of warm, moist air must be lighter than a similar column of cold, dry air. Therefore, a low-pressure area probably forms under the rising air. The Coriolis effect would give a circular motion to the rising air, forming a cyclone. Adiabatic cooling of the spiraling, rising air would cause condensation and the formation of clouds, which in turn would release energy. The released energy increases the velocity of the winds, and a storm develops.

hurricanes—small-diameter cyclones

A **hurricane** is a cyclone with a small diameter and a very great pressure gradient. Such a great change of pressure within a comparatively small diameter produces winds that exceed 75 miles per hour and are capable of causing tremendous destruction.

Thanks to the efforts of a small number of courageous fliers and scientists, we now know much about hurricanes. The hurricane hunters of the United States Air Force, the Navy, and the National Weather Service fly out into the tropical Atlantic Ocean, where hurricanes form. Once they locate a hurricane, they fly with it and plot its course and wind speeds. This information is radioed to the Hurricane Center in Miami, Florida, which can then broadcast a hurricane warning days before the storm reaches land.

Today weather satellites also maintain a constant watch for hurricanes. They are able to spot forming hurricanes even when the

wind and weather

This rare satellite photo of three Pacific storms was taken in 1976. The tropical depression in the center is flanked by Hurricane Kate, at left, and another depression at right which later became Hurricane Liza.

storms are still far out at sea, and to monitor their growth and the direction of their course.

Hurricanes are rarely more than 300 miles in diameter. Within the doughnut-shaped storm is an eye, or center, from 5 to 20 miles across. In the eye of the hurricane the winds are almost totally calm, and no rain falls. Barometric pressure in the eye may fall as low as 27.6 inches (937 mb). Outside the eye, in the body of the storm itself, the winds reach velocities of between 75 and 200 mph.

Although the circling winds inside a hurricane may blow at these high velocities, the hurricane itself moves much more slowly along its path, generally at a speed ranging from 5 to 20 mph. As a result, most people in its path can be warned far enough in advance to take precautionary measures. Windows can be shuttered or boarded up, supplies of food and water can be stored, preparations can be made for loss of electric power, and those who live in low-lying regions near the seashore can move to higher ground inland.

Our hurricane season reaches its peak in August and September. Because the equatorial heat belt lies almost entirely north of the equator, most cyclonic maritime storms are generated in the Northern Hemisphere. Hurricanes start in the Atlantic Ocean between North Africa and the West Indies. They begin when, for some unknown reason, a disturbance known as a **tropical depression** forms in the belt of the trade winds. This disturbance is carried along by the currents of air as an eddy is carried in a river.

Air begins to flow into the disturbed region from all sides, and as it does so, it rises. Clouds form, releasing heat energy which helps to power the storm. Still more air flows into the region, and the Cor-

hurricanes in action

wind and weather

A Kansas Highway Patrol trooper took this series of photos in 1971, detailing the formation of a tornado. The storm's growth can be followed from the mammatocumulus cloud at upper left to the fully-developed funnel at lower right.

iolis effect causes it to rotate in a counterclockwise direction. The rising, humid air is cooled, producing heavy rain. Through a complicated chain of events, heat energy is changed to kinetic energy, causing the high winds of the hurricane.

As it gains strength, the hurricane sweeps across the Caribbean Sea, and usually heads for the North American mainland. It is preceded by a **storm surge,** a sudden rise in sea level that may be as much as 20 feet or more. This sudden rise may occur in a period of only a few minutes, and is a major cause of the damage that accompanies hurricanes.

A hurricane is actually a heat engine working off the energy received from the sun. It takes heat energy from the tropics and deposits it in the temperate zones, where less solar energy is received. In this way, hurricanes help to distribute the solar energy over the globe. If scientists ever become successful in quenching hurricanes as soon as they form, this distribution of solar energy will not take place. The consequences are unclear, but it is possible that in one sense hurricanes are "safety valves" which provide a release for excess solar energy.

wind and weather

tornadoes

Of all storms, **tornadoes** are the smallest and by far the most violent. These storms are confined almost exclusively to the United States, although occasional **twisters,** as they are often called, have been reported in other parts of the world. Tornadoes may occur in any month. They are most common, however, during the spring and early summer, when the mT air in the Mississippi Valley may meet a still-powerful polar air mass. The highest incidence of tornadoes is concentrated in the midwestern states of Kansas, Oklahoma, Missouri, and Texas.

The method of formation of a tornado is similar to that of a thunderstorm. The result, however, is a cyclone even more intense and with a much smaller diameter than a hurricane. Most tornadoes are about 600 feet wide, but some have reached diameters of as much as half a mile, and still others are less than 100 yards wide. The path of the storm is usually only a few miles long, but one twister, in 1925, traveled more than 200 miles before dissipating.

The winds within a tornado reach extremely high velocities. The rising air may move at more than 100 mph, and horizontal winds blowing into the twister may reach speeds of 300 mph. The actual speed of the whirling winds cannot be directly measured but must be estimated. Speeds of 120 mph have been recorded before the measuring instruments blew away.

The immense destructiveness of a tornado is the result of two factors. First, all the energy is concentrated in a small-diameter storm. Second, the rapidly-rising air inside the funnel—where the eye would be in a hurricane—creates such a sudden drop in barometric pressure that as the tornado passes over a building, air rushes out of the building with great violence, and the building actually explodes.

predicting the weather

One of the best ways to watch the movement of weather patterns across the country is to study the daily weather maps published in most newspapers. These maps show the temperature, the wind direction and velocity, the precipitation occurring at various weather stations, and the locations of highs, lows, and fronts. These daily maps are prepared by the National Weather Service and are simplified versions of the detailed maps prepared for use by professional meteorologists.

The symbols used on weather maps to report data are easy to learn. As you can see on the map in the accompanying figure, atmospheric pressure is shown by isobars—lines along which the atmospheric pressure is the same. From the pattern of isobars, you can lo-

IN FOCUS:
killer tornadoes: the american nightmare

On April 3, 1974, a series of more than a hundred killer tornadoes whirled viciously through thirteen states between Mississippi and Canada. By the time the front that had spawned the twisters finally dissipated, hundreds of people were dead and thousands more had been injured. Property worth hundreds of millions of dollars had been destroyed. In the wake of the storms lay the debris of thousands of buildings, exploded by the force of the atmospheric pressure within them as the passage of a tornado suddenly lowered outside pressure by inches within a few seconds. Blades of straw embedded in fence posts and telephone poles, and steel girders twisted into massive knots, provided grim evidence of the storm's fury. The small city of Xenia, Ohio, was in ruins, almost totally obliterated by one of the largest of the tornadoes.

The United States has the highest frequency rate of tornadoes of any country in the world. The conditions that generate these storms arise when warm, moist air moving up from the Gulf of Mexico meets a cold polar air mass from Canada and a wave of dry air moving eastward from the southwestern desert areas. This creates a highly unstable region along which a squall line forms. Massive thunderstorms spawned along the squall line become the "parents" of tornadoes.

Usually, but not always, tornadoes strike during the hours of late afternoon or early evening in the months of April, May, and June. A children's classic of 1900, *The Wizard of Oz*, opens with a remarkably accurate description of the signs that preceded the tornado which carried Dorothy out of Kansas and on her way to fantasy land.

"Tornado weather" generally begins with unusual warmth and mugginess for the time of year. As the storm gathers, the sky often takes on a murky greenish color, and surface winds become abnormally calm. Dogs and other domestic animals may exhibit peculiar behavior; birds stop singing and often go to roost. Distinctive mammatocumulus clouds form—low-hanging convex shapes that precede the appearance of the characteristic black tornado funnel. The emergence of the funnel cloud is accompanied by a deafening roar that has been compared to the sound of hundreds of freight trains or jet planes.

Scientists are making an increasingly detailed study of the causes and mechanics of tornado formation, in an effort to improve prediction methods. On a radarscope, the approach of a squall line and the thunderstorms that accompany it can be followed. The radar picture shows that the "parent" thunderstorm about to give birth to a tornado curves into a formation called a hook, shaped like the numeral 6. At the edge of this hook, a tornado often forms. Unfortunately, radar can "see" only the hook and not the twister itself. However, the presence of a radar hook is considered an automatic signal for the issuance of a tornado warning.

Dr. Theodore Fujita, a leader in the field of tornado theory and prediction, suggests that successful prediction in the future will be based on a combination of satellite observations and radar readings. Weather satellites cannot spot tornadoes themselves because of the heavy cloud cover, but they can "see" the anvil-shaped clouds that Fujita believes give rise to twisters.

Fujita has come to the conclusion that the occurence of severe tornadoes follows a cyclic pattern. He believes that there is a so-called "tornado triangle" in which tornadoes are most likely to form, and that the position of this triangle changes from year to year as the result of long-range climatic patterns.

But is there a more effective and accurate method of prediction than radar? Newton Weller believes that there is. Weller's method is based on the fact that tornadoes are typically accompanied by an unusually bright and blue type of lightning. His technique uses an ordinary television set to detect the presence of this unusual lightning.

First, the television set is turned on, and the contrast control is adjusted to make figures on the screen appear as silhouettes. Next, the set is turned to channel 13 (whether or not it is carrying a picture) and the brightness control is adjusted to make the screen almost completely dark. The final step is to switch to channel 2. Now you are ready to detect an approaching tornado.

According to Weller's theory, if a tornado is on the ground within 15 or 20 miles of your location, the abnormally bright lightning that goes with it will cause the dark television screen to glow brilliantly. If this should happen, Weller says, there is no time to stare at the screen—take cover immediately!

Weller's technique has been the target of much skepticism, although it has been reported successful by a number of people who have practiced it. Recently, a large mail-order house has offered for the first time a tornado-detecting device based on the same general principles.

One thing is certain—an accurate method of prediction is badly needed. Dwellers in the middle western "Tornado Alley" region are, of necessity, competent shirt-sleeve meteorologists. But no meteorologist is perfect; and those who have experienced the devastation of a tornado will welcome any new method of prediction that might give them accurate advance warning of these ferocious killer storms.

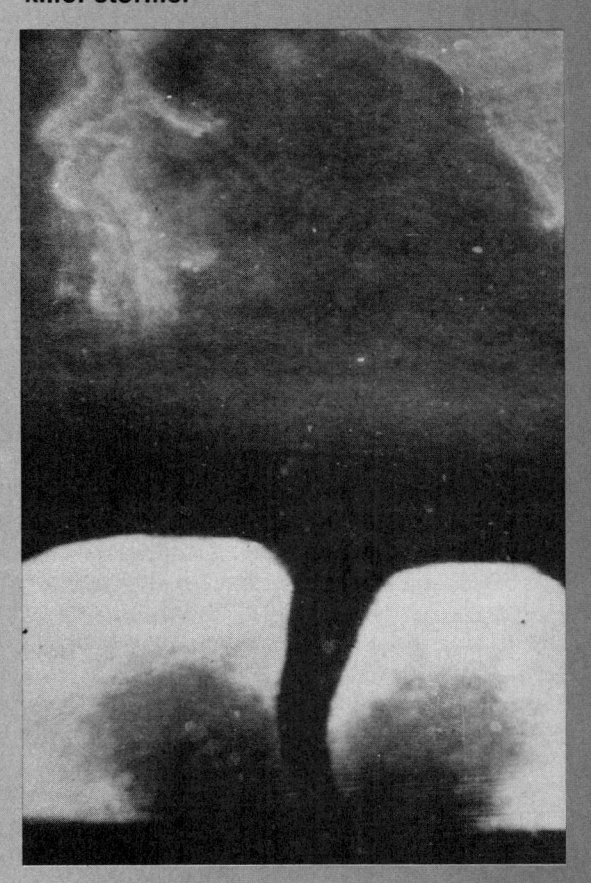

wind and weather

cate the centers of highs and lows. At the end of each isobar is a number showing the pressure in inches of mercury. Generally, above 30 inches is considered high pressure and below 30 inches is consid-

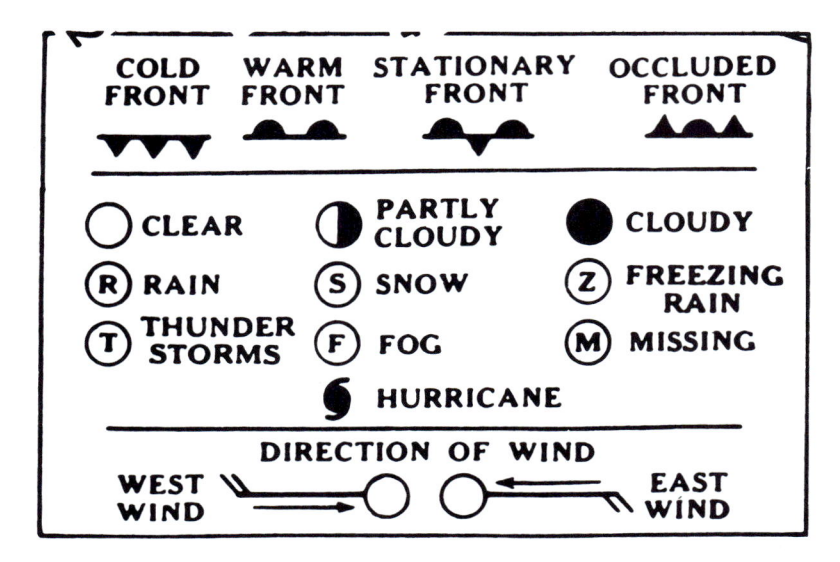

These standard symbols are used in the legends of most newspaper weather maps.

ered low pressure; however, *high pressure* and *low pressure* are relative terms.

Fronts are indicated by heavy solid lines, with symbols on one or both sides of the lines. Look at the legend to learn the symbols for the four types of front. Each weather station is represented by a circle. The symbols within the station circles indicated the condition of the sky when the data were collected. The number next to the station circle is the temperature in degrees Fahrenheit at the time the station reported.

Attached to each station circle is a **wind arrow,** which points in the direction from which the wind is blowing. Attached to the arrow at an angle may be one or more short lines. As you can see from the legend, a very short line represents a wind velocity of about 5 mph. A longer line represents a wind velocity of about 10 mph. By combining long and short lines, the wind velocity can be reported within 5 or 6 mph of true velocity. For a wind of 55 mph, a small flag is used.

Suppose we analyze five days of winter weather in the United States, by carefully examining the weather conditions on January 22, 1968, at 1 P.M. Eastern Standard Time. The most important weather structure on the map is the low-pressure area that lies over the Great Lakes region. A cold front extends from the center of this low, southwestward to Kansas. There it turns abruptly northward along the face of the Rocky Mountains. In Montana the front is stationary. A little snow is falling in the mountains of Colorado and in the region just south of James Bay, directly north of the word LOW. There are sever-

analyzing the weather

"EIGHTEEN HUNDRED AND FROZE TO DEATH"

The most notable thing about the summer of 1816 was that it never really arrived. Although some warmish weather did finally reach the northeastern United States by August, most inhabitants of New England felt that this was hardly adequate compensation for the heavy snow that had fallen in June, killing crops and shrivelling the young leaves of trees and shrubs. Disgruntled old-timers, who had never seen anything like it, promptly dubbed that year "eighteen hundred and froze to death."

Even when we take into consideration the fact that weather observation in the early nineteenth century was by no means a science, and that tales about unusual weather tend to grow taller during the telling, there is still no doubt that the summer of 1816 must have been a pretty dismal affair. Weather records that have survived indicate that 1816 richly deserved its popular name — "The Year Without A Summer."

Everything seemed pretty normal until June, although there had been a spell of unseasonable warmth around the end of May. Then, early in June, winds suddenly shifted to the northwest, and New England was blanketed by a wave of arctic air. Temperatures in some places dropped as much as 50°F during a twenty-four hour period. Severe frosts reached as far south as Virginia, and snow fell over much of New England and as far west as Ohio. One Vermont newspaper reported that ". . . on the night of [June] 7th and morning of the 8th a kind of sleet or exceeding cold snow fell . . . and measured in places where it drifted 18 to 20 inches in depth." Temperatures during the rest of the summer averaged well below normal, although farmers managed to sow and harvest a belated crop.

What caused the Year Without A Summer? Can it happen again? Some scientists believe that the frigid summer of 1816 was somehow related to sunspot activity. Another theory suggests that a change in upper air currents prevented a large polar air mass from retreating to Canada as usual during the summer months. Perhaps the most plausible explanation is that volcanic ash from the violent explosion of Mt. Tamboro in Indonesia a year earlier remained suspended in the atmosphere and thus blocked out an unusual amount of solar radiation. Whatever the cause, it is perfectly possible that we may someday experience another "summerless" year, and that skiing and sledding will be popular sports for the Fourth of July holiday.

wind and weather

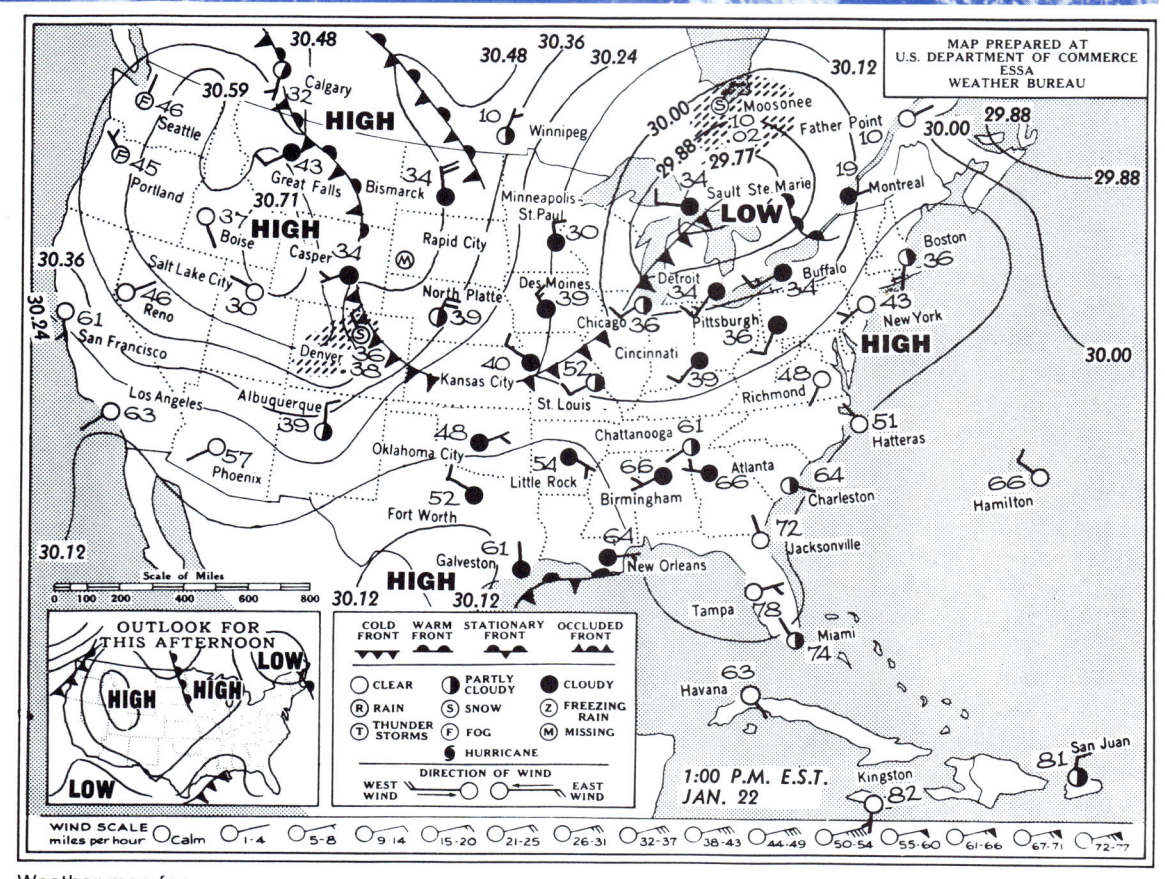

Weather map for January 22, 1968.

al highs outside of the front line. A strong high is centered over the Idaho–Montana border. Watch this particular high as we trace the way the weather moves across the continent from day to day. Along the Atlantic Coast you will see a mild ridge, representing an elongated high-pressure region.

At 1 P.M. the next day, January 23, the cold front has moved eastward and lies along the Appalachian Mountains as a shallow, narrow low, known as a trough. Rain and snow blanket the east because the cold air behind the front has chilled the moist air from the Atlantic Ocean. The stationary front that was in Montana yesterday now extends along the front of the Rockies and across the plains into Texas, where it joins the cold front. The intensifying high in the Great Basin has pushed the front line a little eastward. Notice the new cold front developing in the British Columbia–Washington border region. To see the effect of a passing cold front, watch the temperatures in New Orleans and Miami during the next few days At the time this map was prepared, both were on the warm side of the cold front.

wind and weather

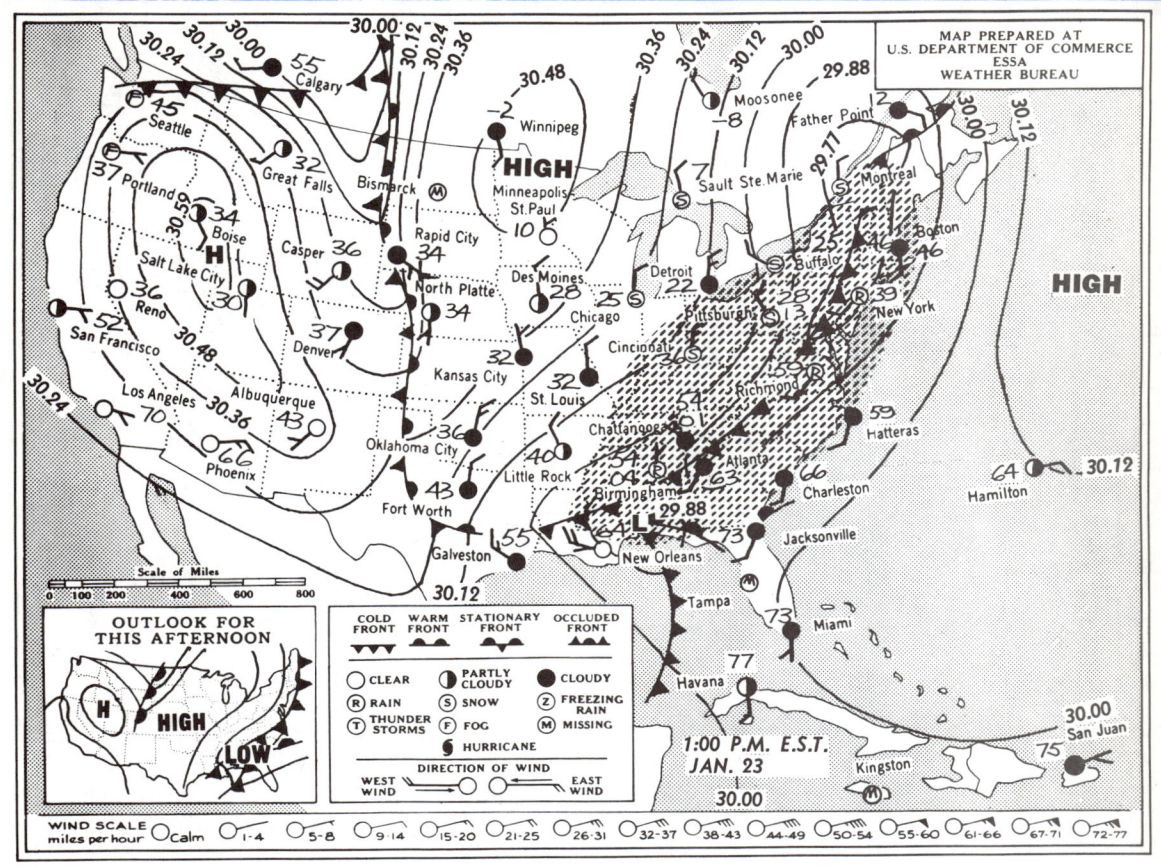

Weather map for January 23.

Now check the map for January 24. The cold front has swept past New Orleans, dropping the temperature from 64°F to 43°F, but has not yet reached Miami. The rainy low that blanketed the Atlantic Coast on the 23rd has moved out to sea, and the Coriolis effect is carrying it northeastward between the Atlantic Coast and Bermuda. The pressure gradient of the low is getting steeper. Compare the spacing of isobars around the low on January 24 and 25. Notice the "tail" on the front. This is an occluded front. What happened to the temperature in Miami as the cold front passed it?

Now compare the situation in the Northwest on January 25 with that on January 20. Notice the wet weather in the Puget Sound region, and the low pressure north and northwest of the Great Lakes. Another winter storm is advancing on the Midwest and the Atlantic Coast. A ridge has developed just west of the Appalachian Mountains. If this high-pressure region intensifies, it will force the low far into Canada as it flows eastward. The afternoon outlook, shown in the lower left corner of the map for January 25, suggests that the high-pressure region will do just that. To learn whether or not the

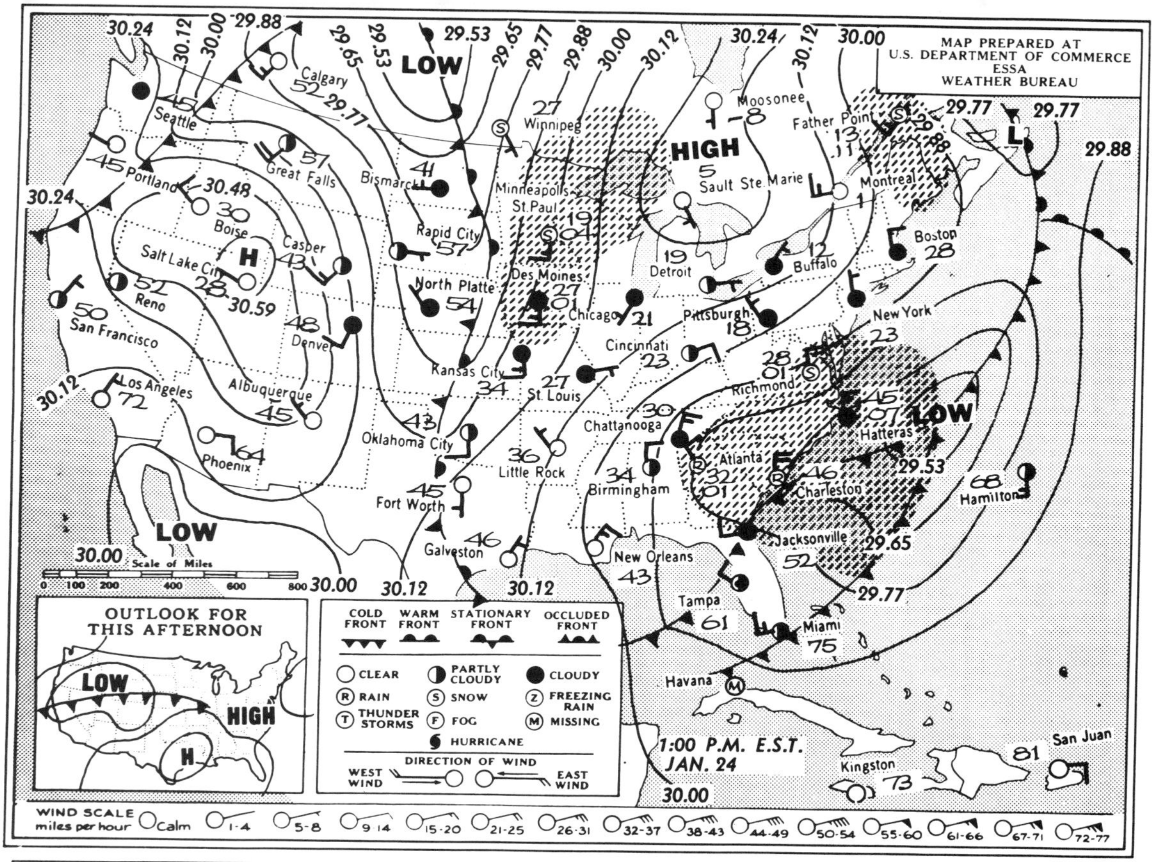
MAP PREPARED AT
U.S. DEPARTMENT OF COMMERCE
ESSA
WEATHER BUREAU
LOW
HIGH
H
LOW
LOW
Calgary
Seattle
Portland
Great Falls
Bismarck
Winnipeg
Moosonee
Father Point
Montreal
Sault Ste. Marie
Minneapolis
St. Paul
Boise
Casper
Salt Lake City
Reno
San Francisco
Rapid City
North Platte
Des Moines
Chicago
Detroit
Buffalo
Boston
Pittsburgh
New York
Cincinnati
Denver
Kansas City
St. Louis
Richmond
Los Angeles
Albuquerque
Chattanooga
Atlanta
Hatteras
Oklahoma City
Little Rock
Birmingham
Charleston
Phoenix
Fort Worth
Hamilton
Jacksonville
Galveston
New Orleans
Tampa
Miami
Havana
Kingston
San Juan
Scale of Miles
0 100 200 400 600 800
OUTLOOK FOR THIS AFTERNOON
COLD FRONT
WARM FRONT
STATIONARY FRONT
OCCLUDED FRONT
CLEAR
PARTLY CLOUDY
CLOUDY
RAIN
SNOW
FREEZING RAIN
THUNDER STORMS
FOG
MISSING
HURRICANE
DIRECTION OF WIND
WEST WIND
EAST WIND
1:00 P.M. E.S.T.
JAN. 24
WIND SCALE miles per hour
Calm
1-4
5-8
9-14
15-20
21-25
26-31
32-37
38-43
44-49
50-54
55-60
61-66
67-71
72-77

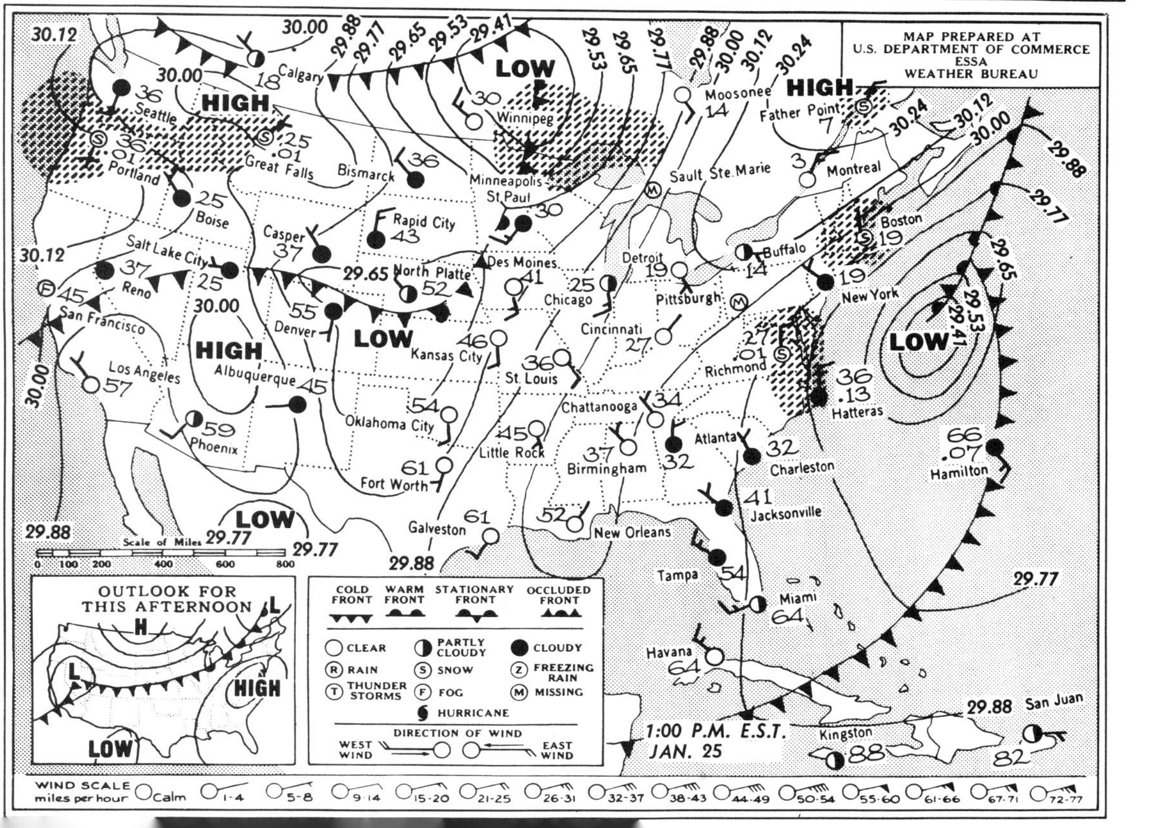
MAP PREPARED AT
U.S. DEPARTMENT OF COMMERCE
ESSA
WEATHER BUREAU
LOW
HIGH
HIGH
HIGH
LOW
LOW
LOW
Calgary
Seattle
Portland
Great Falls
Bismarck
Winnipeg
Moosonee
Father Point
Montreal
Sault Ste. Marie
Minneapolis
St. Paul
Boise
Casper
Salt Lake City
Reno
San Francisco
Rapid City
North Platte
Des Moines
Chicago
Detroit
Buffalo
Boston
Pittsburgh
New York
Cincinnati
Denver
Kansas City
St. Louis
Richmond
Los Angeles
Albuquerque
Chattanooga
Atlanta
Hatteras
Oklahoma City
Little Rock
Birmingham
Charleston
Phoenix
Fort Worth
Hamilton
Jacksonville
Galveston
New Orleans
Tampa
Miami
Havana
Kingston
San Juan
Scale of Miles
0 100 200 400 600 800
OUTLOOK FOR THIS AFTERNOON
COLD FRONT
WARM FRONT
STATIONARY FRONT
OCCLUDED FRONT
CLEAR
PARTLY CLOUDY
CLOUDY
RAIN
SNOW
FREEZING RAIN
THUNDER STORMS
FOG
MISSING
HURRICANE
DIRECTION OF WIND
WEST WIND
EAST WIND
1:00 P.M. E.S.T.
JAN. 25
WIND SCALE miles per hour
Calm
1-4
5-8
9-14
15-20
21-25
26-31
32-37
38-43
44-49
50-54
55-60
61-66
67-71
72-77

wind and weather

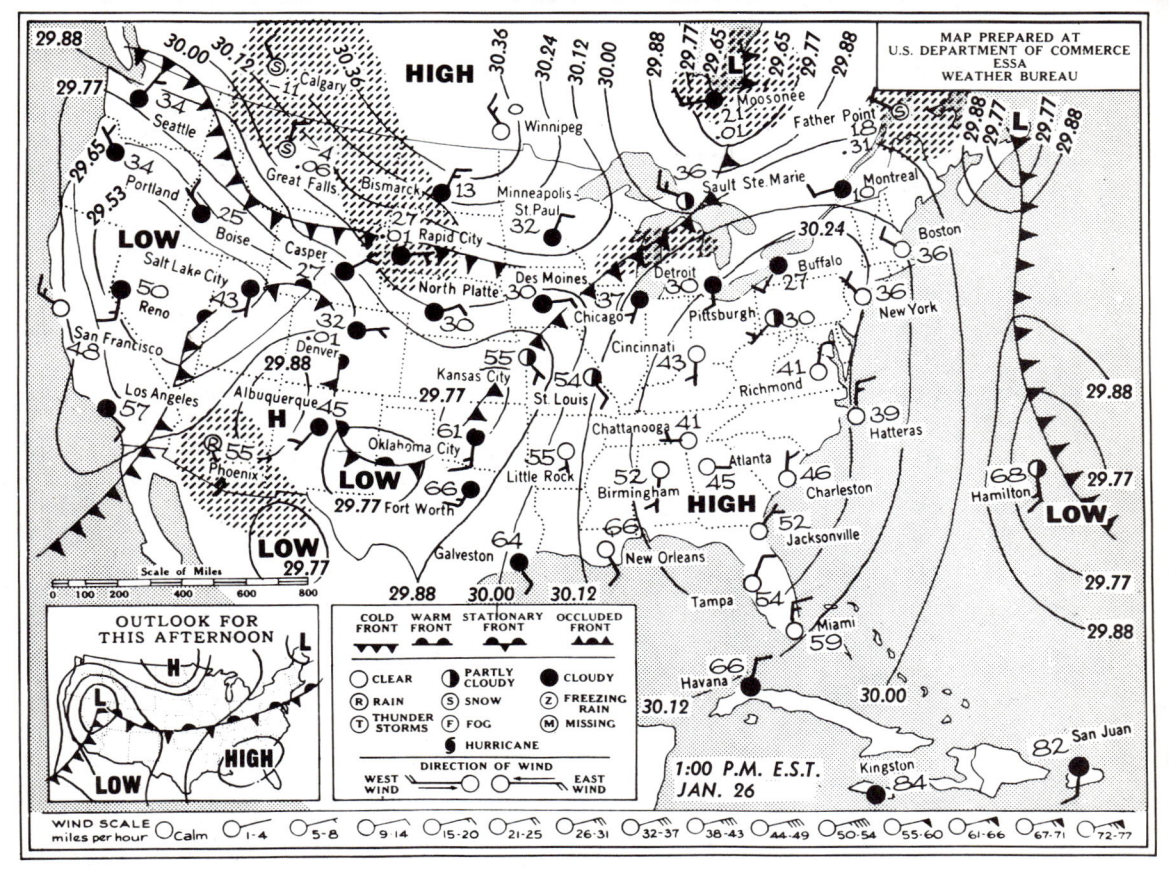

meteorologists were correct in their prediction, look at the map for January 26.

   They were right! The bad weather was forced to go around the Appalachians and out to sea via the St. Lawrence River valley.

(*Facing page*) Weather maps for January 24 and 25.
(*Above*) Weather map for January 26.

An artist's conception of a hurricane-hunter plane, flying through highly turbulent air in an attempt to penetrate to the calm "eye" of the storm.

wind and weather

SUMMARY

Water plays an important role in most types of weather. Liquid water changes into water vapor by evaporation. The rate of evaporation is dependent on wind and the amount of water available.

Humidity is the water vapor content of air. It may be either relative or absolute. An increase in humidity usually means that bad weather is on the way. Condensation occurs when water changes from vapor into its liquid state.

Convection is the upward motion caused by the buoyancy of heated air. The instability or stability of air is determined by the ease with which an air mass can release energy; this is based on factors such as moisture content and temperature.

Fogs and clouds are made up of water droplets suspended in the air. Clouds are usually classified as either stratus or cumulus; many variations are possible within these general categories. Cumulonimbus clouds are the familiar thunderheads.

Precipitation is the process in which water condenses in the air and falls back to earth. Tropical rains probably form by coalescence. It has been theorized that temperate zone raindrops form as ice pellets, which melt before they reached the ground. Snow, sleet, and hail are other familiar forms of precipitation.

Air masses may be either polar or tropical. The point at which two air masses meet is called a front. Storms usually occur at the front between a warm and a cold air mass.

The thunderstorm is the commonest of the violent storms, and usually does not last very long. Thunderstorms are accompanied by lightning, a form of static electricity that moves between clouds or from a cloud to the earth. Texas northers and New England northeasters are other storms which last longer. Like all storms, they are caused by a collision between two air masses.

A cyclone is a system of winds blowing in toward a low-pressure area. An anticyclone is a high-pressure system from which winds circle outward. Hurricanes are small-diameter cyclones which produce winds in excess of 75 miles per hour. New forecasting techniques enable hurricanes to be predicted several days in advance. Tornadoes are very small but extremely violent cyclones, with winds that may reach hundreds of miles per hour. Although they rarely last for more than an hour, the damage they cause is far greater than that of any other storm.

REVIEW QUESTIONS

1. What conditions are necessary for condensation to occur?
2. Distinguish between stable and unstable air.
3. Describe two theories of raindrop formation.
4. Describe at least three types of cooling.
5. What is a front? What happens along it?
6. Distinguish between a cyclone and an anticyclone.
7. Describe the formation and breakup of a typical thunderstorm.
8. What type of front is most likely to produce stormy weather? How does it form?
9. Explain why the combination of heat and high humidity can be so dangerous.
10. What is saturation?
11. Describe the sequence of events when lightning strikes the ground.
12. List three types of precipitation and explain how each is formed.
13. Distinguish between absolute and relative humidity.
14. Describe the formation of a hurricane.
15. What is a polar outbreak?
16. Why can temperature inversions create a health problem?
17. Explain why the North American continent is subject to such a wide variety of spectacular weather.
18. What causes a New England northeaster to form?
19. Describe the two general types of cloud and two subtypes of each.
20. What is a tornado? Where does it occur most frequently?

PART 4

OCEANOGRAPHY

CHAPTER THIRTEEN

THE OCEANS

When Abraham and Lot, back in early Old Testament times, found the desert getting too small to hold both of them, Abraham proposed a friendly solution. One of them would remain on the high, arid grazing grounds; the other would take his flocks and herds and move down into the fertile, well-watered lakeside area around a cluster of towns known as the "cities of the plain." Lot could have the first choice. Being nobody's fool—and perhaps wondering if senility had finally caught up with his respected father-in-law—Lot took what was obviously the better option. He moved down to the plain and settled in the pleasant little city of Sodom.

As any reader of the sequel knows, Lot turned out to have made the wrong choice. Sodom and its sister city of Gomorrah now lie under several feet of water, probably submerged by a great earthquake sometime around 1900 B.C. The Biblical story tells of a rain of fire and brimstone, so perhaps a volcanic eruption accompanied the earthquake. In any case, the cities of the plain are long gone, and it is hard to believe that the land where they stood was ever green and fertile. For this is the shore of the Dead Sea, one of the most desolate spots on earth.

The Dead Sea gets its name for good reason. It is so salty—over eight times more so than ordinary sea water—that nothing but bacteria can live in it. Only a few tough, salt-loving plants grow on its shores. Yet billions of cubic feet of fresh water flow into it every year, chiefly from the swift-running Jordan River to the north. The problem is that they cannot flow out again. For the Dead Sea lies in a rift valley—a place where the earth's crust has sunk far below the level of the surrounding land—and its shores are nearly 1,300 feet below ordinary sea level. There is simply nowhere for its waters to go. So they evaporate in the fierce desert heat, leaving behind the small salt content found in even the freshest water. Gradually the salt has accumulated, until by now the Dead Sea is so salty that a swimmer cannot sink in it if he tries.

the oceans

The Dead Sea will never be a friendly place. But its very harshness has sometimes been an asset. The Wilderness of Engedi, where David took refuge from the jealous wrath of King Saul, lies on its western shore. Masada, the hilltop fortress where a thousand Jewish Zealots held off a Roman army fifteen times their number for nearly two years, overlooks its southern reaches. And in a scattering of caves near its northern end were preserved for nearly two thousand years the fascinating fragments of parchment and papyrus now known as the Dead Sea Scrolls.

Today a kibbutz occupies the land around Engedi where David and his men once hid. With proper irrigation, the land even here can be made to bear fruit. If old Lot were to return a few years hence, he might be tempted to repeat his once-disastrous choice, and move down again to a new, green city of the plain.

structure of the ocean

The earth we live on is indeed the watery planet. Oceans cover 71 percent of its surface. This can be said of no other planet in our solar system. In fact, Mars, Venus and the rest do not have anything even remotely resembling an ocean. And this difference is vitally important to us, for the oceans are central to the existence of life. All fresh water originally comes from the sea. All forms of life depend on water.

Oceanography—the study of the oceans—is a young science. It dates from the 1850s, when Matthew Maury of the United States Navy began to compile sailing charts from the information that had been gathered over centuries. This information consisted of records of water depth, winds, currents, and tides. The next great step forward was a survey of physical, chemical, and biological properties of the oceans. This research was undertaken from 1872 to 1876 by the scientists aboard a British vessel, H. M. S. *Challenger*, on an expedition around the globe.

During the International Geophysical Year of 1957–58, giant strides were made in learning about the oceans. But despite many new discoveries, scientists believe we still know less about the deep than we do about the moon or Mars. More information will be vital in the future. The oceans will become increasingly important as a source of food and minerals, a place for recreation, and a watery highway for growing international trade and commerce.

We speak of "oceans," but in reality there is only a single ocean. All the basins of the sea are connected, and ocean water flows freely between them. For the sake of convenience, we divide this world ocean into separate regions; for example, our maps show the

the oceans

Piston corers, such as the one shown here, are an important aid to our understanding of the composition of the ocean floor. The cores they bring up show differentiated layers of sediments.

Atlantic, Pacific, and Indian Oceans. The Arctic and Antarctic Oceans also are distinguished in popular usage. In a sense, we can think of the continents as "islands" emerging from the surface of these oceans.

The deepest parts of the oceans are farther below sea level than the highest mountains are above sea level. The average depth of the oceans is about 3.7 kilometers, whereas the average height of the continents is barely 0.8 kilometer. The total area of the oceans is more than 224 million square kilometers. They contain more than 528 million cubic kilometers of water. If the earth's surface could be smoothed out so that there were no irregularities above or below the seas, ocean water would cover the entire earth 3,650 meters deep.

Oceanographers recognize three layers of the ocean. These layers vary in temperature, salinity, and dissolved oxygen. They include a shallow surface **mixed zone**, a **transition zone**, and a **deep zone**. The surface layer makes up only about the top two percent of the ocean's

layers of the ocean

the oceans

waters. The transition zone contains another 18 percent, and the deep zone makes up the remaining 80 percent. However, these figures are only approximations. Depending upon the latitude and the season, thickness and temperature of the surface layer can vary.

The three layers of the ocean receive very different amounts of light. For example, only 40 percent of the sunlight that strikes the ocean surface penetrates below one meter. Only two percent reaches a depth of 30 meters. Thus, the greater part of the ocean is a dark, cold world. The amount of sunlight that penetrates the water has a significant impact on the animal and plant life that exists at various depths. One cupful of surface water may contain millions of microscopic plants or swarms of tiny animals, each no larger than a dust speck. In contrast, below depths of about 180 meters no plant life exists, and animal life assumes different forms and habits. Some creatures roam at different levels, depending on whether it is day or night.

ocean temperatures

At the surface, water is warmed or cooled. In the equatorial region the input of solar energy is high, and much of it is converted to heat. In the high latitudes the input of solar energy is low. Therefore, surface temperatures of sea water vary from the equator to the poles. However, as we will see later, ocean currents distribute some of the warm, equatorial surface water to the colder regions. This is one of the reasons why the higher latitudes do not become colder and colder.

Temperatures at the surface can range from 16°C in the Persian Gulf to −15°C in the Arctic or Antarctic. The cold water in high latitudes does not freeze because of its salt content. With few exceptions, water temperatures decrease with depth. The vertical temperature patterns differ, depending on the latitude. The temperature pattern for a higher latitude is about the same from top to bottom.

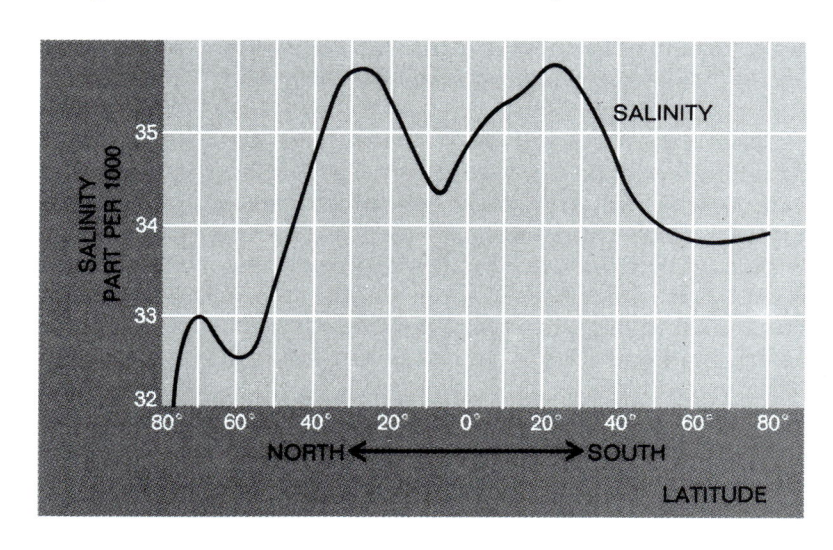

The surface salinity of sea water shows slight variations from north to south. This change is due to differences in precipitation and evaporation.

the oceans

salinity

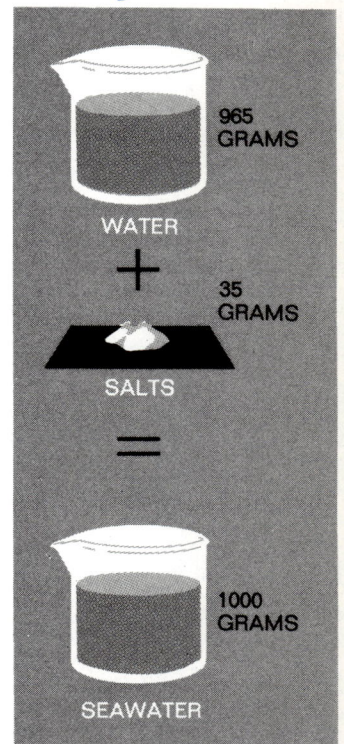

If you have ever swallowed water while swimming in the ocean, you know that sea water tastes salty. It should. If you were to evaporate 30 kilograms of sea water, about 1 kg of a white salt mixture would remain. Most of this mixture—about 78 percent—would be sodium chloride or common table salt. The other 22 percent would be made up mostly of sulfate, magnesium, calcium, potassium, and bicarbonate. Minute amounts of other elements and compounds have also been discovered in ocean water. In fact, the ocean probably contains at least trace amounts of all of the earth's naturally-occurring elements.

If the oceans were formed by rain water, which is fresh, how did the salts get there? Many of them were carried by rivers and streams from weathered rocks on the continents. Outgassing—the emission of water and dissolved gases from the earth's interior through volcanic eruptions—released still more. Others were produced by biological activity in the sea.

Salinity is measured as the weight of dissolved salts in 1,000 grams of water. The average salinity of sea water is about 35 grams per 1,000 grams of sea water (parts per thousand). Throughout most of the oceans, salinity ranges from 33 to 37. Where frequent rain or snow dilute the ocean, salinity is low. Where evaporation is high, salinity is high. Salinity reaches 41, for example, in the Red Sea, a dry region. Precipitation and stream runoff are high in the Baltic Sea, and salinity there is only about 7.

Why haven't the oceans become solid beds of salt over time? The reason is that minerals are removed from sea water as well as added. Ocean creatures build shells and skeletons out of some of them. Other material falls to the bottom as sediment. Humans have been extracting common salt from the sea since the latter part of the Stone Age. At present, the United States obtains its entire supply of magnesium and about 75 percent of its bromine from the sea.

In 1,000 grams of sea water there are approximately 35 grams of salts, mostly sodium chloride.

ocean currents

A **current** may be defined as a large body of water flowing in a certain direction. *The ocean is constantly in motion* may sound like a line from a nursery rhyme or popular song, but it accurately describes the nature of the sea. For centuries, sailors have been aware of currents and their effects.

Ocean currents exert a major influence on climate. They also distribute dissolved minerals and food for marine animals around the globe. Currents move horizontally, vertically and at all depths. When we study them broadly and over a long period of time, we can see that some currents form an orderly pattern of movement. However, when different currents are studied simultaneously on a short-range basis, we notice a great deal of irregularity. In this respect, ocean currents might be compared to the atmosphere, which contains both long-range climate patterns and short-range weather changes.

Sailors have measured and recorded currents and winds for hundreds of years. In the shipping lanes, where traffic is heavy, currents were accurately mapped because they were observed so frequently. In the areas that were seldom traveled, which include most of the ocean, there was not enough information to chart currents with any degree of accuracy.

Think of the problem of measuring a current from a ship moving along with the current. Unless you are within sight of land, you have no direct method of measuring your speed or direction. However, a radio navigational aid called **loran** (from **LO**ng **RA**nge **N**avigation) does help in locating the position of a ship. If the position of a drifting ship is determined every hour, then speed and direction of its movement can be determined. Of course, the wind as well as the current will influence the drift of the ship. You can see that there are many problems in making current measurements.

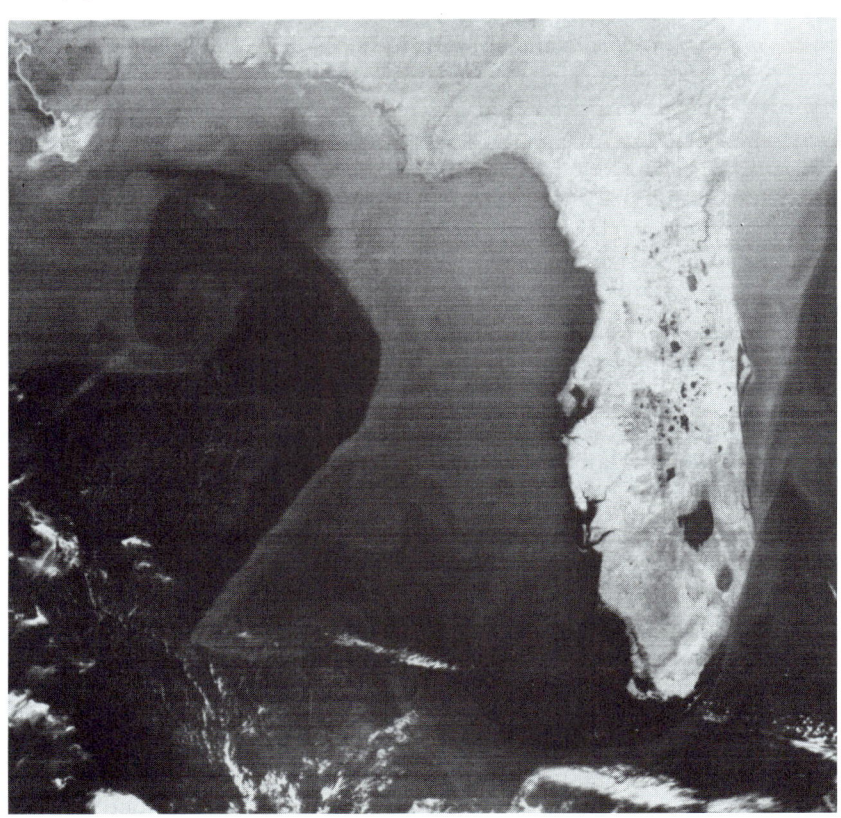

A loop current in the Gulf of Mexico shows clearly in this satellite photograph. The exact size and position of such a current is subject to continual change.

the oceans

Dyes and radioactive materials have been released in the water and their movements tracked. So have buoys that emit sound waves. Still another technique involves measurement of the faint electric current produced by a mass of water moving through the earth's magnetic field. In some cases the movements of large water masses can be detected by measuring temperature, salinity, and oxygen content. Finally, oceanographers have set up test tanks in which the currents of the ocean can be simulated.

Now, after this brief introduction to currents, it is clear that we must ask a basic question: what makes water move in the first place? Exactly what forces are involved, and how do they work?

gravity and density

Gravity will move water, but by itself it cannot create an ocean circulation. However, when its force is combined with the sun's energy, gravity is a strong mover of water. We saw earlier that some of the sun's energy heats ocean water. Water gains heat in the tropics and loses it in the higher latitudes.

In order to see how heating, cooling, and gravity can cause an ocean circulation, let us set up an imaginary model of an ocean that covers the whole globe. For the sake of simplicity, we shall also imagine that there are no winds and that the earth does not rotate. Our tropical ocean would be heated at the surface by energy from the sun. Our polar oceans would lose heat at the water–air boundary.

Since the tropical water would be warmer than the polar water, it would be less dense, and a given mass of it would thus occupy more space. In order to occupy more space, it would have to expand upward against the atmosphere, which would offer less resistance than the colder, denser polar water. In this way, the level of the tropical ocean would become a little higher than the level of the polar oceans. Therefore warm tropical ocean water would flow down-

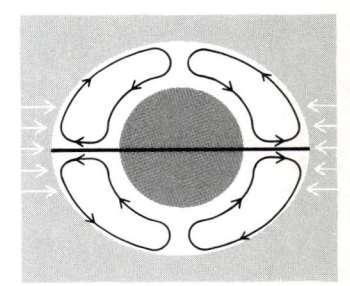

The model above shows the general ocean current patterns that are caused by heating at the equator and cooling near the poles.

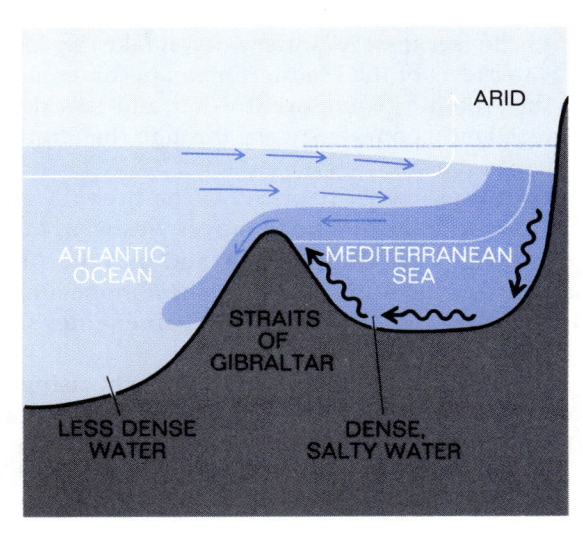

The circulation pattern of ocean water in an arid climate is shown at left. Why is the water of the Mediterranean denser than that of the Atlantic?

hill toward the polar oceans.

The movement of sea water from the equator toward the poles would also be influenced by the heavy rainfall in the tropics, which would help to raise the sea level at the equator. This rainfall would add water to the surface of the ocean faster than the water could spread out horizontally.

In the same ocean model, cold air would absorb heat from the surface of the polar oceans. The surface water would become colder and therefore more dense. When this happened, the surface water would sink, because it would have become denser than the water below the surface. At some depth below the surface, polar water would flow toward the tropical ocean. At the surface, water would be moving in the opposite direction, as we have already seen.

Thus, in our "hot–cold" model of the ocean, a difference in density due to a difference in heating and cooling produces a polar–equatorial circulation. The salinity of water also affects its density; the greater the salt content, the denser the water. A difference of salinity between two areas will produce a very slow and deep current. Such a difference in salinity between two areas occurs because evaporation of water makes a sea saltier, and heavy rainfall makes it less salty.

The Mediterranean Sea connects with the Atlantic Ocean. Since the Mediterranean is located in an arid climate, dry air removes more water from the sea than it gives to it. In fact, the Mediterranean loses an average of 70,000 tons of water every second to the air above it. The salinity of this sea is about 39 parts per thousand, whereas the average salinity of the Atlantic is only about 35 parts per thousand. Why should this difference create a current between these two bodies of water?

When the dense, salty water is cooled in the winter, especially in the eastern end of the Mediterranean, it becomes even denser and sinks. Since it is denser than Atlantic water, it takes up less space. As a result, the water level of the Mediterranean becomes slightly lower than that of the Atlantic. Therefore, fresher and less dense surface water from the Atlantic pours eastward through the Straits of Gibraltar into the Mediterranean.

In the meantime, what happens to the heavy Mediterranean water? Flowing close to the bottom, it moves westward out of the sea, over a ridge known as the Gibraltar sill, and into the Atlantic. Submarine commanders have made use of these opposite currents to drift in and out of the Mediterranean with their engines silenced.

the force of wind

Let us set up another model of the earth, using the global winds—the trade winds and the prevailing westerlies—as the only driving force for water. When a light breeze blows over the ocean, there is very little friction between the air and the water. But as the wind increases in velocity, it causes the water surface to ripple.

the oceans

When ripples appear, the wind presses on the side of the ripples and the water starts to move with the wind.

How fast and how deep down the water moves depends on three factors: the speed of the wind, the length of time the wind blows in a certain direction, and the distance of open ocean over which it blows. The last factor is known as **fetch**. In general, the speed of the water is only about 2 percent of that of the wind. For this reason, wind-caused currents in the open ocean are not very swift.

If our imaginary "wind" model of the earth possessed no continents, the global winds would produce broad surface currents flowing around the earth. There would be a westward current in the region of the trade winds and an opposite current in the region of the westerlies. On the actual earth, there is a region where no continent blocks wind or current. The westerlies in the far Southern Hemisphere, below the tip of South America, make a circular pattern around the earth. This wind pattern produces a broad, shallow current, called the **West Wind Drift,** that circles the Antarctic continent.

Adding the continents to our "wind" ocean model would give us a fairly accurate representation of the major surface currents of the oceans. The currents produced by the trade winds become the **North and South Equatorial Currents.** In the western margins of most oceans, these currents run into land, which turns them north or south. As they flow north or south along the coast, they are gradually turned eastward by the Coriolis effect. Eventually, the currents reach the latitudes of the westerlies, which drive them eastward. In the Northern Hemisphere these currents are called the **North Atlantic Drift** and the **North Pacific Drift.**

What happens to the ocean water between the trade winds and the westerlies? These areas have variable, light winds. In the Northern Hemisphere the bodies of water in these areas rotate slowly clockwise. The trade winds push the southern part of the rotating water westward. The westerlies push the northern part eastward. Such rotating water is called a **gyre.** The Sargasso Sea is the gyre in the North Atlantic Ocean. Warm surface water piles up in the gyres because of the Coriolis effect. That is, as the winds drive the water into motion, water slips to the right of the direction of the wind. In this way, water is pushed into the center of the gyres by the trade winds and the westerlies.

As the currents around the gyre flow poleward between the continent and the gyre, they become narrow and deep. The Gulf Stream, the Kuroshio, and the Agulhas are examples of such currents. They lie on the western sides of their respective oceans.

On the eastern side of the oceans, water is generally flowing toward the equator as part of the eastern side of the gyre. Here the currents are very slow, wide, and shallow. Although they are not fast and deep, they do transport a great deal of water because of their breadth. The Canary and Benguela currents in the eastern side of the

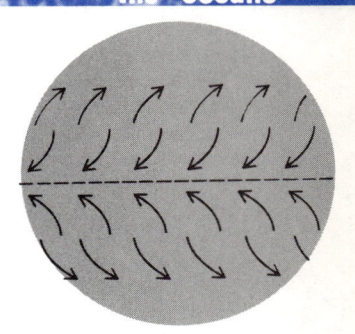

The effect of winds in producing ocean currents is shown in the model above. The center line represents the equator.

In the satellite photograph below the Gulf Stream is clearly visible, flowing roughly parallel to the Atlantic Coast of the United States.

Atlantic Ocean and the California and Peru currents in the Pacific Ocean are examples. These currents are relatively cool, wide, and shallow. They seldom exceed a speed of 1.6 km per hour.

We have already discussed the area called the doldrums, which lies between the northern and southern trade winds. This is the area of the **heat equator**—just north of the geographical equator. Here air is generally rising, and the winds are weak and variable. Therefore, there is no force to drive water. In the Atlantic the trade winds pile up water along the northeast coast of South America, making a "hill" of water. The water on the northern slope of this "hill" flows north into the Caribbean. The water on the southern slope flows south and becomes the Brazil Current.

In the area of the doldrums, where there is no prevailing east wind, water flows downhill eastward, back toward Africa. This is known as the **Atlantic Equatorial Countercurrent.** In the Pacific there is also an equatorial countercurrent, which is even better developed than the one in the Atlantic. In the Indian Ocean there is an equatorial countercurrent, south of the equator, that varies with the monsoon winds.

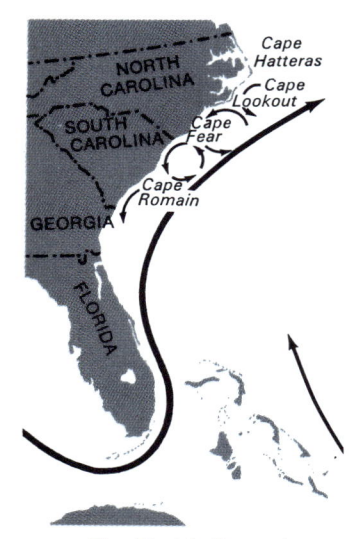

The only great ocean current along the shores of the United States is the **Gulf Stream,** perhaps the mightiest of all ocean currents. Its source is the broad surface of the equatorial Atlantic. We have seen how the trade winds drive the surface waters westward. The northeastern coast of South America guides most of the water of the equatorial currents into the Caribbean Sea. However, even in the Caribbean there is no well-defined, fast-flowing current. Pushed on by the trade winds, the broad, slow surface flow is concentrated and funneled into the straits between Yucatan, Mexico, and the western tip of Cuba. From there it runs into the water of the Gulf of Mexico, which turns the current eastward.

By this time the current is known as the **Florida Current,** which flows up along the east coast of Florida. The fastest part of the current is usually about 29 km offshore, where it averages about nearly 5 km/hr (about 150 centimeters per second). As it sweeps past Miami, the Florida Current has 1,000 times the flow of the mighty Mississippi River. Technically, it is not called the Gulf Stream until it passes Cape Hatteras in North Carolina.

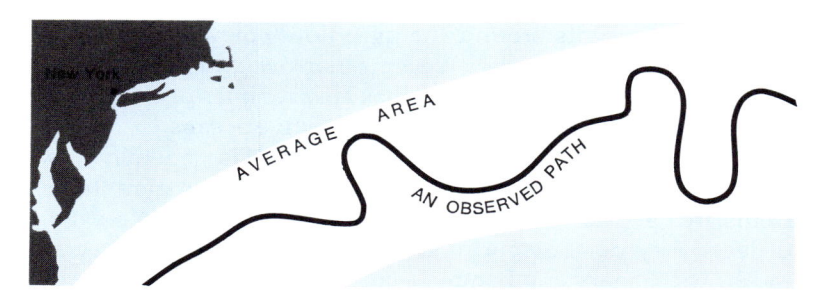

The Florida Current originates in the Gulf of Mexico. When it passes Cape Hatteras, it becomes known as the Gulf Stream.

The Gulf Stream meanders along its length, often forming eddies. Although this huge ocean current continuously changes its position, it can usually be found within a well-defined area.

After the Gulf Stream takes a northeasterly course east of Cape Hatteras, it does not stay in a fixed path. It meanders like a river. Some of the meandering loops work themselves downstream. Others break off from the main current, forming great detached eddies of water. North and west of the Gulf Stream the coastal water is cold, having come south from between Labrador and Greenland. South and east of the Gulf Stream the water is warm, about 25°C (76°F).

The Gulf Stream is thus a very intricate flow of warm, tropical water with cold, arctic water flowing southward along its west side. Its pattern is complicated by changing, cold countercurrents, meanders, and branches. It is not possible to drift on the Gulf Stream and get a free ride to Europe. Oceanographers have tried to do this, only to find themselves caught in a countercurrent flowing in the wrong direction.

In the northern part of the North Atlantic, the Gulf Stream acts as a kind of rim—or even a "dam"—for the Atlantic gyre. Thus it keeps warm surface water from bathing the shores of Greenland. At the same time, however, it carries warm water to the western coast of the British Isles and Norway. Perhaps if this "dam" along the Sargasso Sea were not there, even more warm water would spill over to the European coast.

deep circulation

We have already seen that the density of ocean water depends on its temperature and salinity. The colder or saltier the water, the denser it is. Around Greenland and Antarctica, water is cooled to freezing during the winter. When water freezes, it gives off its dissolved salts. These in turn make the surrounding water saltier and therefore denser. This cold, salt-heavy water that is denser than the water around it slowly sinks and spreads throughout the oceans. Remember that our "hot–cold" ocean model suggested such a circulation.

It has been estimated that the slow convection of this heavy water is only about 20 kilometers per year. Such a movement is far too slow to be detected by current-measuring instruments of any kind. Scientists must make inferences based on comparative densities, salinities, and water pressures at different depths and at different places. They know, for example, that water loses some of its oxygen as it sinks to deeper levels. The oxygen disappears gradually when it becomes involved in chemical reactions with dead microorganisms. Thus the amount of dissolved oxygen in a sample of water taken from the deep gives scientists a rough idea of the age of that water.

We have learned that in a few places cold water leaves the surface and moves downward, carrying oxygen with it. In other places, to balance this, water must rise. Below-average surface temperatures prevail off the west coasts of South America and Africa. Where is the

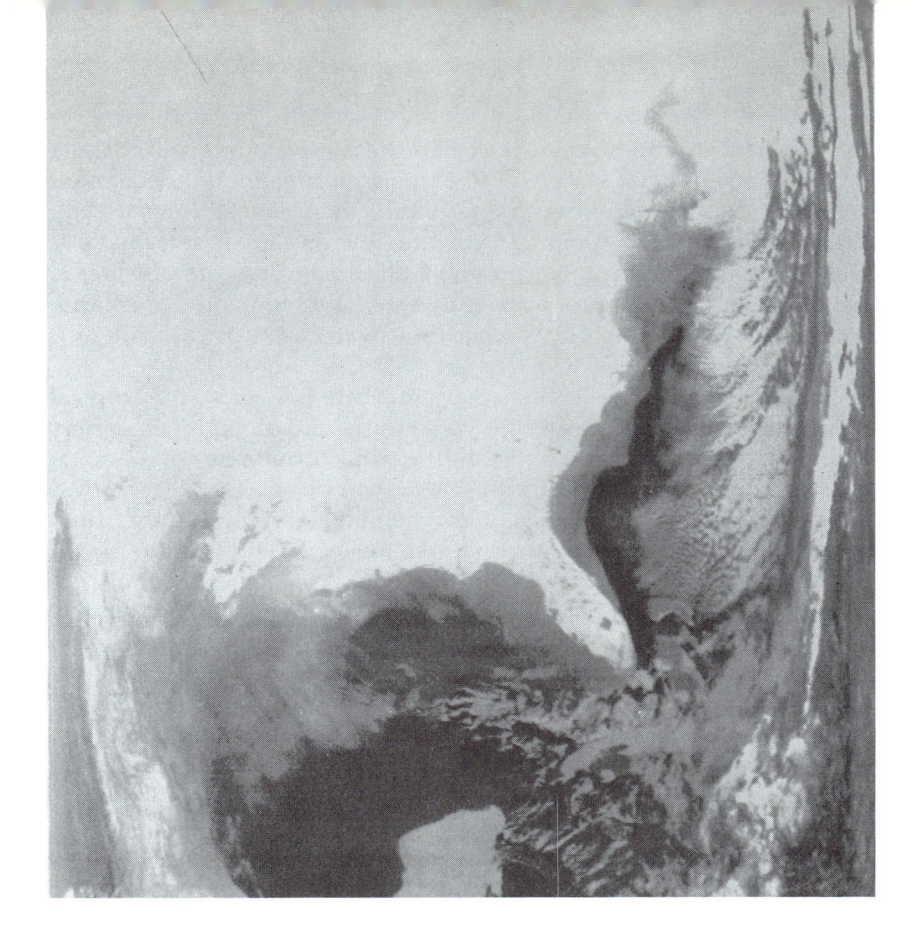

Here the Gulf Stream appears dark, as photographed from a satellite. Surveys from satellites have been very helpful in charting the positions of major ocean currents.

cold water coming from? In this case, the trade winds blow warm surface water away from the coast. Cool water from about 200 meters below the surface then creeps up along the bottom near the coast. This upward movement of water is called **upwelling.** Water that has upwelled is cooler than the water it has replaced. Usually it is also rich in minerals necessary for plant growth. For this reason, areas of upwelling support a great deal of marine life.

Learning more about the circulation of the oceans is important to our understanding of climate, fishing, and the effects of dumping wastes into the sea. Dumping of atomic wastes without certain knowledge about the movements of the seas, for example, could prove a serious danger to marine life and people as well.

waves

All of us have looked at an ocean, a lake, or a pond and noticed that the surface of the water is rarely still. It is usually disturbed by waves. They may be tiny ripples that lap gently at the shore of a pond

the oceans

A turbulent sea off the craggy coast of Maine. The treacherous northern Atlantic has long been a major obstacle to the New England fishing economy.

or giant breakers that crash on the ocean shores. What are these waves? How are they different from each other? What causes them to form? These are questions that we must answer before we can explain the effects of the sea on the land.

Waves involve movement, and movement involves energy. Physicists define a wave as a disturbance or vibration that moves progressively through a medium. Notice that the definition does not mention energy. Energy causes the wave, and energy is carried by the wave, but it is not a part of the wave. When waves move across water, the originally calm, level surface of the water is changed to a surface that moves up and down. The up-and-down movement is the disturbance. The fact that we can see a wave move across the water shows that it is moving progressively. The medium is the water.

The high point of a wave is called its **crest** and the low point is

The parts of a wave are shown here. Compare our discussion of water waves with the discussion of light waves in chapter 2.

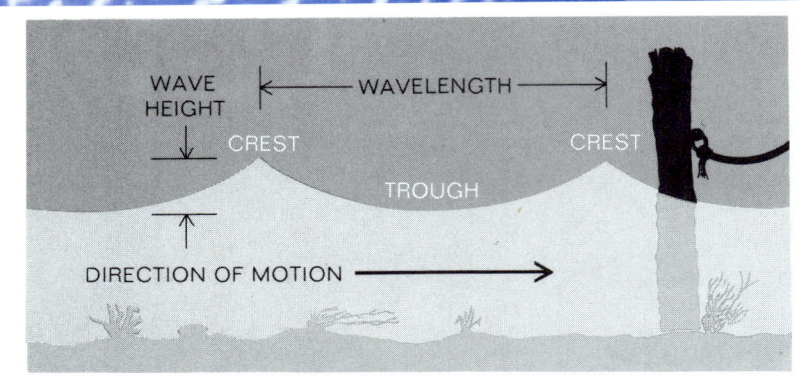

called a **trough**. The distance from one crest to the next, or from one trough to the next, is the wavelength. The vertical distance from the top of the crest to the bottom of the trough is the **wave height**.

The number of waves that pass a fixed point in a second is called the **frequency**—five per second, for example. Wave frequency can be expressed in another way. This is the **wave period**, the time it takes for one complete wave to pass a fixed point. The wave period is usually measured in seconds. If you know the wavelength and the wave period, how can you determine the speed of the wave?

The movement of a wave makes it seem that the water at the surface is moving in the direction the wave is traveling, but this is not what is really happening. An object floating on the surface of the water does not travel along with a wave. Instead, the object just bobs up and down in the water as the wave passes. It also appears to move a little forward as the crest of the wave passes under it. Then it moves a little backward as the trough passes. In fact, it has a circular motion. This means that the particles of water making up the wave must also be moving in a circle. The particles of water at the top of the circle form the crest of the wave. When the same particles are at the bottom of the circle, they form the trough of the wave.

The more energy is transferred to the water, the larger the wave grows. The larger the wave, the larger must be the **circles of motion**. This has been demonstrated in the laboratory by measuring the radii of the circles of motion of objects being moved by waves. This evidence suggests that the size of the circles of motion is closely related to the amount of energy transferred to the water by the wind or other force. The more energy is transferred, the larger the circles of

The wave moves forward, but the cork and the water remain in the same position unless they are moved by the force of blowing wind.

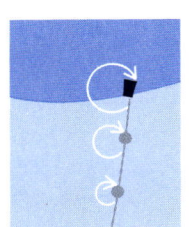

motion at the surface become, and the deeper they go beneath the surface.

kinds of waves

Imagine two people holding a rope stretched between them. One jerks the rope to the left and right, or up and down, and you see a wave travel along the rope. We could call such a wave a **shake wave.** The physicists' name for it is a **shear wave.**

Can we produce waves by another method? Place a spiral spring on the floor and stretch it slightly. The toy called a Slinky is just right for this demonstration. Compress, or squeeze together, a section of the extended spring somewhere near the middle. Then quickly let go of the compressed section. What happens? You see a wave of compressed loops travel the length of the spring and return. Physicists call this a **pressure wave.**

In both cases disturbances are moving progressively. Both are waves, but the pressure wave moving in the spring is different from the shear wave moving in the rope. In pressure waves the principal motion is forward-and-backward progressive motion.

Two different kinds of waves are caused by pressure on water. One is like that in the Slinky. There, the pressure you applied compressed part of the spring. When you released the pressure, a wave traveled through the coiled spring. Sound waves are pressure waves of this sort.

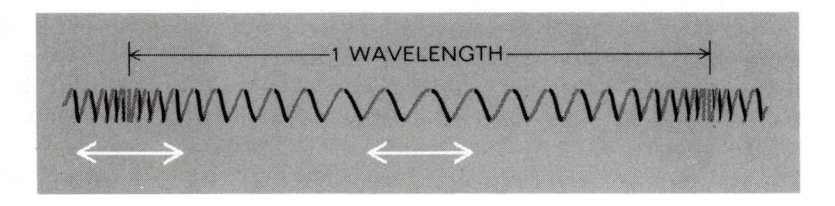

A diagram of a pressure wave. Seismic P-waves, which were discussed in chapter 5, are pressure waves.

The other kind of pressure wave can be seen at the surface of the water. Wind pressure raises the water into ripples, and then gravity pulls the raised water back down. You know from observation that the pressure–gravity waves travel in water. As we noted earlier, their speed is related to, but slower than, the wind that causes them.

Waves appear to the onlooker to be simple phenomena, but to the oceanographer they are an extremely complex subject for study. Research in wave theory requires a firm knowledge of mathematics, physics, and computers. Scientific findings about wave action have many practical applications. For example, knowing how waves act can better prepare us to protect our beaches against the effect of winter waves. Too often in the past, the selection of suitable beach and resort sites has been based only on aesthetic and political considerations rather than on a scientific appraisal of environmental matters, including wave action.

the oceans

causes of waves

In the Slinky experiment you started a wave by pushing together some of the spirals of wire. The same thing will happen if you stretch a section of the Slinky. When you release the part of the spring that you pulled apart, a wave system moves through the Slinky. Thus, it appears that either pushing or pulling on a medium may cause a wave.

What could push or pull on water to start a disturbance moving? If a boat moves through the water, it pushes water out of the way. Do waves form from this pushing? If you don't know the answer, push your hand through some water.

Try some experiments. Fill a pan with water and let it stand until the surface is quiet. With your mouth just above one edge of the pan, blow across the surface of the water. Do waves form? What natural action have you imitated? The most common water waves are caused by the wind. These are called **wind waves.**

When two fluids, air and water in this case, are moving at different rates and come in contact, there is turbulence. Air does not move smoothly along the surface of the ocean. Instead, some air twists into **eddies.** In an eddy, some of the air moves upward and some downward. This produces changing pressure on the surface of the water. Fluids respond almost instantly to small changes in pressure. The downward-moving eddies of air depress part of the water surface and cause other parts to rise. The depressions and bulges in the water surface are the beginnings of wind waves.

The raised areas of the water surface offer resistance to the flowing air. The pressure of the air is generally uniform on the raised water surface. On the lee side of a wave — the side away from the wind — the air is more turbulent than on the windward side. Therefore, the wave moves in the direction the wind is blowing.

As long as the wind continues to blow, waves will grow and move before the wind. They will grow until the wind pressure can build them no larger. Three factors determine the size of a wave. They are the same ones we saw in our discussion of currents: the wind velocity; the length of time the wind has been blowing; and the distance over which the wind affected the water, or the fetch.

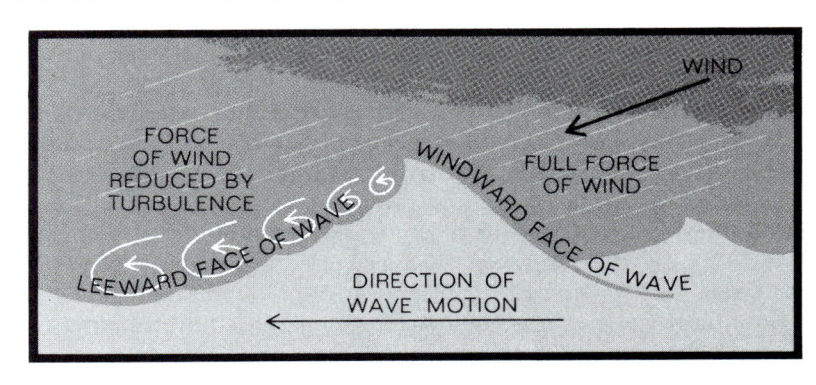

Unequal pressure of wind on water causes the surface of the water to ripple. As the wind continues to blow, waves are formed.

As soon as the wind that is building a wave dies down or the wave moves out of the range of the wind, the wave's height decreases and its wavelength increases. Because of its momentum, the wave continues to move across the surface of the ocean. But it now has a very low wave height and a very long wavelength. Waves that are no longer being built up by the wind are called **swells.** They will continue to move until friction converts all the kinetic energy of the moving water to heat.

In a swell, energy is being used up, and no new energy is being gained from the wind. Swells are considered old waves. But young waves receive energy from the wind faster than they expend it. That is why they grow. In young waves, there is a relationship between wavelength and wave height. The greatest height that a wave can have is about one-seventh of its wavelength. Otherwise, the wave becomes too steep and will break, causing a **whitecap.**

waves and storms

Ships and people have been destroyed by waves throughout the history of seafaring. But it is around the shorelines that waves have done their worst damage. Books about the sea are full of stories about the devastating force of waves whipped to a fury by high winds. What happened at Tillamook Rock, Oregon, demonstrates the power of waves. A tall lighthouse is situated at Tillamook Rock. During one of the most violent storms in the region, the waves lifted a 60-kilogram boulder from the coast and heaved it through the roof of the lighthouse. The roof was more than 30 meters above the sea!

How high can a wave get? Gale-force winds (27 to 55 knots) that blow across a long fetch may build waves with wavelengths as great as 300 meters. Such waves theoretically could have wave

Waves roll in and break on the rocky coast of Oregon. Coastlines are in a state of constant change as the result of wave action.

IN FOCUS:
exploring the last frontier

The tiny, oddly-shaped craft settles gently to the floor of the Caribbean. Suddenly, a bright floodlight mounted on her aft section pierces the watery darkness. At the ship's porthole, a human face appears. Then a mechanical arm slowly extends from the craft's bow. It claws at the ocean floor immediately ahead of it, shaking the little ship as it pries loose handfuls of the bottom sediments. What is this strange visitor, over 3,600 meters below the waves? The ship is the submersible *Alvin*, and her mission is to probe the depths of a vast rift in the sea floor that separates two of the earth's gigantic crustal plates. The *Alvin* is a recent addition to deep-sea exploration, an addition that is opening new avenues of research in oceanography.

The ocean has been called the last frontier, one of the few places on earth that still guard their secrets from the prying minds and machines of human beings. But within the last ten years or so, a drive has been launched to uncover these secrets. Valuable resources lie in the depths of the sea, and the technology for reaching them is now available as a spinoff of the space program.

For decades, oceanographers have employed a number of conventional techniques. Surface vessels such as the *Glomar Challenger* have served as floating laboratories for teams of scientists who study the ocean's currents, biology, chemistry, and geology. Fixed platforms are another type of seagoing laboratory, and have enabled researchers to conduct long-term studies.

But some of the most spectacular advances in undersea exploration have come from manned submersibles—pressure-sealed, maneuverable craft capable of carrying researchers into the depths. At present, one of the hardest-working submersibles happens to be the *Alvin,* operated by the Woods Hole Oceanographic Institute. This 7-meter-long mini-sub can accommodate two people in comfort, but occasionally a third has been shoehorned into its tiny pressure sphere. This compartment, constructed of a layer of titanium two inches thick, protects the little ship from pressure to a depth of 3,600 meters. At such depths, pressure is as much as two and one-half metric tons per square inch, enough to make an ordinary submarine look as though it had been through a trash compactor. The *Alvin* maintains radio and sonar contact with her constant traveling companion, the *Lulu,* a catamaran which hovers protectively at the surface.

The *Alvin* has recently starred in two important deep-sea ventures. In 1974 she took part in Project FAMOUS, an operation which probed a portion of the Mid-Atlantic Ridge in an attempt to discover why the sea floor is spreading. In early 1976, the *Alvin* descended to the Cayman Trough, a gash in the sea floor nearly 1,500 km long and 7,680 m deep in some places. This trough separates the westward-moving American plate to the north from the eastward-moving Caribbean plate to the south. The goal of this expedition was to map and discover the nature of the rock structures in the region.

To carry out her missions, the *Alvin* comes equipped with an impressive array of hardware. Drills, hammers, and a manipulative arm ending in a claw are located on the exterior of her hull. These enable the crew to examine and collect any samples of rock, sediments, and undersea life that look interesting. Television and still cameras take thousands of photos of the surroundings. Inside, the passenger compartment is crowded with gauges, electrical equipment, and flashing panel lights. Indeed, riding in this tiny sphere with the clicking, pinging, and humming of the electrical gadgetry has been compared to being trapped inside a gigantic wristwatch.

An even more interesting development in undersea exploration is the habitat, an underwater house located at shallow depths, in which teams of scientists live in a pressurized atmosphere and study the marine environment on intimate terms. Their exploration of the sea involves frequent diving sorties outside the habitat.

The Sealab habitat program, conducted by the U.S. Navy, is well known. Another very ambitious habitat project is Tektite II, which began in the early 1970s and is operated by the Department of the Interior. The goal of the project was to study sediment formations and sea life. A series of eleven-man teams lived and worked a total of 915 man-days at a depth of fifty feet below the surface, near the Virgin Islands. Their habitat consisted of two cylindrical steel tanks, twelve feet in diameter and eighteen feet in height. Each tank was really a duplex apartment. One upper room served as a storage area, while the other contained communications and lab equipment. One lower room held diving equipment and was equipped with a hatch through which the crew could enter and exit. The remaining room was the furnished, plushly carpeted living quarters, where the researchers ate, slept, and listened to music. A support barge tethered nearby supplied the habitat with air, water, and electricity through an umbilical system of hoses and cables.

The habitat's advantage over other deep-sea exploration techniques is that it enables researchers to study the ocean depths on a long-term, day-to-day basis. The success of some of the habitats has prompted predictions of future underwater cities, thriving on the floor of the ocean. But such latter-day Atlantises, if they ever become reality, are probably as far away in time as settlements on Mars.

heights of one-seventh of that, or about 43 meters. A hurricane wind blowing across a long fetch could conceivably create even higher and longer waves. Waves more than 30 meters high have been observed at sea, but they are rare.

Earth scientists are interested in knowing how to calculate the depth to which water is disturbed by waves. Such knowledge helps to explain observations of the sea bottom made in offshore waters. Of course, water molecules cannot be observed directly. Scientists must therefore observe the effect of the motion of water upon visible particles such as mud, sand, and silt. Scientists using scuba equipment have watched and photographed the movement that takes place in a circle of motion.

Scientists can also generate small waves in experimental, glass-sided wave channels. By adding visible material to the water, they can study the motion of water at several depths. From such observations, students of wave theory know that the longer the wave period and wavelength, the deeper the disturbance. They also have learned that there are differences in the actions of young wind waves and swells. In waves, the radii of the circles of motion increase with increasing wavelength. In swells, the radii decrease with increasing wavelength.

As waves approach the shore, the water becomes more and more shallow. Somewhere offshore, the deepest circles of motion will meet the sea bottom. Where the depth equals one-half the wavelength, the wave touches bottom. The wave is then said to **feel bottom.** For example, swells with crests 100 meters apart will feel bottom in 50 meters of water.

As the water in the circle of motion brushes against the bottom, the water rapidly transfers energy to the sediments. This causes the sediments to move. The loss of energy changes the circle of motion to an ellipse. Since the wave loses energy to the bottom, its forward motion slows down. Therefore, the shallower the water, the slower the wave. When a wave feels bottom, its wavelength decreases, and its wave height increases.

The lower part of the ellipse of motion drags on the bottom. This means that the upper part of the ellipse is moving faster than the lower part. What would happen to you if your feet moved forward more slowly than the upper part of your body? You would lose your balance and fall forward. The same thing happens to a wave. The bottom of the wave moves more slowly than the crest. The top of the wave curls forward and crashes downward. When a wave does this, we call it a **breaker.** The wave actually breaks apart. When the wave breaks, it loses its energy rapidly to the material on the sea bottom. This stirs up the sea bottom and moves the sediments. Such energy has been felt by anyone who has been hit by a breaker at the seashore.

HARNESSING TIDES FOR POWER

People have long realized that the tides represent a potential source of clean, relatively unlimited energy. The British Domesday Book of the eleventh century refers to a mill at Dover which was driven by the tides. Remains of tidal pumps and mills, in use during the last century, dot the English landscape. Today, two tidal hydroelectric plans are in operation—one in France and a smaller facility in the Soviet Union. About a dozen other sites are candidates for such plants, including Severn, England; Passamaquoddy Bay at the United States—Canada border; and San Jose, Argentina.

Tidal plants can be built only in locales which have large tidal ranges (the difference in sea level between flood tide and ebb tide). The range at the French facility at Rance on the English Channel, for example, averages about 8½ meters. Rance, completed in 1966, has 24 generators. It is the world's first major hydroelectric plant driven by tides. A 700-meter-long dam spans a narrow estuary, separating the sea from the tidal basin. Sluice gates in the dam are opened to permit the incoming tide to enter the basin, then closed at high tide to trap the water. At low tide, the gates are re-opened to release the water to the sea. As the water enters and leaves the basin, it turns the blades of reversible turbines in the dam, thus generating electrical power.

Such plants provide pollution-free energy which is self-renewing, and therefore virtually inexhaustible. But Rance's $100 million price tag for construction alone disproves the common misconception that tidal power is free or cheap. Off and on over the past half century, a tidal power facility has been proposed for Passamaquoddy Bay. However, the building cost of this huge project—it would require 11 km of dams, 150 water gates, and many maritime locks—far exceeds the cost of power that could be had from other sources. Whether any of the other potential tidal power sites can be developed economically is one of the big question marks in the world's future energy picture.

the oceans
tsunamis

There are forces in nature that jolt the water in an ocean more deeply than wind moves it at the surface. Those include underwater earthquakes and volcanic eruptions, and submarine or coastal landslides. An earthquake may cause a small section of the ocean bottom to drop or to rise. This violent action makes the water drop or rise, causing a wave. Waves caused by earth movements are called **seismic sea waves,** or **tsunamis.** They are very long and low. Their wavelengths are about 200 kilometers and their periods approximate 1,000 seconds.

A tsunami (Japanese for "large waves in harbors") can be coupled with great destruction. One that occurred at Awa, Japan, in 1703 killed more than 100,000 people. Frequently, the arrival of the tsunami is preceded by a pronounced lowering of the sea. That is what happened on April 1, 1946, in the Hawaiian Islands. The surf stopped pounding, and the sea became strangely quiet. Some people who came down to shore to watch the amazing spectacle paid for their curiosity with their lives when the first great wave hit the beach. When the tsunami had ended, it had left behind 173 dead and $25 million in damages.

Yet so shallow are the waves of a tsunami in the open sea that ship captains are often unaware that they are passing. They are less than one meter high, and there can be as much as 200 kilometers

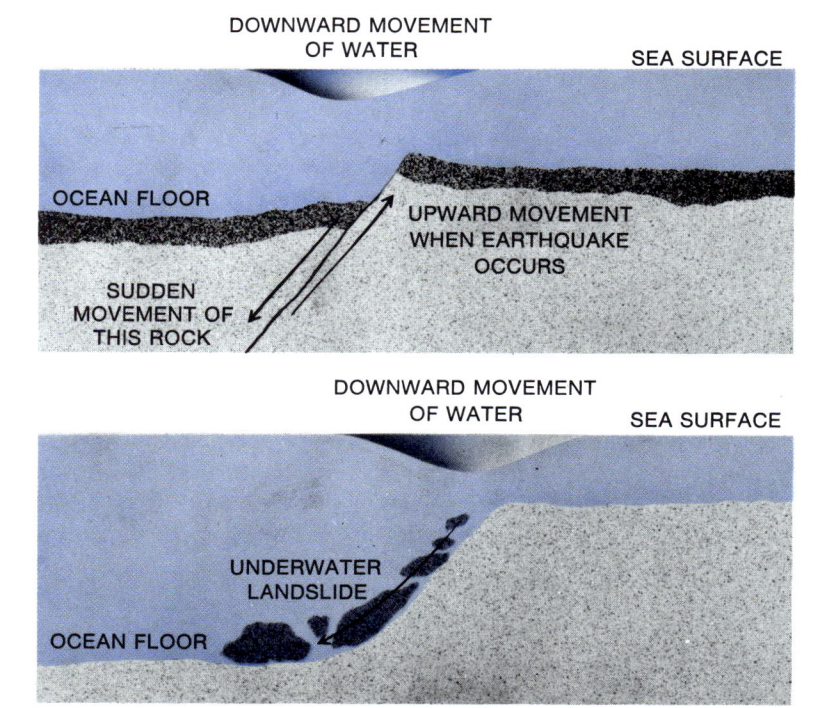

A tsunami occurs when the ocean bottom is violently disturbed by an earthquake, a volcanic eruption, or a landslide.

the oceans

between crests. A captain in a ship off Hilo, Hawaii, in 1946 was thoroughly startled when he saw huge waves breaking in the harbor. He had not noticed anything unusual happening moments before.

tides

Those who live by the ocean are familiar with the daily, rhythmic rising and falling of sea level. Sea captains have long known that they must guide their ships by these changes, known as the **tides.** In certain harbors, the onrush of high tide can crush even large ships against the docks. Captains in some places must wait for high tide before they take on or unload cargo. Unless they move out well before low tide, their ships will be stuck in mud.

Although people have been familiar with tides for as long as they and the sea have existed, it took Sir Isaac Newton to tell them why the tides exist. Simply, the tides exist because of the gravitational pull of the moon on the earth's waters and, to a lesser extent, because of the pull of the sun upon those same oceans.

To understand tides we have to remember that the gravitational force of the sun is exerted on the entire planet. But the force on each region is not the same. Some regions are nearer the sun and are attracted more strongly than regions that are farther away. The side of the earth nearer the sun tries to move towards the sun. The side farther from the sun tries to move away from the sun.

The gravitational pull of the moon draws the earth's oceans toward it in the same way. Waters that face the moon will tend to bulge toward it. Waters in other parts of the globe will flow in the direction of the bulge. This horizontal movement of the water is more important in creating tides than the vertical, upward lift of the moon, because the earth exerts a downward pull upon its waters that is one million times stronger than the upward lift of the moon.

Low tide (left) and high tide (right) on the coast of Nova Scotia.

the oceans

One important point to note is that the "bulges" exist in two places, not one. One bulge exists where the earth faces the moon. Another bulge exists on the side of the earth exactly opposite to where the earth faces the moon. This second bulge is caused by the centrifugal effect of the earth's rotation, combined with the fact that the opposing gravitational pull of the moon is weakest at this point.

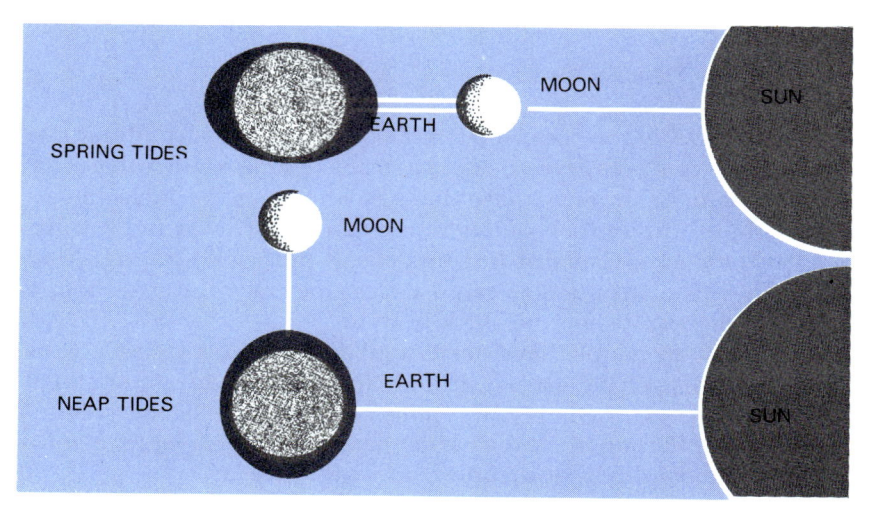

Spring tides occur when the moon and the sun are aligned with the earth. Neap tides occur when sun and moon are at right angles to each other with respect to the earth.

kinds of tides

When the moon and the sun are in line with the earth, as they are during a full or new moon, both exert their gravitational pull on the waters. The highest high tides and the lowest low tides occur at such times. These strong tides are called **spring tides.** When the moon and the sun are at right angles to each other, during the half moon and three-quarter moon, weak tides occur. They are called **neap tides.**

Because of the angle of the moon's orbit around the earth, tides last for different lengths of time in different places. Basically, there are two kinds of tide. One, the **semi-diurnal** (half-day) **tide,** has a period of 12 hours and 25 minutes from one high tide to the next. The other, the **diurnal** (day) **tide,** has a period of 24 hours and 50 minutes. In some places there is a mixture of diurnal and semi-diurnal tides. These so-called **mixed tides** result in two high and two low tides in one day, with a considerable difference between the heights of the morning and evening tides. If you live on the Atlantic coast of the United States, you are probably more familiar with semidiurnal tides. Pacific coast residents are more accustomed to mixed tides.

The Bay of Fundy, which separates Nova Scotia from Maine and the Canadian mainland, is noted for particularly high tides. The range between high and low tide is as much as 15 meters. The head of the bay is much like one end of a bathtub. Water set in motion there will rock back and forth—oscillate—with great vigor for about 12 hours. This 12-hour period of oscillation almost coincides with the period of the tide. Hence, the two forces—oscillation and tide—sustain and reinforce each other. Furthermore, the Bay of Fundy has a narrow opening. Water rushing into the bay from the ocean is compressed as it passes through the opening. Forced into a diminishing amount of horizontal space, it must rise vertically, thus reaching a very high level at high tide.

Tidal current is the horizontal flow of water that accompanies a rising and falling tide. The coastal areas that are alternately submerged and exposed by flood tides and ebb tides are called **tidal flats.** Tidal flats may range from narrow strips to wide areas of land.

The power of the tides will undoubtedly receive increasing attention in the future as a possible source of energy, as supplies of oil and natural gas dwindle. Water wheels driven by the tides were used to grind grain as early as the twelfth century. Tidal power plants for the generation of electrical energy may one day be built wherever there are narrow, enclosed bays and large tidal ranges.

In some places, such as the Bay of Fundy, the high tide flows in as a wall of water. The tidal range here may reach as much as 20 meters.

SUMMARY

Oceanography is a young science, concerned with the study of the oceans that cover 71 percent of the earth. Much of the water that forms the oceans probably came from deep within the earth, early in the earth's history. The average depth of the oceans is about 3.7 kilometers.

The layers of the ocean include a mixed zone, a transition zone, and a deep zone. The three layers receive very different amounts of light, which has a significant effect on the types of life that exist at the different levels.

In addition to large amounts of common salt, sea water contains sulfate, magnesium, calcium, potassium, and bicarbonate. As a result, sea water is much denser than pure water. Salinity throughout most parts of the ocean ranges from 33 to 37 parts per thousand.

The ocean is constantly stirred by deep ocean currents. The movement of these currents is generated by a combination of the Coriolis effect, the wind, and the variation in density and temperature of different regions of the ocean. The only great ocean current along the United States coastline is the mighty Gulf Stream.

A wave is a disturbance that moves progressively through a medium. Waves may be classified as either pressure or shear waves. The most common water waves are caused by wind. The size of a wave is determined by wind velocity, the length of time the wind has blown, and the distance over which the wind has affected the water.

Tides, the daily rise and fall of sea level, exist because of the gravitational pull of the moon and the sun on the earth's waters. Tidal current is the horizontal flow of water that accompanies a rising or falling tide. Swift-moving tidal currents may produce turbulent, dangerous seas.

REVIEW QUESTIONS

1. What percentage of the earth's surface is covered by oceans?
2. Explain how ocean currents affect climate.
3. What is salinity? How is it measured?
4. What three factors affect the movement of wind waves?
5. Trace the path of the Gulf Stream.
6. What factors determine the density of ocean water?
7. Distinguish between wavelength, wave height, and wave frequency.
8. Explain the difference in the way shear waves and pressure waves move.
9. What is a tsunami?
10. What is the cause of tides?
11. What influence does the Coriolis effect have on ocean currents?
12. Why is it not possible to get a free ride to Europe by drifting on the Gulf Stream?
13. What causes a wave to break as it approaches a beach?
14. What is unusual about the tides in the Bay of Fundy?
15. What is fetch?
16. Distinguish between spring tides and neap tides. What is the cause of each?

CHAPTER FOURTEEN

OCEAN AND LAND

Who owns the ocean? Nobody worried much about this back in the time of the Greeks and Romans. The ships of those days were frail little wooden shells that hugged the coastline and made only occasional brief dashes across the open sea. Once out of sight of land, everything was up for grabs—mostly by the ever-roaming pirates.

The question really came up in the seventeenth century, after the invention of gunpowder and the development of cannon. When an enemy ship could stand offshore and lob cannonballs into your city's marketplace, there was good reason for wanting to establish your territorial borders somewhat farther away than the low-tide line. So by the nineteenth century it was generally agreed that a nation had jurisdiction over the waters for a distance of three miles out from its coasts. Three miles was as far as the big guns of the day could fire. Beyond that, the rest of the ocean—the high seas—remained open to everyone. (Of course, the British navy sometimes acted as if it owned even that. A major cause of the War of 1812 was the British practice of stopping U.S. ships on the high seas to search them for British deserters.)

Today the matter has taken on a whole new set of dimensions. The ocean is no longer just something to sail over. It is not even just something to take fish out of. It is a potential source of any number of valuable minerals, and a dumping ground for human trash from sewage to atomic wastes. We are suddenly having to deal with a lot of new problems. For instance: should the mineral wealth scattered on the ocean floor—and the oil and gas under it—belong to anyone who can get it (which means the rich nations), to the nearest nation, or to some sort of international body that would give landlocked nations such as Chad and Bolivia and Switzerland a share of it? Then there is the matter of fisheries. Overfishing has seriously reduced the world's stocks of

ocean and land

important food fish — to the point of giving rise to "fishing wars" among the fishermen of rival national fleets. "Farming" of these fish is possible to restore the stocks, but what nation is going to spend the necessary money unless it can have some control over the harvest? Again, if one nation dumps DDT, kepone, mercury, and other toxic substances into the ocean, these long-lasting toxins are eventually going to turn up in fish all over the globe. They may even interfere with the microscopic plant life — the plankton — that supplies a substantial portion of the oxygen we breathe. Who has the right to regulate this dumping — and how can the regulations be enforced? As for oil spills, we have already seen what they can do, and there is every reason to expect more of them, not fewer, in the years to come.

These are knotty problems — so knotty that conferences sponsored by the United Nations have been trying to resolve them for years. The question of who owns the ocean — once so ridiculous that no one would have thought to ask it — has become more and more crucial with each new discovery we have made about this vast realm of water in which our island continents sit.

the ocean bottom

If we could drain the ocean by pulling a giant plug, a balloonist or pilot flying over the empty basin would see a terrain as diverse as the mountains, plains, and valleys of the continents. Scientists have given unique names to the topographic features beneath the sea to distinguish them from similar features found on land.

Although people have been studying the ocean depths for more than a century, many of its secrets remain hidden. Today hundreds of vessels from many nations are engaged in oceanographic research. They are unlocking many of the sea's mysteries, but much remains to be done, for every new answer raises additional questions.

When we represent dry land on a sheet of paper, we call the result a map. The same sort of representation of the ocean is called a **chart**. Mapping the land is much easier than charting the ocean bottom. How easily could you draw a map of the land if you were in a balloon floating over the top of a solid layer of clouds?

Until the 1900s, charting ocean depths was done with a length of rope or steel cable and a piece of lead. Seamen lowered a heavy lead weight on a piece of line to the bottom. The lead was smeared with a sticky fat, which held and brought back to the surface a sample of the bottom sediments. The line was marked every **fathom** (6 feet, or about 2 meters) to measure the depth to which the lead weight had sunk. It could take as long as an hour and a half to lower the line and even longer to bring it up.

A major advance—the **echo sounder**—had been made by the time of the 1925–27 voyage of the German ship *Meteor.* Echo sounding is based on the rate of speed at which sound travels through water. A device called a **transducer** or **pinger,** on the bottom of a ship or towed behind it, makes a high-pitched noise, or "ping." The sound travels to the ocean bottom, and some of the sound energy reflects back to the vessel as an echo. There it is received by a special listening device called a **hydrophone.** The sound and its echo are recorded by a stylus on a rotating arm which "writes" across a moving strip of paper. The time between the ping and the echo is measured with an accuracy of small fractions of a second, and is recorded on the paper as the depth.

Just as it once was fashionable to believe the earth was flat, it used to be thought that the bottom of the sea was monotonously level and featureless. Today we know better. Oceanographers have distinguished three main regions of the terrain beneath the sea: the continental margins, the ocean basins, and the oceanic ridges. The **basins** are located between the margins and the ridges. They contain a variety of distinguishing features.

"Abyssal" means too deep to be measured, unfathomable. But today oceanographers are able to measure the depths of the **abyssal plains,** which are broad, almost flat areas of the ocean bottom. Over much of their extent they appear to be as featureless as the great Staked Plain of Texas, without the gullies that erosion has cut into the Texas plains. Here and there low, gently sloped hills rise 30 to 400 meters above the plains. Less frequently but much more conspicuously, isolated mountain peaks tower upward.

Abyssal plains are the flattest areas on the surface of the earth. For example, the Argentine Plain found off the coast of Argentina has less than 3 meters of relief in a distance exceeding 1,300 kilometers. Other similar regions are located in the Indian Ocean, southeast of Ceylon, and in much of the ocean basin from Bermuda to the Mid-Atlantic Ridge.

Abyssal plains are found in all the oceans. They are more common where there are no deep **ocean troughs** bordering close upon the continents. These troughs or trenches act as traps for the sediment carried down the slopes. Since the Atlantic has fewer trenches than the Pacific, it contains more abyssal plains.

Among the features of the ocean floor are **oceanic rises.** These are large, gently sloping areas, the highest point of which may or may not be above the ocean's waves. The Hawaiian Islands are at the top of an immense rise from which gigantic shield volcanoes emerge.

Seamounts are volcanic peaks that dot the ocean floor and rise at least 1,000 meters above it. These steep, conical peaks exist in all

ocean basins

This box corer is only one of several similar devices used to retrieve cores of ocean-bottom sediments.

ocean and land

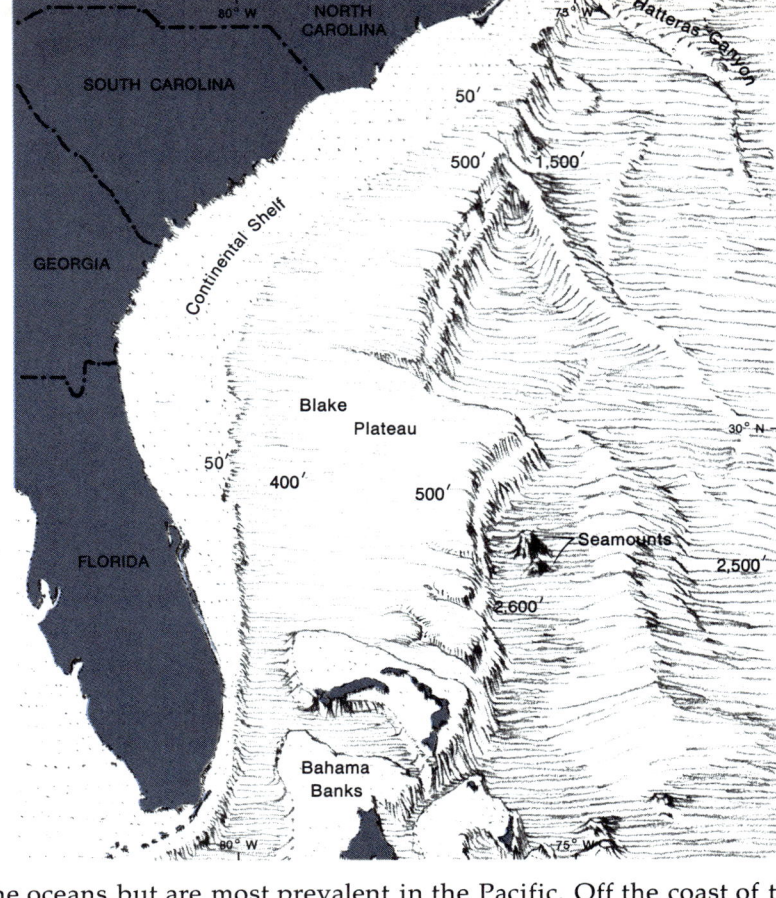

The Blake Plateau, off the coast of Florida, is an unusual underwater tableland, which may have been cleared of sediments by the deep waters of the Gulf Stream.

the oceans but are most prevalent in the Pacific. Off the coast of the Carolinas, seamounts rise so high that their tops just break the surface of the ocean, forming the Bermuda Islands. Similar examples of these exposed submarine volcanoes in the Pacific are the islands of Tahiti and Rarotonga.

Guyots are seamounts with flat tops. They were discovered in the Pacific in 1946 by H. H. Hess, who named them after Arnold Guyot, a Swiss geographer. The best explanation for these structures is that they were once seamounts. At one time their tops reached close enough to the surface of the ocean to be affected by the action of great ocean waves. It is believed that wave action eroded the loose volcanic material that composed the top of the seamount, leaving a flat top. The tops of many, if not most, guyots are now so far below the surface that even the greatest ocean waves do not disturb them. In the Pacific they lie 900 to 1,550 meters below the surface. The largest known guyot has a flat top 56 kilometers wide at its widest point.

We do not know whether much more water has entered the oceans since the waves planed the tops of these submarine moun-

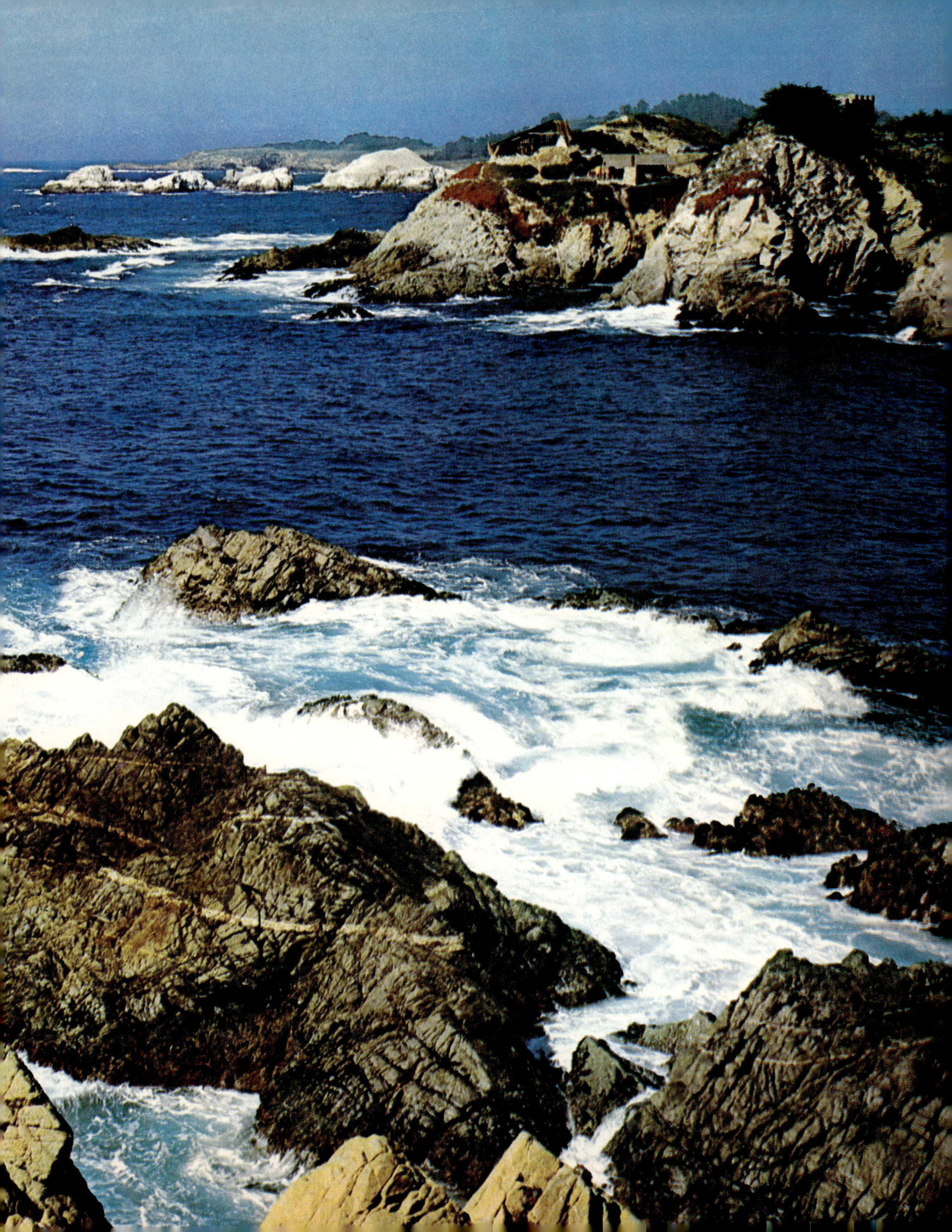

OCEANOGRAPHY

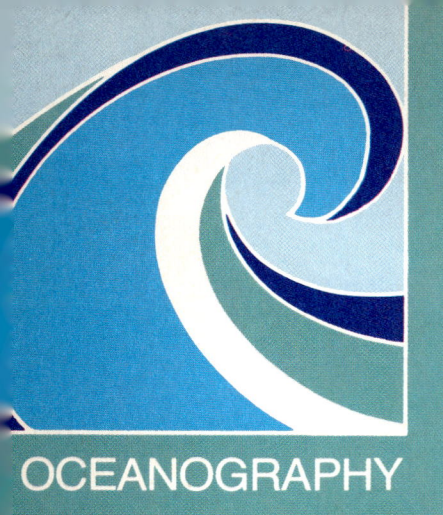

The oceans of our planet are perpetually fascinating. Soon after the dawn of human history men began to go down to the sea in ships. First they braved the surging water in frail rafts or canoes. Later, the Phoenicians learned to use sails on their boats to harness the power of the wind. The Vikings crossed the Atlantic long before Columbus' famous voyage, in ships that now seem laughably small and flimsy. Still later, the clipper ships of the nineteenth century transported merchandise and passengers all over the world. England, the nation that ruled the waves, effectively ruled a good deal of the rest of the world in consequence. Humans had finally conquered the oceans—or had they? Remembering the *Titanic,* we realize that the sea is still the master.

(Opposite page) The ocean is often assisted by other forces in its work of shaping the land. In the Northwest Territory of Canada, a continental glacier extends to the edge of the coastline. As the water continually batters and erodes the shoreline, the icecap moving over the land carves it into a typical glacial landscape.

(Below) A pair of volcanic islands have emerged from the ocean off the coast of Iceland—Surtsey (above) and Little Surtsey. This photograph was taken just after the islands were formed. Little Surtsey has since disappeared, eroded away by the sea. Surtsey itself may eventually meet the same fate.

(Opening page) Wherever the ocean meets the land, there is action. Here on the coast of California, a rocky headland juts into the sea and is battered by surging waves. Twice a day, the rising and falling tides help to shape the coastline. The sea is never entirely still, and thus the point where land and sea meet is always changing.

Beaches can be found in all shapes, sizes, and types of materials. Here, on the eastern coast of England overlooking the English Channel, a sandy beach is abruptly terminated by a portion of the chalk cliffs of Dover that extends right to the water's edge.

As a line of waves approaches the shore, the waves feel bottom and slow down. The line of waves then bends and turns toward the shore, so that the waves finally strike the beach almost parallel to it.

Large breakers suitable for surfboarding form only where the beach is shallow and does not drop off quickly. Such beaches are common along the coasts of California and Hawaii, drawing enthusiastic surfers from all over the world.

This marine terrace at Cape Perpetua, Oregon, must have been formed in shallow water when the land was lower than it is now. Evidence for this is provided by the shells of shallow-water marine animals that can be found on most marine terraces.

Underwater research by oceanographers and marine geologists indicates the presence of large quantities of petroleum under the ocean floor. Here a helicopter is landing on an offshore oil drilling rig in the Persian Gulf.

Along certain coasts the range of sea level between high and low tides is very wide. In the Bay of Fundy in Canada, the range is as much as 21 meters, which creates obvious problems for the ships that dock there.

Many forms of marine life have not changed much over geological ages. These brachiopods date from the Devonian Period. They are similar in many ways to modern shellfish.

This is a modern scallop. Compare it with the fossil brachiopods in the picture above. Terrestrial life no longer resembles its early ancestors very closely. Why have there been relatively few changes in so many sea creatures?

The material of which a beach is made can vary widely. This is a pebble beach. The dark color of the pebbles and the rocks in the background suggests a possible volcanic origin. They may be some type of basaltic rock.

Torghattan Island, which lies off the western coast of Norway, shows evidence that sea level here was once much higher than at present. The tunnel in the center of the island was cut by waves when the sea level was at that height. To the right and left of the tunnel are marine terraces.

The Salinas River flows into Monterey Bay, California. As it passes through the fertile farmland, the river picks up a variety of sediments. When it meets the sea and slows down, it deposits these sediments, discoloring the ocean water.

ridges and trenches

tains, or whether the bottom of the ocean has sunk. Probably a combination of events took place.

Oceanic ridges make up more than 20 percent of the earth's surface. They are found in all of the major oceans. Perhaps the most interesting feature of the Atlantic basin is the mid-ocean ridge which extends the full length of the North and South Atlantic oceans. Some oceanographers believe the Mid-Atlantic Ridge is only a section of a mid-ocean ridge that continues through all the oceans, forming a mountain range almost 64,000 kilometers long. Like the Rocky Mountains, the mid-ocean ridge is not a wholly connected, single range of mountains but a series of ranges that are almost connected.

The part of the mid-ocean ridge that lies in the North Atlantic Ocean is the best known. Individual peaks extend upward about 5,000 meters from the adjacent valley bottom. In one area in the middle of the ocean, the peaks break through the surface to form the volcanic islands called the Azores. Pico Island, one of the Azores, rises about 2,300 meters above sea level.

Some of the better-known parts of the Mid-Atlantic Ridge appear to be formed by two parallel ridges, separated by a very deep valley. To some geologists this valley resembles the great Rift Valley of Africa—deep depressions flanked by steep, high sides. Therefore,

ATLANTIC OCEAN

Above, a cross-sectional view of the Mid-Atlantic Ridge. Below, an artist's concept of the undersea topography around the Ridge.

they call it the **Mid-Atlantic Rift.** The rocks that have been dredged from the flanks of the mid-ocean rift are basalt, a rock that is formed by volcanic action. The Rift Valley of Africa also has many volcanoes within it.

What has caused the mid-ocean ridge? As we saw in chapters 6 and 7, volcanoes on the continents are associated with young and growing mountains. In such regions, huge cracks occur in the crust of the earth. Through these cracks molten rock sometimes oozes to the surface as a slow lava flow, or explodes as a violent volcano. If the mid-ocean ridge is composed of material that has oozed upward through cracks, it is a system of cracks one hundred times as long as any we know on the continents. Most geologists regard the mid-ocean ridges as perhaps the most geologically significant features of the ocean floor.

As spectacular as the ridges are the long, narrow trenches and troughs in which the deepest ocean waters lie. These great gashes in the ocean floor may reach depths of more than 9,200 meters. The longest of the Pacific trenches is in the South Pacific Ocean, bordering close upon the coast of South America. There the east wall of the trench is almost continuous with the west flank of the Andes Mountains. The combined difference in elevation of the earth's crust from the bottom of the Chile–Peru Trench to the summit of the Andes is about 14 kilometers, the greatest vertical distance on earth.

The deepest water in all the oceans is found in the Mariana Trench in the western Pacific. There the floor of the Challenger Deep lies about 12 kilometers below sea level. In January 1960, the deep-sea vessel *Trieste*, piloted by Jacques Picard for the United States Navy, made a successful dive to the bottom of that awesome trench in the earth's surface.

deep sea sediments

Sediments of varying thickness blanket the ocean floor. In the Atlantic, the thickness ranges from 500 to 1,000 meters. But in some deep trenches, which act as traps, sediment can be as deep as 9,000 meters. In the Pacific Ocean, it may measure 600 meters.

The sediment reaches the ocean floor by several means. Some of it is transported by wind, precipitation, and erosion. Part of it is made up of the skeletons or shells of living creatures. Some is produced by eruptions of volcanoes on land or in the sea. And some is carried down the continental shelf or slope by the force of turbidity currents, which we will discuss a little later.

Ocean sediments can be divided into several types. **Lithogenous** debris consists largely of mineral grains brought to the ocean by rivers, wind, melting glaciers, eruptions, and erosion. The remains of microscopic marine animals and plants are the main sources of **biogenous** debris. **Hydrogenous** debris is made up of minerals that crystallize from seawater.

Hydrogenous sediment is likely to draw the most attention in

Alvin prepares for a dive. This tiny submersible has provided us with an immense amount of information relating to several branches of oceanography.

the near future. Part of it consists of nodules, or small, irregular, rounded lumps, of manganese oxide. These are potentially important sources of manganese and small amounts of cobalt, iron and nickel. The greatest concentrations are at depths of more than 3,800 meters. Hence their exploitation will require new, economical techniques for dredging. New methods of processing these materials must also be invented.

Old-fashioned bottom sampling had shown that different parts of the deep sea contain sediments of different kinds. Modern deep coring has confirmed these earlier discoveries and has also shown that the sediments of the deep sea are stratified and varied. The combination of deep coring and radioactive dating tells us that the rate of deposition is very slow in the deep ocean. It is estimated that only about 1 millimeter to 1 centimeter of material is deposited in 1,000 years in those parts of the ocean farthest from land.

At the bottom of the sea is a dark, eerie, cold world with stupendous pressures. No light penetrates here. Maybe that is why the deep-sea squid ejects a spurt of luminous fluid, while its shallow-water relatives shoot out a black, inky substance. Other deep-sea creatures compensate for the lack of light by possessing lights of their own, in the form of luminous cells like those of glowworms or fire-

IN FOCUS:
the coral reef: living geology

When the first Western galleons penetrated deep into the uncharted Pacific Ocean, they returned from each voyage—those that returned at all—laden with the artifacts of the various native cultures they encountered. Weapons, carven wooden idols, necklaces of rare shells, priceless ceremonial robes made entirely of feathers—all of these were eagerly sought as curiosities, later to find their way into the hands of European curio collectors. But of all the things brought back by the Pacific explorers, perhaps the most valuable was coral, which could be used for many kinds of jewelry either by itself or in combination with precious gemstones. Coral soon became a major item of commerce—but despite its value, for a long time the traders who risked their lives to bring home supplies of it had no idea just what it really was.

At first glance coral appears to be a type of rock, and since it is the primary material of which many Pacific islands are composed, it would seem logical to think of it as rock. But coral is no ordinary rock. It is made up of the lime-containing skeletons of billions upon billions of tiny coral animals, left behind when the corals themselves died. The many coral islands that dot the Pacific were formed by accumulations of coral skeletons over geological ages.

Corals are not confined to the Pacific Ocean, although it is here that most of the precious coral used for jewelry is found. However, the animals can thrive almost anywhere if certain special environmental features are present. The water must be clear, warm, and quite salty for them to do well, and these requirements are a consequence of their rather peculiar biology.

The coral animal itself is a soft polyp, related to the hydra and the jellyfish. A strange symbiotic relationship exists between corals and several types of single-celled algae. These algae live in the tissue of the coral and in the space between the coral animal and its external skeleton. The algae use carbon dioxide produced by the coral during respiration to carry out photosynthesis. This process, in turn, supplies the coral with the oxygen it needs. Neither the algae nor the coral can thrive without the other.

Since the process of photosynthesis is dependent on light and temperature, reef-forming corals can live only where the ocean water is free of silt, warmer than 18°C, and shallow enough to allow light to penetrate. These requirements limit most coral growth to the region between about 30°N latitude and 30°S latitude.

Charles Darwin, in 1842, was the first scientist to make a detailed study of the coral islands of the Pacific. On the basis of his observations, he classified them in three categories. One type of island has a fringe of coral growing out from its shore—a fringing reef. A second type is surrounded by a barrier reef of coral some distance offshore. In the third type, the island itself has disappeared completely, and all that remains is a roughly circular reef of coral, known as an atoll.

Darwin suggested that the three types of coral islands represent three different stages in the history of an oceanic island. According to his theory, the island is originally formed by volcanic activity on the ocean floor. Coral animals gradually settle near the shores of the island and build a fringing reef. Because of local isostatic adjustment, the island slowly settles. As it settles, the coral must grow upward to survive, and thus forms a barrier reef. In time, the island may disappear beneath the water, leaving only an atoll encircling a shallow lagoon.

Darwin's theory has been supported, if not proven, by core samples taken from various atolls. At depths of several thousand feet the composition of the cores changes from coral to basalt. This indicates that, as Darwin suggested, the base of the coral island originated as a volcano which erupted from the ocean floor.

A second theory of coral island formation suggests that when the glaciers of the great Ice Ages melted, the water thus released caused the level of the sea to rise by as much as 300 feet. If this occurred, the corals would have had to grow upward toward the surface to reach the light and temperature conditions they required for survival. Below a depth of about 180 feet they would have been killed by the cooler water and the lack of sunlight.

However the coral reefs were formed, it is clear that the formation of each type depends on the ability of the coral animals to grow upward rapidly enough to keep up with the rising sea level. In some places, dead or "drowned" reefs exist, where the rate of growth was unable to keep ahead of the rising waters. The Chagos Archipelago, in the Indian Ocean, is a fine example of such a drowned reef.

Today the continuing growth of coral reefs is threatened by many human activities. Pollution of the clear tropical ocean water, for instance, is constantly occurring as the result of oil spills and raw sewage from passing ships. Great portions of living reefs are often destroyed by careless dredging techniques. Industrial wastes and thermal pollution also take their toll of coral. And every one of these conditions also has destructive effects on the many fish and undersea creatures who inhabit the waters around every coral reef. Attempts are now being made, however, to preserve these unique habitats and leave the tiny coral animals free to continue their million-year island-building task.

ocean and land

A remarkable view of the ocean floor. The sea stars shown are not as fragile as they look, since they can withstand the high pressure of the ocean depths.

flies. Some can turn their "torches" on and off at will as they pursue their prey. Plants cannot live so far away from sunlight. The great depths of the sea are a carnivorous world in which each species is both hunter and hunted.

the continental margin

Now we can turn our attention from the ocean bottom to the shore. Before we can step on dry land—the continents—we first approach the **continental margin** as we come up from the ocean basin. The continental margin consists of three distinct parts: the rise, the slope, and the shelf.

The **continental rise** is a gently sloping region that stretches toward the shore for hundreds of kilometers. Above the rise is the considerably more steeply graded **continental slope.** The average width of the slope is 20 kilometers. Between the slope and the shoreline is the gently graded **continental shelf,** with an average width of 65 kilometers.

We have become familiar with the continental margins in recent years through photographs taken by offshore oil drilling rigs.

ocean and land

The margins, particularly the outer continental shelf, have great economic importance as potential major sources of oil, natural gas, and minerals. And they are teeming with marine life which provides a valuable source of food.

composition

The continental shelf is largely the result of deposition by glaciers, streams, and the growth of coral. During the ice ages, much of the shelf was dry land. Animals roamed, and trees and other vegetation grew. Dredging of the continental shelf has produced skeletal remains of horses, mammoths, mastodons, and other land animals. The discovery that freshwater bogs once existed on the shelves is further evidence that they were once dry land. The shelf is vital to us as a source of energy, sand and gravel, food, and other materials. A growing proportion of the world's petroleum and natural gas is being derived from the shelves.

By now, enough data have accumulated to give us a reasonably good idea of the composition of the surface sediments. The most widespread are particles between 0.06 and 2.0 millimeters in diameter, which we call sand. There is also mud, a finer material than sand. Mud includes silt and clay. One way to determine whether a very fine sediment is mud or sand is to squeeze a handful into a mass and let it dry. If it falls apart easily when it is dry, the sediment is sand. If it holds together, it is mud. The least common of the shelf materials is gravel, which varies in size from the largest sand grains to boulders as much as half a meter in diameter.

On the continental slope, the most common materials are clay-sized particles. Sediment deposition occurs at a slower rate here than on the shelf, and the deposits are not as thick. Plant and animal life are considerably less abundant than they are on the shelf.

The continental rise was formed, oceanographers believe, by sediments swept down from the shelf and slope and piled up at the base of the slope. The deposits of sediment are as much as 10 kilometers thick at the bottom of the slope, and become thinner as the rise slopes outward toward the edge of the ocean basin.

geographic features

Until 1946, the continental shelf was of little interest to anyone except oceanographers and fishermen. In that year the United States claimed the oil and other mineral resources that lie on or under the shelf along its coastlines. This meant there was a need to define the term **continental shelf.** Legally, the shelf extends outward to the 100-fathom depth line. The scientific definition of the continental shelf, however, is "the zone around the continents from the low-water line to the depth at which there is a marked increase in slope to a greater depth." If the shelf could somehow be drained, another 11 percent would be added to the continental land surface.

The width of the shelf varies greatly. Off San Francisco, California, it is only about 48 km. Along the East Coast, the shelf varies in

ocean and land

This is a general profile of the ocean and the coastal region. The width of the continental shelf and the pitch of the continental slope may vary widely from region to region.

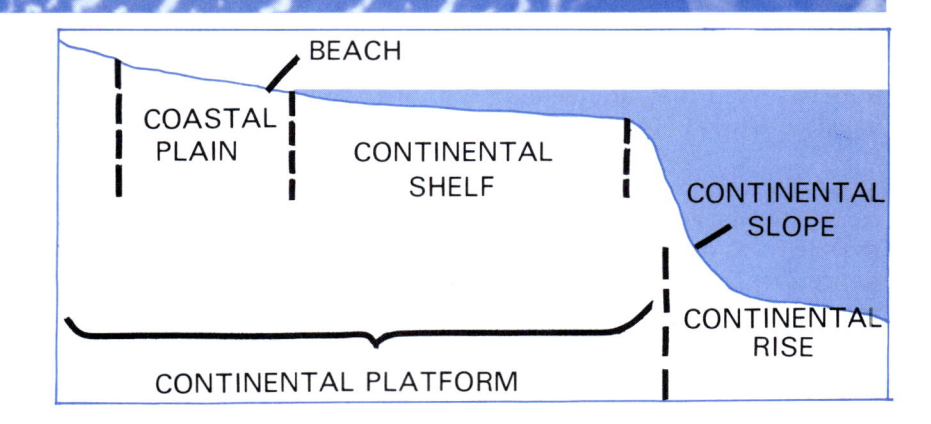

width from about 416 km off Boston, Massachusetts, to about 19 km off Miami, Florida. Just as there is no set distance that the shelf extends seaward from the land, there is also no set depth at which it ends. As a worldwide average, however, the continental shelf changes pitch, or slope, permanently at a depth of about 150 meters (490 feet).

A convenient way to express a pitch is to state it as an angle above or below the horizontal. A vertical canyon wall has an angle of 90 degrees. A steep ski slope might be 37 degrees. The surface of the continental shelf, however, has an average angle of pitch that is considerably less than 1 degree. The average descent of the continental shelf is only about 2 meters per kilometer.

It was formerly believed that the surface of the continental shelf was quite smooth. This is probably the reason why it was called a *shelf*. As sonic methods of sounding replaced the old lead line, more detailed information about the shelf was gathered. We now know that the surface of the shelf in the Gulf of Maine is as rugged as any hilly country on the land. Glacial deposits and underwater terraces, formed when sea level was lower during the ice age, may be responsible for this. Along the southern Atlantic Coast and the Gulf Coast there are extensive areas where the surface of the shelf appears to be relatively smooth. There it may be even flatter than the Great Plains.

The continental shelf off the Pacific Coast is very different from the shelf off most of the Atlantic Coast. It is composed of many basins and ridges—something like those off the Maine coast. At present our knowledge of the western shelf of North America is not so detailed or complete as that of the eastern shelf. This is being remedied rapidly by the oceanographic institutions of the West Coast and of the United States government.

The continental shelves extend at a gentle pitch outward from the land to the place where the sea floor begins to descend more steeply toward the true ocean bottom. The **continental slope** is this steeper edge of the shelf. What do we mean by *steeper*? We have seen that the average pitch of the shelves is considerably less than 1 degree, or about 2 meters per kilometer. The continental slopes drop off a little more rapidly. Their average pitch is from 3 to 4 degrees, or about 70 meters per kilometer. The continental slopes of the West Coast have a steeper-than-average pitch, about 100 meters per kilometer. Where the continental shelf is narrow, there is generally a comparatively steep continental slope.

The **continental rise** begins where the slope ends and edges gradually toward the ocean basin. Its width varies from about 100 to 1,000 kilometers and its average grade is about 4 meters per kilometer.

canyons

Among the most puzzling features of the ocean bottom are the steep valleys that cut across the continental shelves. These are called **submarine canyons** because they compare in cross section with such gigantic river-cut continental features as the Grand Canyon. On the East Coast, the Hudson River Canyon extends from the mouth of the Hudson River 128 km across the shelf and down into the true ocean basin more than 3 km deep. The Hatteras Canyon, off Chesapeake Bay, may be even deeper.

How were these canyons cut—if they were cut? When the canyons were first discovered, it was assumed that no erosion could occur underwater. This assumption made it necessary to believe that the canyons were cut by rivers at a time when the present continental shelves may have stood more than a thousand meters above sea level. However, there is no acceptable way of accounting for such a great lowering of sea level.

Another possibility is that a mixture of mud and water could slip down a slope as a current and erode the bottom, forming a canyon. Such a current is called a **turbidity current.** It is created when sand, mud, and other sediment is dislodged and slides downslope at great speed, eroding what lies in its path. Its action is similar to what would happen if you squirted ink from a fountain pen into a pan of water. The ink, being heavier than the water, would sink to the bottom of the pan and spread out. Often turbidity currents cut across the rise, slicing channels until they lose their power and finally come to a halt on the ocean floor.

An earthquake off the coast of Newfoundland in 1929 helped to establish the existence of turbidity currents. The quake snapped 13 transatlantic telephone and telegraph cables. It first was thought that the tremor itself caused the breaks. Later investigation showed that a sediment-laden turbidity current triggered by the quake had torn apart the cables. Other evidence came from cores taken from the ocean bottom, which contained samples of plants and animals that dwell only on the continental shelves. Their existence in the ocean basin could be explained only by the action of turbidity currents.

ocean and land

the coastal plain

The coastal plain, the beach, and the shorelines are three distinct areas. The **shoreline** can be defined as the line of contact between water and land. The **beach** extends from the low-water line to the highest portion of the shore washed by waves and tides. The **coast** is a larger region that can extend far inland from the shore. For example, the coastal plain that extends for 3,350 kilometers along the Atlantic and Gulf Coasts ranges in width from 160 to 480 kilometers.

Wherever the sea meets the land, there is action. In one place a rocky headland juts into the sea and is battered by the surging storm waves. In another place the smooth sweep of a curving beach of sand is forever shifting as waves wash across it. What is now an island may someday be tied to the land by a broad stretch of sand. What is now a peninsula may be partly flooded by the sea and an island may be formed.

The relationship between the level of the land and the level of the sea is subject to change. Sometimes the land itself moves up or down. At other times, sea level may change when water is added to or removed from the ocean.

The coast of Alaska offers an example of the extreme changes that can occur over a comparatively short time in the contest between land and sea. Only a century ago, a number of bays on the coast were used as harbors. Today these same bays are too shallow for ship traffic, because that part of the continent has risen.

Breaking waves constantly reshape a beach. In winter, storms and their accompanying high tides remove huge amounts of beach material.

ocean and land formation

Many factors can cause changes in sea level. Imagine the continents as being like large rafts of rock floating upon the interior matter of the earth. Add weight to the raft in the form of huge ice sheets or great outpourings of lava and the raft will sink. Sea level, then, will appear to have risen. Result: the land surface of the continent will shrink as water spills further inland. The opposite effect will occur if the "load" is lightened when great ice sheets melt and erosion shrinks the mountains. The land rises, sea level seems to drop, and the coast grows larger.

When the sea rises with respect to the land over a long period of time, the coast is drowned, or submerged. It is then called a **submergent coast.** If the opposite occurs, the exposed sea bottom becomes part of an **emergent coast.** The land that was once sea bottom becomes part of a new coastal plain.

The submergent–emergent classifications can be tricky, however. For example, two emergent coasts may have different features, partly because they were formerly submergent coasts at different time periods. Coasts also are classified according to whether erosion or deposition predominated in shaping them. Lava flows, for example, are depositional land forms. Hills worn smooth by ice sheets and mountains dissected by streams are examples of erosional land forms.

The shaping of the coast is a continuing process. In the scale of geologic time, many of the land sights familiar to us will disappear. For example, William Morris Davis, a geographer, predicted in his essay, "The Outline of Cape Cod," that the Cape will vanish completely, through erosion, in 8,000 to 10,000 years.

Marine terraces are striking examples of the changes caused

Parts of the eastern Canadian coast are subject to an unusually wide tidal range. Here low tide exposes large rocks that are completely submerged at high tide.

ocean and land

by the shifting of land and sea level. Some of the most carefully studied marine terraces in the world are those along the coast of Palos Verdes, California. You can see a series of flat-topped structures, rising in stair-like sequence from the shoreline toward the continental interior. Each of these flat structures is a marine terrace. The highest one is about 300 meters above sea level. In the same area, geologists have discovered some terraces that are 150 meters below sea level. In each case, the terrace must represent a time when waves eroded rock very close to sea level.

Observation of the marine terraces at Palos Verdes led geologists to pose a question. How can we account for the terraces both above sea level and below sea level in deep water? We know that terraces are cut just offshore in shallow water. Could some of the terraces in California have been cut in deep water?

There are two pieces of evidence that tell us they could not. First, we know of no action of the sea that could erode solid rock at great depth. Even the largest waves do not disturb water very deep in the ocean. The second piece of evidence is the shells on the surface of marine terraces. Most of them are shells of shallow-water animals.

How, then, can we explain marine terraces found both above sea level and below sea level in deep water? Geologists believe the marine terraces are evidence of changing sea level in the past. All the terraces were originally formed in shallow water. Those now above sea level must have been formed when the land was lower than it is now. Those now in deep water must have been formed when the land was higher than it is now.

submergent coasts

When sea level rises with respect to the land, one of the first results is the drowning of the river valleys that lead to the ocean. The high ridges that flank the valleys remain above sea level and jut out along the new shoreline. These are called **headlands.** The waterfilled river valleys are called **bays.** Chesapeake Bay in Maryland is an example, as are the bays along the coast of Maine.

After the land has been submerged, the irregular headlands are attacked and eroded by the waves and currents of the sea. Debris eroded from the headlands begins to fill the shallow bays. Given sufficient time, the irregular submerged coastline will be straightened by waves and currents. If the right conditions exist, sandy beaches may form, and the entire coastline may be changed.

Indications that coasts have submerged can be found all over the world. The Connecticut coast in the Clinton area, for example, has submerged nearly 3 meters in the last 3,000 years. Scientists have determined this by studying plants buried below sea level. These plants used to grow on the land edge of salt-water marshes.

emergent coasts

Just as signs of "drowned" land exist, so do signs of exposed land that was formerly covered by the sea. On the coast of Italy at

THE CHANGING SEASHORE

There is endless fascination in the perpetual cycle of change that takes place at the ocean beach. The basic mechanism behind this tug-of-war between land and sea involves an ongoing transfer of energy from the wind to the ocean waves. Along the coastline, energy from the wind transforms the waves into breakers that continuously pound the shore. As the waves thrust forward they are pulled downward by a combination of gravitational force and the resistance of the land, and the water returns to the sea. It is through the interplay of wind and water that the contours of a beach are constantly reshaped.

The erosion and rebuilding of a beach occurs according to a seasonal cycle. The stormy winds of winter diminish the size of the beach, producing towering waves that eat away enormous amounts of sand. Then, in the calmer weather of spring and summer, the beach is built again, as gentler waves deposit new sand along the shore.

Barrier islands are particularly vulnerable to the assaults of the sea. During severe storms the waves may surge over the island, carrying sand toward the shore and pushing the island back toward the mainland. Repeated storms, combined with a gradual rise in sea level, may eventually obliterate a barrier island entirely.

The encroachment of the sea takes a drastic toll of beachfront structures. Small docks may simply collapse into the sea, and houses built on the sand may be swept away by unusually high-surging waters. We are beginning to realize that our own carelessness may contribute to such problems. Human interference, in fact, frequently speeds up the process of erosion, by altering the natural topography of the shoreline. Build a road that cuts through the beach dunes, for example, and one barrier to erosion is automatically removed. Destroy the line of trees that lies around the fringe of a beach, and there will be no more strong, deep-reaching roots to hold the soil in place.

The physical forces that govern the interaction between the ocean and the beach are, so far, beyond human control. In fact, the enormous potential power of wave action has led some scientists to speculate on the possibility that eventually entire continents may be gnawed away by the ocean. But when we remember that beach erosion, however severe, takes place in a general context of equilibrium between land and sea, there seems to be no immediate need to worry about building arks.

ocean and land

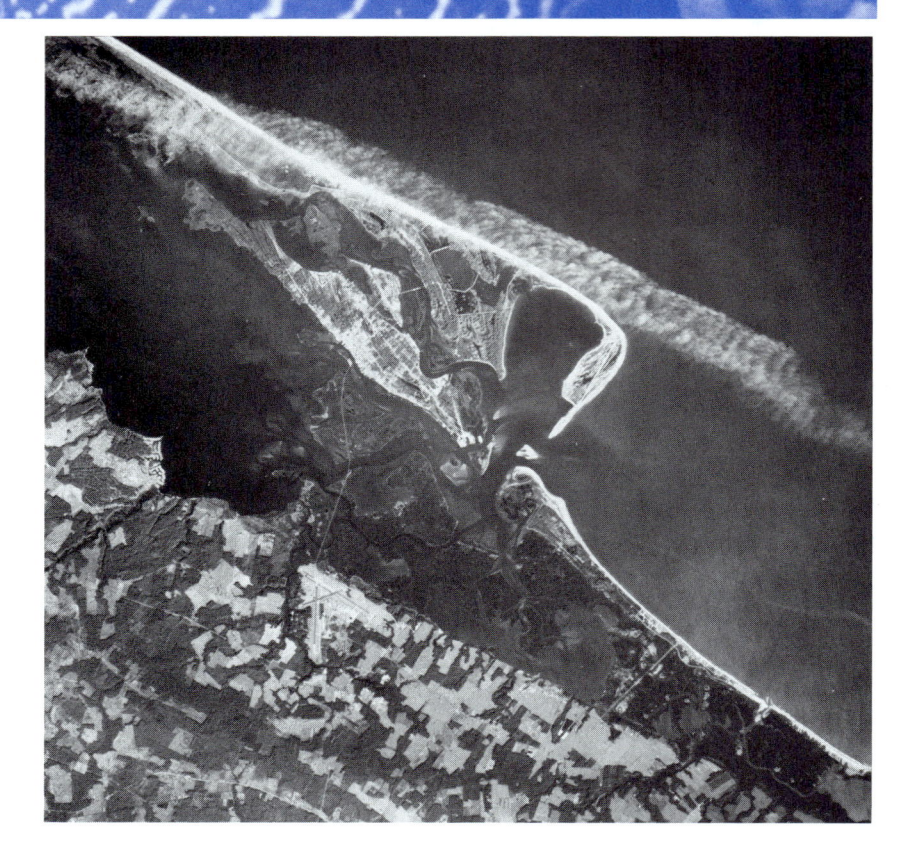

An aerial view of Assateague Island clearly shows the formations created by wave action and shifting of sand. Some of these bars disappear entirely after a time, and then are rebuilt, so that the shape of the island is not constant.

Pozzuoli, a little north of Naples, stands the Greek temple of Serapis. The columns of the temple have small holes in them made by boring clams. The holes are about 3 to 5 meters above the level of the Mediterranean Sea. They are clearcut signs of an emerged coast.

We also know that the modern coastal plain of Georgia was under water in Cretaceous time. This is indicated by the fossils contained in the rocks. In the southwestern corner of the state, about 3,000 meters of marine sediments have been deposited on the basement crystalline rocks.

The **fall line** indicates about where the coastline was during the Cretaceous Period. North of the fall line, the surface is composed of the metamorphosed rocks of the Appalachian Mountains. The fall line marks the point where streams that begin in the Appalachians and flow eastward toward the Atlantic leave the hard rock of the mountains and encounter the softer sedimentary rocks of the coastal plain. At that juncture, the streambeds erode faster, producing waterfalls and rapids.

The age of shorelines of submergence affects the ways humans earn their livelihood. The youthful shoreline with its splendid bays fosters fishing, shipbuilding, and other sea-related activities. On the other hand, the mature shoreline of submergence lacks the natural harbors needed for shipping.

The sedimentary rocks at the bottom of a rock core taken from the coast of Georgia are of Cretaceous age—about 135 million years old. This is the same age as that of the rocks at the surface near the fall line, about 280 kilometers away. The uppermost rocks in the core are about 18 million years old. Above them are unconsolidated sediments, mostly sands and clays.

When the rocks in the core are compared with those on the surface of the coastal plain, we find they are related. For instance, the uppermost rock in the core is a sandstone. Slightly inland from the coast this same rock appears at the surface, eroded into gentle hills. A little further inland, the surface rock changes to the same kind found in the second layer in the core.

As we proceed still farther inland, we find that the surface rocks are the same as those that appear deeper and deeper in the core taken from the coast. Finally we reach the steep fall line. Here, at the limit of the coastal plain, the surface rocks are the same as those found at the bottom of the core—Cretaceous sediments. Continuing on, we come upon the **oldland,** an area of still older rock lying inland from the coastal plain.

The fall line is about 300 meters above present sea level. It is doubtful that the oceans have lost that much water since Cretaceous time. It seems quite probable from geologic evidence that the Appalachian Mountains have risen about 300 meters in the last 100 million years. This uplift also raised the sea bottom at the coast above sea level. Therefore, the Georgia coastal plain and other emergent shorelines are exposed sea bottom.

the beach

To the earth scientist a **beach** is a moving deposit of material between land and water. The material is being moved along the shore and along the offshore bottom by the action of the waves. We usually think of a beach as a broad expanse of fine, white sand bordering the ocean. But this is only one kind of beach. Beaches form an almost continuous fringe along the continents and islands. They may be less than a meter or more than a hundred meters wide. This depends on the steepness of the land where it meets the sea. In a few places, where steep cliffs continue below water level, there may be no beach at all.

Beaches are formed and changed by combinations of waves, currents, and winds. From tide to tide or day to day the changes are small. From season to season and year to year, however, they can be

ocean and land

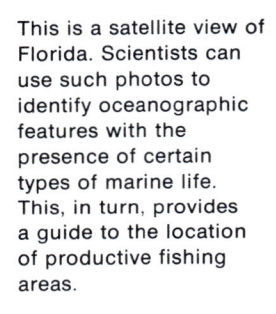

This is a satellite view of Florida. Scientists can use such photos to identify oceanographic features with the presence of certain types of marine life. This, in turn, provides a guide to the location of productive fishing areas.

much more visible. Perhaps the greatest changes are noticed after a severe storm.

The materials of which beaches are made range from fine clay to large boulders. There are **sand beaches, pebble beaches, cobble beaches,** and mixtures of all three. A **shingle beach** is built of small, flat stones arranged by waves into a pattern like that of shingles on a house. When a beach is built of silt or mud, it is called a **mud flat.**

The familiar sandy terrace shoreward of the area covered at high tide is called the **berm.** In the direction of the sea, beaches extend outward as far as ordinary waves move the sand or other sediment. Geologists have found that this distance is about 10 meters below the low-tide level. The seaward side of the berm, which is constantly washed by the waves, is called the **beach face.** Offshore and parallel to the shoreline are ridges of sand called **bars,** which can be seen only at exceptionally low tides. Landward, the beach extends to the edge of the permanent coast, which may consist of sand dunes or cliffs.

There is a constant exchange of sand between berm and bar on most beaches. In winter the waves are more powerful. They bring sand ashore, but their backwash is so powerful that they remove more sand than they deposit. Winter waves create offshore bars with this sand, at the expense of the berm. In warmer, calmer weather, the opposite happens. Smaller waves rebuild the berm at the expense of the bar. Summer bathers then can sun themselves on a conveniently wider berm.

Beach sediments usually come from a distant place. Rivers carry sediments from the land, and these are sorted and distributed by currents, tides, and waves. In glaciated areas, such as the coast of New England, waves work over the glacial till. In lower latitudes many beaches consist of fragments of shells and broken coral.

wave and current action

Did you ever wonder why waves approach a beach almost parallel to it? Even when the wind is at right angles to the beach, the waves roll in almost head on. When waves move toward a beach at an angle, the end of the wave front nearest the shore enters shallow water before the rest of the wave. Because the shore end of the wave feels bottom, it slows down. When that part slows down, the faster-moving part gets ahead of it, thus bending the entire wave front toward the shore.

If you live along the West Coast, you have seen long, rolling swells. The prevailing westerlies over the North Pacific have thousands of kilometers to produce this swell. That is why surfing is so good on the West Coast. The surf from the breaking swell varies from beach to beach, depending on the slope of the bottom. To have large breakers, you need a shallowing beach. Breakers do not form where beaches drop off quickly.

How does the bottom affect the formation of breakers? Imagine a beach with offshore ridges and a valley perpendicular to the shore. The wave front of the swell does not feel bottom along the valley. Therefore it hits the beach without forming a large breaker.

But look what happens to the part of the swell that passes over the ridge. Because the ridge is shallow, the swell feels bottom far from shore. When the wave feels bottom, it slows down and refracts, or changes direction. Notice that this happens on both sides of the ridge.

Surfing is largely confined to the West Coast, since breakers can form only where a beach drops off gradually.

ocean and land

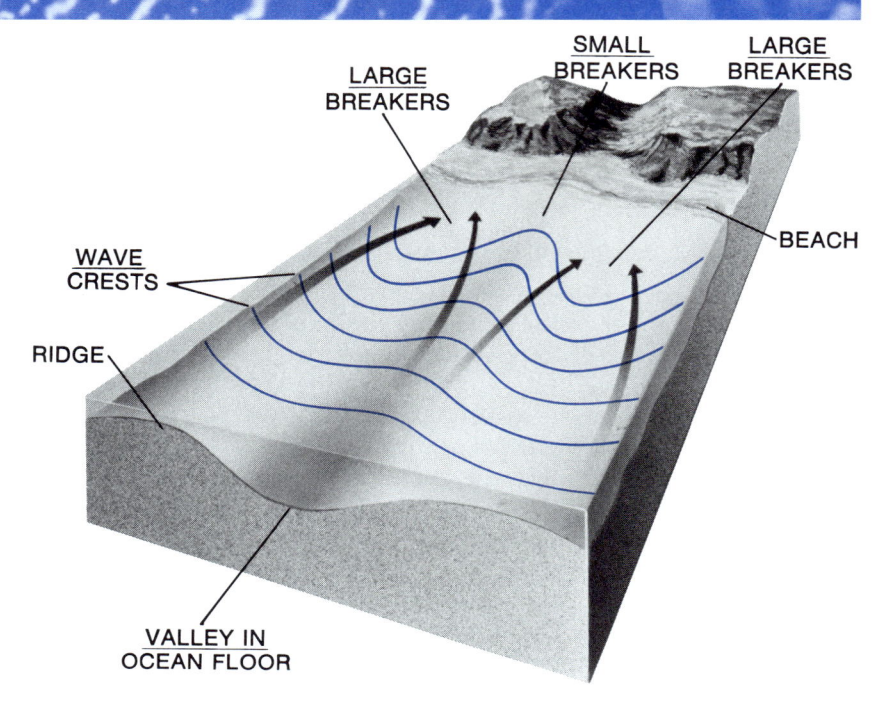

[Diagram labels: LARGE BREAKERS, SMALL BREAKERS, LARGE BREAKERS, BEACH, WAVE CRESTS, RIDGE, VALLEY IN OCEAN FLOOR]

The topography of the ocean bottom along a coast determines the size of the breaker a wave can form.

Waves have a certain amount of energy. As a wave moves from deep to shallow water, it slows down, and its wavelength decreases. This leaves it with excess energy, which increases its wave height.

When the swell moves over the ridge, therefore, it not only bends, but increases in height. The ridge acts as a magnifying lens. It focuses the growing waves on a section of the beach where the ridge meets the land. At that point the breakers are enormous. Farther along the beach, where the underwater valley meets the land, the breakers are small.

The high points or headlands along a coast, when followed out to sea, become underwater ridges. The ridges focus the waves on the headlands. The focused waves batter the headlands and gradually erode them. The eroded material is carried into quiet bays and deposited, eventually forming a sandy beach.

Watch the waves move toward you at the shore. How does the water return to the sea? This depends on the angle at which the waves approach the shore. Water returns to the sea by moving under the incoming wave. This is the **backwash,** and it can be strong enough in heavy surf to pull you off your feet. However, it ceases to pull a few seconds before the next wave breaks.

Waves often approach the shore at a small angle. They also roll off the beach at a small angle and start a current flowing along the shore—the **longshore current.** That is why you often drift along the shore when you are swimming. If it is rapid enough, a longshore cur-

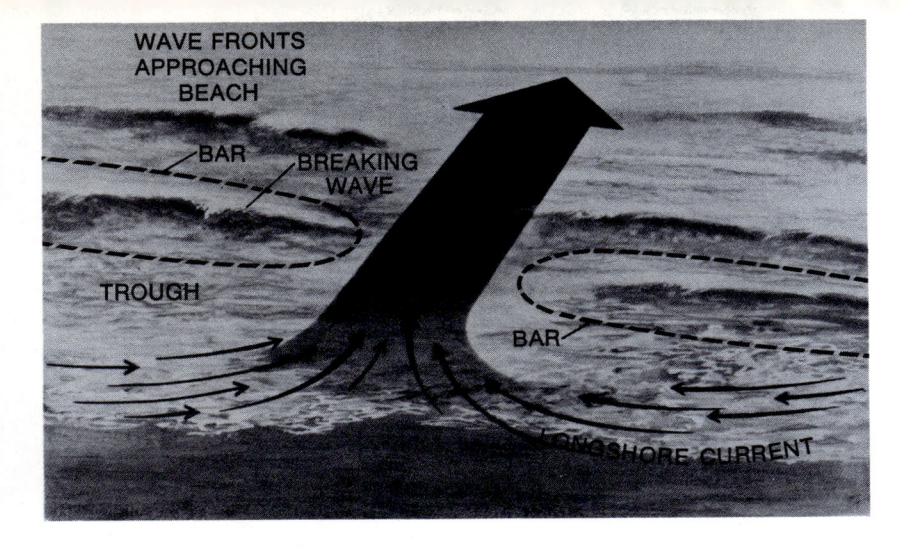

Longshore currents are usually caused by the wind. The rip current flowing seaward is often formed when a channel exists between offshore sand bars.

rent moves sediment along with it. Called **longshore drifting,** this movement is believed to be largely responsible for the building of parts of bars below the water surface.

Longshore currents sometimes feed into a stronger current flowing more or less directly out to sea. This is the **rip current,** and it can be dangerous. However, it is usually quite narrow, so if you are caught in a rip current, you can usually get out of it by swimming parallel to the shore. If you try to swim straight toward shore, you will be fighting the full force of the current.

barrier islands

As waves break offshore, the bottom sediments are tossed forward. Then, when the waves slow down, the material is deposited closer to shore. Some is returned to deep water by backwash and rip currents, but some remains where it was deposited, forming a bar. As this deposit grows, it begins to interfere with the free flow of waves toward the shore. The sea becomes shallower where the bar is growing. Waves feel bottom in that area and tend to break there. Then the bar grows faster and faster.

Eventually, enough material is deposited to form a bar that is above water at low tide. Waves with a greater-than-average wave height will deposit even more sediment on the bar. In time, the bar will be built well above the average sea level. Large bars become known as **barrier islands.**

The water between the barrier island and the mainland is protected from the waves of the open ocean. It is quiet water, which rarely develops large waves. This sheltered body becomes known as a **lagoon.** The sediments being carried to the sea by streams will usually be deposited in the lagoon. Eventually, they may fill the lagoon and turn it into a marshy mud flat. Plants that can grow in brackish, or slightly salty, water will trap more sediment and help build the level higher. If enough deposition takes place in the lagoon, a strip of dry land may develop and connect the barrier island to the mainland.

ocean and land

other sand formations

The only requirement for a barrier island is plenty of loose sediment offshore and medium-sized waves. Such islands form much of the United States coastline, including the coast from the Rio Grande in Texas, around the Gulf of Mexico, and all the way up the Atlantic shore to Long Island. However, there are only small bits of this kind of coastline on the Pacific Coast.

A half-dozen major United States cities are built on barrier islands, including Galveston, Texas, and Atlantic City, New Jersey. These islands may be many kilometers from the mainland and some of them are dozens of kilometers wide. Sandy hills or dunes on the islands may be nearly 30 meters high.

The headlands of a submergent coast are battered and eroded by waves. The sandy sediments produced by the action of the waves are moved by local currents. These currents tend to be strong near the headlands, which jut into the sea. In the bays the currents are weaker. On the sheltered side of the headlands the current loses its force and drops its load. As a result, an underwater bar of sand is formed. If this bar grows to the surface and is attached to the land, it is called a **spit.** If longshore currents form a spit into a curve, it is called a **hook,** such as Sandy Hook, New Jersey.

Sometimes a spit reaches partly across the mouth of a bay. It is then called a **bay-mouth bar. Bay barriers** are spits that completely block the mouths of bays. Many bays can be kept open only by dredging. Do you see how barrier islands, bay-mouth bars and bay barriers straighten a coastline?

Looking down into an underwater habitat. These scientists are conducting various psychological tests related to the problems of living in such close quarters. The results will be used in planning space laboratory missions.

ocean and land

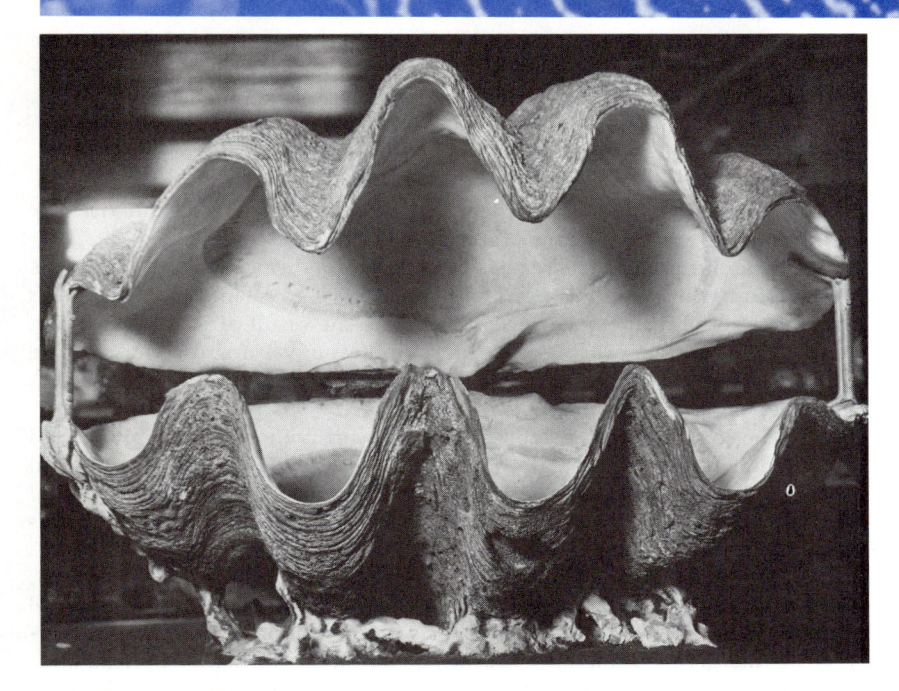

Mammals, such as whales, are not the only marine creatures that grow to enormous size. This clam, which lives in warm Pacific waters, is more than a meter across.

The bottom of McMurdo Sound, in Antarctica, has been the subject of much oceanographic research. Here we see some of the odd types of sea life that live beneath the cold Antarctic waters.

SUMMARY

A map of the ocean bottom is called a chart. Modern methods of charting the ocean bottom include the use of echo sounders, tranducers, and submersible craft which can descend to a depth of more than three kilometers.

Oceanographers have divided the undersea terrain into three main regions: the continental margins, the ocean basins, and the oceanic ridges. The abyssal plains are extensive, almost flat areas of the ocean bottom. They are most common where there are no ocean troughs. Oceanic rises are gently sloping areas which may reach above sea level.

Seamounts are volcanic peaks that rise from the ocean floor. A guyot is a seamount whose top has been flattened by wave erosion. Oceanic ridges are found in all of the major oceans, and are the result of volcanic activity.

The ocean floor is covered with sediments of varying thickness. These sediments may be lithogenous, biogenous, or hydrogenous. The rate of sediment deposition in the deep ocean is very slow.

The continental rise slopes gently upward to the more steeply graded continental slope. Between the slope and the shoreline is the continental shelf, largely the result of deposition by glaciers, streams, and the growth of coral. Submarine canyons, probably carved by turbidity currents, cut across the continental shelves.

The coastal plain is made up of the shoreline, the beach, and the coast. Coastal areas are constantly changing as the result of variations in sea level and the action of wind and water. Coasts can be classified as either submergent or emergent. Submerging coasts may lead to the formation of headlands and bays.

Beaches are formed and changed by waves, currents, and winds. They may be sandy or stony, or a combination of both. A beach can be divided into a berm, the sandy terrace shoreward of the high tide line, and the beach face, which is constantly washed by the waves. Offshore sand bars can only be seen during very low tides. Sand is exchanged between berm and bar as a result of wave action.

When a wave moves from deep to shallow water its wavelength decreases and it becomes higher. Water returns to the sea by moving under the incoming wave, forming a backwash. It may roll off the beach at a small angle, creating a longshore current. This, in turn, may feed into a stronger rip current, which can be dangerous to swimmers.

Higher-than-average waves may deposit so much sediment on a bar that it becomes a barrier island. The sheltered area of water between the barrier island and the mainland is called a lagoon.

REVIEW QUESTIONS

1. What do we call a map of the ocean bottom?
2. Describe an abyssal plain. Where is it most commonly found, and why?
3. Distinguish between a shoreline, a beach, and a coast.
4. What are the three main regions of the undersea terrain?
5. Describe an ocean trench. Where is the deepest known example?
6. What is a seamount?
7. Name and describe the three divisions of the continental margin.
8. How is a barrier island created?
9. Distinguish between a berm and a beach face.
10. How are oceanic ridges formed?
11. What is a submergent coast?
12. Describe several kinds of beaches?
13. What is a guyot, and how is it formed?
14. What are some of the indications that a coast is emergent?
15. Name and describe three types of ocean sediments.
16. What is so unusual about deep-sea life forms?
17. Describe the conditions necessary for breakers to form.
18. What is a continental shelf, and how does it form?
19. What is a backwash?
20. Explain the formation of marine terraces.
21. How is a submarine canyon created?
22. Explain the difference between a spit and a hook.

PART 5

ASTRONOMY

CHAPTER FIFTEEN

THE SOLAR SYSTEM

"The king is dead; long live the king!" Good King Edward had died early in 1066. Having no son, he had left the throne of England to the powerful Harold, Earl of Wessex. But across the waters there were other claimants to the throne—Harold Fairhair of Norway, and William, duke of the great duchy of Normandy in France. In September King Harold had to march northward with his army to stop an invasion by his Norse rival. Scarcely had he defeated and killed him when word came that Duke William had landed on the southern coast. By forced marches the king rushed south, gathering such men as he could to replace the veterans killed in the battle against Fairhair. At dawn on October 14 the two armies met, a few kilometers northwest of Hastings. Arrows flew in clouds, mounted knights charged thundering into the melee, the terrible two-handed English battleaxes cleaved flesh and bone and steel. In late afternoon Harold fell, slain by a Norman arrow; as night fell the leaderless English broke and fled. On Christmas Day William, Duke of Normandy, was crowned as King William I of England.

If the victors' account of the matter is to be believed, the Normans had divine help in their campaign. Sometime in 1066 a brilliant comet had appeared in the skies over Europe. In those days everyone knew that a comet foretold great changes—plagues, droughts, wars, the death of kings, the fall of nations. William, a clever psychologist as well as a good general, saw his opportunity. He spread the word that this particular comet was a sign of his own impending victory and Harold's utter defeat. The Bayeux tapestry, a contemporary portrayal of the affair, gives the Norman viewpoint. It shows a cluster of Englishmen gazing at the comet in obvious alarm—beaten by their own fear before the battle ever began.

How terrified the English soldiers really were by the comet remains an open question. There were plenty of other

the solar system

reasons that could explain their defeat. But the belief that comets portend terrible things lasted for centuries more. In 1456 another great comet passed over Europe—the same year the splendid Christian city of Constantinople fell to the Turks. Many people were sure there was a connection.

There was—but not the one that people were looking for. The two comets were, fact, one and the same. Now known as Halley's Comet, it visits the earth every 76 years, and its next return is scheduled for 1986. Few people today will see it as a sign of impending disaster; but the fiery head and long, glowing tail in the night skies will still be enough to make many look up and wonder at this strange visitor from the far depths of space.

an overview of the solar system

Our earth is only one member of a group of celestial objects held together by the gravitational attraction of the sun. We call this family of objects the **solar system** (from the Latin *sol*, "sun"). The sun contains nearly 99.9 percent of the mass of the entire system, and is self-luminous—that is, it shines by its own power. In addition to the sun, the solar system contains the nine planets and their satellites, as well as comets, meteors, and the asteroids or minor planets.

Table 15-1 shows that most of the matter not contained in the sun is in the form of planets. Not all planets are equally massive. The earth, for example, has about 3 millionths the mass of the sun. Mercury, the least massive planet, has about 1/18 of the earth's mass. Jupiter, the most massive, has 318 times the earth's mass, and even this is only about 1/1,000 the mass of the sun.

Table 15-1. Masses of solar system objects

Object	Percentage of Mass
Sun	99.86
Planets	0.135
Satellites	0.00004
Comets	0.0003
Asteroids	0.0000003
Meteors	0.0000003
Interplanetary Matter	0.0000001

The planets travel in elliptical orbits around the sun, as we saw in chapter 1. Except for Pluto, their orbits lie in very nearly the same plane. Pluto, however, orbits at an angle of 17° to that plane. Table 15-2 gives some of the details about the planets and their orbits.

the solar system

Table 15-2. Properties of planets

Planet	Average distance from sun (km)	Rotation time	Length of year	Diameter (km)	Average density (g/cm^3)	Number of satellites
Mercury	5.7 × 10^7	59 days	88.0 days	4,800	4.8	0
Venus	10.7 × 10^7	249 days	224.7 days	12,160	4.9	0
Earth	14.8 × 10^7	23.9 hrs.	365.3 days	12,756	5.52	1
Mars	22.7 × 10^7	24.6 hrs.	687.0 days	6,720	3.95	2
Jupiter	77.7 × 10^7	9.8 hrs.	11.9 yrs.	142,400	1.34	13
Saturn	1.4 × 10^9	10.2 hrs.	29.5 yrs.	118,000	0.69	9
Uranus	2.9 × 10^9	10.8 hrs.	84.0 yrs.	48,000	1.36	5
Neptune	4.5 × 10^9	15 hrs.	164.8 yrs.	44,800	1.30	2
Pluto	5.9 × 10^9	6.4 days	248.4 yrs.	5,760	?	0

Mercury has the smallest orbit, averaging about 57,600,000 km (36 million mi) from the sun. The orbit of Pluto is a hundred times larger. Thus we see that diagrams on a textbook page cannot be drawn to scale; they can only indicate the general arrangement of the solar system.

Distances within the solar system are measured in millions or billions of kilometers, figures not familiar to us from our everyday lives. For comparison, consider that it takes about 12 days to count to a million (one count per second), and 30 years to count to a billion. The Metroliner, which travels between New York and Washington, D.C. in three hours, would take 200 years to reach the sun and 2,000 years to reach Saturn.

To make such large numbers easier to manage, astronomers use the earth's distance from the sun (150 million km, or about 93 million mi) as a standard length. This is called the **Astronomical Unit (A.U.).** The distance of Mercury from the sun thus becomes 0.4 A.U., and the distance of Pluto from the sun is about 40 A.U.

Six of the planets are known to have satellites or moons orbiting them. A satellite is much smaller than the planet it orbits. The most massive known satellite is Neptune's Triton, and the least massive is one of the moons of Jupiter, an irregular chunk of rock about 8 km across. Our own moon is one of the largest satellites, and is also the only one that we have studied in great detail.

Besides the planets and their satellites, the solar system contains thousands of minor planets known as **asteroids,** which are largely confined to the region between the orbits of Mars and Jupiter. The largest asteroid is about 800 km in diameter, but most of them are very much smaller. Bodies smaller than the asteroids also orbit the sun, some of them mere particles of dust. They can be observed only if they enter the earth's atmosphere, become very hot, evaporate, and leave a glowing trail in the sky. These objects are **meteors,** often called shooting stars.

Comets are frozen chunks of rock, water, and gases, about a kilometer across, which travel in elongated elliptical orbits around

Halley's Comet photographed on June 6, 1977; its next appearance is awaited in 1986.

the solar system

Drawing shows how much larger the sun is than any of the nine planets. The rings of Uranus have only recently been discovered.

Comet West, photographed on March 12, 1976, appeared about 25 times the size of the moon. Notice its two distinct tails.

astronomical measurements

the sun. When a comet approaches the sun it becomes warm and the gases evaporate, forming a spectacular "tail" that can stretch for a billion kilometers. After rounding the sun, comets return to the depths of space, perhaps to reappear only after thousands of years.

How do we know the distances to the various objects in the solar system? Fortunately, there is a method which enables us to find the distance to an object without actually going there. The method was invented by the Greeks, and is called **triangulation**.

Suppose we want to determine the distance to the moon. Two well-separated points on the earth, A and B, are chosen. The distance between the points becomes one side of a triangle, and is called the **baseline.** The directions in which the object lies when viewed from the two ends of the baseline are then noted. The two directions are two angles of the triangle. If we know one side and two angles, we

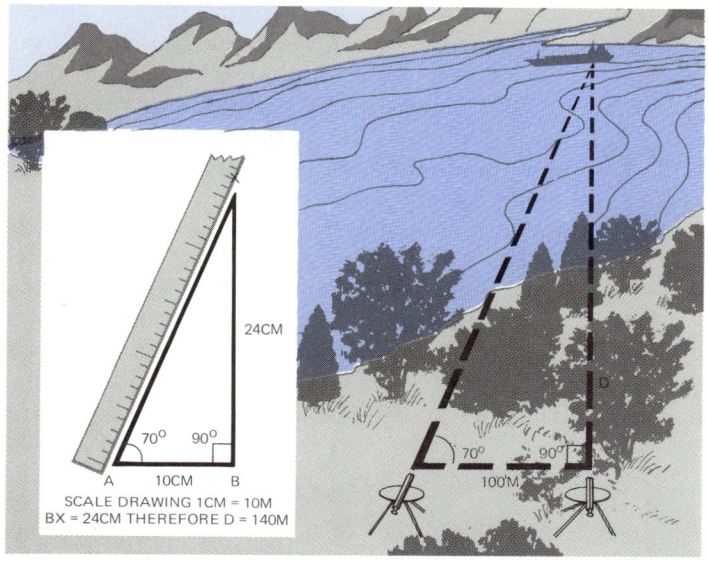

the solar system

Distances to unreachable objects can be measured by triangulation. Inset box shows how the distance is calculated using a scale drawing.

can then draw the triangle to a suitable scale, and work out the distance to the object mathematically.

In the case of the moon, simultaneous observations are made from two widely separated observatories. The direction of the moon can be found very accurately by comparing its position with respect to the distant background stars, which are so far away that their directions remain virtually unchanged when they are viewed from different points on earth. Now we can calculate the distance to the moon: it is about 350,000 km (240,000 mi).

Radar can also be used to determine distances to the moon and nearby planets. Radar signals traveling at the speed of light are bounced off the object, and the reflected signal is then received on earth. The time lag between broadcast and reception of the signal is measured very accurately. The distance to the target can then be calculated, since we know that the speed of light is 300,000 km/sec. The average distance to the moon turns out to be 348,403 km. Because the moon's orbit is elliptical, the actual distance varies within a range of about 30,000 km.

Below left is the 1000-ft radio telescope, built into a sinkhole amid the hills at Arecibo, Puerto Rico, which is used to determine planetary distances.

Laser beams from earth returned to us by this special reflector on the moon enable us to determine the earth–moon distance accurate to a few cm.

the solar system

the Kelvin temperature scale

Below, a photograph shows pairs and groups of sun spots on the sun on December 21, 1957, a year of sunspot maximum.

Totally eclipsed sun, below right, photographed from Green River, Wyoming on June 8, 1918, shows corona and prominences.

The Fahrenheit and Celsius temperature scales are convenient for everyday use, but in one sense they are arbitrary. For example, 0°C is not the lowest temperature that a body can attain; heat can still be extracted from it at 0°C. It can become still colder and have a temperature "below zero." When all the heat has been extracted from a body it is as cold as it can possibly get. Its temperature is then —273°C. Scientists call this temperature **absolute zero.**

A system that puts zero degrees at the true absolute zero is called the **Kelvin** scale of temperature, and is generally used by astronomers. (Lord Kelvin was a British physicist whose researches led him to the notion of the absolutely cold object.) In the Kelvin scale, water freezes at 273K and boils at 373K, and absolute zero is 0K. We can easily convert Celsius temperatures to the Kelvin scale. The Kelvin temperature is simply the Celsius temperature plus 273. In writing it, we do not use the ° symbol. We write 309K rather than 309°K.

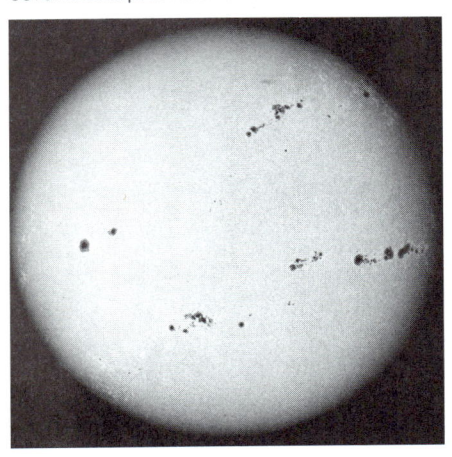

the sun

The sun is the most conspicuous astronomical object, at least from our earthly viewpoint. It is also the only one essential to life on earth, since it provides nearly all the earth's energy. As we saw in earlier chapters, it is the moving agent behind winds, storms, and other weather phenomena, as well as the primary source of the energy contained in fossil fuels. Variations in the amount of energy radiated from the sun may have caused the ice ages. Outbursts at its surface can interrupt long-range radio communications, as shortwave radio buffs are likely to notice during the next few years of expected high solar activity. For these reasons, we shall look first at the external aspect of the sun and its effects on the earth.

the solar system

general properties of the sun

The sun is a gigantic ball of hot gas with a mass 330,000 times that of the earth. Its radius is 109 times that of the earth, and its volume is over a million times greater than the earth's. But its density is only 1.41 g/cm^3, less than a third that of the earth.

The sun radiates energy in all directions, mostly in the form of visible light. The earth, as we saw in chapter 11, intercepts only 2 billionths of the energy, and reflects back half of what it does intercept. Still, the earth receives a thousand times more energy than human use requires. Several billion high-powered cars could be kept going at full speed by the solar energy received on earth.

the outer layers of the sun

The sun seems to have a sharp and definite edge, so it is not meaningless to say that its diameter is 1,288,000 km. But it has no solid surface, and investigation shows that gases extend well beyond the normally recognized **edge,** or **limb,** of the sun.

We can observe only the atmosphere of the sun, which can be divided into three layers. The innermost of these, called the **photosphere** or "globe of light," is bright and opaque; this is what we see when we casually look at the sun. The photosphere is less than 300 km thick. Temperature at the top of it is around 4,500K, increasing to about 7,000K at the bottom.

The **chromosphere** or "globe of color" surrounds the photosphere. It is transparent and not very bright, making it difficult to observe although it is almost 30,000 km thick. The chromosphere can be observed telescopically during a total solar eclipse, during which the moon comes between the sun and the observer and blocks out all of the photosphere. Then, when the overwhelming light put out by the photosphere is blocked, we can see that the chromosphere shines with a pinkish light.

The chromosphere can also be observed by photographing it through a filter that allows only certain wavelengths of light to pass through. For example, a filter may be chosen which allows only red light to pass through. In this light the photosphere is very faint but the chromosphere is bright. A photograph taken in this way is called a **filtergram.**

Three pictures of the same region around a sunspot. Picture at right is in white light, center picture is a filtergram in the red light of hydrogen, at left is a picture in the light of calcium.

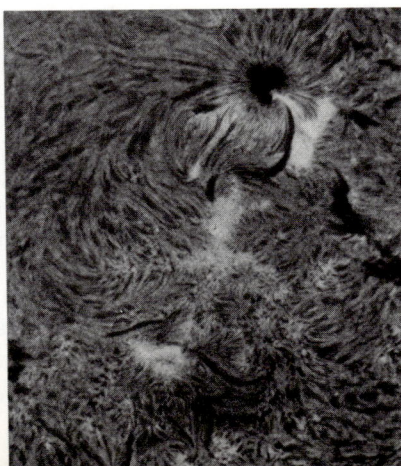

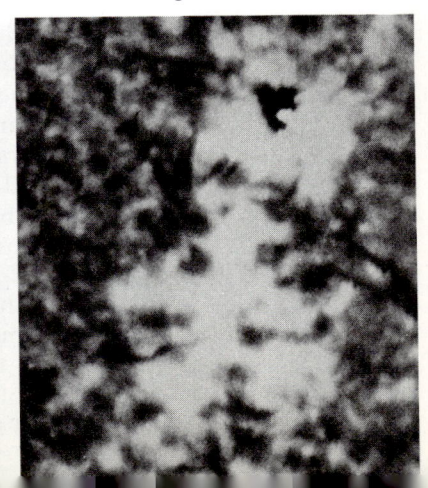

the solar system

The temperature of the chromosphere has been estimated at around 100,000K. This surprisingly high temperature is due to turbulent motions in the underlying layers of the sun. These motions begin to expend themselves in the chromosphere, heating it up as a consequence. We can compare this process to ocean waves that begin to break as they approach the shallow regions near a beach.

Surrounding the chromosphere and merging with its spiky outer layer is the **corona** or "crown," which often extends more than a million kilometers from the sun. During a total solar eclipse the corona can easily be seen shining with a pearly white glow, about half as bright as the full moon. When the sun is not eclipsed, the corona can be viewed through a **coronagraph,** a telescope with highly polished lenses in which a black disk blocks out the image of the sun.

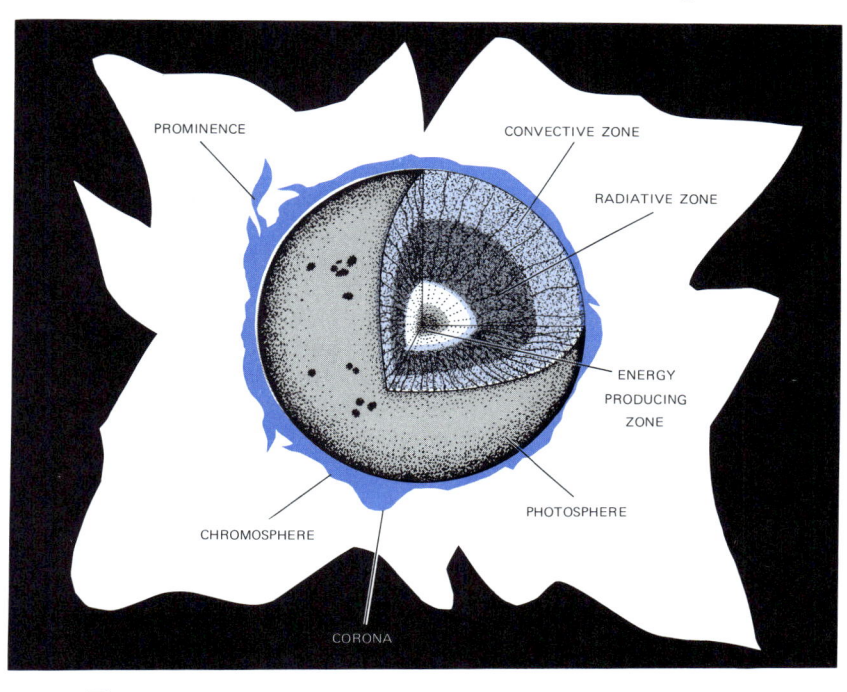

Almost all the energy of the sun is produced at the very center. It slowly trickles out and escapes from the photosphere in the form of light. The corona is very hot and is continually expanding away from the sun, producing the solar wind.

The most interesting thing about the corona is its extreme heat—about 3 million K. It is heated by the same mechanism that heats the chromosphere. The hot corona possesses enough energy that it can expand away from the sun. In the early 1960s, observations made from spacecraft showed that a "wind" of particles, mostly protons and electrons, flows from the sun and extends far beyond the earth. This **solar wind** is actually the expanding corona. Its velocity seems to vary according to what is happening at the sun's surface.

the sun's rotation

The way the sun rotates enables us to deduce something about its internal structure. Modern physics has demonstrated that objects such as the sun must be gaseous. If we construct solid or liquid bodies of increasing mass, we find at first that the radius is larger if

the solar system

the mass is larger. But when we reach masses of about 1/1,000 that of the sun, such as Jupiter, we find that adding more mass produces a *smaller* object. If we could construct a solid or liquid body with a mass equal to the sun's, it would be smaller than the earth, because such great masses have very strong internal gravitation fields that compress them into small volume. Strange as it may seem, the only way to construct a large object like the sun is to make it out of a hot gas.

This is intriguing, but we should note that Galileo did not need modern physics to show that the sun was not a solid. While observing the sun with his telescope in 1610, Galileo noticed that its surface was not clear but had dark spots. He found that these **sunspots** traveled with the sun's rotation, and that those in different regions traveled at different speeds. He concluded, therefore, that different regions of the sun rotate at different speeds. The equator seems to rotate once every 25 days, but a region near 30° latitude takes 2½ days longer, and the poles appear to rotate once every 35 days. This clearly implies that the sun is not a solid.

One thing, incidentally, should be stressed here: looking di-

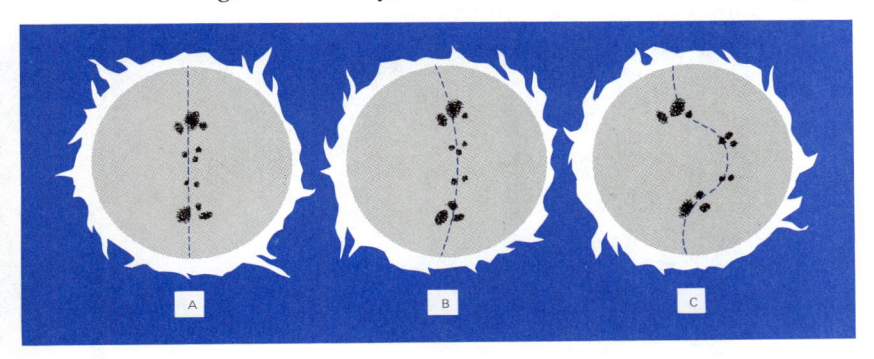

Sunspots in a straight line in A. In B, one revolution later, the spots nearer the equator have pulled ahead of those nearer the poles. In C is the situation after yet another rotation.

rectly at the sun, either through a telescope or with the naked eye, may permanently damage your vision. One safe way to observe the sun is to view it through two layers of *exposed* photographic film which has been developed. It is also safe to wear dark sunglasses and look at the reflection of the sun on an unsilvered piece of glass, or on still water. Galileo did not look directly at the sun through his telescope, but used the instrument to project a large image of the sun.

solar activity

At first glance, the surface of the sun appears smooth and quiet. But closer observation with special instruments reveals that the sun is seething with activity, the effects of which are felt even at the earth's surface. This activity takes a variety of forms; here we can mention only a few of the most important.

sunspots

Sunspots are now known to be broad, shallow depressions in the photosphere. They range from 1,000 km to 50,000 km in diameter, and usually have dark centers with lighter borders. They have a temperature of about 4,000K, almost twice that of the filament in a household bulb. They appear dark only because they are surrounded by material which is about four times brighter. The lifetime of a sunspot may be a few hours or many months.

Hot gas constantly flows out of a sunspot and cools down as a result of expansion. The behavior of the gas is determined by the sunspot's most important characteristic—its magnetic field, which may be as much as 8,000 times greater than that of the earth. (Fields of this strength have been produced on a small scale in laboratories. In such a field you cannot hold onto a wrench or a screwdriver; it is torn from your grasp and flies to the magnet producing the field.) In a sunspot the magnetic field and the gas interact in complicated ways. Gas is normally not responsive to a magnetic field, but a hot gas contains atoms broken up into fragments with positive and negative charges. It is electrically conductive, and so it can interact with the magnetic field.

As we saw in chapter 5, a magnet has two poles, called North and South. Since sunspots have magnetic fields, they always occur in pairs, each member of the pair possessing one of the poles. Very often up to a hundred sunspots may appear in complicated groups, with complex magnetic structures.

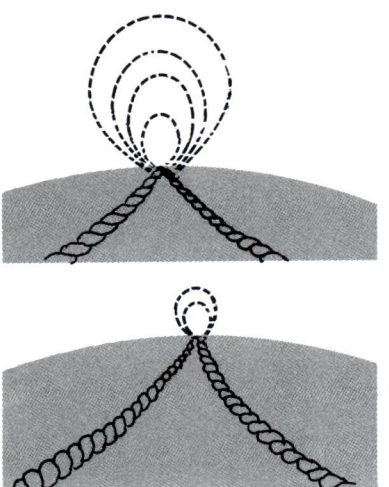

Complex interactions between the solar magnetic field and the ionized gas of the sun result in sunspots and other surface activity. Photos show intricate structure within a sunspot.

One theory of sunspot formation is based on the overall structure of the sun's magnetic field, which is about eight times stronger than that of the earth. According to this theory, the field is stretched into peculiar shapes because different parts of the sun rotate at different speeds. The magnetic field eventually develops small holes through which gases escape and are seen as sunspots.

Sunspots are not equally likely to be found at all times. They seem to go through a cycle, reaching a maximum number about every 11 years on the average. But this 11-year period is not constant;

the solar system

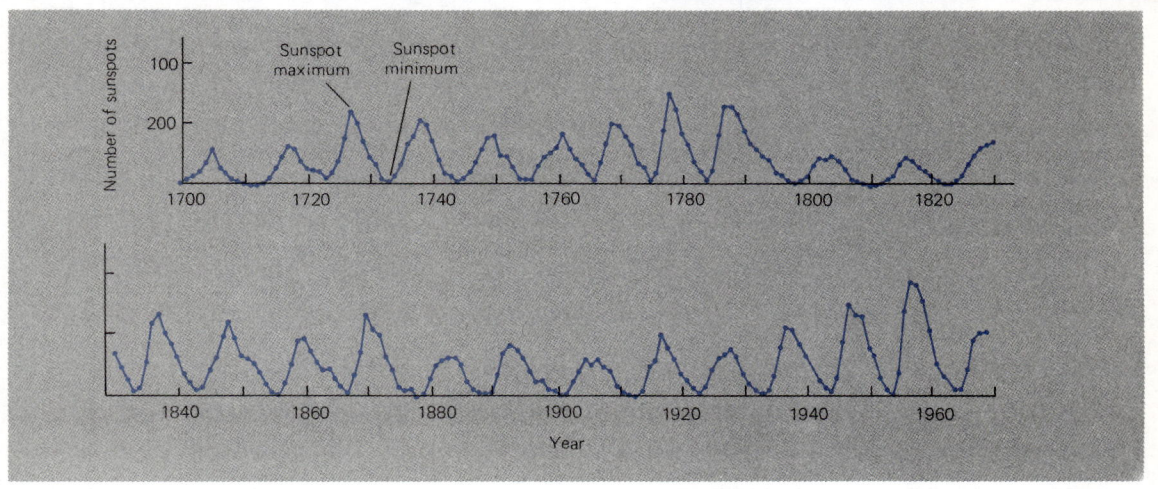

The sunspot cycle with an approximately 11-year cycle may affect climate on earth.

it may vary between 8 and 18 years. The sunspot cycle is not well understood as yet, but it is clearly a symptom of deep changes going on inside the sun. At the minimum of the cycle there are few sunspots; at the maximum more than a hundred may appear. There are indications that our weather may be affected by the sunspot cycle, as we will see a little later.

other surface phenomena

Three other sorts of events are associated with sunspots. **Faculae** are regions of high temperature in the photosphere and chromosphere. They appear before sunspots and often persist after the spots have disappeared.

Prominences are structures that extend out from the chromosphere and into the corona. They can easily be seen during total solar eclipses, and can also be observed with coronagraphs and filtergrams. Eruptive prominences are particularly spectacular, shooting outward from the sun at velocities up to 1,000 km/sec. They may extend a full solar radius from the sun, and they always occur near magnetic regions. Prominences have a complex structure, resulting from the interaction of the hot gas with the magnetic field.

Flares appear as intensely bright spots on the sun's surface, which reach maximum brightness in about 5 minutes and fade away after 1 to 3 hours. They vary in diameter from 5,000 km to many thousands of km. Near the maximum of the sunspot cycle, flares occur every few hours.

When a large flare occurs, the explosive wave it creates travels out from the sun, reaching speeds of up to 1,500 km/sec and creating hordes of charged particles. If the flare is directed at the earth, the particles reach us in about two to three days. The earth's magnetic field traps them, and they move toward the magnetic poles. Some of them penetrate to the upper atmosphere, producing the colorful auroral displays often known as "northern lights."

the solar system

At right, the development of a solar flare is seen in photographs taken two days apart.

The largest eruptive prominence ever observed, on June 4, 1946, became nearly as large as the sun in one hour. Within a few hours it had disappeared.

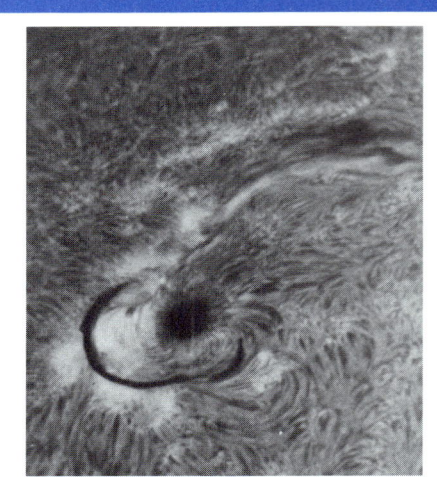

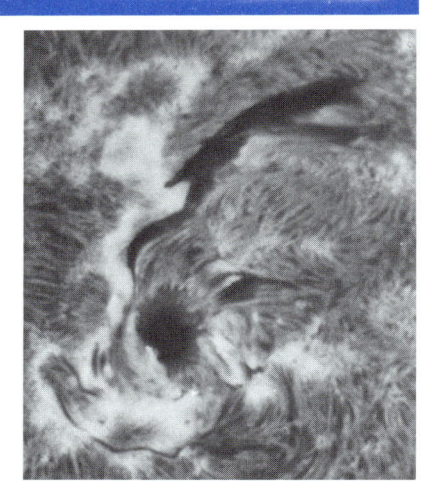

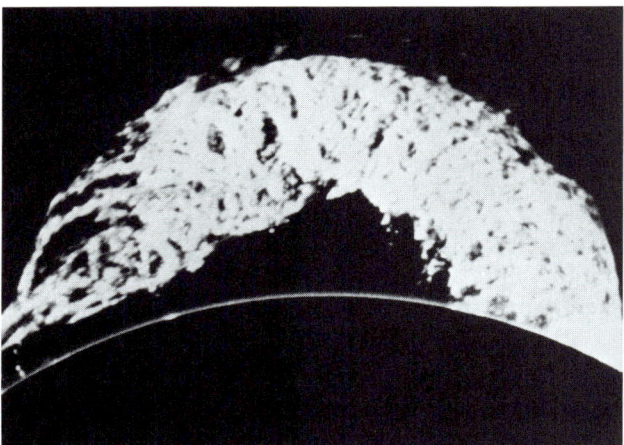

At right is a large flare extending beyond the edge of the sun.

Flare activity seems to modify the earth's climate significantly. There has been much controversy about the exact nature of the interaction between solar activity and our environment on earth. We have found that if the sun has 100 sunspots the earth's surface temperature during that period rises by about 1K. But little else is known, and to this day the exact relationship between solar activity and climate remains a very interesting puzzle.

planets and their orbits

The modern view of planetary orbits originated with Kepler, Galileo, and Newton, whom we discussed in chapter 1. Kepler did not successfully explain why the planets moved in their particular orbits.

the solar system

This was left to Galileo, who investigated the properties of masses and forces, and to Newton, who may have been the greatest scientific genius of all time. As we saw earlier, Newton proposed a new, more accurate version of Kepler's third law, which took into account the masses of the attracting bodies and could be applied to any situation where one body was in orbit around another. Now it was possible to determine the mass of the sun by observing the orbital size and period of a planet revolving around it, and to calculate the mass of a planet by observing the orbit and period of one of its satellites.

Newton's theory was very successful because he did not attempt to explain everything about the solar system. For example, he did not try to explain the sizes of the planetary orbits, assuming that they were the result of the original accident (or design) of creation. Newton claimed only to explain how things were getting along—in other words, he tried to define the rules of the game. But it is one thing to explain to a friend the rules of baseball, and quite another to explain why the New York Mets are three runs ahead in the bottom of the ninth inning!

Bode's law

The distances of the planets from the sun are not entirely arbitrary. A mysterious relationship appears to govern the sizes of planetary orbits. Known as **Bode's Law,** this relationship is based on a discovery made by Titius of Wittenberg in the eighteenth century. Titius began by taking the numbers 0, 3, 6, 12, 24, 48, 96, etc. He added 4 to each one and divided the resulting number by 10. The numbers he thus arrived at appeared to represent orbital sizes and relationships very well.

Table 15-3. Bode's Law

Titius Numbers		Planet	Actual Mean Distance from the Sun in A. U.
(0 + 4)/10 =	.4	Mercury	.387
(3 + 4)/10 =	.7	Venus	.723
=	1.0	Earth	1.000
=	1.6	Mars	1.534
=	2.8	Asteroid Ceres	2.77
=	5.2	Jupiter	5.203
=	10.0	Saturn	9.539
=	19.6	Uranus	19.539
=	38.8	Neptune	30.6
=	77.2	Pluto	39.4

In 1781 Uranus was discovered, and its location seemed to conform to Bode's Law. There was still some concern about the apparent gap between Mars and Jupiter, but when Ceres, the first asteroid, was discovered in 1801, it filled this gap.

Neptune and Pluto do not conform to Bode's Law, an indication that it is not really a "law" but an equation that expresses a relationship. The significance of Bode's Law is not clearly understood. It

IN FOCUS: solar activity and the earth

For centuries the sun appeared to be a spotless ball of light. When Galileo discovered sunspots in 1611, however, quite a few people began to wonder whether they affected the earth, and how. Many connections between sunspots and events on earth were suggested; even today, statistics have often been misused to demonstrate relationships between solar activity and practically everything. Of course, most of these alleged relationships are ridiculous. But there is growing evidence that the sunspot cycle does affect the climate on earth.

The 11-year sunspot cycle has been very steady in recent times, but this was not always the case. Around the end of the nineteenth century, E.W. Maunder examined historical records of sunspot observations dating back as far as 250 years. He discovered that in 1640 the number of sunspots observed declined. Between 1645 and 1715, almost none were recorded. This low point of sunspot activity is now called the *Maunder minimum*.

Was this shortage of sunspots real, or were people simply not seeing them? By the middle of the seventeenth century, many observers possessed telescopes better than Galileo's. In those days, the appearance of a sunspot was a very special event, and an active search for them was always in progress. But during the period from 1672 to 1704, not one spot was observed in the northern hemisphere of the sun. We can only conclude that the spots were really missing.

Maunder reasoned that such a clear-cut change in solar activity might have profound effects on conditions on earth. In 1894 he assembled his findings in a paper, published under the title of *A Prolonged Sunspot Minimum*. It attracted little attention from the scientific community, so Maunder tried again in 1922. He published a second paper on the same subject, which was also largely ignored. After his death a few years later, it appeared that Maunder's hypothesis was doomed to obscurity.

Gradually, however, other data were being compiled which tend to support Maunder's belief that the effects of solar activity could be observed on earth. The Maunder minimum, for example, coincided with a minimum in the number of auroras. These "northern lights" are produced by interaction of the upper atmosphere with charged particles expelled from the sun during violent solar events associated with sunspots. Only 77 auroras were recorded all over the world during the 70 years of the Maunder minimum. For a 37-year period, not a single aurora was seen. But when sunspots reappeared in 1716, so did the aurora. People in northern Europe, who had quite forgotten about auroras, watched them with a good deal of apprehension. Edmund Halley, the great British astronomer, observed an aurora in 1716 and wrote a paper about it. It was the first he had seen in his 60 years of life.

Two other points support Maunder's hypothesis. First, total solar eclipses observed during the Maunder minimum showed that the sun had a small corona with no coronal streamers, indicating little solar activity. Second, the Maunder minimum falls within the coldest years of the

"Little Ice Age," a period during which temperatures all over the world were unusually low. On the other hand, during the twelfth century Europe was unusually warm, and the amount of solar activity during this time is known to have been unusually high.

At the time Maunder's second paper was published, a botanist named A.E. Douglas was investigating a method of dating trees by the rings found in their trunks. Each year a tree adds a new layer or ring to its trunk. The thickness and structure of this layer can tell a scientist many things about the climate that year. Thick rings, for example, usually mean rain and sunshine, while a thin ring may mean that the weather was cold and dry.

In 1920, Douglas reported that tree rings showed a pattern of variations which repeated itself in cycles of 10 to 20 years; it seemed to be the same for trees all over the world. When he began to work with very old tree samples, however, Douglas noticed that the pattern was absent for the latter part of the seventeenth century. This suggested that the world's climate had remained steady for an unusually long time during those years. But Douglas could think of no reason for this until he happened to read Maunder's second paper. Then he guessed that the sunspot cycle was somehow connected with the earth's climate.

Douglas' work has now been improved on by application of radioactive dating techniques to tree rings. Trees absorb radioactive carbon-14 from the air and incorporate it into their structure. C-14 is produced from nitrogen by cosmic rays from space, and cosmic rays in turn are affected by the earth's magnetic field. High solar activity modifies the magnetic field so that fewer cosmic rays reach the earth's atmosphere. As a result, there is less C-14 in the air during years of intense solar activity. This is reflected in the low C-14 content of tree rings formed during those years. Conversely, rings formed during the 70 years of the Maunder minimum have an extremely high content of C-14, a further indication of extremely low solar activity during the period.

In addition to the variation in sunspot activity, many scientists now believe that the actual luminosity of the sun also fluctuates. To confirm this hypothesis, however, we need to observe the energy output of the sun very accurately, and this cannot be done from our position under the earth's variable atmosphere. Such observations will have to be conducted from space, which will be an expensive proposition. Placing an observation device on the moon during one of the Apollo explorations would have been an ideal solution. Unfortunately, this was not done, and so the question remains unanswered.

undoubtedly tells us something about the process which formed the planets. This significance is heightened by the discovery of similar relationships between the satellites of Jupiter and Saturn and the planets they orbit. But what Bode's Law *does* tell us is still unknown.

the Roche limit

One more law concerning orbits should be mentioned here. What would happen if the moon were closer to the earth? The tidal forces exerted by the earth on the moon would be stronger, and would try to "stretch" the moon. If the moon were close enough, these forces would be great enough to overcome the gravitational forces that hold the moon together. The moon would be torn apart, and its fragments might orbit the earth in a system of rings. The closest distance that a satellite can approach to a planet without being torn apart is called the **Roche limit.** The Roche limit for the earth is about 16,100 km.

the terrestrial planets

If we ignore Pluto, about which very little is known, the eight other planets fall into two groups. Mercury, Venus, and Mars are called **terrestrial planets,** because they resemble the earth in size, density, and composition. The outer planets—Jupiter, Saturn, Uranus, and Neptune—form another group which are similar among themselves, and are known as the **Jovian planets** (from Jove, an alternate name for the classical god Jupiter). As we have noted, these two groups of planets are separated by the asteroid zone that lies between the orbits of Mars and Jupiter. All the terrestrial planets have now been visited by spacecraft from earth. Successful landings have been made on Venus and Mars. Information about these planets is rolling in faster than it can be assessed, and many new discoveries have been made.

Mercury

Named after the swift messenger of the gods, **Mercury** is a small world, only 5,600 km in diameter. It orbits the sun at an average velocity of 48 km/sec, at an average distance of about 58 million km.

As seen from the earth, Mercury is never more than 28° away from the sun, and can be viewed only just before sunrise or just after sunset. Through earth-based telescopes it looks like a pinkish disk with a few shady markings. Mercury's slow rotation is hard to detect visually, but it becomes apparent when we bounce radio waves off the planet's surface. In 1962 an analysis of the reflected signals revealed that Mercury rotates once every 58.64 days, about two thirds of its 88-day orbital period. This slow rotation is a consequence of the tidal forces exerted on Mercury by the sun.

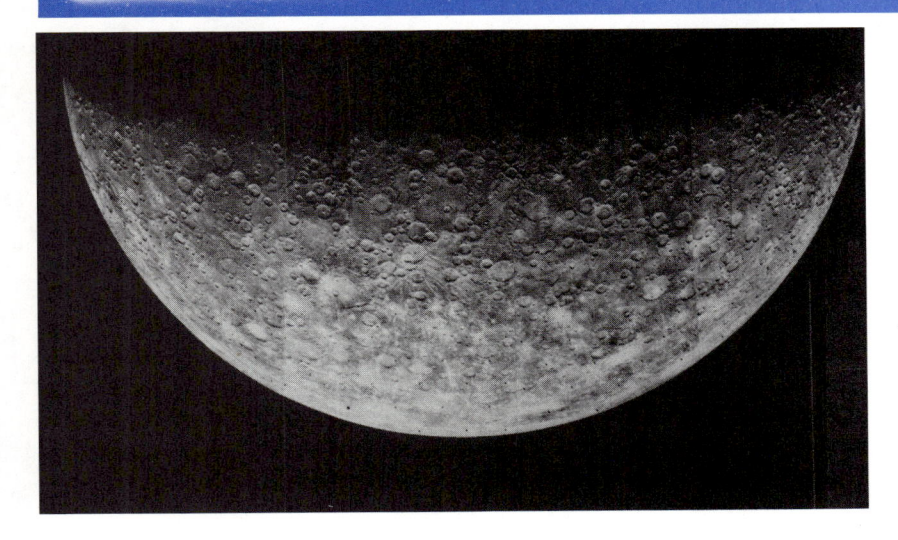

the solar system

Photo-mosaic of Mercury from Mariner 10 as it approached the planet.

Mariner 10 periodically encounters Mercury. On its third encounter, before it ran out of steering gas and lost contact, it sent back this detailed view of the fractured and ridged plains with large craters.

Most of our knowledge of Mercury is based on photographs and other information transmitted by Mariner 10 as it flew past the planet on March 29, 1974. The spacecraft had already flown past Venus about a month earlier, and the gravitation of that planet had been used to deflect it toward Mercury. The path of the Mariner was carefully tracked, and by observing the amount of deflection scientists were able to calculate Mercury's gravity to be about one-eighth that of earth.

Photographs show that Mercury is a barren, cratered world, rather like the moon. The surface is as dark as coal and reflects only a sixteenth of the light falling on it. There is a trace of hydrogen around the planet, but this appears to be an effect of the solar wind rather than a true atmosphere. Mercury has a faint magnetic field, probably due to a solid iron core. Such a core is indicated by the planet's high density, 5.2 times that of water. However, this theory is still unconfirmed.

Mercury appears to be "two-faced," since one hemisphere is much smoother and more free of craters than the other. There seems to be no surface evidence of tectonic activity, such as grabens, which would indicate faulting. The planet is both very hot and very cold. On the daylight side the temperature can reach 700K, high enough to melt lead and zinc. On the night side, with no atmosphere to retain the heat, the temperature can drop to 100K, cold enough to freeze carbon dioxide.

Venus

The planet that most resembles the earth is **Venus,** named after the goddess of love. At its closest approach to us it is only 42 million km away; no other planet comes closer. At its brightest, it is brighter than any other celestial object except the sun and moon. On dark nights Venus can cast a shadow. It is easy to locate, and often

the solar system

As seen from earth, Venus goes through a cycle of phases.

On February 6, 1974, a day after Mariner 10 swung by Venus on its way to Mercury, it took this ultraviolet photograph of the cloudy planet from a distance of 720,000 km.

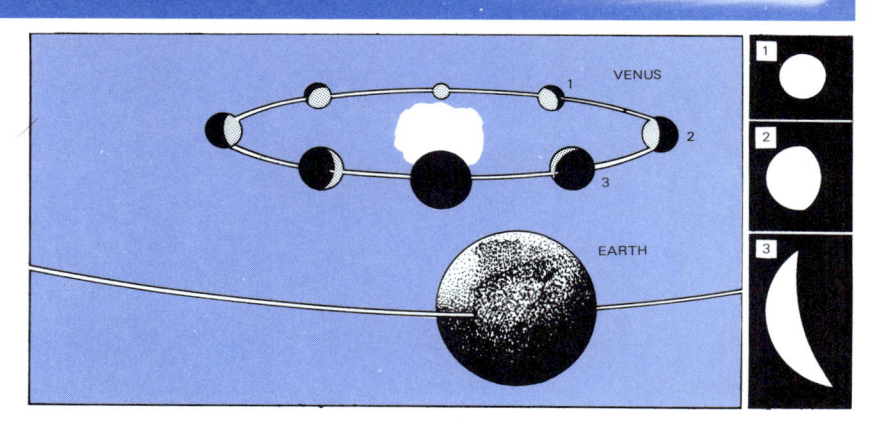

appears as a bright "star" in the east, before sunrise, or in the west, after sunset. It is never more than 48° away from the sun.

Although Venus approaches us so closely almost nothing was known about it until recently. There are two reasons for this. First, when it is close to the earth Venus appears as a crescent, and lies in a direction near that of the sun. It is up when the sun is up, setting just after sunset, or rising just before sunrise. Second, the entire surface of the planet is shrouded by a permanent cloud layer, which conceals all surface features. Since it is not possible to see any landmark that turns with Venus on its axis, the length of the Venusian day cannot be determined by visual observation. The following description draws on information compiled by radar and space probes.

The diameter of Venus is 96 percent that of earth; its volume is 86 percent that of earth; and its density is 5.23 times that of water, compared to 5.52 for earth. This leads us to suspect that the interior of Venus must resemble that of earth, with a crust, a mantle, and a molten core. If this is the case, Venus must be a geologically active planet. Some evidence for such activity became available quite recently from radar studies of the surface. Exceptionally fine radar images were obtained which rival optical photographs of the moon from the earth.

The most striking surface feature is a trench 1,400 km long and as wide as 150 km in places. It is very like the Great Rift Valley of East Africa, and is a definite indication of tectonic activity. There are also mountain ranges whose shapes suggest fault displacement. One peak, 400 km across at the base and cratered on top, may be a gigantic shield volcano.

Besides these features, two interesting regions have been discovered. One is a large basin about the size of Hudson Bay. The basin seems to be surrounded by a blanket of debris, such as would result from the explosive impact of a large meteor. Near this basin is an area named Maxwell, which reflects radio waves well and is almost certainly the result of a large lava flow. Maxwell is crossed by parallel ridges hundreds of kilometers long, which are clearly due to

THE ATMOSPHERE OF VENUS

Earth's sister planet, Venus, is a sister of a different complexion. The hot Venusian surface is covered by a dense atmosphere of carbon dioxide, containing clouds of sulfuric acid. The atmosphere is so dense that it affects light in very peculiar ways. We know, of course, that light rays are bent when they pass through atmospheric layers of different densities. On earth this effect is small, and permits us to see only a very short distance beyond the horizon. But on Venus, the atmospheric density increases so much near the surface that the corresponding effect is much greater. Unfortunately, the atmosphere is hazy and the visibility poor. But if it were transparent, an explorer on Venus would be able to see all the way around the globe. Theoretically at least, he could stand at any point and look at the back of his own neck!

Scientists have always found it puzzling that a planet resembling the earth in so many ways should have such a very different atmosphere. So far, a complete explanation has not been arrived at, but there are some interesting hints. For one thing, the atmosphere of Venus is composed almost entirely of carbon dioxide. Although there is as much carbon dioxide on earth as there is on Venus, most of it is dissolved in the oceans or locked up in the form of carbonates.

One possible explanation for this difference between the planets is that the young earth was cool enough to have oceans. Carbon dioxide dissolved in the oceans, and over time was converted into carbonates by the marine life forms that evolved. But since Venus was warmer than earth, the water in its atmosphere remained in the form of water vapor. This vapor trapped heat through the greenhouse effect, and raised the temperature still further. No oceans were able to form on Venus, and its carbon dioxide remained in the atmosphere. Eventually, the water vapor escaped into space, leaving Venus a very dry planet. As the sun becomes hotter during the next few billion years, it is possible that the earth's oceans will evaporate and the two planets will become not just sisters but twins.

the solar system

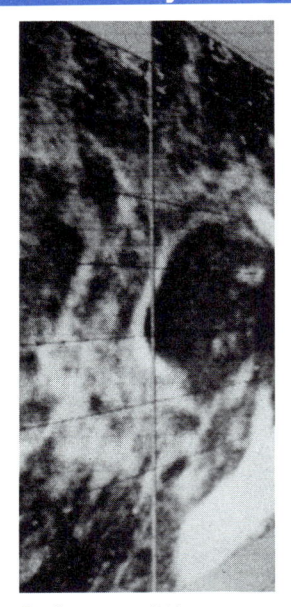

Radar map of Venus indicating tectonic activity.

Venera landings

Below is the sandy slope on which Venera 10, the Russian spacecraft to Venus, landed.

tectonic activity. There can be no doubt that Venus is a dynamic planet.

The many similarities between Venus and earth serve only to emphasize the many differences. One is the way in which Venus rotates. Radar studies have shown that Venus rotates once upon its axis every 243.1 earth days. Surprisingly, the rotation is clockwise, in the direction opposite to that in which Venus and all the other planets revolve around the sun. Furthermore, Venus always presents the same face to the earth at the moment of closest approach. This suggests that the tidal action of the earth has played a role in determining the rotation of Venus, but at present no one knows how the earth could have affected Venus to such an extent.

Conditions at the surface of Venus are very different from those on earth. Writers used to imagine lush forests, sparkling lagoons, and delightful people under those dense clouds. But in reality, the surface of Venus is more a Hell than a Heaven. It is extremely hot—around 750K both day and night. This high temperature is due to the greenhouse effect which we discussed in chapter 11. Carbon dioxide in the Venusian atmosphere allows energy from the sun to reach the surface, but traps the heat radiated outward by the planet. The atmospheric pressure is about 100 times that on earth, and the atmosphere is unbreathable. About 95 percent of it is carbon dioxide; the remainder is made up of argon, nitrogen, and traces of oxygen. Wind speeds are low. Occasionally there is a sulfuric acid drizzle.

Despite the hostile surface conditions, Soviet scientists have landed two Venus probes that survived long enough to transmit pictures. The picture from Venera 9, which landed on a steep slope, covers only a few meters. It shows jagged rocks, which must be young because erosion would have rounded the corners of older rocks. Soviet scientists believe that these rocks were formed by recent tectonic activity. Venera 10 landed on a sandy surface, and the picture extends for about a kilometer.

The Venera missions provided astronomers with a number of surprises. For example, the light is unexpectedly bright at the surface, as bright as an overcast day on earth. Furthermore, the famous clouds are more like a haze, and do not extend below 49 km from the surface. Most astronomers now believe these clouds consist of sulfuric acid. Some observations by Mariner 10 support this belief, and it is hoped that future landings by advanced United States spacecraft will provide more details. All indications are that if we were to deliberately design a hostile planet, we could not do much better than Venus.

the solar system

Mars

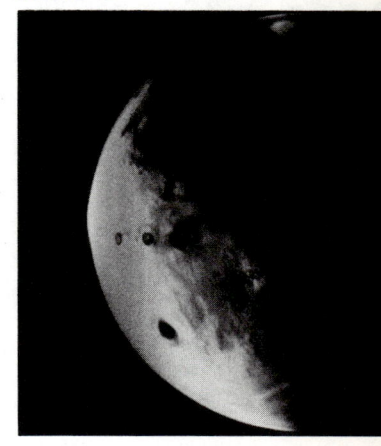

Mars, the red planet named after the god of war, has excited more curiosity than any other. Mars is comparatively small, with a diameter one-half and a mass one-tenth that of earth. Since its atmosphere is thin and it approaches to within 55 million km of the earth, it has been very well observed.

For more than a century people have speculated about intelligent life on Mars. These speculations probably began with Secci's telescopic observation of streaks on the planet which he called *canali* (Italian for "channels"). The word, incorrectly translated as "canals," gave rise to the notion of a canal-building race of Martians.

In the years that followed, various observers reported seeing canals. Others disputed these observations, and the dispute was not settled until Mariner 4 flew past Mars in 1965. This spacecraft and two others, Mariner 6 and Mariner 7, photographed a bleak, cratered surface, with no canals or bodies of water. There seemed to be little probability of a Martian civilization. But if what these craft and others which followed have reported has made Mars less interesting to ethnologists and romantics, the planet has become far more interesting to astronomers and geologists.

Mars with its gigantic volcanoes, as seen from Viking 1.

atmosphere and internal structure

The Martian atmosphere is thin and cold, with a surface pressure less than 1 percent that at sea level on earth. The temperature ranges between 293K and 198K. Carbon dioxide is the main atmospheric gas, but argon, hydrogen, oxygen, and water vapor are also present. High-altitude clouds are common. Those at 45 km are probably frozen carbon dioxide, and those below 30 km are probably frozen water vapor.

Dust storms, sometimes covering the whole planet, are stirred up by winds blowing at hundreds of kilometers per hour. The dust clouds reach enormous heights, and the scouring effect of the wind-blown dust is the principal cause of erosion on Mars. The red dust suspended in the atmosphere makes the Martian sky an unearthly pink.

Mars has a density only 3.97 that of water. The interior is believed to be less differentiated than that of the earth. Observations by Mariner 9 indicate that Mars may have a partially melted core. The mantle is thought to be denser and less well separated from the core than the earth's. The interior temperatures are probably not hot enough to melt nickel or iron.

Thin clouds of ice in the atmosphere of the cratered planet Mars.

surface features

Mars is heavily cratered by meteor impacts. The craters have been eroded by the atmosphere, and most of the smaller ones have been obliterated. The largest craters are very old, and are concentrated in the Southern Hemisphere. The Northern Hemisphere contains large smooth areas, some of them covered by drifting dust. In places, high winds have piled the dust in dunes.

The most surprising features of Mars are the thousands of winding channels which resemble dry river beds. Within the chan-

the solar system

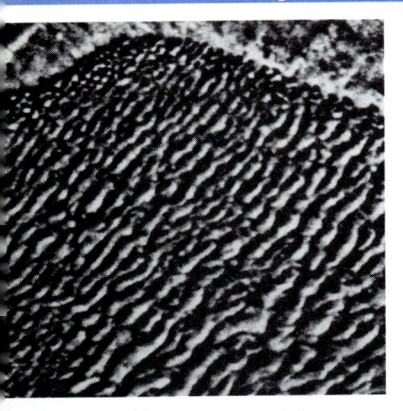

Above are sand dunes on Mars produced by strong winds. Above center, are sinuous channels almost certainly cut by flowing water. Above right is a picture from Viking Orbiter 1 showing braided channels, islands, and grooves in what must once have been a river bed.

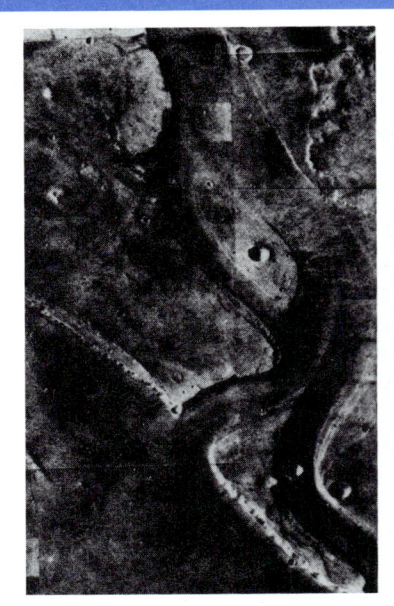

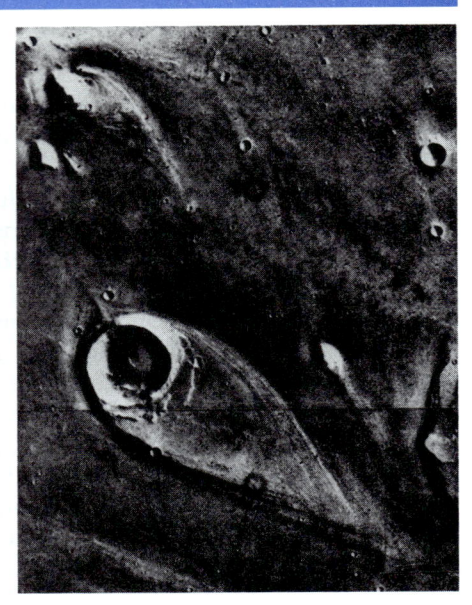

nels are braided structures, little deltas, and islands. These structures must have been cut by water, or a similar substance. In some places the erosion pattern shows that the water must have flowed in a turbulent stream. But where did the water come from? One theory suggests that the polar ice caps contain quantities of water. The tilt of the Martian axis is altered periodically by the gravity of Jupiter. When the tilt increases, hotter summers might cause the water to melt and flow in rivers.

Other striking Martian surface features include gigantic canyons and volcanoes. Some of the so-called canals are actually huge canyons, much larger than any on earth. One of them, nicknamed Feather River, stretches for 5,000 km. At places it is 150 km across and 6 km deep, completely dwarfing the Grand Canyon. These Martian canyons are clearly the result of tectonic activity.

Matching the size of the canyons are the huge shield volcanoes that have been formed by repeated lava flows. Olympus Mons, for example, is at least three times larger than Mauna Loa, the largest shield volcano on earth. The other volcanoes are smaller, but still larger than anything on earth.

Martian soil

We now understand why Mars is red. Its color is believed to be due to limonite, which is formed from an iron oxide called hematite by the action of water. It is interesting that the Martian rocks themselves are not colored. Instead, they have a colorful outer coating, a "desert varnish" produced by water that soaks into the rock and causes the color to form from the outside in.

The surface rocks are silicates, and appear to be basaltic rather than granitic. At least six kinds of rock were found around the Viking

the solar system

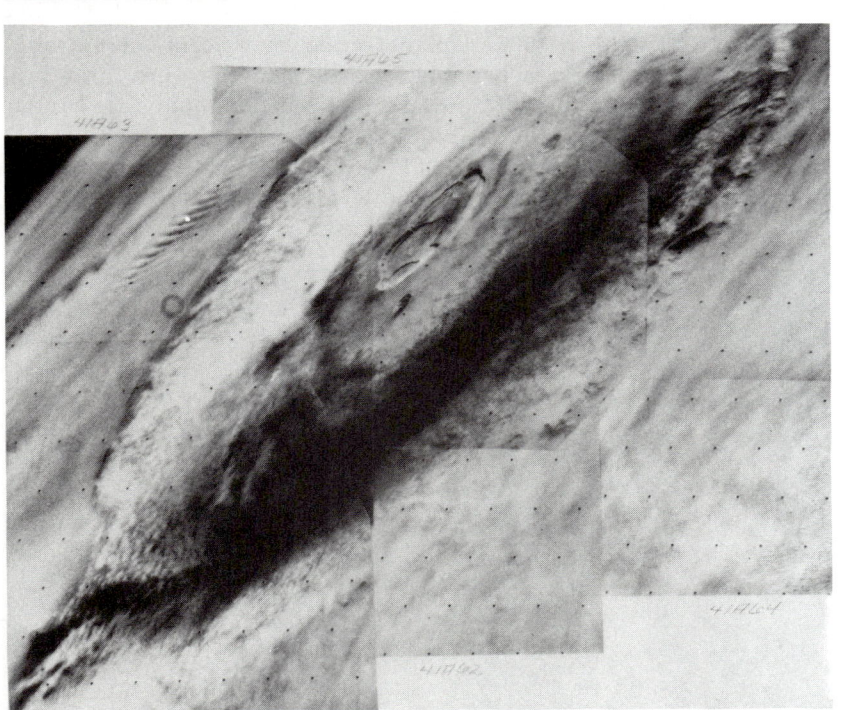

419

This huge equatorial canyon on Mars was probably produced by faulting and then carved into complex shapes by the wind, as well as by periodic freezing and thawing of underground ice.

At left is Olympus Mons, a gigantic shield volcano whose 25 km height dwarfs any mountain on earth. In this Viking Orbiter 1 photograph, clouds reach up nearly to the top. Only the volcanic crater, 80 km across, is exposed.

Below, the rock-strewn surface of Mars photographed by Viking Lander 1.

the solar system

is there life on Mars?

the moons of Mars

Below is a detailed picture of Phobos showing a set of parallel grooves that may have resulted from the tidal stretching exerted by Mars.

1 lander. They did not seem to have come from nearby craters, but may have been transported to the site by flowing water.

Digging experiments performed by the Viking landers indicate that the Martian soil is rather firm and is covered by a thin crust. The texture below the surface is like that of a wet beach, although no water is now present. These observations indicate that the soil was wet at some time.

No Martian civilization has been discovered. But might there be some form of life on Mars? In theory, this is possible: some forms of antarctic lichen have been grown under conditions similar to those on Mars. To investigate this question, the Viking landers were equipped to analyze the Martian soil for any recognizable signs of life.

Preliminary results are puzzling. Some life-indicating reactions have been detected, but further experimentation is needed before a conclusive answer can be given. In this way, as in others, Mars promises enough problems to keep scientists happy for a long time.

Mars has two tiny satellites, **Phobos** (fear) and **Deimos** (panic), discovered in 1877. A planet-wide dust storm was in progress when Mariner 9 reached Mars, so its cameras were turned on the satellites instead of on the surface. The Martian moons are irregular in shape; Phobos, which is twice as large as Deimos, measures 22 km by 15 km. The orbital period of Phobos, 7.5 hours, is less than one Martian day, making it the only known satellite in the solar system that rises in the west and sets in the east.

At right, Phobos, the inner satellite of Mars, from Viking Orbiter 2.

the solar system

Pluto

Named for the god of the underworld, **Pluto** is often considered a terrestrial planet, but so far we know very little about it. Discovered only in 1930, Pluto is the remotest of the known planets. It is believed to be a small, cold world, covered with frozen gases. Even in the largest telescope it appears as a faint, star-like object rather than as a disk. Pluto's size has been roughly estimated from its brightness; it is not much more than 5,000 km in diameter. Its brightness varies regularly over a period of 6.39 days. From this we can deduce that it rotates, but its other major features remain a mystery.

the Jovian planets

Beyond the orbit of Mars are the asteroids, and beyond the asteroids lie the giant planets: Jupiter, Saturn, Uranus, and Neptune. They contain 99 percent of the mass in the planetary system, and are similar in many ways. For example, they all have low densities, comparable to that of water. Saturn is actually less dense than water; if it could be dropped into an ocean large enough to hold it, it would float. The density of the Jovian planets closely resembles that of the sun. This implies that they are made up chiefly of hydrogen and helium, since there is no other way to explain such low densities.

Observations show that the Jovian planets are giving out heat. Jupiter, for example, gives out twice as much heat as it receives from the sun. The interiors of these planets are probably quite hot. Most of their heat is given out as invisible infra-red radiation; the visible light that we observe is entirely reflected sunlight.

The Jovian planets possess cloudy atmospheres, containing ammonia, methane, and other chemicals. No surface can be seen beneath the clouds, and it is possible that no solid surface exists. Theoretical models indicate that these planets are almost entirely liquid, a theory that has been verified in the case of Jupiter. Like the other Jovian planets, Jupiter rotates rapidly and thus bulges at the equator, which causes its gravitational field to differ from one place to another. Pioneer 10, as it flew by Jupiter, measured the gravitational field very accurately. It turned out to be just what we would expect to find if Jupiter's interior were liquid.

Thirty of the 33 known satellites in the solar system belong to the Jovian planets. Some of them are very small, only tens of kilometers across, but others are almost as large as Mercury. The larger satellites have densities and surface reflectivities intermediate between those of the terrestrial planets and the Jovian. These satellites seem to consist of rocky material together with ices of ammonia, methane, and water.

the solar system

Jupiter

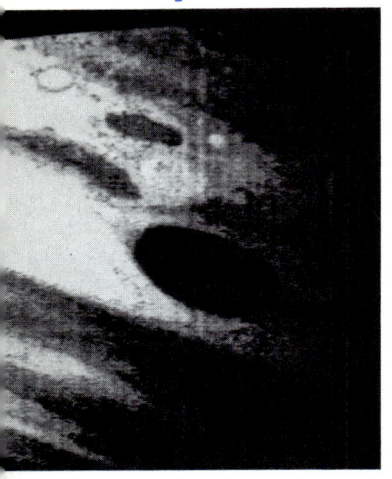

The largest and most massive planet is **Jupiter,** named after the king of the gods. Its diameter is 11 times that of the earth; together with its high reflectivity, this makes it a bright object in the sky.

Jupiter has been observed telescopically since 1610, when Galileo turned his telescope on it and discovered four of its satellites. In the last few years Pioneer 10 and Pioneer 11 have transmitted pictures as they flew by the planet. Jupiter appears to be crossed by permanent bands, consisting of light-colored **zones** and darker **belts.** These bands evidently belong to the atmospheric layers rather than to a solid surface, since observation shows that the equatorial bands rotate more quickly than those at higher latitudes. The Jovian atmosphere has a complex structure. The belts, for example, are lower and warmer than the zones, a pattern reminiscent of the trade winds on earth.

Interaction between the zones and the belts produces eddies, which can be seen as spots. The most famous of these is the Great Red Spot, which was discovered in 1665 and has persisted to this day. Pioneer 10 discovered that the Great Red Spot is a gigantic oval hurricane, 14,000 km across and 35,000 km long—large enough to hold a dozen earths. Meteorologists are intensely interested in the Great Red Spot, and hope that it may shed some light on the formation of hurricanes on earth. Understanding how it has been able to maintain itself for more than 300 years might help us to predict and possibly to control hurricanes.

Jupiter's Great Red Spot imaged by Pioneer 11 at about a million km from the planet.

inside Jupiter

The interior of Jupiter is as interesting as its surface. Jupiter's volume is about 1,300 times that of the earth, but its mass is only 318 times as great, so it is therefore only a third as dense as the earth.

Various models of Jupiter have been calculated on the basis of all the observations of the planet. According to one model, the banded appearance of Jupiter is due to various gases found in the outer layers of the atmosphere. The color of Jupiter is probably due to clouds of ammonium hydrosulfide, a poisonous gas that smells like both ammonia and rotten eggs. Below this visible layer, hydrogen is compressed into a dense fluid. At a depth of 3,000 km, the pressure is 100,000 atm and the temperature about 7,000K, causing hydrogen to become liquid. At a depth of 20,000 km the pressure is 5 million atm and the temperature 10,000K. Hydrogen under these conditions becomes metallic and can conduct electricity. In the center of the planet, at a depth of almost 70,000 km, there is a small core containing iron and other heavy elements, at a temperature of about 30,000K.

Some people have called Jupiter a "failed star," but it would have to be at least 100 times more massive in order to become a star. Only then would its center heat reach the millions of degrees required to start the nuclear reactions that power stars, a process we will discuss further in chapter 16. The source of Jupiter's energy, in fact, may simply be the heat left over from its formation.

the solar system

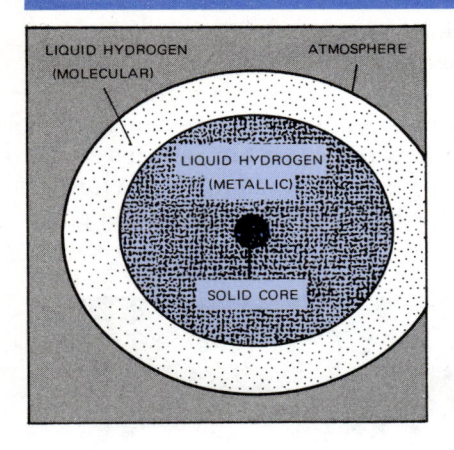

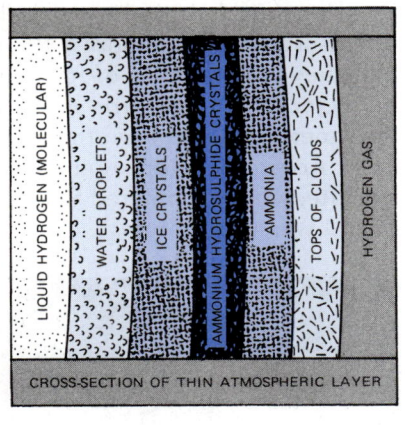

A well accepted model of Jupiter's interior. Although the atmosphere is thin compared to the planet's radius, it nevertheless consists of several layers.

The metallic hydrogen in Jupiter's interior is probably the source of its magnetic field, which is about eight times stronger than that of the earth. The magnetic field traps particles of the solar wind, which interact with the satellite Io to create bursts of radio waves. Bursts of waves also appear to come from the surface of Jupiter at intervals of about two seconds, each containing as much energy as a hundred hydrogen bombs.

Is there a possibility of life on Jupiter? Between the inner heat and the outer cold is a region of moderate temperatures, in which organic chemicals are abundant and lightning bolts crash through the turbulent gases. Such conditions are believed to have contributed to the formation of life on earth. Although Jupiter has no solid surface, it is theoretically possible that some form of life adapted to existence in the atmosphere may have evolved.

Thirteen satellites are known to orbit Jupiter, and the presence of another is now suspected. Four of them are very large, with masses approaching that of Mercury. The other nine are much smaller, and may be asteroids that were captured by Jupiter's strong gravitational field.

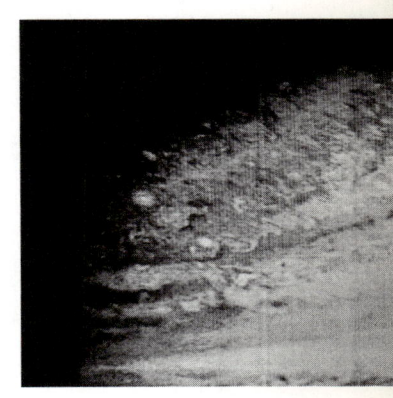

Photograph of Jupiter's cloud tops as Pioneer 11 flew past, 40,000 km above them.

Saturn

Named after the father of Jupiter, **Saturn** is the second largest planet. Its mass is about a third of Jupiter's, and its volume about half. The two planets are believed to be very similar in structure, but Saturn's smaller mass produces a lesser degree of compression in its interior gases. This explains both its low density and its very elliptical outline. Saturn has ten major satellites. The largest, Titan, has a permanent atmosphere, and atmospheric pressure at its surface may approach that of the earth.

Saturn is one of the two planets known to have rings; its rings can be seen through even a small telescope. Through larger telescopes, four separate rings can be distinguished. They occupy a region between 80,000 km and 136,000 km from the center of the planet.

The rings are believed to be billions of tiny satellites composed of frozen water and ammonia. Recent radar studies appear to confirm that they are made up of icy chunks no more than a meter or so in diameter. The outer parts of the ring system orbit more slowly than the inner parts, and even the outermost ring is within Saturn's Roche limit. According to one theory, the rings may be remnants of a former satellite, which wandered within the Roche limit of the mother planet and was torn to bits.

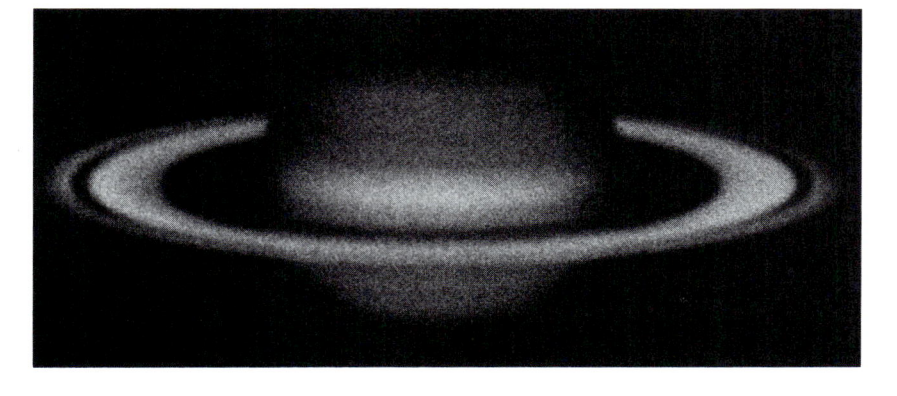

This photograph of Saturn from the earth will become obsolete when Pioneer 11 sends back a close-up of the ringed planet in 1979.

Uranus

Uranus, named after the forefather of the Titans, was discovered in 1781. Its mass is about fifteen times that of the earth, and it rotates quite rapidly. The density of Uranus is slightly greater than that of Jupiter. This is an indication that although hydrogen and helium are its primary constituents, Uranus possesses a greater percentage of heavy elements than Jupiter. Its average distance from the sun is about 2.8 billion kilometers, giving it an orbital period so long that it has completed only two and a half orbits since its discovery.

Uranus has five moons, one of which was discovered as recently as 1948. Until early in 1977, however, no one had suspected that Uranus also possesses five rings, which seem to be very much like those of Saturn. The four inner rings appear to be only a few kilometers wide, but the outermost ring may be as much as 95 kilometers in width. The rings cannot be viewed through terrestrial telescopes. They were discovered by observers flying in a specially-equipped laboratory aircraft.

Another intriguing thing about Uranus is that it rotates on its side. Its equator is inclined at an angle of 82° to its orbit. Both its moons and the newly-discovered rings orbit in the plane of its equator, so that if we could have a close-up view of Uranus, it would probably look much like Saturn tipped over sideways.

Neptune

Not very much is known about **Neptune,** the eighth planet from the sun. It is physically very similar to Uranus, and orbits at an average distance of 4.5 billion kilometers from the sun. Since its discovery in 1846, Neptune has yet to complete one orbit.

the solar system

A peculiarity of Neptune concerns its two satellites. **Nereid,** the smaller, has a diameter of less than 100 km. It revolves around Neptune in an extremely elongated orbit. Nereid approaches to within about 2.4 million km of the planet, and then recedes to more than 16 million km.

Triton, Neptune's second satellite, is the most massive in the solar system. It orbits Neptune in a direction opposite to that of the planet's rotation. Some astronomers believe that Triton's peculiar orbit can be explained by assuming that Pluto was once a satellite of Neptune. They believe that it may have interacted with Triton and thus been flung out of Neptune's family. In the process, they suggest, Triton's motion was reversed.

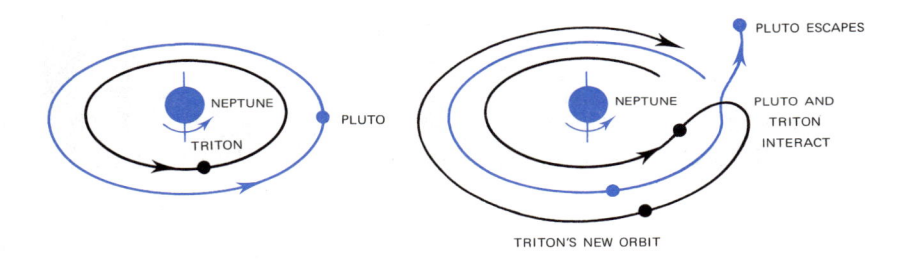

Pluto might have been a satellite of Neptune which escaped after interacting with Triton; this may explain how Triton's orbital motion came to be reversed.

This theory has by no means been fully accepted, and some questions remain. But it does have some attractive features. It explains three things: the fact that Pluto's orbit sometimes takes it within the orbit of Neptune; the presence of little Pluto beyond the Jovian giants; and Pluto's slow rotation, which might have been tidally slowed down when it was Neptune's satellite.

the moon

The moon is our nearest celestial neighbor, and our only natural satellite. It has been studied telescopically since the beginning of the seventeenth century. On July 19, 1969, the astronauts of the Apollo 11 mission landed on the moon, and it has since been successfully explored by five other Apollo missions. No other extraterrestrial body has been examined so closely.

The Apollo missions greatly increased our knowledge of the moon. Besides performing experiments on the surface, the astronauts brought back hundreds of kilograms of lunar rocks and soil, which have been analyzed by laboratories around the world. We now have a fairly detailed picture of the moon's history.

In one way, the moon missions have been disappointing.

the solar system

The rotating earth carries an observer around with it; enabling us to observe the moon from slightly different directions. Thus we can see a little around the edge of the moon.

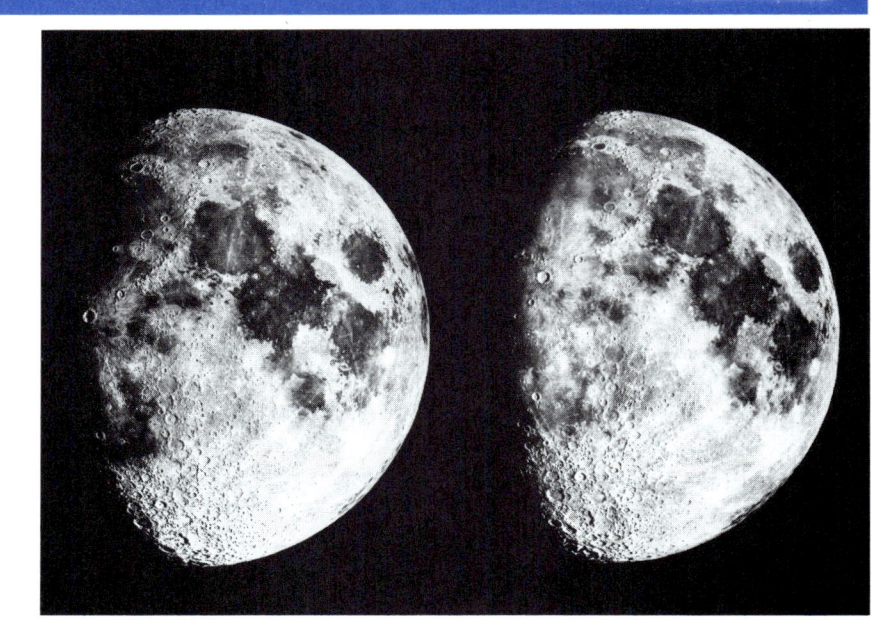

Some scientists had hoped that the moon would be covered with unchanged material from the period before the formation of planets. This could have cleared up many of the mysteries about the formation of the solar system. But no such material was discovered; all the lunar rocks have been transformed by igneous processes during the moon's early history, and not one of them seems to represent the state of the moon as it was first formed.

the motion of the moon

The mutual gravitational attraction between the earth and the moon keeps them in orbit around each other. The orbit is actually around their **balance point,** the center of mass. Because the earth is 81 times more massive than the moon, the center of mass is 81 times closer to the center of the earth. It lies 1,600 km beneath the earth's surface.

The moon travels along its elliptical orbit around the earth at 3,700 km/hr, and has an orbital period of 27.3 days. In the same period of time it rotates once upon its axis. Slightly more than half the surface becomes visible over its orbital period, because its axis is slightly tilted and because it does not travel at a uniform velocity along its orbit. But 41 percent of the surface is never visible from earth. The hidden part was first observed in 1959 by the Soviet Luna 3. Since then, the entire surface of the moon has been throughly mapped by the American orbiters.

It is not coincidental that the moon's period of rotation is the same as its period of revolution. According to one theory, the moon and the earth once rotated much faster than they do now. The earth's gravitational pull raised rock tides in the elastic moon, slowing down

the solar system

its rotation. Today the moon is permanently distorted; it is slightly egg-shaped, the small end pointing toward the earth.

The rotation of the earth continues to slow down at a rate of about .002 seconds per day every century. As it slows, the orbit of the moon becomes larger. Eventually, the moon will take 50 days to orbit the earth instead of the 27 days it does now.

The moon is not luminous; moonlight is really the reflected light of the sun. At different times all areas of the moon are lit by the sun. But at any one instant only half the moon is bright; the other half is dark. As the moon orbits the earth we see varying amounts of the light side and the dark side. When the moon lies in the direction of

phases of the moon

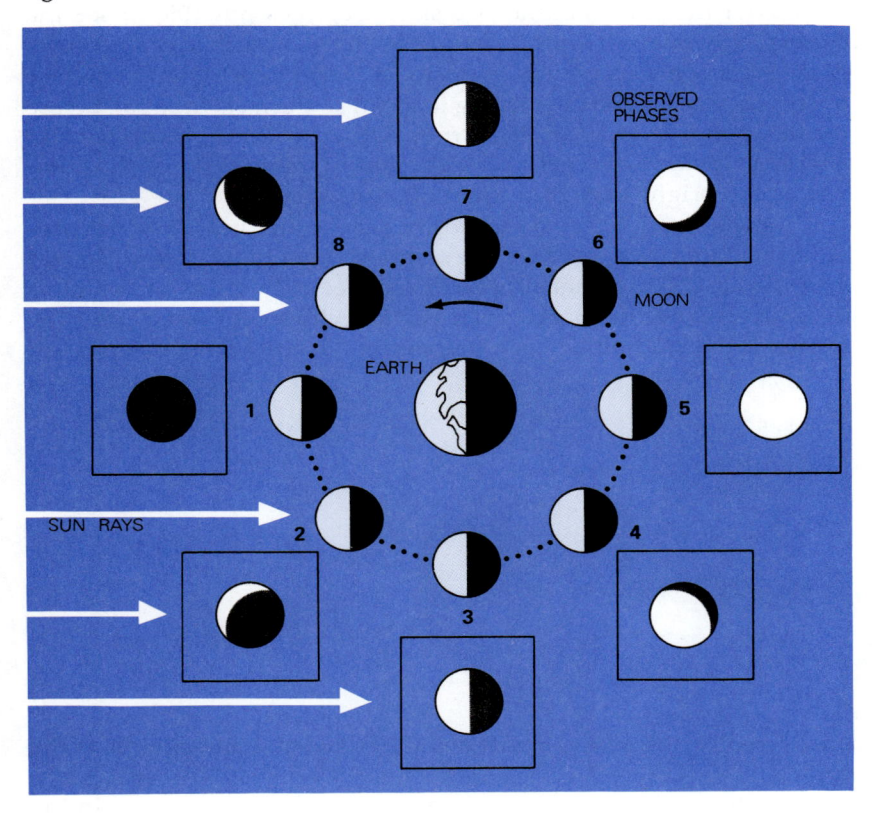

Figure explains the phases of the moon in terms of its position in its orbit around the earth.

the sun we see none of the lighted side, and we say the moon is **new**. As it proceeds in its orbit we see progressively more of the light side, which appears as a thin crescent that gradually grows. When the moon lies in the direction opposite that of the sun, we see all of its light side, and we say that it is **full**. As the cycle continues we see less and less of the moon's light side until it becomes a crescent, and then new again. The entire cycle of phases takes about 29.5 days, about 2.2 days longer than its orbital period. The extra time is needed because

the solar system

eclipses

during the same period the earth has been traveling along its own orbit around the sun. The direction of the sun as seen from the earth has changed, and the moon must travel an extra 1/12 of its orbit to lie in this changed direction.

The diameter of the moon is about 1/400,000 that of the sun. It is also 400,000 times closer to us. For this reason the sun and moon both appear to be the same size in the sky. This sets the stage for one of the most spectacular of astronomical events—the **total solar eclipse.**

When the orbit of the moon takes it to a point almost exactly between an earth-based observer and the sun, the moon completely blocks out the sun's luminous disk. It becomes possible to see the corona, only one-millionth as bright as the sun itself. The sky grows dark. Stars come out, birds go to roost, and scientists and amateur observers work feverishly, trying to glean all the facts they can before the moon moves away on its journey. Others without scientific tasks simply enjoy the unforgettable sight. But in minutes the spell is broken as the bright edge of the sun appears again.

Even though they occur nearly every other year, total eclipses are considered rare because they appear total only from a limited viewing area. Observers at other places see only a part of the sun's disk covered by the moon. Such **partial eclipses** do not permit the corona to be studied. If even one percent of the photosphere is left

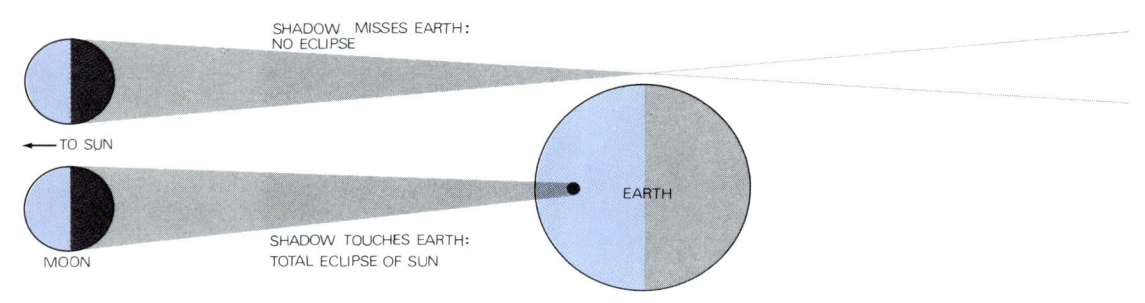

Not every new moon results in a solar eclipse. In most cases the moon's shadow misses the earth.

uncovered by the moon, it radiates 10,000 times as much light as the corona.

A **lunar eclipse** occurs when the full moon moves within the shadow of the earth, and thus cannot be illuminated by the sun. The observer sees the moon darken. It does not disappear completely from sight but glows a dull red, because sunlight is reflected by the earth's atmosphere onto the moon's surface. Solar eclipses occur when the moon is new and lunar eclipses when the moon is full. But the moon is not eclipsed every time it is full because the plane of its orbit is at an angle to the earth's orbit around the sun. The moon usually travels above or below the shadow of the earth. The complicated cycle of solar and lunar eclipses repeats itself, almost exactly, ever 18 years and 10 days.

When the moon's edge passes in front of a star, the star blinks out almost instantaneously, an indication that the moon possesses no atmosphere. If the moon had an atmosphere the star would fade out gradually instead. The lack of a lunar atmosphere, which has been confirmed by moon explorations, is easy to explain.

You will remember from chapter 11 that a gas consists of molecules in motion. If the molecules in a planet's atmosphere move fast enough to reach escape velocity, the gravity of the planet can no longer hold them, and they leave the planet forever. The escape velocity from the earth is 11.9 km/sec; from the moon it is only 2.35 km/sec. Obviously, the earth can hold onto some molecules that the moon cannot.

the moon's missing atmosphere

Gas	Time for 50% of gas to escape
Hydrogen	A few weeks
Helium	A few weeks
Water	Several thousand years
Oxygen	Several hundred thousand years
Nitrogen	Several hundred thousand years

Table 15-4. Escape rates of gases from the moon

We have seen that hydrogen and helium are light enough to escape entirely from the earth's atmosphere. The moon, with its lower escape velocity, has lost not only its hydrogen and helium but also its other gases. Volcanic vapors are often released in small quantities from the moon's surface, but they soon disperse. Because there is no atmosphere the sky is jet black as seen from the surface. The stars shine steadily, without twinkling, even while the sun shines in the sky.

Because the lunar surface is exposed to space, the solar wind strikes it unhindered and makes it very dark. The surface reflects only 7 percent of the sunlight falling on it. Meteors, too, strike the surface constantly. The surface is covered with sand and rubble as a result of this ceaseless bombardment, since even the smallest meteors, which would be stopped by an atmosphere, can cause damage.

There is very little erosion on the moon other than that performed by the meteor bombardment. There is no weather on the moon, nor is there water. Neil Armstrong's famous footprint may survive for millions of years, perhaps long after every sign of the human race has vanished from the earth.

The absence of liquid water on the moon is connected with the absence of atmosphere. For water to exist in liquid form it must be under pressure. A reduction in atmospheric pressure causes water to boil at a lower temperature. On the airless moon, any water that once existed would have long since boiled away.

Astronaut Neil Armstrong's famous footprint on the moon.

the solar system
the lunar surface

craters

The moon has bright and dark regions. People have given imaginative names to the resulting pattern, of which the most familiar is the "Man in the Moon." In this respect, the moon has been an astronomical "ink-blot" test!

Telescopic examination reveals that the moon is covered with craters, which are more common in the light-colored areas. The dark areas appear to be smoother, and many of them are roughly circular in shape. Early observers, struck by their smoothness, called them **maria** (Latin for "seas." The singular form of maria is **mare.**) The light areas were called continents or **highlands.** These names are still used, even though there is no water in the maria and the highlands are not necessarily higher than the surrounding surface.

The craters of the moon range in size from microscopic to hundreds of kilometers across. Some of the largest have walls that tower thousands of meters high. The largest craters are found on the far side of the moon. But why are there craters on the moon to begin with?

A few of the moon's craters may be the result of volcanic activity, but the majority seem to have been formed by the impact of massive bodies on the moon's surface. Like the earth, the moon is constantly colliding with chunks of matter that abound in interplanetary space. An object 500 meters across, striking the surface at a velocity of 50 km/sec, can release as much energy as hundreds of millions of tons of TNT. The explosive collision produces a large crater, and often expels matter to great distances, creating a striking pattern of **rays** around some craters.

Why is the earth not cratered like the moon? For one thing, our atmosphere stops the smaller meteors, and erosion wears down the meteor craters that do form. But what about craters such as Crater Bailey on the moon, which is 300 km across? Such a crater on earth would last for hundreds of millions of years.

One possible answer might be that most of the large craters were formed when the moon was young. That phase ended four billion years ago, and since then only much smaller objects have collided with the earth and the moon. Since most of the earth's surface is only about three billion years old, larger craters from the early bombardment have been obliterated.

A newer theory proposes that the meteor bombardment about four billion years ago may have been directly responsible for the formation of the earth's first ocean basins. According to this theory, deep basins were formed by meteor impacts, providing a place for the water on earth to collect. Igneous activity beneath the ocean basins caused uplift and rifting of the ocean floor. As the rifts were widened by sea floor spreading, lithospheric plates were formed, and plate tectonic activity began. (For a review of plate tectonic mechanics, see chapter 8.)

The Apollo 11 Lunar Module returns to the orbiting Command Module after its stay on the moon.

the solar system

maria

The lunar maria are plains of dark basalt, which reflect only 6 percent of the light that falls on them. Maria cover about a third of the earth side, but are almost absent on the far side. It is believed that they originated as basins formed by meteor impacts, which were subsequently filled with lava.

Maria are relatively flat, but close examination reveals many features. **Wrinkle ridges** are raised structures hundreds of meters high and many kilometers long. **Domes** are swellings, perhaps caused by magma under pressure. **Rilles** are river-like grooves, either straight or winding, that usually appear in regions of the maria that border the highlands. The rilles may have been caused by lava flows. **Scarps** are fault-like discontinuities in the maria, which may have formed as the lava cooled.

Rilles on the moon are river-like structures cut by flowing lava in the moon's past.

highlands

The highlands of the moon reflect up to 18 percent of the light falling on them, and are usually at a greater elevation than the maria. We know that the highlands are older than the maria, because the lava flows that created the maria appear to have flowed over and covered the adjoining highlands. Photographs of these highland regions show that the hills seem to have been sculpted by material ejected by the impacts that formed the maria.

Additional evidence for the great age of the highlands comes from radioactive dating of material brought back to earth. Rocks from the highlands are definitely older than rocks from the maria. Some of them are almost as old as the material of meteorites, which can be up to 4.6 billion years old.

Highlands cover two thirds of the near side of the moon and virtually all of the far side. This difference is significant. Obviously, the maria on the near side formed after the moon's rotation had been

the solar system

slowed by tidal forces. Besides this, the difference is almost certainly related to the thickness of the moon's crust, which is much thicker on the far side than on the near side. The thinness of the crustal layer on the near side might have facilitated the lava flows that made the maria.

lunar geology

Although rock samples brought back from the moon have not provided any explanation for the formation of the solar system, they have nonetheless proved very informative. The surface of the moon is covered by soil and rocks formed by continuous meteor bombardment. The dusty soil that covers the entire surface of the moon to depths of up to 20 meters sticks to everything, and seriously interfered with many of the instruments carried by the astronauts. Microscopic glass spheres, formed by local melting caused by meteor impacts, are mixed with the soil. Farther away from the point of impact the soil is welded into a light breccia. Still farther away, the impact has shattered existing rock into fragments.

Most of the lunar samples have been soil and small pebbles from the surface. But since the constant meteoric bombardment churns up the surface of the moon, some of this material must have come from depths of at least several meters. One exception to this is a sample brought back from the famous Split Rock, a rock which some-

Astronaut Harrison H. Schmidt, now Senator from New Mexico, stands beside the famous Split Rock during the Apollo 17 mission to the moon's surface.

how became dislodged from its hilltop location, rolled down the slope, and split into several parts. This enabled one of the astronauts to get a sample from inside a large rock.

moon rocks

There are many kinds of moon rocks, including several different types of basalt. All of them are igneous, and were formed without contact with water. None of them contains any sign of fossilized life forms.

Mare basalt is a dark volcanic rock found in the maria. It contains more iron and less sodium than most earth basalts. A type of breccia known as **KREEP norite** is found in the highlands. It is named for its high abundance of potassium (K), rare earth elements (REE), and phosphorus (P). It has less iron and magnesium than earth rocks. KREEP norite is believed to be older than the mare basalts, and may have been produced by partial melting of the lunar surface.

The most abundant variety of moon rock—anorthosite—is light-colored and coarse-grained, consisting mostly of large plagioclase feldspar crystals. Although this is the most common rock on the moon, it is rather rare on earth, and is usually found in regions where fractional crystallization of magma has occurred. This suggests that large parts of the moon were once molten, a possibility that is difficult to accept. If such melting had taken place, the peculiar shape of the moon would have been smoothed out, and it would be more nearly spherical.

What little we know about the moon's interior comes from experiments left on its surface by the Apollo astronauts. The density of the moon is 3.134 times that of water, or a little more than half the density of the earth. Lunar density as a whole is only slightly greater than the density of the rock samples from the surface. It is possible that the entire moon is made up of material similar to that at the surface, but the higher overall density may simply be due to compression of the material at great depths.

Seismic stations set up by each of the Apollo landing missions are still reporting, and their reports indicate that the moon is seismically very quiet. Occasionally a meteor strikes the surface, or a crater slumps. Other than this, most of the seismic activity occurs in one region, 800 km below the surface. It appears to be triggered by the tidal effect of the earth, since most moonquakes occur when the moon is at its closest to the earth. Each moonquake releases very little energy, only about a billionth as much as an earthquake.

Since moonquakes are so feeble they are not of much use in probing the moon's interior. NASA scientists have tried to compensate for this by crashing spent spacecraft onto the surface of the moon and analyzing the transmitted vibrations received at the seismic stations. But this method cannot be used to study the deep interior. In fact, there seemed to be no way of doing this till fate lent a hand.

In 1972 a one-ton meteor crashed into the far side of the moon, causing vibrations which traveled through the moon and were picked up by seismic stations on the near side. The S-waves that should have passed through the central regions of the moon were apparently damped out, an indication that the moon has a small,

Photomicrograph of lunar rock containing plagioclase, pyroxine, and black ilmenite.

the interior of the moon

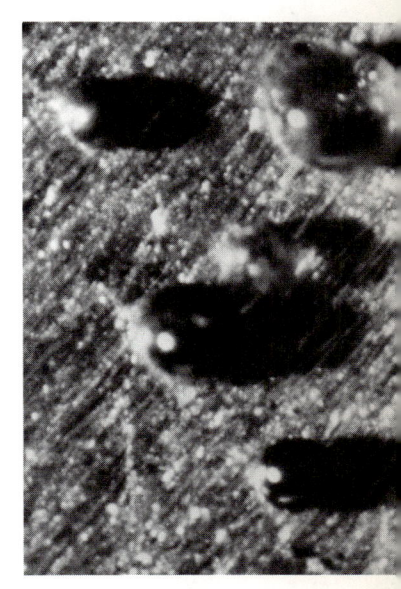

Tiny glass spheres 1/50 cm in diameter in lunar soil sample.

the solar system

plastic core. Further analysis showed that the core is surrounded by a deep, strong mantle. Above the mantle is a crustal layer about 60 km deep.

the origin of the moon

As we noted earlier, lunar exploration has failed to explain the origin of the earth–moon system. Many theories have been advanced and later disproved. It is still not possible to select among the theories that remain, but it seems likely that the earth and the moon formed from the matter left over after formation of the sun. (We will discuss the sun's formation in detail in chapter 16.)

According to this theory, the formation of the earth and the moon did not use up all the material in their region of the solar system, but left material in the form of large boulders. As we have seen, these boulders crashed onto the surfaces of both the earth and the moon, a bombardment which continued for hundreds of millions of years. It is possible that a little less than four billion years ago the moon passed close to the earth. Their mutual gravitational attraction would have ravaged both their surfaces, but the moon would have been more strongly affected by tidal distortions. Perhaps, at this time, parts of the moon were torn away and later fell back, creating the mare basins. Tidal interaction made the moon's orbit larger and less elliptical. The earth's rotation gradually slowed, and eventually the earth–moon system settled into its present form.

visitors from space

If we watch the sky on a clear, dark night we often see the moving points of lights called "falling stars" or "shooting stars." In reality, they are not stars at all. They are meteors, small solid bodies that collide with the earth's atmosphere and are vaporized as a result. The light that we see comes from the hot vapor. Larger meteors that manage to pass through the atmosphere and reach the earth are called **meteorites.**

meteors

Most of the meteors we see are no larger than a grain of sand. A pea-sized body can produce an exceptionally bright meteor. These tiny objects collide with the atmosphere at velocities up to 72 km/sec. The resulting friction first heats and then vaporizes them. Most incoming meteors have been completely vaporized by the time they reach an altitude of 60 km. Careful watchers can detect luminous trails that linger for seconds after the passage of a meteor.

By analyzing the light emitted by meteors we can determine their composition. It varies from almost pure metal—iron rich in nickel—to entirely stony material. The highly metallic kind are called **iron meteorites,** and the purely stony kind are known as **stony mete-**

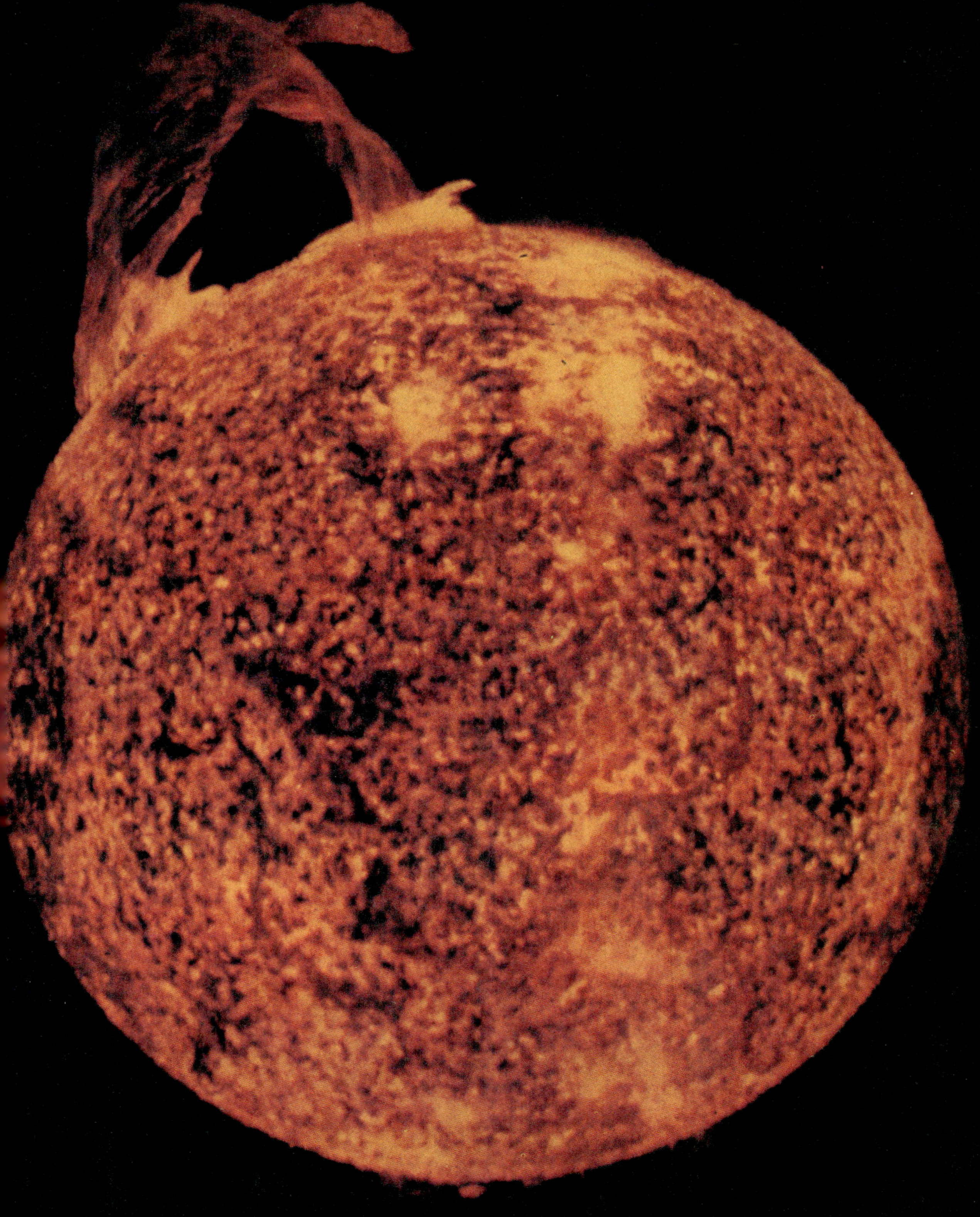

ASTRONOMY

Excellent color pictures that have become available only in recent years have added to our delight in astronomy. Visiting spacecraft have transmitted color pictures of Mars and Jupiter back to us; in 1979, Pioneer 11 should send us spectacular views of Saturn. Meanwhile, special techniques have solved the problems encountered when ordinary color plates are used to photograph extremely faint astronomical objects. Most astronomical objects are far too faint to show color when observed visually with a telescope, and it is especially exciting to have color pictures of these objects. Besides their beauty, color photographs can be quite instructive; for example, two populations of stars are clearly distinguishable in the color photograph of the Galaxy in Andromeda, as they never would have been in a black and white photograph.

The Ring Nebula is the best known planetary nebula. Astronomers believe that giant stars lose their outer layers in the process of becoming white dwarfs. The resulting planetary nebula is an expanding shell of gas exited into radiating visible light by the invisible ultraviolet light from the extremely hot dwarf star in the center. The red light is from hydrogen and ionized nitrogen.

This gigantic prominence, extending half a million km from the sun's surface, was photographed on December 19, 1973, from Skylab in earth orbit above the obscuring atmosphere. The photograph, made in the ultraviolet light of ionized helium, shows the intricate veiled structure that results from the complex interactions between the gas, the sun's gravity, and the local magnetic field.

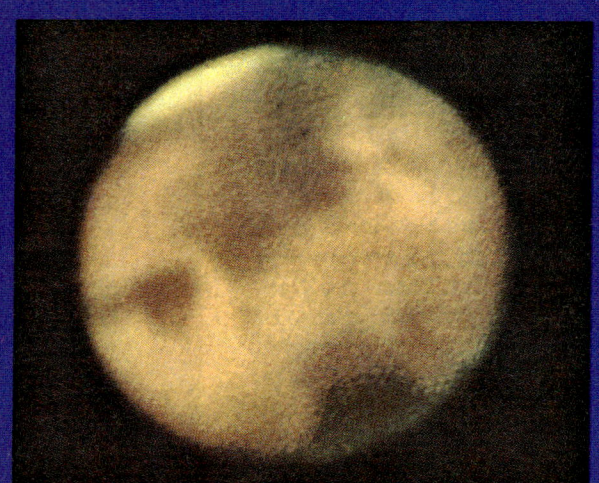

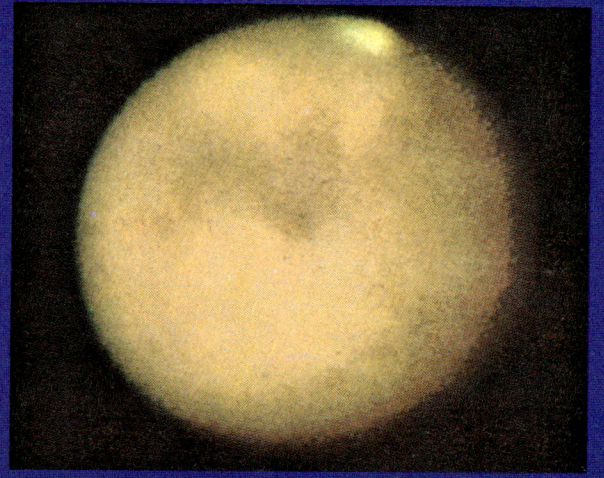

Photographs from earth clearly show the seasonal changes that occur on the Red Planet. Notice how the ice cap changes.

The Great Red Spot on Jupiter, imaged by Pioneer 11 on December 2, 1974 as the planet's powerful gravity flung the fragile American spacecraft towards its 1979 rendezvous with Saturn. Each of the smaller spots seen in the picture is a storm larger than the earth. Radio waves speeding at 300,000 km per second took more than half an hour to bring this picture to earth.

Beyond the sampler scoop of Viking Lander 1 is the red-orange surface of Mars strewn with a variety of angular rocks. The rocks are coarse-grained; most are light-colored, but there are some dark ones. Large blocks, the size of a human, can be seen on the horizon where the red soil meets the strangely pink sky. The ground near the bottom of the picture has been seared by the lander's rocket blast during landing.

Astronaut Owen K. Garriot undertakes a "space walk" outside Skylab 3 to fit a solar shield to help shade the Orbiting Workshop. NASA's Skylab was launched in May 1973. It carries several experiments; more than fifty major research projects have been performed by the three three-manned crews that have visited it. The astronauts have been able to photograph objects using wavelengths that do not penetrate the earth's atmosphere. The ultraviolet photograph of the sun on the first page of this color insert is an example.

The Great Nebula in Orion is a vast cloud of gas. The ultraviolet radiation from the hot, newly-formed stars excites the gas into emitting visible light. Star formation is believed to be going on in this region. Infrared telescopes have revealed what appear to be protostars. Radio telescopes have detected clouds containing various molecules.

The Trifid Nebula in Sagittarius contains hot young stars, excited gas, and dust. The intense light from the young stars exerts sufficient pressure to push the gas outward. The dark lanes are cooler clouds, rich in dust. The light in the red region is emitted by ionized gas; in the blue region light from the embedded star is being scattered by grains of interstellar dust.

The Great Galaxy in Andromeda is a spiral galaxy that is slightly larger and more massive than our own. Notice how the nuclear bulge containing older, reddish stars differs in color from the spiral arms containing hot, blue, relatively young stars. The two satellite galaxies are small ellipticals. Light from this galaxy takes nearly 2 million years to reach us.

orites. The intermediate varieties, in which masses of stone are imbedded in metal or masses of metal in stone, are called **stony–iron meteorites.**

The composition of the stony parts of meteorites is just about the same as that of some of the basic igneous rocks found on earth — olivine and pyroxene. But a few very odd stony meteorites appear as blackish granular masses, rich in carbon compounds. There has been a recent controversy over these meteorites. Some scientists believe that they are bits of sedimentary rock, and that the carbon compounds were derived from life on another planet.

Most meteors appear to have orbits similar to those of comets, and may, in fact, be debris from disintegrating comets. This theory is supported by the observation that when the earth passes through the orbit of some comets, there is increased meteoric activity. For example, the earth encounters the orbit of Biela's comet every October 20. On that night, large numbers of meteors appear to move away from a point in the constellation Orion. Such **meteor showers** can be observed on other dates throughout the year. During an average shower the observer can see about 25 meteors every hour, but sometimes unusually heavy showers produce hundreds of thousands of meteors during the same period. By contrast, the normal non-shower rate is 7 meteors every hour.

While some meteors may be cometary in origin, others, called **fireballs,** are not. These have orbits similar to those of asteroids. They enter the atmosphere at relatively low velocities and burn up at low altitudes. A fireball can be luminous enough to make the night sky as bright as day. Some fireballs explode with a crackling noise as

While Josef Klepsta was photographing the Andromeda Galaxy at the Prague Observatory on September 12, 1923, a bright meteor, usually called a "fireball" or a "bolide," crossed the field of the camera, resulting in this unusual picture.

meteorites

they pass overhead. Those that survive their passage through the atmosphere finally strike the earth.

Where do meteorites come from? Two theories have been accepted by most scientists. First, most meteorites originate in our own solar system. Second, they probably have much in common with the asteroids that lie between Mars and Jupiter.

A meteorite cut and polished.

It has been suggested that one kind of stony–iron meteorites, known as **chondrites,** may be small masses of solar system material that never accumulated to form a planet. Chondrites contain tiny, rounded pieces of olivine, pyroxene, or both, embedded in an iron–nickel mass. Other scientists believe that meteorites are fragments of an exploded planet that once orbited the sun between Mars and Jupiter. Radioactive dating has determined that most meteorites are about 4.6 billion years old, although some are younger. Recently, several scientists have suggested that meteorites of different ages each resulted from the disintegration of a different asteroid at a different time.

meteor craters

There are probably many meteor craters on the earth's surface. In the United States a small group of craters can be found at Odessa, Texas. Many large and small craters have been discovered in Canada, by studying aerial photographs used to map the country. Perhaps the most famous of these is a very large crater at the northwestern tip of Labrador. The best-known crater in North America, however, lies about 32 kilometers west of Winslow, Arizona.

The crater near Winslow looks like a low, flat-topped hill from a distance. But instead of a flat summit there is a hole about 300 meters deep and 1,300 meters in diameter, deep enough to hide a 60-story building. This curious depression was first brought to the attention of the scientific world early in the twentieth century. For several decades there were many arguments about its origin. Close examination, however, reveals that the crater was formed by an explosion caused by a collision between the earth and a meteor. It is now known as Meteor Crater.

Measurements of gravity in Meteor Crater indicate that far below the surface there is a mass of at least 3 million tons — possibly a sphere of iron more than 150 meters in diameter. Smaller pieces of

the solar system

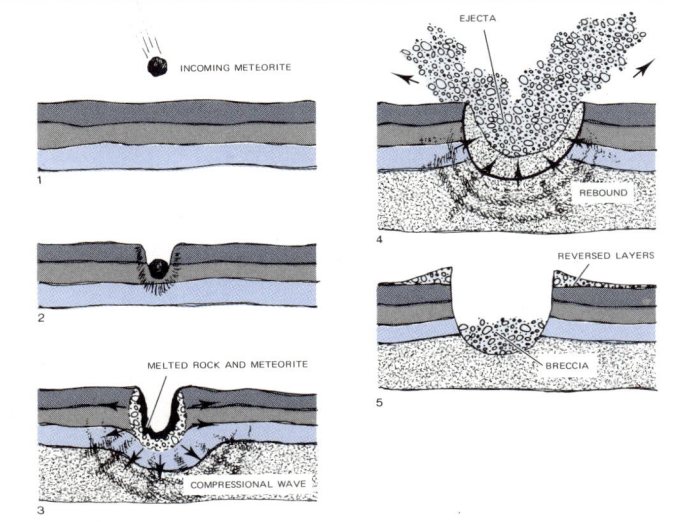

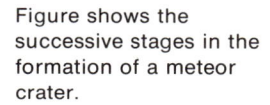

Figure shows the successive stages in the formation of a meteor crater.

The Barringer meteor crater in Arizona is about a mile across.

iron are scattered about the vicinity of the crater. These show the erosion characteristic of meteorites that have passed through the atmosphere.

To conclude our discussion of visitors from space, we should mention a spectacular event which occurred early in this century. On June 30, 1908, an immense explosion took place over northern Siberia, between the Lena and Yenisey rivers. The explosion blew down the pine forest for 65 km around, and was accompanied by a great flash of light. A year later, when the site was studied by scientists, depressions up to 50 meters in diameter were found.

Careful search failed to reveal any meteoric material. To this day, no one is really certain what caused this catastrophe, but there is a clue. At the time of the explosion a very small comet was close to the earth, and this comet has since disappeared. Some scientists therefore believe that the explosion occurred when the head of the comet entered the earth's atmosphere.

the solar system

SUMMARY

The solar system includes the sun, the planets, their satellites, asteroids, comets, and meteors. The sun contains most of the matter in the solar system. The combined mass of the planets is only about 1/100 that of the sun.

Distances to planets are measured by triangulation. This involves observing the planet from two different points that are well separated. A more recent method of finding distances uses radar.

The sun is a gigantic ball of gas, of which we see only the three outer layers: the photosphere, the chromosphere, and the corona. The surface of the sun has various observable features including sunspots, flares, and prominences. The gravitational attraction of the sun keeps the planets in their orbits, following Newton's laws. Bode's law states the relationship between the sizes of the planetary orbits. The Roche limit determines the closest point a satellite can approach its mother planet without being torn apart by tidal forces.

The terrestrial planets include Mercury, Venus, and Mars. Mercury, closest to the sun, has almost no atmosphere and is cratered like the moon. Venus resembles the earth in mass and density but has an atmosphere 100 times more dense. Mars is only about half the size of the earth, and has an extremely thin atmosphere.

Beyond Mars are the Jovian planets, composed mostly of hydrogen and helium. Jupiter is the largest, with a banded appearance and a strong magnetic field. Saturn and Uranus are surrounded by rings consisting of chunks of ice. Uranus and Neptune are very similar to each other. Pluto, the outermost planet, may once have been a satellite of Neptune.

The moon is earth's only satellite. Its surface consists of the older highlands and the more recent maria. Craters caused by meteor bombardment cover the moon. Information obtained from the small amount of seismic activity on the moon indicates that it has a crust and a partially molten core.

Meteors are small bodies that enter the earth's atmosphere and burn up due to the resulting friction. They are called meteorites if they strike the earth. Most meteorites are stony, but a few are almost pure metal.

the solar system

REVIEW QUESTIONS

1. What are the various types of objects that make up the solar system?
2. Describe the three layers of the sun that can be observed from the earth. How can we observe the corona of the sun?
3. How do we know that the sun rotates? How do we know that it is not a solid?
4. What are sunspots, and how do they affect the earth?
5. What forms of energetic activity can be observed at the surface of the sun? How do they affect the earth?
6. How does the tidal action of the moon affect the earth? What effect does the tidal action of the earth have on the moon?
7. How was the rotation period of Mercury discovered?
8. Why is the surface of Mars red? How does its surface differ from that of the earth?
9. Why do we think that Mars once had flowing water? What might have happened to it?
10. What is the Great Red Spot of Jupiter?
11. Why do some astronomers think that Pluto may once have been a satellite of Neptune?
12. How does the far side of the moon differ from the near side?
13. Why is nearly half the moon's surface hidden from earth-bound observation?
14. How were the craters on the moon formed? Why were similar craters not formed on the earth?
15. How did meteors originate?
16. Most meteors are of the stony variety. Why do most of the meteorites in museums belong to the metallic type?
17. Describe the three basic types of moon rocks.
18. How can we study the interior of the moon? What chance event led to our increased understanding of the moon's interior?
19. Why is a total solar eclipse so much more interesting than a partial solar eclipse?

CHAPTER SIXTEEN

THE UNIVERSE AND BEYOND

Fascinated, the time travelers watched the scene unfolding before them. Millions of miles away, but clear in the ship's viewer, the vast cloud of glowing gas spun ceaselessly in the black depths of space. As it turned, its center began to bulge, then spread out into a broad, flat disc. The disc separated into rings; the rings condensed into globules revolving around the great central mass. From their ship, the time travelers watched it all, shortening billions of years into a few hours of experience by an occasional touch on the time levers of their control panel. At last they switched off the viewer and looked in silence at each other. No one seemed to be in a hurry to talk—not after actually watching the birth of a solar system.

Fantasy, of course. It will be a long time yet—if ever—before human beings can really watch such things in such a way. And yet, in a sense, it may be that we can, even now, watch the birth of worlds. In June, 1977, astronomers of the National Aeronautics and Space Administration announced the discovery of a star that seems to be in the process of forming planets.

The star, known by the unromantic name of MWC 349, is very young, probably no more than a thousand years old. It is about ten times as large as our sun, and far more massive. It is surrounded by a wide, intensely glowing disc, somewhat like the rings of Saturn but far larger. The material in the disc seems to be steadily spiraling inward toward the central star, so the disc will probably be altogether gone in another century. But astronomers believe that some of the disc material may be condensing into planets. If so, then, in a sense, the star provides us with a sort of real-life time machine. We cannot go back and look at the beginnings of our own solar system, but we can look at the beginnings of another one that must be, in many ways, very like our own.

Unfortunately, this new baby solar system will probably

the universe and beyond

not live to grow up. The larger and more massive a star is, the faster it burns out. This one is slated to last only about a hundred million years, compared to our own sun's probable lifetime of ten billion years. This won't even give its planets time to solidify and cool down, let alone evolve any life forms. So our remote descendants will never be signing up for spaceship tours to the exotic worlds of MWC 349, with their soaring rosy mountains, their beautiful golden trees, and their charming, fairy-like inhabitants. Oh, well — better luck next time.

observing stars

Throughout human history people have been intrigued by the apparently unchanging patterns of stars in the sky, and have given special names to various groups of stars. In the Northern Hemisphere these **constellations** are generally named after mythical gods and heroes — Orion the hunter, Aquarius the water bearer, and Pegasus the winged horse. In the Southern Hemisphere, the names of the constellations reflect the interests of northern sailors who named them — Telescopium, Sextans, and Volans the flying fish. Modern astronomers have assigned entire regions of the sky to each constellation, so that every star belongs to one constellation or another.

The rotation of the earth makes the sky appear to rotate. As a consequence, stars rise and set. Only the constellations above the horizon can be seen at any given time, and some constellations in the southern sky can never be seen by northern observers, because they do not rise above our horizon. Furthermore, as the earth travels in its orbit the sun appears to lie in different constellations. These special constellations make up the **Zodiac.** When the sun is in one constellation of the Zodiac, that constellation and others nearby are above the horizon only when the sun is up, and so cannot be observed.

The stars of a constellation are seldom related to each other. We see them as patterns only because they lie in the same general direction from the earth. One star of a constellation may be relatively close to the earth and another very distant. Observers of the constellations over thousands of years have asked the same questions over and over again. What is a star? Why does it shine? Now we are able to answer this question, and the answer is partly based on our understanding of the observable properties of stars.

A fanciful drawing of the constellation Orion from an old star atlas.

brightness

Even to the naked eye some stars are obviously brighter than others. In a scientific context, the term brightness has a very precise meaning. The brightness of a star is the amount of its light that falls

the universe and beyond

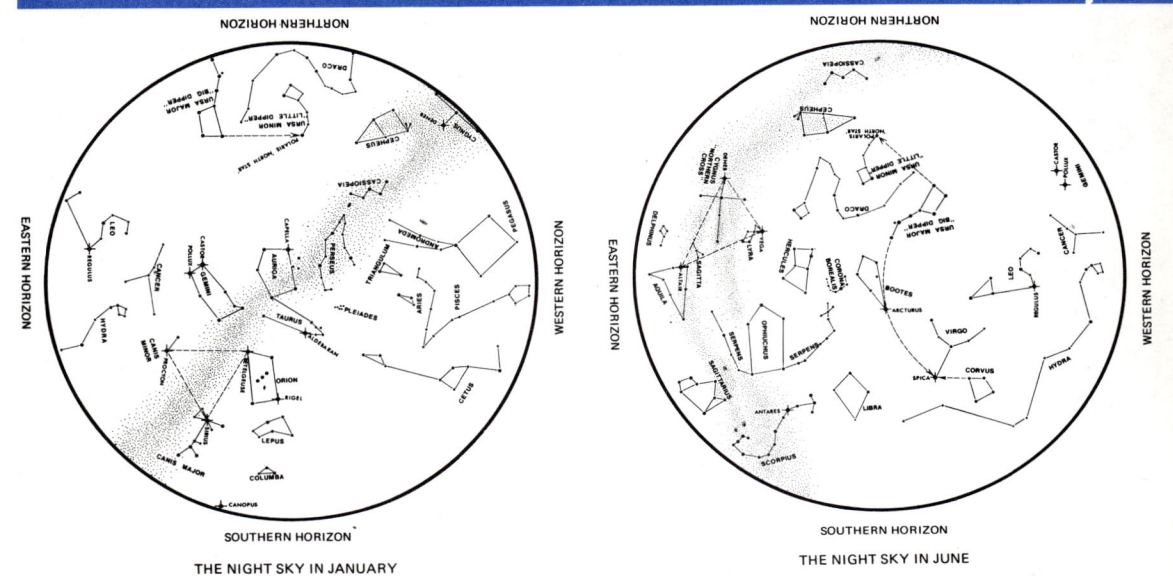

THE NIGHT SKY IN JANUARY

THE NIGHT SKY IN JUNE

The sky in January and in June. The maps are correctly oriented when held above the head, facing down.

every second on one square centimeter of the earth's surface. A star appears brighter than another star if more of its light falls on this unit of the earth than does the light from the second star.

Brightness alone does not tell us much about a star. We know, for example, that a lamp may appear bright either because it is close to us or because it is very powerful. Is a star bright because it is very near, or because it is radiating a lot of light? To answer this question, we must know both the star's distance from us and the way in which distance affects brightness.

measuring stellar distances

The obvious way to determine the distance to a star would seem to be by triangulation—the same method we used for the moon in chapter 15. However, a difficulty arises here. When we looked at the moon from two different points on earth, it appeared in different positions relative to certain stars. This difference in position of an object seen from different angles is known as the object's **parallax**. But stars are too far away to show a measurable parallax when observed from even the most widely separated points on earth—a maximum separation of about 13,000 km. A wider separation is needed, and we can find it in the earth's orbit, which has a diameter of about 300 million km. By observing a star when the earth is at different points in its orbit, we can achieve a wide separation of viewpoints.

Even with this amount of separation, the shift in a star's direction is very small—less than a thousandth of a degree. Astronomers usually call *half* the change in angle the parallax of the star. Such small angles are measured in seconds of arc (written ''). One degree is equal to 3,600''.

Mathematical calculation shows that if the parallax of a star

the universe and beyond

were 1″ the star would be about 30 trillion km away. Astronomers call this distance a **parsec (pc)**, derived from "**par**allax **sec**ond." The smaller the parallax, the farther away is the star. If the parallax is ½″ the distance is 2 pc. If the parallax is ⅕″ the distance is 5 pc, and so on. Light, which travels a distance equal to seven times around the earth in a single second, takes about 3¼ years to travel 1 pc. The distance light travels in a year is called a **light year (ly)**. A parsec is therefore equal to 3¼ ly.

To understand the immense distances between stars, consider that our nearest star neighbor, Alpha Centauri, is 1.3 pc away. Most stars, however, are more than 200 pc away, and their parallaxes are too small to be measured. At 200 pc the parallax is only ¹⁄₂₀₀″, the limit of measurement. Determining the distance to a star farther away than 200 pc requires more sophisticated methods than parallax.

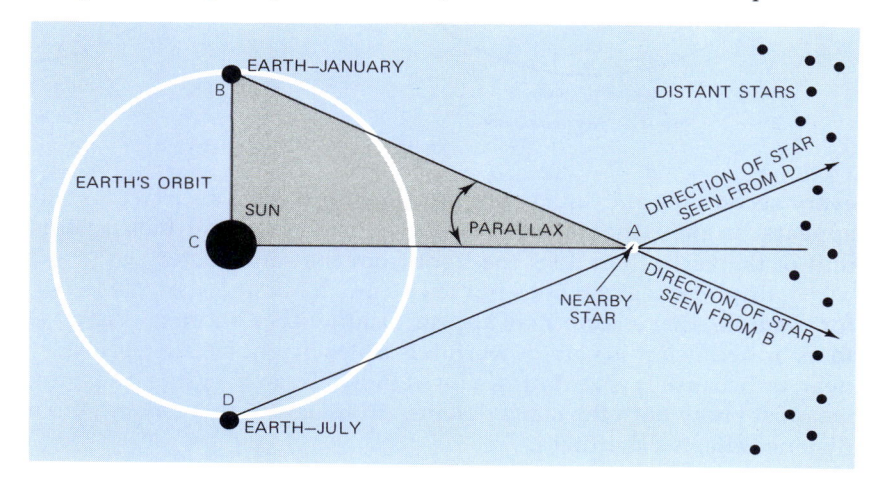

The nearby Beehive Cluster of stars in the constellation Cancer can be observed with binoculars.

The distance to a star is determined by measuring its parallax, using the diameter of the earth's orbit as the baseline. This method works only for the nearest stars.

Table 16–1. The nearest stars.

Star	Distance (parsecs)	Luminosity (sun = 100)
Proxima Centauri	1.31	.005
Alpha Centauri A	1.35	150.
Alpha Centauri B	1.35	40.
Barnard's Star	1.81	.044
Wolf 359	2.35	.0017
Lalande 21185	2.52	.53
Sirius A	2.65	2300.
Sirius B	2.65	.19
Luyten 726-8A	2.72	.0064
Luyten 726-8B	2.72	.004
Ross 154	2.9	.04
Ross 248	3.16	.01
Epsilon Eridani	3.3	27.

luminosity

One of the things we know about light is how distance affects the brightness of a light source. To illustrate: if the distance to a star were doubled, its brightness would be reduced $2^2 (= 4)$ times. If the distance were increased by n times, the brightness would be decreased n^2 times. Thus, if we know the distance to a star we can allow for the decrease in brightness, and can calculate the star's light output per second, or **luminosity**. When we do this, we find that stars are rather like the sun. Some are ten thousand times fainter, and others are brighter by the same factor, but most of them are enough like the sun that we can consider them the same type of object. Stars, then, are suns—or better yet, the sun is a star. Knowing this, we can use our detailed knowledge of the sun to help us understand stars.

colors of stars

Stars differ in color as well as in brightness, but only the brighter stars have noticeable color. Even then, "blue" stars are only slightly bluish and "red" stars are only slightly reddish. One reason for this is that even bright stars are faint by ordinary standards, and the human eye is not able to distinguish the color of faint light.

One way of measuring a star's color precisely is to photograph the star twice, first using a photographic plate that responds to blue light and then with a plate that responds to yellow light. If a star is blue, the photographic image on the first plate will be intense and the image on the second plate will be faint. If the star is yellow or red, the reverse will be true. By carefully comparing the images on the two plates, we can establish the exact color of the star.

But what does the color of a star tell us? First, we must understand how hot bodies radiate light. As the temperature of an object is increased, it first glows a deep red and then, progressively, it glows orange, yellow, white, and blue. At every temperature all wavelengths, or colors, of light are emitted, but most radiation occurs at one particular wavelength. In a relatively cool body, most radiation is emitted at wavelengths corresponding to red. In very hot bodies, most radiation is emitted at a wavelength corresponding to blue.

This gives us a basis for determining the temperatures of stars. Red stars, for example, are obviously rather cool. Judging from their color, their surface temperature is about 3,000K. Yellow stars, such as our sun, have surface temperatures around 6,000K. White stars, such as Sirius, have temperatures of about 10,000K, and some blue stars exceed 30,000K.

Still another property of a hot body enables us to determine the size of a star. Each square centimeter of a hotter body emits much more radiation than each cm^2 of a cooler body. If we raise the temperature of a body, not only does its color change but also it radiates more light.

Now, suppose we apply this to stars. If two stars have the same luminosity, but one is blue and the other is red, which star is

the universe and beyond

Cool stars radiate much less light per square cm than hot stars. Therefore stars of the same luminosity but different surface temperatures will differ in size. A star whose temperature is 3,000K will be about 16 times larger than a star of 12,000K.

larger? The red star is obviously cooler than the blue star, and so each cm² of its surface must radiate less light than each cm² of the blue star. But since their luminosities are the same both stars are radiating equal amounts of light. It is clear that the red star's surface must con-

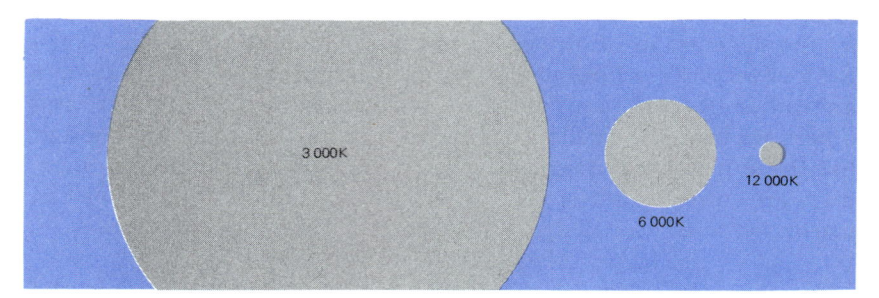

Table 16–2. The sizes of stars.

Star	Diameter (Sun = 1)	Type
Betelgeuse (variable)	500 750	Supergiant
Aldebaran	45	Giant
Arcturus	23	Giant
Antares	640	Supergiant
Scheat	110	Giant
Ras Algethi	500	Supergiant
Mira	420	Supergiant
Canopus	39.0	Giant
Sirius A	1.76	M-S
Vega	3.03	M-S
Altair	1.65	M-S subgiant
Procyon	2.15	M-S subgiant
Fomalhaut	1.56	M-S
Regulus	3.8	M-S
Sirius B	.022	White dwarf
40 Eridanus B	.018	White dwarf
Van Maanen's star	.007	White dwarf

tain many more cm² to make up for the fact that each cm² radiates less light—in other words, the red star must be larger. Using mathematics, we can calculate the size of a star if we know its color and luminosity.

spectra of stars

We can learn even more about stars by breaking starlight up into its various colors, using a **spectrograph.** Light from a star is passed through a prism, which bends the shorter wavelengths more than the larger. This spread of colors, in order of wavelength, is called a **spectrum.** By examining a photograph of the spectrum, called a **spectrogram,** we can tell what wavelengths are present in that star's light.

Atoms radiate or absorb only certain specific wavelengths of light. The spectra of most stars show evidence of such absorption. When a certain wavelength has been absorbed, that part of the spectrum looks like a dark line, known as an **absorption line.** Every element produces a characteristic pattern of absorption lines.

The presence of strong absorption lines of an element does not necessarily mean that that element is particularly abundant in the star. The ability of atoms of a gas to absorb radiation depends on the temperature of the gas. It may happen that one star has a temperature just right for a certain element to absorb radiation, but in another star the temperature is wrong. For example, hydrogen is the most abundant element in most stars, but only those at around 10,000K show strong hydrogen lines.

Once the spectrum of a star has been interpreted we know both the star's temperature and its composition. Hydrogen makes up about 70 percent of the mass of most stars. The rest is made up of helium and small percentages of other elements.

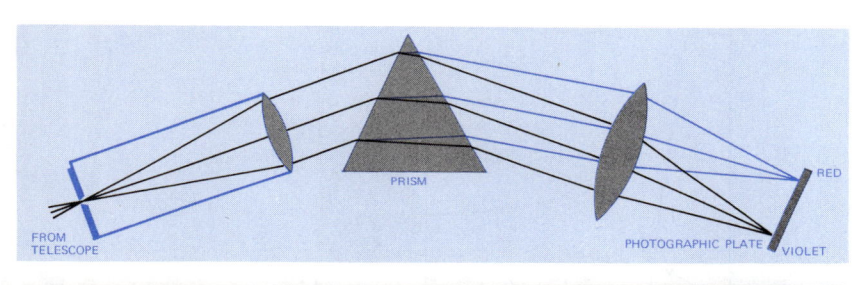

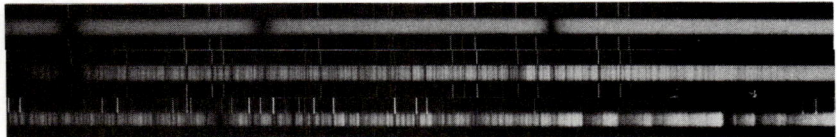

Drawing shows a simplified prism spectrograph. The glass prism separates the various colors of light. Below the drawing are spectra of a hot star, a star like the sun, and a cool star. The hot stars show the few but distinctive lines of hydrogen; the sun-like star shows metallic lines; the cool star has dark bands showing the presence of molecules.

masses of stars

The next step in understanding stars is to find out how massive they are. We can determine a star's mass by observing its gravitational effect on another body. Single stars are too far apart to affect each other measurably, but not all stars are single. In many cases, two stars are in orbit around each other. These are called **binary stars,** and more than half the stars may belong to binary systems. If Jupiter had been a hundred times larger than it is, it would have been a star and would have formed a binary system with the sun.

Applying Newton's laws, we can calculate the masses of binary stars by observing the sizes and periods of their orbits. Most stars have masses between 0.1 and 10 times the mass of the sun. The stars in a binary system are usually too close to be distinguished, and are seen as one star. In a few cases, however, the two stars periodically eclipse each other. By studying the way their light varies

the universe and beyond

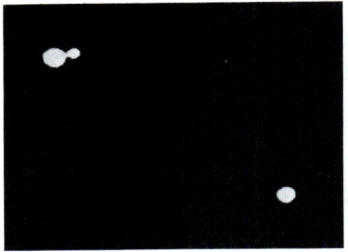

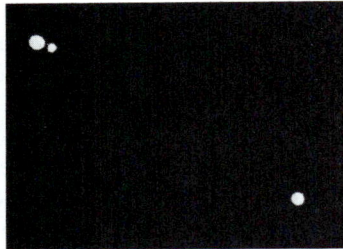

Above, photographs of Kruger 60, a binary star, covering a fourth of its 44-year period.

At left, a diagram of the eclipsing binary Algol. An eclipse occurs every 2.8673285 days when the bright star is eclipsed by the dim companion. When the dim star is eclipsed by the bright star there is only a small dip in the brightness, called the secondary eclipse.

Table 16–3. The range of properties of stars.

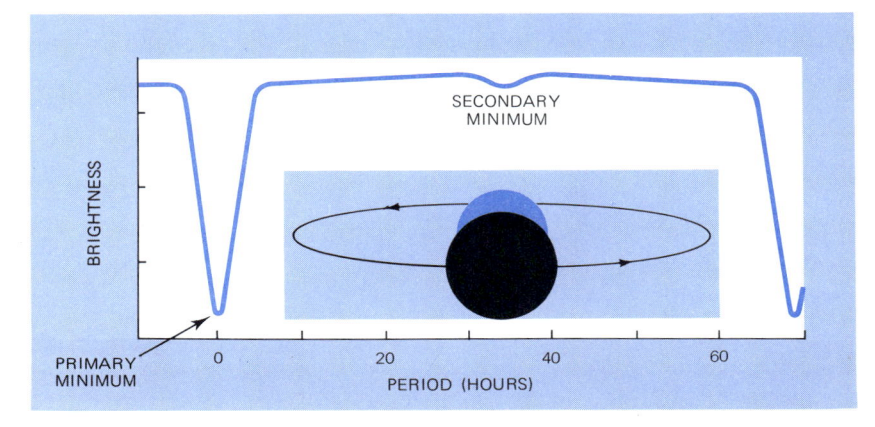

Property	Range
Luminosities	1/1000 – 10,000
Mass (Sun = 1)	1/10 – 10
Radii	1/100 – 500
Temperatures	3000K – 30,000K

during an eclipse, we can calculate the sizes of both stars in an eclipsing binary.

putting it all together

Once scientists have learned the properties of a group of objects, they next ask if there are any relationships among the properties. To find these relationships, they may plot a graph of one property against another. The accompanying figure shows the weights of students in a class plotted against their heights. The distinctive pat-

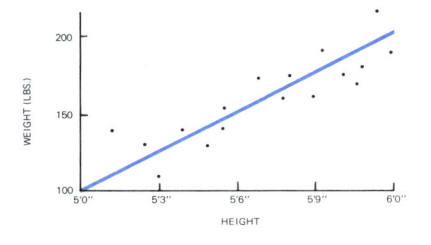

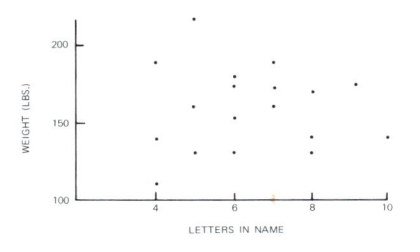

The weights of students plotted against their heights shows an obvious correlation. Their weights, as one may expect, are in no way related to the number of letters in their names.

tern shows that a student's weight is closely related to his height. (Of course, there are a few exceptions.)

In the accompanying figure a star's luminosity is plotted against its mass. We can see a clear relationship. A star as massive as the sun is as luminous as the sun, but a star $1/10$ as massive is only $1/1,000$ as luminous. The graph shows that a star 10 times as massive as the sun is 5,000 times as luminous.

This tells us something about the lifetimes of stars. Stars radiate energy, so they must have some energy source, or "fuel." We can assume that a more massive star has more of this fuel. Since luminosity is the rate at which the star radiates energy, it is thus the rate at which the fuel is being consumed. If we know how much fuel there is *and* the rate at which it is being consumed, we can calculate how long it will last.

Consider a star of 10 solar masses — that is, 10 times as massive as the sun. It has ten times as much fuel as the sun. But, as we saw above, it is using it up 5,000 times as fast as the sun, so its lifetime can only be $1/500$ that of the sun. In general, more massive stars use up their fuel and burn out more rapidly.

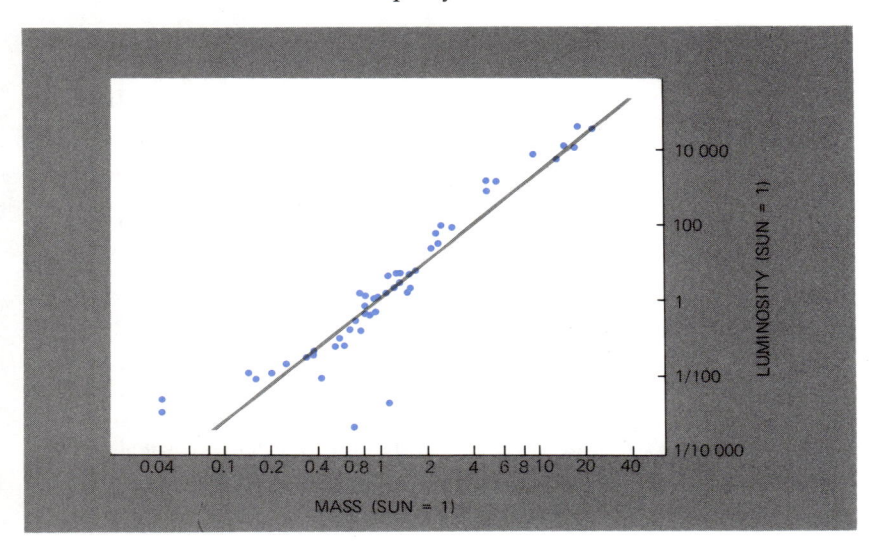

mass–luminosity relationship

The mass–luminosity relationship is obeyed by 90% of the stars. Stars of greater mass have much greater luminosity.

Hertzsprung–Russell diagram

The next step is to plot the luminosity against the temperature (or color). This forms a **Hertzsprung–Russell (H–R)** diagram, named for the scientists who first plotted this relationship. In the accompanying H–R diagram, we can see that stars are distributed in a definite pattern. About 90 percent of the stars fall roughly along a diagonal, called the **main sequence (m-s).** The hotter, bluer stars are very luminous, and the cooler, redder stars are weakly luminous.

Most of the stars that are not m-s stars lie in the lower lefthand corner. They are hot, but weakly luminous, so they must be very

the universe and beyond

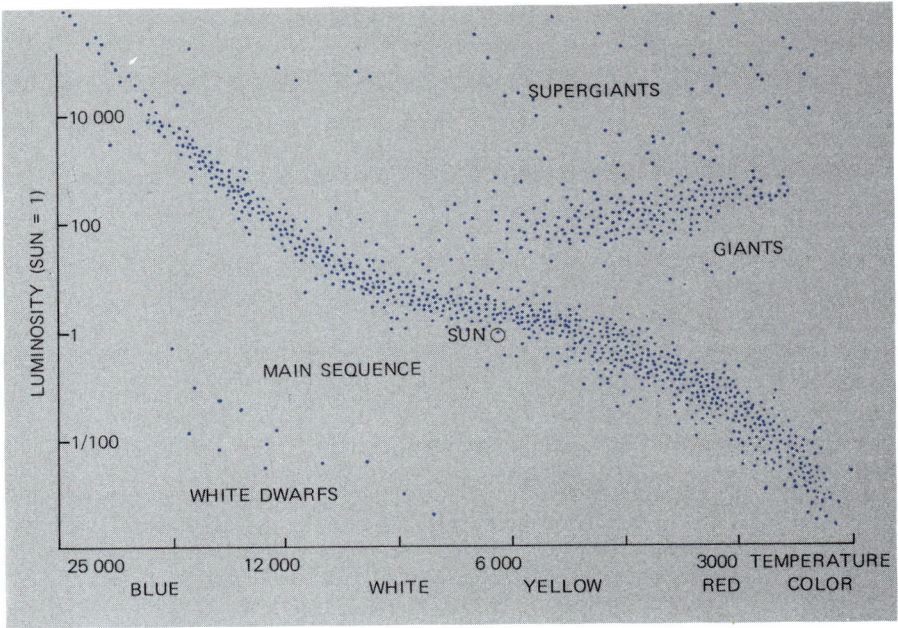

Above is the Hertzsprung–Russell diagram. The H–R diagram is a plot of the luminosity of a star against its color. On the diagrams stars separate themselves into general groups. The three main divisions are main sequence stars, giants, and white dwarfs.

Table 16–4. The brightest stars.

Star	Constellation	Distance (parsecs)	Luminosity (sun = 1)	Type
Sirius	Canis Major	2.7	24	M-S
Canopus	Carina	55	1,500	Giant
Alpha Centauri	Crux	1.3	1.5	M-S
Arcturus	Boötes	11	110	Giant
Vega	Lyra	8.1	55	M-S
Capella	Auriga	14	160	Giant
Rigel	Orion	250	18,000	Supergiant
Procyon	Canis Minor	3.5	7.2	M-S subgiant
Achernar	Eridanus	20	220	M-S
Beta Centauri	Centaurus	90	3,800	Giant
Altair	Aquila	5.1	11	M-S subgiant
Betelgeuse	Orion	150	14,000	Supergiant
Aldebaran	Taurus	16	100	Giant
Alpha Crucis	Crux	120	3,400	Sungiant
Spica	Virgo	80	2,400	M-S
Antares	Scorpio	120	5,400	Supergiant
Pollux	Gemini	12	41	Giant
Fomalhaut	Piscis Austrinus	7.0	14	M-S
Deneb	Cygnus	430	50,000	Supergiant
Beta Crucis	Crux	150	6,000	Subgiant

small. These are known as **white dwarfs.** A few stars lie in the upper right corner. They are cool and red but very luminous, so they must be very large. These are called **red giants.** Above them are even larger stars, the rare **supergiants.** In the next section we will look at the structure of some of these stars.

the universe and beyond

stellar structure and evolution

Stellar astronomy attempts to trace the lifetime of a star, from its formation to its extinction. This requires an understanding of the structure of stars. At first sight this seems like an impossible task, but in fact we do understand stars, perhaps better than any other astronomical objects. We can trace their evolution in detail. Although gaps exist in our knowledge, the gaps are being filled up.

There are two reasons for our spectacular success in this field. First, stars are gaseous; their internal activity produces observable effects on their external appearance. Second, stars are so numerous that we can observe various types at various stages of their lives. Our present theory of stellar structure combines the sciences of physics and astronomy. By understanding stellar structure, we can understand stellar evolution.

the energy of stars

In our discussion of stellar energy, we shall take our own sun as an example, since it is a typical m-s star. As we saw in chapter 15, the sun's interior is a hot gas, reaching a temperature of 15 million K at the center. Heat flows out from the sun, escaping into space at the surface.

The material of the sun obstructs radiation very effectively, because its temperature and density cause its gases to become opaque. The light energy inside the sun does not escape easily. It is absorbed and re-emitted after it has traveled only one centimeter. Energy produced at the center of the sun undergoes a trillion such scatterings, and thus takes about a million years to reach the surface. (If unobstructed it would have reached the surface in three seconds.) As a result, the sun has a large amount of energy. If its energy source were turned off, little change would be noticed on earth for thousands of years.

source of the sun's energy

The sun is the only star whose age we know directly. It appears to be about 5 billion years old, and its luminosity has remained fairly steady. Considering its great age, we must ask what energy source could have kept the sun going for such a long time.

The source of the sun's energy cannot be a chemical fuel, such as coal. Such fuels could keep the sun going for only a few thousand years. Gravitational energy has also been suggested. We know that water falling toward the center of the earth can be used to produce electrical energy. Similarly, the sun could contract—fall toward its center—and release energy. But calculations show that this could have kept it going for only about 50 million years.

the universe and beyond

Obviously, a better source of energy is needed, and we know that such a source is available. As we saw in chapter 1, matter can be converted into energy. The conversion of only one gram of matter can release enough energy to power a large city for a full day. On earth, the most spectacular example of such a conversion is the hydrogen bomb. The sun and the stars, in fact, are controlled hydrogen bombs, which do not go off all at once but burn for billions of years.

How does a star convert matter into energy? We know that the sun is mostly hydrogen, and that the temperature at its center reaches 15 million K. At this temperature, hydrogen nuclei travel at very high speeds, crashing into one another and combining to form heavier, more complex nuclei. The most common process converts four hydrogen nuclei into one helium nucleus. The four hydrogen nuclei together are 0.7 percent more massive than the helium nucleus thus formed. The surplus mass is converted into pure energy, which powers the sun and all m-s stars.

As the hydrogen "burns," the helium "ash" accumulates at the center. The structure of the star is altered. Eventually, it evolves into a red giant. Using the enormous computers that have now become common, we can trace the evolution of a star by calculating what happens when more helium accumulates in the center. We can now see how stars are formed and how they evolve.

formation of stars

Highly luminous stars are almost always found near vast clouds of interstellar dust and gas. As we have seen, such stars have very short lifetimes. They could not have formed more than a few million years ago, and so cannot have moved far from their birthplace. This suggests that there is a connection between interstellar clouds and the birth of stars. Perhaps all stars have formed out of such clouds.

It has been demonstrated that a large uniform cloud of gas does not remain uniform. It breaks up into clumps, curdling like sour milk. The clumps, about 0.1 ly across, then collapse under their own gravity. At first the collapse occurs rapidly, and the energy released

Diagram shows the successive stages in the formation of a protostar from the galactic gas.

the universe and beyond

escapes. Eventually the star-to-be, or **protostar,** becomes opaque, trapping the energy and heating up as a consequence. With this heating, the internal pressure increases and the collapse slows down greatly. The star now becomes visible, many times more luminous than the sun.

The energy lost by the star is made up from its gravitational energy. Because it is radiating so much energy, its internal pressure decreases and it again contracts slightly, heating up the interior until it reaches around 10 million K. At this point nuclear reactions begin, and hydrogen begins to be converted into helium. The star has tapped a rich source of energy. It no longer has to contract to meet its energy needs, and will not have to change for a long time. Stars of different masses settle down at different points in an H–R diagram, forming the main sequence. A star spends about 90 percent of its lifetime in the main sequence.

So far, our discussion has neglected the effects of rotation. We know that when a large, slowly rotating body contracts, its rotation speeds up. Most interstellar clouds rotate slowly, perhaps once in a million years. But they must contract to a millionth of their size to become stars, and this contraction greatly speeds up their rotation.

As rotation speeds up, the protostar flattens out. At some point, matter is flung out of the equatorial regions to form a disc. The material of the disc cools down and solidifies into small objects called **planetesimals,** which fall together to form the planets. As the planets form, they sweep up the rest of the planetesimals and gas. Satellites were quite possibly formed in the same way. As we noted in chapter 15, planetesimals that were not captured in this way are probably the asteroids and meteors that we observe today.

The size of a planet, according to this theory, depends on several things. It depends on the amount of matter available at the region where the planet formed. It also depends on the temperature of the region, since temperature strongly affects the way in which matter solidifies.

This theory of planetary formation predicts that there should be other planetary systems besides our own. But such systems cannot be observed directly. Their existence can only be theorized on the basis of the small effects they produce on the stars around which they orbit. Some observations suggest that three of the ten nearest stars have planetary companions.

As we noted earlier, an m-s star accumulates helium in its center. When this helium core has grown to 12 percent of the star's mass, the star begins to evolve rapidly. Its size increases a hundred times. Its surface cools, but its larger size makes it very luminous; it has become a red giant. For stars such as the sun, this phase occurs after an m-s stay of 10 billion years. Our sun has 5 billion years left before this happens. More massive stars, however, evolve much more rap-

formation of planetary systems

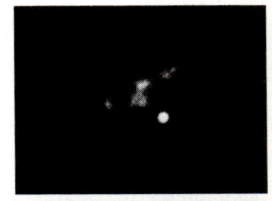

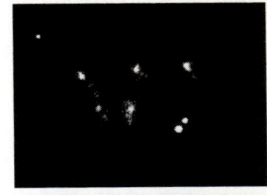

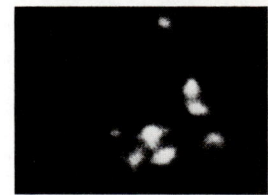

Herbig-Haro objects may be newly formed stars.

red giants

the universe and beyond

idly; if the sun were only 10 percent more massive, it would be preparing right now to become a red giant. The most massive stars stay on the main sequence only for a million years, and then evolve into supergiants.

As a star evolves into a giant its interior becomes hotter. When the temperature reaches 100 million K, helium begins to be transformed into carbon and oxygen. In the most massive stars the temperature can reach billions of degrees, and carbon and oxygen can "burn" to form heavier elements. This does not happen in stars like the sun.

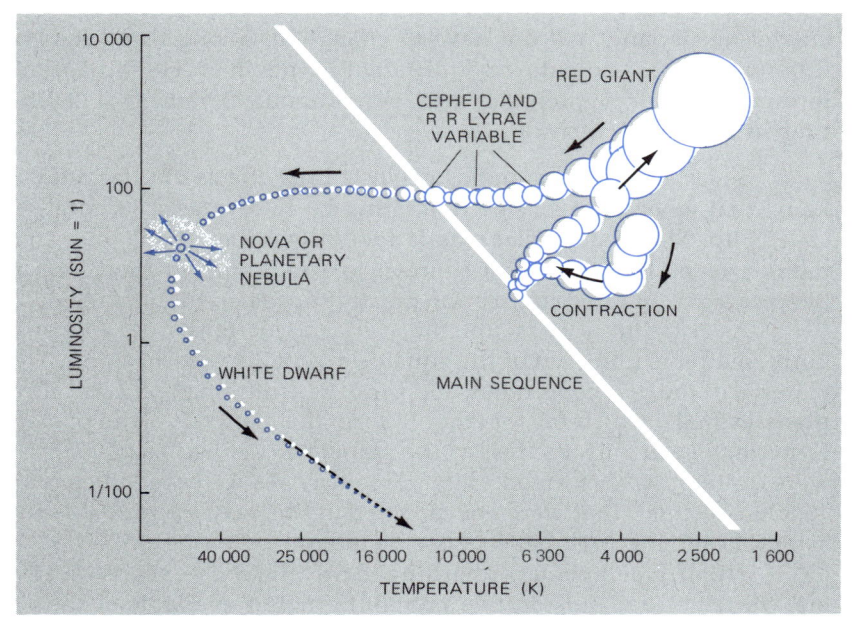

Evolutionary path of a star slightly more massive than the sun. During its giant phase the star is 100 times its main sequence size; it is 1/100 its main sequence size when it becomes a white dwarf. The diagram gives only a hint of these changes. Drawn to scale a red giant would be about a foot across.

white dwarfs

By the time helium begins to burn, a star is near the end of its life. It is now very luminous, and its energy budget is very high, but helium is not an efficient fuel. Burning 1 g of helium produces only $1/10$ as much energy as burning 1 g of hydrogen. Soon the star has lost its energy source; it begins to cool and contract.

A star without energy sources contracts until its density becomes about a million times that of water. Such stars are called **white dwarfs,** and are hot with stored energy. They cool off over a billion years, becoming cold **black dwarfs.**

The first white dwarf to be discovered was Sirius B, the faint companion of Sirius. It is about as massive as the sun, but its volume is a million times less. The matter in Sirius B is very dense; one teaspoon would weigh several tons. Other white dwarfs have since been discovered, and it is now believed that 10 percent of all stars in our galaxy may be white dwarfs.

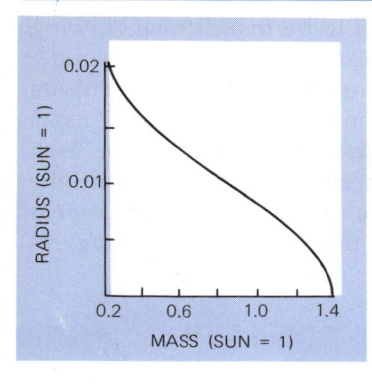

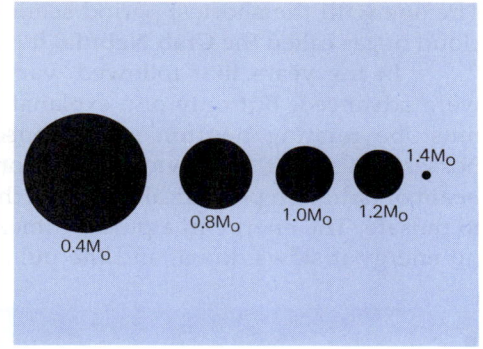

the universe and beyond

The more massive a white dwarf the smaller it is. Around about 1.4 solar mass the size of a white dwarf would tend to zero. White dwarfs of greater than this mass cannot exist.

neutron stars, pulsars, and supernovas

When matter becomes compressed to an even greater extent than in a white dwarf, it turns into a dense gas of neutrons. According to astronomical theory, such a **neutron star** would be about as massive as the sun, but it would be only about 30 km in diameter. A single teaspoon of matter from such a star would weigh a billion tons, more than 10,000 Empire State Buildings.

Neutron stars were objects of theory and speculation until 1969. Then researchers detected regular pulses of radio waves coming from fixed points in the sky. The sources of these are now called **pulsars**. At first, though, the waves were thought to be signs of extraterrestrial intelligence, and were nicknamed **LGMs** (for Little Green Men). Soon other pulsars were detected, pulsing at different rates.

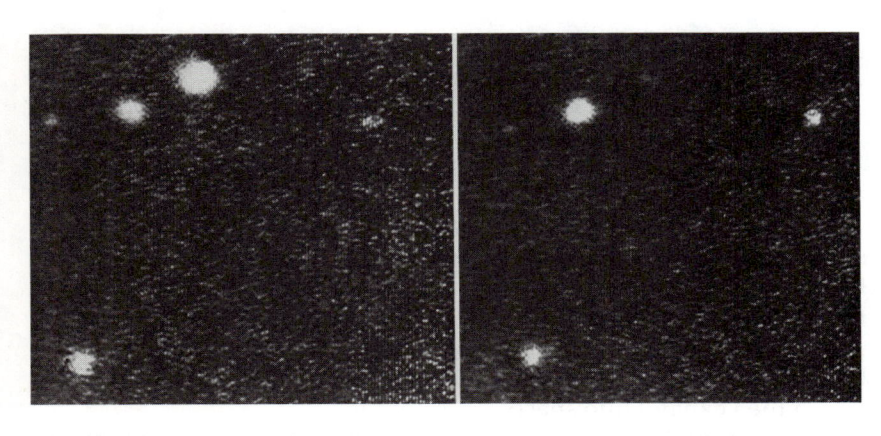

The pulsar in the Crab Nebula photographed when it is "on" and when it is "off". Below is the pulsar going through a cycle of changes.

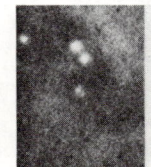

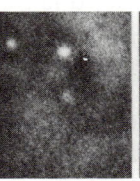

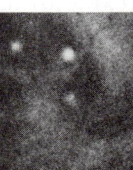

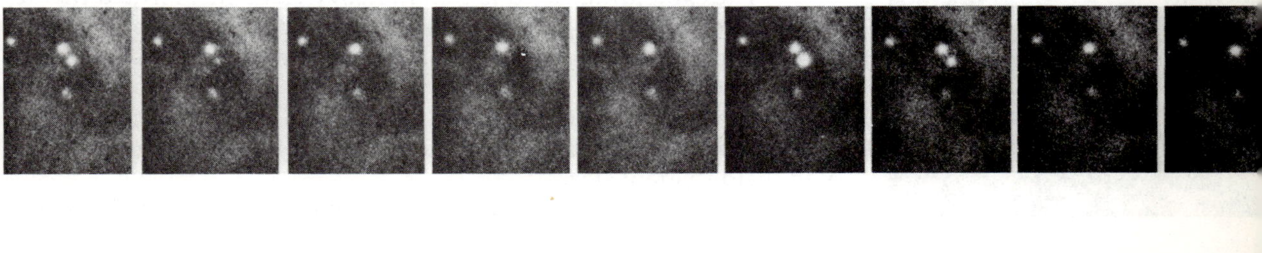

the universe and beyond

The one with the shortest period seemed to be in a diffuse glowing cloud of gas called the **Crab Nebula.** It pulsed 30 times every second.

In the years that followed, various explanations for pulsars were advanced. But only one explanation now seems to fit. Pulsars must be rotating neutron stars, denser than white dwarfs, each possessing a spot from which radiation emerges. As they spin, the beam of radiation periodically sweeps the earth, making them appear to pulsate. The energy in a pulsar comes from its rotation. As it gives up energy it slows down, and the pulses come less often.

Pulsars are believed to be rapidly rotating neutron stars. Bright spots on them radiate light like searchlight beams. In some instances the rotating beam periodically sweeps over the observer on earth, making the pulsar visible. There may be pulsars whose beams miss us. The light is believed to arise out of the interaction of the magnetic field of the pulsar with the surrounding gas.

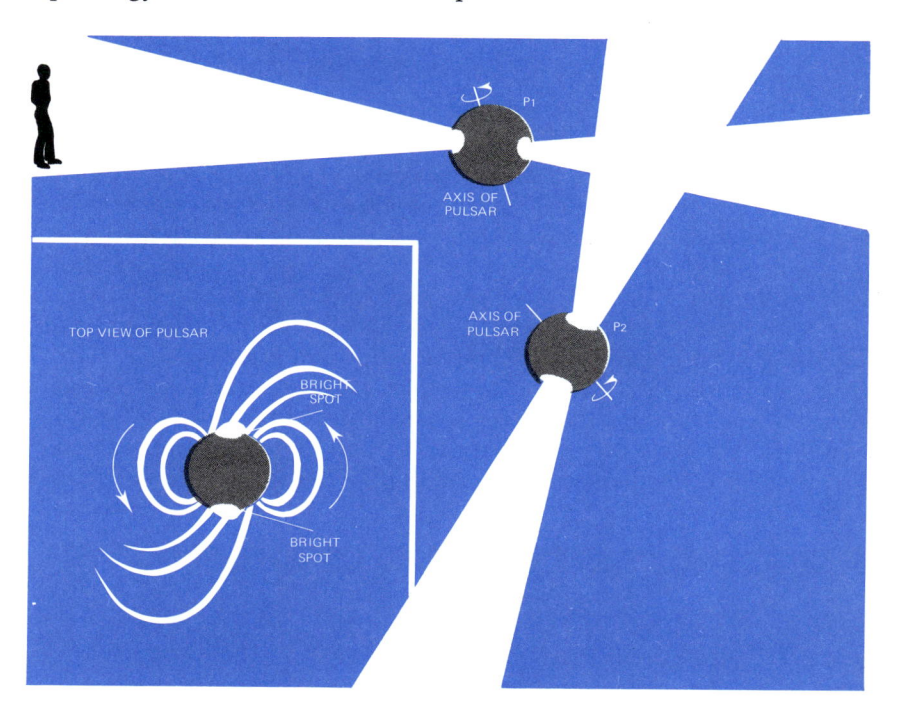

Below, the Crab Nebula, a supernova remnant, photographed in red light.

The Crab Nebula is the remains of a gigantic explosion called a **supernova.** Massive stars are believed to end their lives in this way, but how a supernova occurs is not well understood. According to one theory, the interior regions of an evolved star collapse to form a compact neutron star, rotating thousands of times per second. The collapse releases enough energy to expel the outer layers of the star, which then expand into space.

The expanding cloud is rich in heavy elements, formed inside the neutron star and in the supernova explosion. These elements "contaminate" interstellar clouds, and the stars and planets that form from these clouds are rich in heavy elements. Our sun is one such "second-hand" star. Many of the elements in our bodies, the book you are reading, and the chair you are sitting in were originally created inside a giant star.

The birth of a supernova is the most spectacular of stellar

events. The star's luminosity increases up to a billion times, and then declines after a few days. Supernovas occur in our galaxy about once every hundred years, but only a third of them are visible from earth. The most recent supernova was discovered by Kepler in 1604, and was bright enough to be seen during the day.

Neutron stars have an upper mass limit, probably not more than 2 to 5 solar masses. If their mass exceeds this limit, they begin to collapse. There is no way this collapse can be halted. The material of the star falls together to form an infinitely dense, infinitesimally small object. It is called a **black hole** because it is invisible; not even light can escape from it.

To understand this, we must consider how the escape velocity would change if an object like the sun collapsed. It would increase as the object became smaller and its surface gravity became more intense. The escape velocity from the sun is 600 km/sec. At white dwarf size, the escape velocity would be 6,000 km/sec. At neutron star size,

a more extreme ending

Drawing gives impression of the relative sizes of the sun, a white dwarf, a neutron star, and a black hole.

it would be 100,000 km/sec. By the time the sun collapsed to a diameter of 6 km its escape velocity would be more than the speed of light (300,000 km/sec). Light could no longer leave it, and it would disappear from sight. It would become a black hole.

Although black holes are invisible they exert gravitational force on other objects, so a black hole might be detected if it were a member of a binary system. The mass of the invisible member could be calculated by examining the motion of the visible member.

Black holes are likely to be sources of high-energy radiation. Matter falling into the intense gravitational field of a black hole would be greatly compressed, and nuclear reactions could be set off by resultant heating, producing observable bursts of X-rays and gamma rays. Many astronomers think that the X-ray source Cygnus X-1 is a black hole, and other likely candidates have been reported. However, the observational evidence for black holes is not conclusive.

KEEP AWAY FROM BLACK HOLES!

At the moment we die, some believe, all of our life will flash before our eyes. While this may be just folklore, modern physics describes an equally fascinating process. According to Einstein's General Theory of Relativity, a man falling into a black hole will find all of the time-to-come flashing before his eyes.

In Einstein's theory, gravity is simply the bending of space and time. One aspect of this bending is that time slows down within a strong gravitational field. As seen by a distant observer a man will age more slowly, a clock will run more slowly, and an atom will vibrate more slowly in the region near a massive object. This effect has actually been observed on earth, although the earth's gravity is so weak that extremely sensitive techniques are needed. The slowing of time should be more pronounced near massive, compact objects such as white dwarfs and neutron stars. And near a black hole, it should be quite spectacular.

Suppose we consider Steve and Frank, both carrying clocks. Steve stands far from a black hole, while Frank falls toward it. From Steve's point of view, everything about Frank slows down as he approaches the event horizon of the black hole—the point at which light can no longer escape from it. At first one second of Frank's time is little more than a second of Steve's time. Closer to the event horizon, however, it becomes an hour of Steve's time. Closer still, Frank's second is a year of Steve's time, and then a million years. No matter how long Steve watches, he will never see Frank fall past the event horizon. Steve, we might say, is always on this side of eternity.

Frank, though, is not. From his own point of view, he falls past the event horizon very quickly. As he approaches it, he sees the external universe speed up to a frenzied blur. Cepheid variables seem to pulse more and more rapidly until they are mere flickers of light. Spiral galaxies spin like pinwheels. As Frank goes through the event horizon, all of time in the external universe passes before him. Once through, however, he is on the other side of eternity. He cannot come back to our universe because there is no *when* for him to come back to!

the universe and beyond

galaxies

Stars are grouped into **galaxies,** vast structures sometimes half a million ly across and containing up to a trillion stars. Each star within a galaxy is gravitationally attracted by the others, so the galaxy remains compact and does not disperse. Some galaxies also contain interstellar dust and gas, which can amount to 25 percent of the total mass of the galaxy. The space between the galaxies is relatively empty, with only a few stars or star clusters that have escaped from some galaxy.

With the naked eye we can see three galaxies, not including our own. **M31** or the **Great Galaxy** appears as a hazy patch in the constellation Andromeda. (Of course, it is much farther away than any of the stars in the constellation.) The **Large** and **Small Magellanic Clouds,** smaller than M31, can be seen by viewers on or south of the equator. Besides these, our own galaxy appears as a bright band of light stretching across a dark sky from north to south. Many other galaxies can be observed through a telescope. At least a billion galaxies can be seen in photographs taken with the largest telescopes.

our galaxy

Telescopes reveal that our galaxy is composed of billions of stars. Its milky appearance accounts for its popular name: the **Milky Way.** (The word "galaxy" comes from the Greek *Galaktikos*, meaning milky-white.) From our position within the galaxy it is hard to

Wide-angle photograph of our galaxy, the Milky Way. Notice the dark lanes that seem to be relatively free of stars. They result from clouds of obscuring interstellar matter that block light from the stars lying beyond them.

the universe and beyond

structure of our galaxy

observe its overall structure, rather like trying to tell the shape of a crowd you are standing in. A further complication is that the galaxy is filled with obscuring matter between the stars, which often blocks our view.

Modern ideas about the galaxy are built up of various bits of information carefully pieced together. Much of our information comes from observations using infra-red or radio waves, which are not significantly obstructed by interstellar matter. Observations of other galaxies have also been helpful in understanding our own.

Viewed from the outside, our galaxy would look rather like M31. Its most prominent feature would be a bright disc of stars about 30,000 pc (100,000 ly) in diameter but only 1,000 pc thick. The disc has a central bulge where the thickness increases to about 2,000 pc. The outer, thinner part of the disc is not continuous, like a hat brim, but is composed of a number of winding spiral arms.

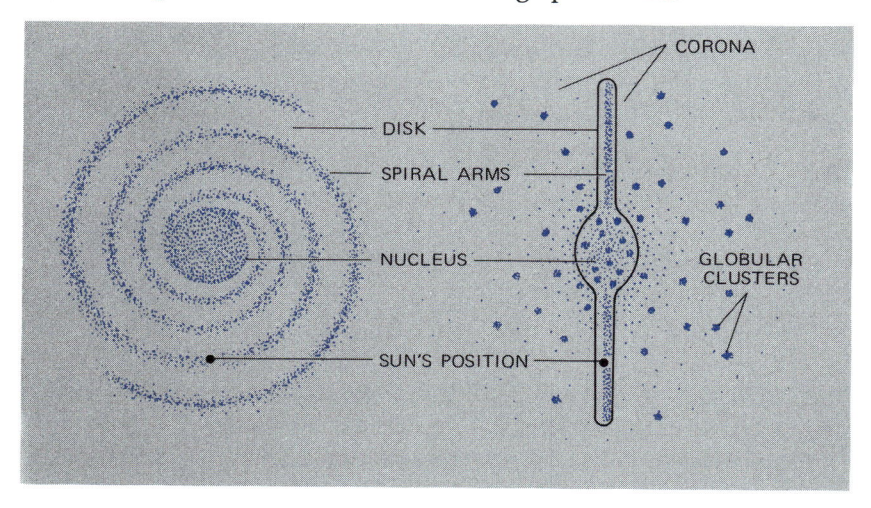

The sun is located in the disk of our galaxy. The nucleus cannot be seen visually because of obscuring matter in the disc. There is a spherical "halo" or "corona" of stars around the disc and nucleus.

The disc of the galaxy contains a wide variety of stars—red giants, white dwarfs, and m-s stars of assorted masses. The presence of m-s stars from 5 to 15 times as massive as the sun is especially significant, because they must have formed very recently. Their existence proves that star formation is still going on in the disc.

Most star formation takes place in the spiral arms. Spiral arms do not contain very much more matter than the rest of the disc, but they are abnormally bright. This brightness can be explained by the fact that they contain large numbers of highly luminous, short-lived stars. As we might expect of regions where star formation is occurring, the spiral arms are rich in dust and gas, which interfere with optical observations.

At the center of the disc is the nucleus of the galaxy. It is only 100 pc across, but has a mass equal to that of 100 million suns. Obscuring matter is thickest in the direction of the galactic nucleus,

the universe and beyond

which is therefore invisible from the earth. However, the nucleus emits infra-red and radio waves, so it can be observed using instruments sensitive to these. The nucleus has a core only 1.5 pc across, probably a very dense cluster of stars. The stars may collide with each other, setting off vast explosions. In fact, violent activity in the nuclei of galaxies appears to be the rule rather than the exception.

Surrounding the nucleus and the disc is the spherical **halo** of the galaxy, with a diameter of about 30,000 pc. It consists almost entirely of low-mass stars, with little dust or gas. Star formation in the halo must have stopped a long time ago. All the massive stars have lived out their short, bright lives; only the dim, low-mass stars go smoldering on. They are difficult to observe and to count at these great distances, so the mass of the halo is not certain. It was once believed that the halo made up only about 10 percent of the galaxy's mass. Now, however, it appears that up to 80 percent of the mass may be in the halo, and only 20 percent in the disc.

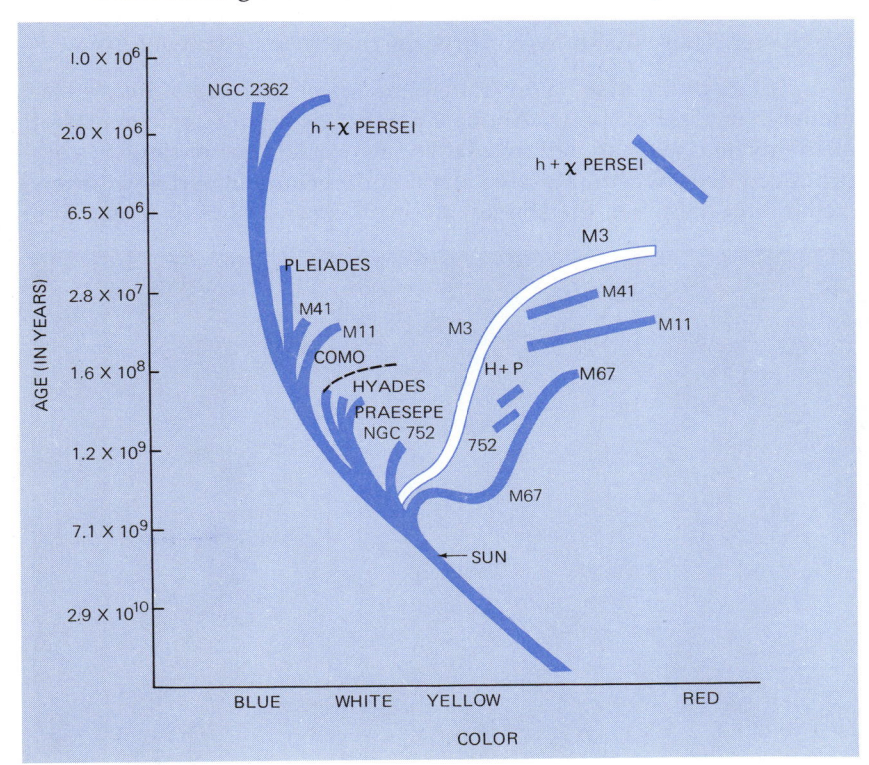

The fact that stars are often found in clusters is very useful to astronomers. For one thing, we can assume that they are all at roughly the same distance from us. (This is similar to the situation of a motorist in San Francisco wanting to know the distance to Manhattan. He calls up the AAA and is told that it is 4,000 km. The same

A composite of the H-R diagrams of several clusters. In the oldest clusters stars of lower luminosity have begun to evolve. In the youngest, even stars of high luminosity are still in the main sequence. The h and χ cluster in Perseus (photographed above) is one of the youngest clusters. Which would you say is an old cluster? M3 is a globular cluster.

star clusters

the universe and beyond

populations of stars

answer would have been all right if he had wanted to know the distance to Brooklyn. After he had driven 4,000 km, the extra couple of km to Brooklyn would be insignificant.)

If we assume that stars in a cluster are at the same distance, we can also assume that distance has affected their brightness in the same way, and that a brighter star is more luminous than a fainter star in the same cluster. We can then plot diagrams that resemble an H−R diagram, even though we do not know the actual distance to the cluster. These diagrams confirm the theory of stellar evolution.

Stars in the disc have a much greater percentage of heavy elements than the halo stars. About 4 percent of their mass consists of elements heavier than helium. As we saw earlier, heavy elements are produced deep within massive stars, and become mixed with interstellar material when the stars explode at the end of their lives. This enriched material, in turn, forms a new generation of stars. The halo stars must be an earlier generation because they have fewer heavy elements. The most massive stars of this generation synthesized the heavy elements that we find in stars of the disc. Stars rich in heavy elements are said to be **Population I** stars; those with few heavy elements belong to **Population II**.

In the accompanying figure is ω Centauri, one of the 120 **globular clusters** discovered in the halo. It is made up of Pop II stars. Globular clusters are compact and spherical. They contain hundreds of thousands of relatively low-mass stars, and this leads us to believe that globular clusters are among the oldest objects in the galaxy. Such clusters have also been detected around other galaxies.

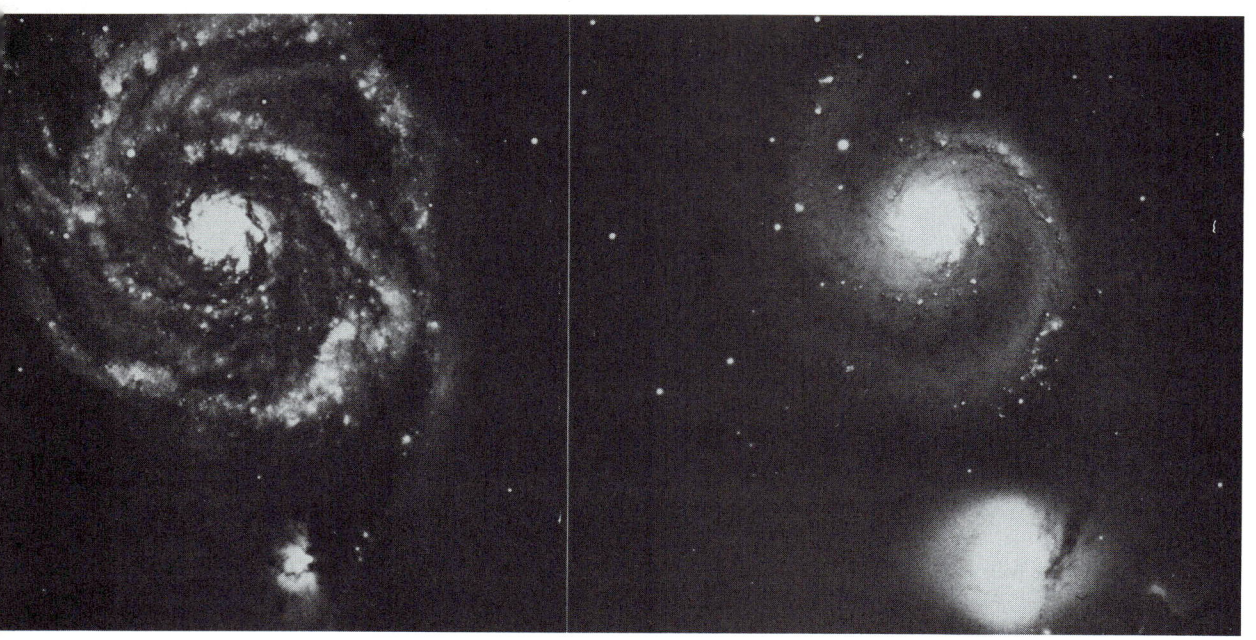

The spiral galaxy M51. The photograph at left in blue light emphasizes the spiral arms with their young, hot blue Population I stars. The red photograph emphasizes the Population II stars that are concentrated toward the galactic center.

the universe and beyond

Open clusters, containing fewer stars and with a looser structure than globular clusters, are found in the disc of the galaxy. Some contain hundreds of stars, but most have only a few dozen. The stars in an open cluster are Pop I, and some of them are very luminous; both of these facts tell us that open clusters are relatively young. All the stars in a cluster have the same composition, and were probably formed from the same cloud of gas.

We can trace the motions of stars in our galaxy through careful observation. Our sun is located in one of the arms of the galactic disc, about two thirds of the way out to the edge and approximately 10,000 pc from the center. Gravitational attraction by all the other mass in

Above left, the brightest globular cluster is Omega Centauri. It may contain a million stars. The Butterfly Cluster above is an open cluster. Both objects are in the southern sky.

motion of stars

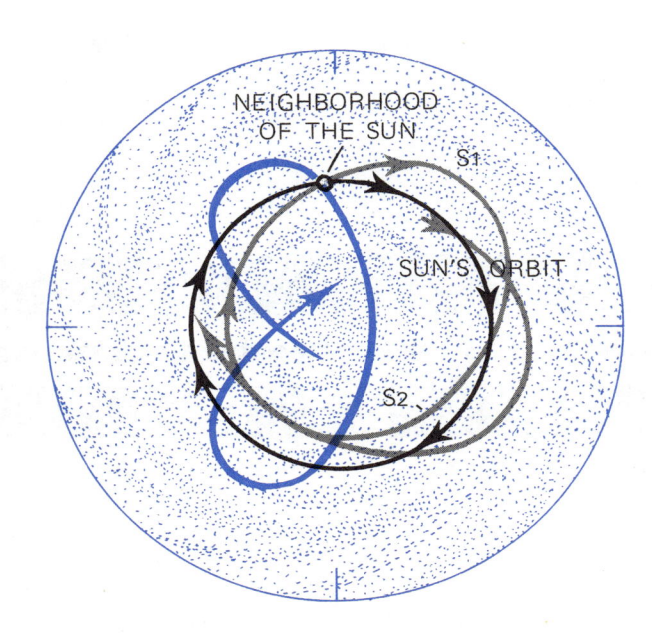

The calculated orbits of two Population II stars that are now in the solar neighborhood are, in general, elongated orbits. Population I stars, such as the sun, have nearly circular orbits.

the universe and beyond

the galaxy acts to pull the sun toward the center of the galaxy. Under the influence of this force, it travels in a vast circular orbit around the galactic center. It must travel 63,000 pc to complete one orbit in a period of about 240 million years.

Other Pop I stars also travel in circular orbits. But the orbits of Pop II stars have complex shapes which seldom repeat themselves.

the Doppler effect

Determining the velocities of stars is a complex process. Let us look at one useful method of finding the velocity at which an object is approaching or receding from us.

We have all noticed that the siren of a fire engine is shrill when it is approaching us, but lower-pitched when it is going away. A similar effect, called the **Doppler Effect,** exists in the case of light. Consider an object emitting light of a particular wavelength. If the object is approaching, the wavelength of its light appears shorter than it really is. If the object is receding from us, the wavelength of its light appears longer.

The apparent change in wavelength depends on the velocity at which the object approaches or recedes. For example, if it is receding from us at a velocity of 2 percent the speed of light, its wavelength

Diagram shows how the motion of the source with respect to the observer produces a Doppler shift. Below we have two spectrographs of Arcturus taken when the earth, in its orbit, was approaching the star and when it was moving away from it. Notice how the Doppler effect shifts the spectral lines to shorter wavelengths in the first case and to larger wavelengths in the second.

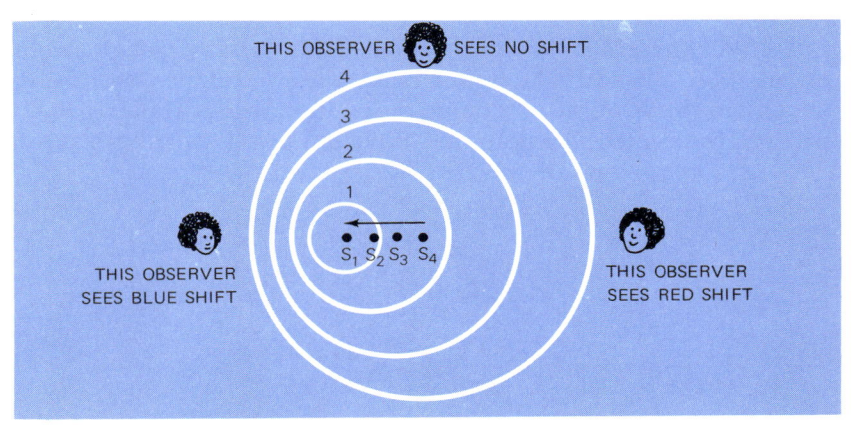

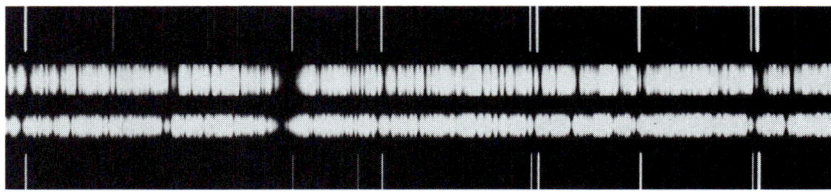

appears to be lengthened by the same 2 percent. The reverse is also ture: if we can tell that the wavelength of its light has been lengthened by 2 percent, we know how fast the object is receding. But how can we tell this? In the case of stars and galaxies, we may recognize the spectral lines of a particular element. We know from laboratory experiments the wavelengths that these lines should have. If they apthe universe and beyond

pear to be longer or shorter than expected, we know that the star is receding or approaching. If the wavelengths seem to be longer, we call this a **red shift,** meaning that the wavelength has shifted toward the red end of the spectrum.

radio astronomy

Radio astronomy has become a major branch of astronomy. It began with the accidental detection of radio waves from the center of the galaxy in 1931. But most of the progress in this field came much later, as an outgrowth of the electronic advances made during World War II.

Radio telescopes are very sensitive radio receivers which employ giant dish-shaped reflectors to gather and focus radio waves. The focused radio waves are collected, amplified, and analyzed. Most radio telescopes operate at wavelengths ranging from 1 meter down to about 2 mm. Astronomical objects emitting radio waves can be scanned by aiming the reflector at them. However, radio telescopes cannot distinguish between two objects that lie in roughly the same direction. This problem can be solved by making simultaneous observations with two or more widely separated radio telescopes.

Radio telescopes at Owens Valley Radio Observatory, California. Set on tracks, they are movable.

interstellar matter

Radio astronomy has been particularly useful for detecting the presence of interstellar matter. The first evidence for **interstellar gas,** however, was found during observations of the spectra of stars. When light passes through clouds of gas, certain wavelengths are absorbed. The resulting spectral lines are added to the spectral lines of the star, and the two sets of lines can often be distinguished. If the cloud of gas has a velocity different from that of the star, the spectral lines become Doppler-shifted by different amounts. This provides information about the velocity of the cloud.

The density of the gas in interstellar regions is only about 1 to 100 atoms per cm^3, about one billion billion times less dense than air. But the total mass of this gas may be as much as 20 percent of all the mass of the galaxy.

For the most part, of course, the interstellar gas cannot be seen.

IN FOCUS: hello out there!

The question of whether life exists outside of our small planet is still unresolved. Humans have speculated about the possibility for centuries, but only in the last decade have advances in Space Age technology enabled us to begin the search for extra-terrestrial life.

What form is extra-terrestrial life likely to take? Where are we most likely to find it? Scientists often use the phrase "life as we know it." Life as we know it is based on the chemistry of carbon. Carbon has the ability to link up with other elements to an almost unique extent, forming giant molecules such as DNA and the proteins, which are the basis of life on earth. Silicon shares some of this ability, but scientists have concluded that silicon-based life is virtually impossible.

If we focus our speculations on carbon-based life, then we can begin to predict the conditions under which it might be found. (Admittedly, this is a limited viewpoint. But without such a limitation we could get lost in wild imaginings of magnetic life forms in the galactic clouds, or neutrino-eating monsters in the core of the sun.)

First of all, the development of carbon-based life requires an environmental temperature of around 300K, roughly the temperature of a warm day. Furthermore, some sort of liquid must be present to facilitate biochemical reactions. Water seems to be best for this purpose, but liquid ammonia and liquid methane might possibly serve under conditions in which the temperature is lower and the atmospheric pressure high. These liquids, of course, are likely to be found only on planets or very large satellites of planets, such as Saturn's moon Titan.

Few scientists today believe that intelligent extra-terrestrial life exists in our solar system. It is possible that even unintelligent life may not exist, although the question is still open. Experiments performed by the Viking spacecraft that landed on Mars have so far proved inconclusive. (Photograph shows the small trench dug by the sampling arm of the spacecraft in order to get a sample of soil for the biology experiment.) There are some indications that life has never existed on Mars. It has been suggested that there may be life on Jupiter, but most people rule out the possibility of intelligent life. For the most part, scientists have centered their search for intelligent life on planets that may be circling other stars of the galaxy.

Our closest neighboring star, Alpha Centauri, is more than 3 ly away, and most of the stars in our galaxy are much farther. How can we search for life at these great distances? Not by visiting the stars ourselves, at any rate. According to Einstein's theory, nothing can travel faster than light. Thus, a round trip to a star-and-planet system 200 ly away would take more than 400 years for a spacecraft traveling at 99.5 percent the speed of light. To accelerate even a tiny 10-ton object to this velocity would require as much energy as our entire technological society uses during a period of about 500 years. Voyages to the stars, therefore, are unlikely in the foreseeable future.

Another possibility might be unmanned space probes, which would be put into orbits around stars possessing planetary systems. The probes would be programmed to detect signs of civilization and transmit the information back to earth.

It appears that the cheapest and quickest way to contact extra-terrestrial beings is by radio. It is also the only method that we are capable of employing at present. The first attempt to detect broadcasts coming from extra-terrestrial civilizations was made in 1960 by Frank Drake of the National Radio Astronomy Observatory. He entitled this attempt Project Ozma, after the fictional Princess of Oz (a far-away land inhabited by strange creatures). Drake aimed a radio telescope at two suitable stars, both about 11 ly away. However, no communications were received from extra-terrestrial beings, and later attempts have also been unsuccessful.

Recently some scientists have suggested setting up a huge collection of radio telescopes to look for signs of extra-terrestrial civilizations. Project Cyclops, named after the one-eyed giant in the *Odyssey,* would consist of 1,500 radio telescopes, computer-controlled to work together as one giant telescope. The cost of the project would be about ten billion dollars. With this project we would not have to wait for civilizations to beam special signals to us. It should be possible for us to pick up their local transmissions, such as radio and television broadcasts.

One of the most imaginative attempts to communicate with intelligent beings outside our solar system is already on its way toward the depths of space. In 1972 the Pioneer 10 spacecraft was launched. It explored the vicinity of Jupiter and, accelerated by the gravity of that planet, became the first product of human technology to leave the solar system.

Pioneer 10 will probably wander through space for many millions of years. On the chance that it might be snared by members of an intelligent extra-terrestrial race, NASA officials placed a message on the spacecraft, in the form of a gold-etched plaque attached to its hull. The plaque shows a model of the hydrogen atom, which is used to establish standards of length and time. These standards, in turn, are employed to show the position of our solar system at the time of launch with relation to the position of fourteen pulsars. The solar system itself is also depicted on the plaque, with a line showing the origin and trajectory of the spacecraft. Standing in front of an outline of the spacecraft are the nude figures of a man and a woman. The man's right hand is raised in the universal human gesture of peace.

Whether the intelligent extra-terrestrials will correctly interpret the gesture of peace is uncertain. We have asked a question of the universe. Now, we can only wait for an answer.

the universe and beyond

At right, a region of diffuse nebulosity around the star Gamma Cygni.

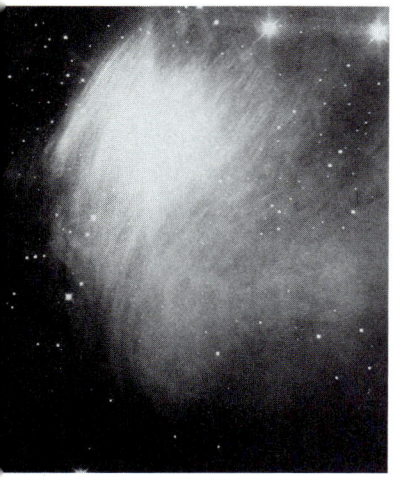

Cloud of reflecting matter around the star Merope, one of the Pleiades. A reflecting nebula of this sort shines with light reflected from nearby stars. Merope is only passing through this region of dust and gas, it was not born here.

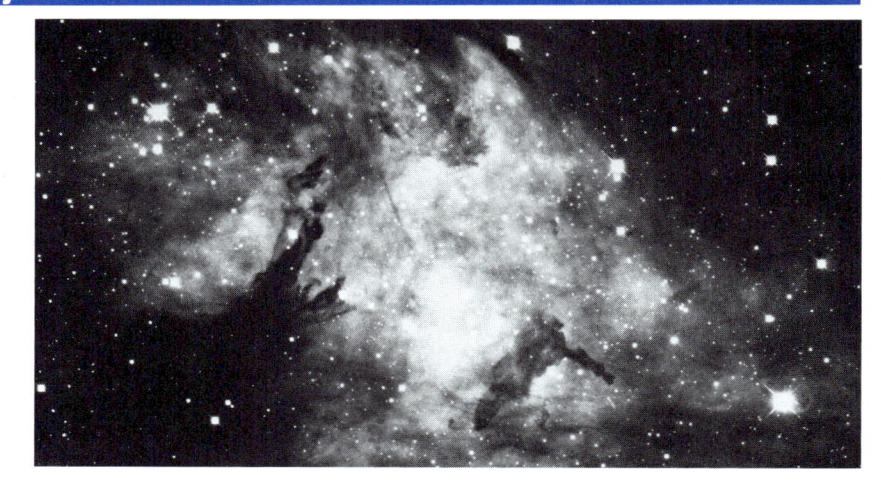

An exception to this is the **emission nebula**. Emission nebulas are vast clouds of gas that often surround hot, bright stars. ("Nebula" means cloud.) The gas is usually left over from star formation. Ultraviolet radiation emitted by the hot stars excites the gas, which in turn emits visible light.

In 1951, radio waves from interstellar hydrogen were detected at 21 cm wavelength. The discovery of these radio spectral lines was a major event. Now the galaxy could be mapped, using the radiation from hydrogen, even where visible light was blocked. Interstellar hydrogen seems to be concentrated in a flat layer less than 100 pc thick. Since the 21 cm radiation is a spectral line, its Doppler shift can be measured. This tells something about the velocity of the hydrogen clouds producing the line. The 21 cm radiation has also been detected from several other galaxies.

Interstellar space also contains small, solid particles of **interstellar dust**. The dust obstructs light much more effectively than gas, and obscures objects that lie beyond it without producing spectral lines. The light from distant star clusters is greatly dimmed by this dust. Interestingly, different wavelengths of light are differently affected. Red light is least dimmed, and blue light is dimmed the most. Infra-red radiation is very little affected, and radio waves are not affected at all.

The interstellar dust probably consists of very small solid grains, with diameters of about $1/10,000$ cm. It is thought that the grains are carbon—in other words, the interstellar dust is essentially soot. Some astronomers believe that each carbon grain is coated with a layer of ice.

Recently a novel theory has been presented to explain the origin of these interstellar grains. It suggests that carbon grains are formed in the outer layers of certain stars. Some red giants are abnormally rich in carbon. These same stars are often abnormally cool, perhaps only 2,000K. At the temperatures of most stars, carbon exists

the universe and beyond

The famous Horsehead Nebula is a large region thick with obscuring dust. It is seen against a glowing mass of gas beyond it.

as a gas. But in these unusual stars, carbon precipitates into grains, a process similar to the formation of snow from water vapor.

interstellar molecules

After the 21 cm line of hydrogen was discovered, few other radio spectral lines were found. It was known that molecules could emit radio frequency radiation, but no one had looked for molecules in space. Prevailing theories held that it was impossible for them to form in space, and that even if they did form, ultra-violet radiation was sure to destroy them.

But in 1968 a team of physicists began to look for interstellar ammonia. Sure enough, they soon detected radiation from ammonia coming from the direction of the galactic center. The astronomical community was startled, and rather reluctant to act on the information. Most astronomers agreed that unusual conditions at the galactic center might produce unusual events, even molecule formation. But did this justify committing expensive resources to a search for molecules elsewhere?

Then water vapor was detected at three different points in the galaxy. This aroused more interest, and other astronomers joined the search. More than 40 kinds of interstellar molecules have been discovered so far, and new ones are being found every week.

The molecules themselves are not new; most of them are quite common. Examples include carbon monoxide (CO), an urban pollutant; formic acid (HCOOH), which makes an ant's bite sting; and ethyl alcohol (C_2H_2OH), the essential chemical in your martini. In

the universe and beyond

space, these molecules exist in uncommon situations. A whole new chemistry had to be developed to explain their behavior under conditions that cannot be duplicated in the laboratory.

How are the molecules able to survive in space? The most widely accepted theory suggests that interstellar dust grains play an important role. Atoms of gas first stick to dust grains, and then nearby atoms link up to form molecules. The formation of each molecule releases energy, which kicks it loose from the grain.

Molecules are always found in large, dense clouds of dust and gas. The dust shields the molecules from the ultra-violet radiation in space. Molecular radiation has now been detected from all over the disc of the galaxy, as well as in other galaxies.

external galaxies

In the 1840s Lord Rosse, an Irish nobleman and amateur astronomer, set up the largest telescope then in existence. Its most important part was a metal mirror almost 2 meters in diameter. This would be considered a large instrument even today. Its large size made it a problem to mount, and it was unwieldy and difficult to point. Moreover, the metal mirror tarnished rapidly. Despite these shortcomings, however, the telescope soon scored a major success. When it was aimed at a nebula in the constellation Canes Venarici, the nebula was seen to have spiral arms extending out from the center. Within the next few years 14 other **spiral nebulas** were found.

Spiral galaxy is seen edge-on. Note layer of obscuring matter in its disk. The halo is too faint to be seen in the picture.

In the early part of this century the 60-inch and 100-inch telescopes were put into operation at Mt. Wilson, near Los Angeles. Photographs taken with these telescopes revealed hundreds of spiral nebulas. Could these be galaxies like our own? They seemed very large: inside many of them individual stars could be distinguished. Still, many astronomers, skeptical of the observations, were reluctant to believe that nebulas were galaxies. As we know, some nebulas are clouds of gas—the emission nebulas—and the general opinion was that all nebulas were such clouds. Stars observed in the spirals were dismissed as bright spots of gas.

External galaxies are much larger than gas clouds within our galaxy. But the sizes of the nebulas could not be estimated because their distances were not known, and they were much too far away for the method of parallax. Astronomers could not determine whether they were small nearby objects or large distant ones. A method of measurement had to be found, and the first person to find one was Edwin Hubble, an astronomer at Mt. Wilson.

cepheid variables

Hubble's method is based on the properties of a peculiar type of star. Some stars change in size periodically, and their brightness varies or pulsates, as a result. Delta Cephei is a yellow supergiant star that pulsates once every five days. When at maximum luminosity it is twice as bright as at its minimum. Other such stars have been discovered, possessing different luminosities and periods. These stars are known as **cepheid variables.** The period of a cepheid is closely related to its average luminosity. Delta Cephei, for example, has a 5-day period and is 600 times more luminous than the sun. A cepheid with a 10-day period is about 2,000 times as luminous as the sun.

In 1923, Hubble discovered faint but unmistakable cepheid variables in the nearest spirals. By observing their periods, he could determine their luminosities. It became apparent to him that these cepheids, though they appeared faint, were really highly luminous — and therefore that they must be very far away. Obviously, the spirals containing the cepheids were large objects at a great distance — galaxies in their own right. Today these nebulas are known as **spiral galaxies.**

Hubble's classification of galaxies

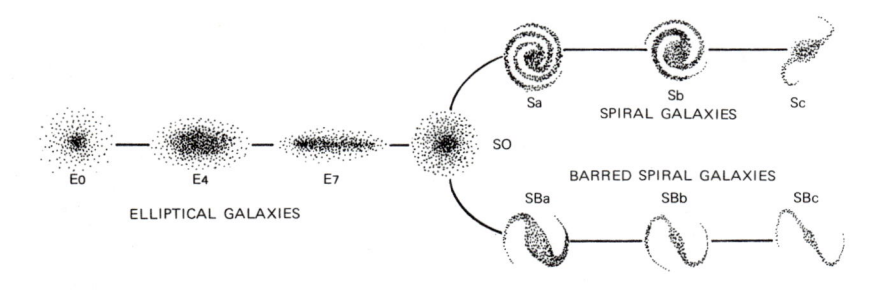

Hubble observed and classified the external galaxies, and we still use a modified version of his scheme of classification. Spiral galaxies make up about two-thirds of all observed galaxies, but there are other kinds as well. The most common are the elliptical galaxies

The various types of galaxies were classified by Hubble, resulting in this well-known diagram.

and the irregular galaxies.

Most spiral galaxies weigh about the same, and all of them seem to contain dust and gas. This is reasonable to expect, because the spiral arms are regions of star formation. Spirals contain bright young stars, and as a class they are the brightest of galaxies. All spirals have Pop I stars in the disc and Pop II stars in the halo. Their spiral arms may be wound tightly or loosely. About one-third of the spirals have a bar in the center, and are known as **barred spirals.**

Elliptical galaxies, so called because of their smooth elliptical

the universe and beyond

Above, three types of spirals. Star formation appears to be still taking place in the least tightly wound spiral.

shape, make up about a third of the observed galaxies. Ellipticals are composed of Pop II stars. They contain little dust or gas, so it is almost certain that star formation is not going on within them. In the nearby ellipticals many globular clusters can be observed.

The smallest ellipticals, known as **dwarf ellipticals,** are only slightly larger than globular clusters. Only the nearest of these can be observed; the distant ones are too faint to be seen. Dwarf ellipticals may be the most common type of galaxy. On the other end of the

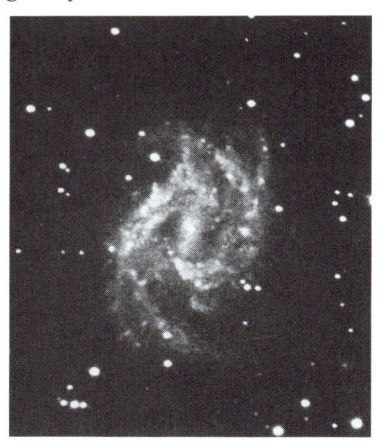

Three types of barred spirals, going from tightly wound at left to loosely wound at right.

scale are the **giant ellipticals,** some of which may contain tens of times the mass of our own galaxy. Giant ellipticals are very luminous, but in general are much dimmer than spirals of the same mass. The spiral's superior luminosity is due to its hot, luminous newly-formed stars.

After the spirals and ellipticals we come to the **irregular galaxies,** which make up only about 3 percent of the galaxies observed. They contain dust and gas, and many of them have luminous Pop I stars indicating that star formation is taking place. The Magellanic Clouds we mentioned earlier are irregular galaxies.

the universe and beyond

At far left is a small elliptical companion of the Andromeda galaxy. At left is the Large Magellanic Cloud, an irregular galaxy which is a nearby companion to our galaxy. This galaxy in the southern skies was first reported to northern scientists by the great explorer Magellan.

As if the galaxies themselves were not large enough, more then half of them belong to clusters. Clusters of galaxies occupy unbelievably huge volumes of space, some having diameters of millions of parsecs. The number of galaxies in a cluster varies. The largest clusters appear spherical, and contain at least 1,000 galaxies, mostly elliptical. X-ray emissions from such clusters have been detected.

clusters of galaxies

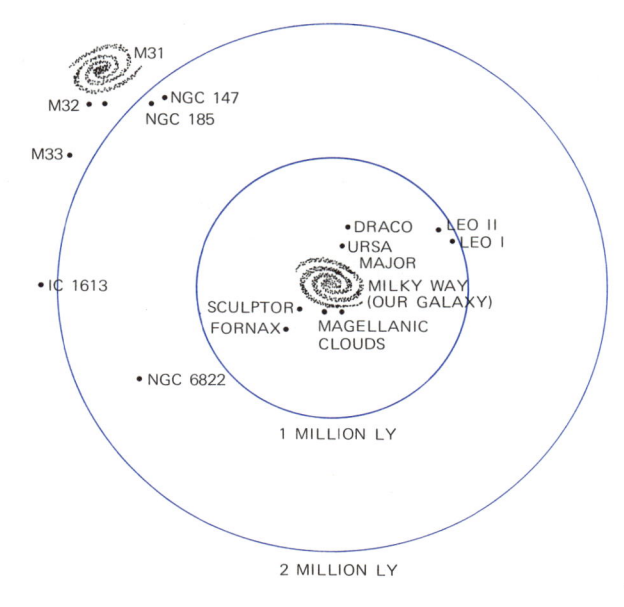

The Local Group of galaxies dominated by M31, the Andromeda galaxy, and our galaxy. Two other massive members have recently been detected using infra-red telescopes. The obscuring dust in our galaxy prevents us from observing them visually.

the universe and beyond

A cluster of galaxies in the constellation Hercules contains several types of galaxies. Most of the galaxies are spirals. Notice how some neighboring galaxies have distorted each other gravitationally.

The smaller clusters are irregularly shaped, and sometimes seem to be made up of even smaller clusters. Irregular clusters contain both ellipticals and spirals. Our galaxy belongs to a small cluster called the **Local Group,** which consists of at least 17 members. The Local Group is dominated by our galaxy and the Andromeda galaxy; the other galaxies cluster around them. The exact boundaries of the Local Group are uncertain. Two other large galaxies, recently discovered by infra-red photography, may belong to the Local Group.

galactic violence

As we mentioned earlier, violent events seem to be the rule in many galaxies. In some, the energy released is stupendous, many times that radiated by our own galaxy. These events appear to have both external and internal causes.

An example of externally caused violence takes place when two galaxies collide, a rather infrequent occurrence. The stars them-

the universe and beyond

selves do not collide; instead, the galaxies pass through each other like swarms of bees. However, gas clouds in the galaxies collide, producing various forms of energy. Some strong radio sources might be galaxies in collision.

In many galaxies, violence is internal. The irregular galaxy M82 is one of the nearest galaxies, about 3 million pc away. Photographs of M82 show gigantic streams of material extending thousands of parsecs above its disc. There is evidence that these streams contain a good deal of dust, which reflects light from the nucleus and

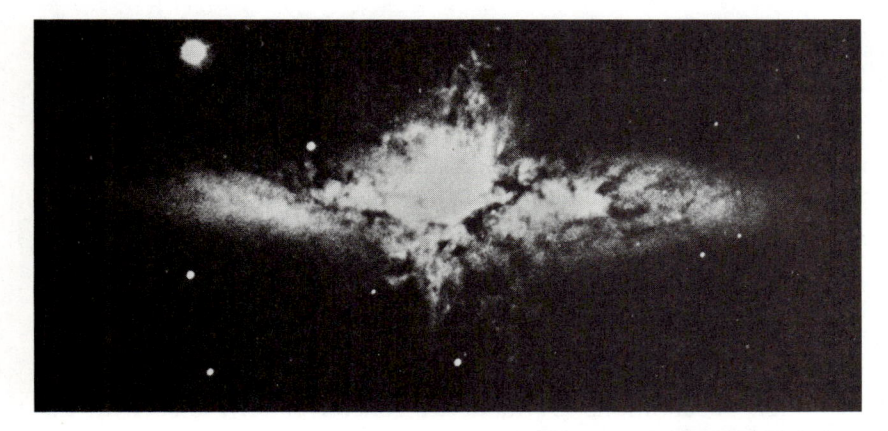

Light from the exploding galaxy M82 takes about ten million years to reach us. Notice the streams of matter extending away from the galaxy.

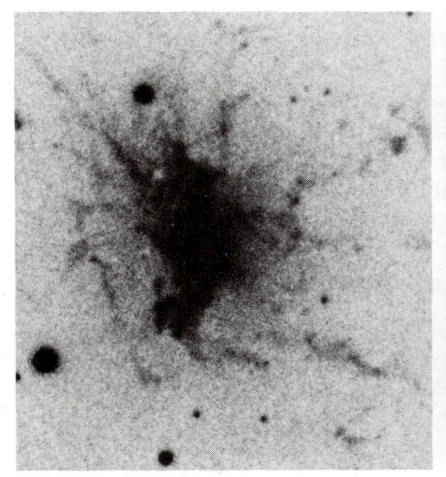

becomes visible. The streams of dust and gas are moving at the unimaginable speed of 1,000 km/sec. All the observations of M82 suggest that a gigantic explosion must have occurred in the nucleus about a million years ago.

There are also other galaxies in which gases are expanding away from the nucleus. These are known as **Seyfert galaxies,** named after their discoverer. They may resemble M82 in many ways, but are distinguished by very bright nuclei, sometimes tens of times brighter than our galaxy. The nuclei are relatively small and appear starlike

At far left is the Seyfert galaxy NGC 1275 seen face-on, with clear indications of explosive activity in its center. Above, gravitational interaction between galaxies spews out long "tails" of matter.

the universe and beyond

in telescopes, but are very often up to a billion times more energetic than stars.

Seyfert galaxies constitute about 2 percent of all galaxies and are very interesting to study. Some astronomers believe that all galaxies are Seyferts for a small percentage of their lives. It is possible that they may be related to the mysterious quasars, which we will discuss a little later.

cosmology

At this point we have looked at most of the important objects in the universe. In earlier chapters, we studied our home planet—Spaceship Earth, as someone has called it. In the last two chapters we have studied the planets, the stars, the galaxies, and the clusters of galaxies. But **cosmology** is the study of the universe itself.

By definition, the **universe** includes all that exists, has existed, and will exist. Cosmology attempts to explain the nature of the universe. Such an attempt is full of difficulties, one of which has to do with the uniqueness of the universe. Consider how a science is built up. First, observations are made. A general pattern is seen in the observations. The pattern is refined into a theory, which is used to make predictions. Finally, the predictions are checked out by further experiment or observation.

In studying stars, as in studying everything else, we use this comparative method. By observing similarities and differences between stars we have discovered how a star works, how it lives, and how it dies. The comparative method works for studying galaxies also. But this method fails completely for the universe, since we cannot observe any other universe. We can speculate about how the

Hale telescope on Mt. Palomar has probably contributed more to cosmology than any other. At far right an observer sits in the observing cage, with the Hale telescope's 200-inch mirror seen beyond.

universe would be if some aspect of it were changed, but we have no way to verify our speculations by experiment or observation. And this is only one of the difficulties in cosmology—there are many others.

the universe and beyond

surveying the universe

As we try to get an overview of the universe, we shall measure distances in light years. When we say that a galaxy is 1 million ly away, we know at once that we are seeing it as it was a million years ago. Some galaxies are 10 billion ly away. Light that we now receive from them started on its long journey before the earth was formed. In one sense, when we look out into space we are looking back into time. Speaking of distances in light years helps us keep this fact in mind.

To get a better idea of the size of the known universe we can construct a mental scale model. In our model, we make our galaxy the size of a dessert plate. Within it, the sun is about the size of an atom. Near the sun, almost too close to be distinguished, is Alpha Centauri, another atom. Across the room is the Andromeda galaxy, about the size of a frisbee. The other members of the Local Group are saucers, coins of various sizes, and match heads, all within the room. M82 is a little less than 30 meters away, across the street. NGC 1068 is two houses down the block. The nearest rich cluster of galaxies, the Virgo Cluster, is in the next block. The quasar 3C 273 is 8 kilometers away. The edge of the known universe is 40 km away; we could drive to it in half an hour.

Distance	Light travel time
New York to San Francisco	1/100 seconds
Earth to moon	1-1/4 seconds
Sun to earth	8 minutes
to Pluto	5 hours
to Alpha Centauri (nearest star)	4 years
to Pleiades (nearby open cluster)	200 years
to M13 (globular cluster)	30,000 years
to the center of our galaxy	33,000 years
to M31 (galaxy in Andromeda)	2 million years
to M82 (exploding galaxy)	10 million years
to NGC 1068 (nearest Seyfert)	30 million years
to the Virgo Cluster	60 million years
to 3C273 (first quasar discovered)	3 billion years
to the most distant cluster known	7 billion years
to the most distant quasar known	15 billion years (?)

Table 16—5. Distances to various astronomical objects.

Hubble's law

The extraordinary distances to far-away galaxies and clusters of galaxies are often estimated using a law discovered by Hubble. When Hubble studied galaxies spectroscopically, he found that the light from most of them was red-shifted, an indication that they are moving away from us. Further investigation revealed a curious relationship. The more distant a galaxy is, the greater its red shift. Obviously, the farther a galaxy is from us the faster it is traveling away. Moreover, there is a consistent relationship between its speed and its distance. This relationship is called **Hubble's Law (H).**

The value for H is 17 km/sec/million ly. In other words, a galaxy a million light years away is traveling away from us at 17 km/sec. Another, at a distance of 2 million ly, is traveling away from us at

the universe and beyond

twice the speed, 34 km/sec. A third, 10 million ly away, is receding from us at 170 km/sec.

We can also reverse this argument. It is hard to measure distances to really distant galaxies, but relatively easy to measure their red shifts. Suppose a galaxy is receding from us at 170 km/sec. We can use Hubble's Law to find its distance—10 million ly from us. One distant cluster of galaxies is traveling away from us at a startling 17,000 km/sec. It is calculated, therefore, to be at a distance of 1 billion ly.

Of course, when we use Hubble's Law this way, we are assuming that it holds true at great distances, which is not at all certain. Much controversy among astronomers centers around this issue. But Hubble's Law will have to do until some better method comes along.

Hubble's Law is probably the most important discovery in cosmology, with many fundamental implications. One implication is that the universe seems to be expanding, dispersing without end.

This implication seemed preposterous to many astronomers during the 1930s. They could not challenge Hubble's observations, which were plain enough. But some of them questioned his interpretation of the observations. The red shift, they argued, did not indi-

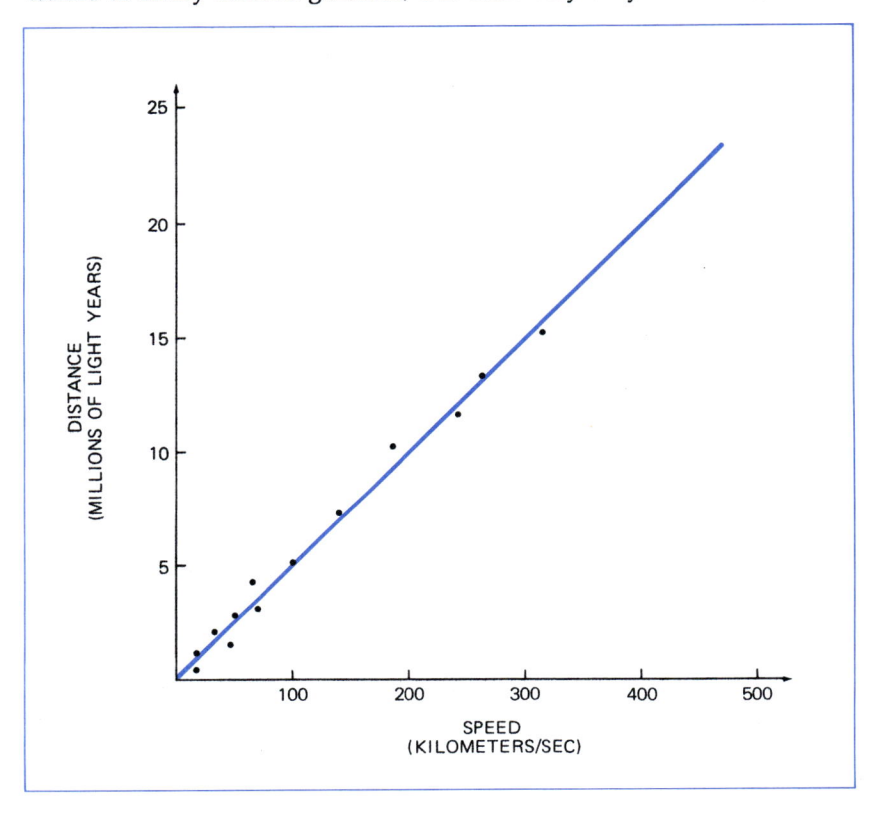

Hubble discovered that distance to a galaxy appears to be related to the rate at which it is receding from us. Diagram at right shows Hubble's relationship.

implications of Hubble's law

cate velocities at all. It was due to light getting "tired" on its long journey to us, losing energy and becoming redder the longer it traveled. This ingenious theory was not accepted by most scientists, since it calls for an entirely new theory of light. This is not to say that it may not have been the right theory. But explanations based on known laws of physics are normally considered stronger than those that call for brand-new laws.

Hubble's Law and his interpretation were also accepted for another reason. They fitted in with a theory of the universe proposed by Albert Einstein. Einstein's universe was constructed around his theory of space–time curvature.

When Einstein's theory was published, around 1920, one of the few persons who understood it was Sir Arthur Eddington, who established the present theory of stellar structure. Eddington calculated that Einstein's universe would not remain still. If it were disturbed ever so slightly it would begin to expand.

Eddington wanted to verify his calculations, but at that time there were few observations to support his ideas. It was true that nebulas were known to have red shifts, but no one knew how far away they were. Eddington belonged to the minority of astronomers who believed that they were other galaxies. Then Hubble came along and solved two problems. First, he demonstrated that nebulas are galaxies. Then he established his law relating red shift to distance.

Hubble's Law seems to imply that the galaxies are rushing away from us. Are we at the center of the universe? In our century such a geocentric view is frowned on. No theory that places the earth in a privileged position is considered acceptable.

But Hubble's discovery does not necessarily imply that we are at the center of the universe. Consider a balloon being slowly inflated. The skin of the balloon represents the universe. Several bugs are resting quietly on the surface of the balloon. They represent the galaxies. As the balloon inflates, each bug finds that the other bugs are moving away. The nearest bug moves away the most slowly, while the farthest moves away the most rapidly—Hubble's Law in action. Each bug observes the same overall situation around itself. In the same way, each galaxy in an expanding universe finds every other galaxy moving away from it. In each case, the motion follows Hubble's Law.

Until fairly recently it was believed that only one kind of geometry was possible—the geometry of Euclid. In **Euclidean** or **plane geometry,** various familiar facts are always true. For example, the angles of a triangle add up to 180°. The area of a circle increases by four times when its radius is doubled. These assertions can be clearly demonstrated on a flat surface.

But other geometries are possible, and were first investigated by mathematicians during the nineteenth century. In **Reimanian**

Sir Arthur Stanley Eddington, eminent British physicist and astronomer, made important contributions to several branches of astronomy and pioneered the study of stellar structure.

some cosmological ideas

the universe and beyond

geometry, the angles of a triangle add up to more than 180°. A circle of twice the radius has less than four times the area. This geometry can be demonstrated on a sphere.

In another form, called **Lobachevskian geometry,** the angles of a triangle add up to less than 180°. A circle of twice the radius will

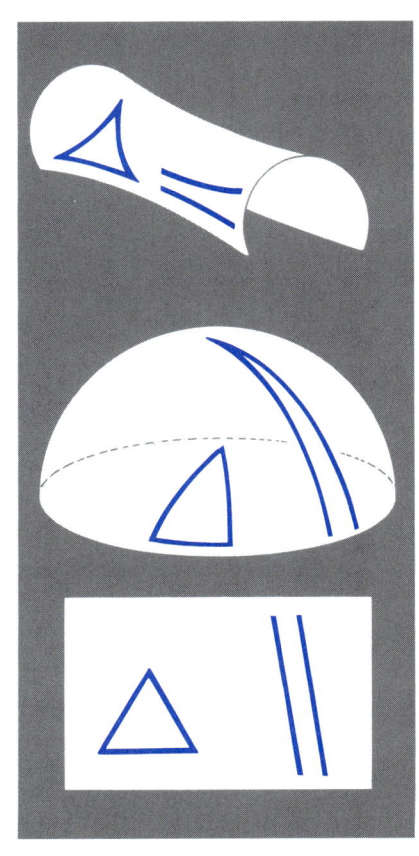

Diagram at far right displays various types of geometries. At right, balloon with dots representing galaxies illustrates the expansion of the universe.

have more than four times the area. This geometry can be demonstrated on a saddle-shaped surface.

We speak of Euclidean geometry as plane or "flat." The other geometries we call curved. The accompanying figure shows how these geometries differ when applied to two dimensional space. It is very difficult to imagine non-Euclidean geometries in three-dimensional space, but mathematicians have been able to explore their consequences. And it is clear that real space does not necessarily have to conform to the rules of Euclidean geometry.

Einstein's theory of the universe combines space and time into a four-dimensional **spacetime.** The geometry of this spacetime is flat only in places far away from matter. The presence of matter causes spacetime to curve, and the curving makes itself felt as gravity. This is extremely difficult to visualize, and making calculations about spacetime is a very complex process. But we can mention the results of one calculation. It states that the greater the density of matter in the universe, the greater is its curvature. According to this calculation, if the density of matter were great enough, the universe would curve and close in on itself. It would have no boundary, but it would be finite. A person traveling in a straight line would come back to where he started, just as a person who travels in a straight line on the surface of the earth will come back to his starting point.

The question remains whether the universe contains enough matter to close it. No one knows. Astronomers, like most other people, are cautious in their dealings with infinity. They would prefer a closed, finite universe. But some recent observations seem to show that the universe was not designed with their tastes in mind. It is, very possibly, open and infinite.

two principles

It is not really possible for us to comprehend the immensity of the universe. The galaxies within it are mere specks, which appear to be distributed endlessly without any sign of stopping. From our position in one galaxy, we can make good observations of only a tiny part of the universe. If we try to generalize the structure of the rest of it on the basis of our limited view, we are forced to make various assumptions.

Cosmologists of one school of thought regularly make an assumption called the **Cosmological Principle.** According to this principle, a person anywhere in the universe would see the same overall situation if he looked around — that is, the universe is uniform.

We make this sort of assumption very often. Looking at a photograph of a Russian power station we might say, "Oh, those are the electrical wires." In saying so, we have assumed that copper conducts electricity in Russia just as it does in America. But sometimes this principle can lead to error. We have only to remember the story of the blind man who touched the tail of an elephant. Applying the Cosmological Principle, he concluded that the rest of the elephant was also like a rope.

During the 1950s three British astronomers proposed a stricter principle. They called it the **Perfect Cosmological Principle.** It states that an observer, anywhere *and at all times,* would see the same overall situation — that is, the universe is unchanging as well as uniform. Since the universe is assumed to be unchanging, it cannot have a beginning or an end. Various astronomers have constructed theories based on this principle. But observations made in recent years suggest that the principle does not apply to the universe.

the big bang theory

According to the **Big Bang** theory, the universe appeared about 20 billion years ago. It was infinitely dense and infinitely hot. It expanded rapidly, and in only seconds it became a raging, chaotic fireball of radiation. During its first 2,000 years the universe contained more radiation than matter. As it expanded it cooled, and much of the radiation was converted to matter. Cooling continued, and about a million years after its formation the universe reached a

the universe and beyond

temperature of about 3,000K. At this temperature, matter that had been in the form of electrons and protons became neutral hydrogen. The radiation and matter stopped interacting. Both cooled down separately.

Then matter began to clump together. The clumps collapsed to form galaxies. As they collapsed, further clumping took place. First, clumps the size of globular clusters were formed. These, in turn, broke up into individual stars. In some galaxies, some of the matter collapsed into a disc before it could break up into stars. These galaxies became spirals. In others, the gas was used up before a disc could form. They became ellipticals.

the steady state theory

Because it assumes that the universe had a beginning, the Big Bang theory violates the Perfect Cosmological Principle. The **Steady State** theory, however, does not. According to this theory, the overall structure of the universe remains unchanged. The theory gets around the observed expansion of the universe by saying that the universe is infinite—and, of course, when an infinite universe expands it remains just as infinite as it was.

But what about the density? If the galaxies are moving apart, space must be getting emptier. The Steady State theory deals with this problem by asserting that matter is continuously created. Of course, it is impossible to test this notion by any observation. The idea of continuous creation requires a whole new physics. Hence, like the idea of tired light, it did not impress very many scientists when it was first advanced.

But there are more serious reasons to believe that the Steady State theory is incorrect. According to this theory, the universe must have looked the same in the past as it does now. We can check this by observing distant objects. (Remember that when we look far out into space we are looking far back in time). In the 1960s astronomers discovered quasars, which we will discuss in the next section. Careful observations indicate that quasars were far more numerous in the past than they are now, which is a clear contradiction of the Steady State theory.

Still another objection has been raised. Remember that when matter began to form galaxies, according to the Big Bang theory, the radiation continued to disperse. If this theory is the correct one, we should be able to observe radiation coming from all directions—and we did, in 1965. The Steady State theory cannot explain this observation. As far as most astronomers are concerned, this clinched the case for the Big Bang theory.

the discovery of quasars

In England, where the weather constantly interferes with optical observations, radio astronomy is highly developed. The Mullard Observatory in Cambridge is one of the leading radio observatories in the world. In the late 1950s Cambridge astronomers completed a

the universe and beyond

survey of the sky. The results of the survey became the *Third Cambridge Catalogue of Radio Sources.* As soon as the catalogue was available, many astronomers tried to get photographs of the radio sources listed, but this proved to be difficult. Radio telescopes are not very directional, and the 3C catalogue was therefore not able to say just where a radio source was. It could only indicate a region in the sky and say that the radio source was somewhere in that region.

A photograph of one such region by the 200-inch telescope at Mt. Palomar, near San Diego, California, contained hundreds of star images. The problem was to decide which of these objects was the radio source. In this case, it was easy. One particular object in the photograph looked very strange; it appeared to be an exploding galaxy, or perhaps the remains of a supernova. Astronomers identified it as a radio source. In most cases, however, the regions contain only star-like objects or faint galaxies.

Then, in 1962, the moon eclipsed the radio source 3C273. The event was observed using a large radio telescope in Australia, and the exact moment of eclipse was noted. Since the moon's position is very accurately known, the precise position of 3C273 could now be calculated. Next, astronomers at Mt. Palomar aimed the 200-inch telescope at that spot, and discovered what appeared to be a faint blue star. In a similar manner the radio source 3C48 was also found to look like a faint blue star. This was very puzzling. Astronomers could not understand how a star could be a radio source.

In 1963, Maarten Schmidt used the 200-inch telescope to obtain a spectrogram of 3C273. The spectrum looked very peculiar. The spectra of stars usually have dark lines, but the spectrum of 3C273 had bright lines. Furthermore, Schmidt could not identify the lines in the spectrum.

On February 5, 1963, Schmidt had a sudden insight. He realized that the spectrum of 3C273 was the familiar spectrum of hydrogen, but red-shifted by an enormous amount. It indicated that the object was traveling away from us at 45,000 km/sec. If Hubble's Law held true, it was almost 3 billion ly away. Obviously, 3C273 could not be a star at all. It was at least a galaxy, and one of the most distant galaxies known.

Since then, other similar objects have been discovered, all of them with large red shifts. Some are receding from the earth at up to 270,000 km/sec—almost 91 percent of the speed of light. Since they are not real stars, but only look like stars, they were called **quasi stellar radio sources,** a name which has been shortened to **quasar.** Over 200 quasars are now known.

If quasars obey Hubble's Law, they must be extremely distant and extremely luminous. Calculations show, in fact, that quasars are hundreds or even thousands of times more luminous than a normal galaxy. They are by far the most luminous objects known.

The luminosity of a quasar is often variable. The brightness of

Quasar 3C273 shows a faint streak in this long-exposure photograph. In general, quasars are difficult to distinguish from stars, except by studying their spectra.

the nature of quasars

the universe and beyond

3C273, which has been under fairly regular observation for about a hundred years, varies with a rough period of about 13 years. During all those years it was considered a blue star, and its special character was not suspected. Another quasar became twenty times brighter over a period of 3 months, and still another changed its brightness in a week.

The rapid variability of quasars tells us something important about them. It has been demonstrated that an object which varies rapidly in luminosity cannot be very large. For example, an object that varies in a month cannot be more than a light month across. Many quasars vary within this short period, and must therefore be about a light month across, or even less. This, of course, makes them much smaller than galaxies, which are tens of thousands of light years across.

How can an object which is a million times smaller than a galaxy be a thousand times brighter? So far, no one knows. Many astronomers have offered explanations for quasars. Some, for example, have tried to show that the red shifts of a quasar do not really indicate great distances. None of these explanations, however, has been entirely successful, and quasars remain a great mystery.

the 3K cosmic radiation

According to the Big Bang theory, a million years after its formation the universe became cool enough for electrons and protons to form hydrogen. Before this happened, the universe obstructed radiation. It was opaque. When all the free electrons had combined with the protons, the universe became transparent. Thereafter, radiation and matter hardly interacted at all.

When radiation stopped interacting with matter, the temperature was around 3,000K. The radiation resembled that emitted by a household bulb, which is a 3,000K hot body. As the universe expanded, the radiation was stretched out. Since the universe has expanded to a thousand times what it was then, so has the radiation. It now resembles radiation emitted from a very cold body, one whose temperature is only 3K. The radiation is no longer visible light; it is radio waves.

The 3K cosmic radiation was discovered in 1965. In a sense, it is the most distant thing that we have observed. Even the largest quasar has a red shift which is only $1/200$ that of the cosmic radiation. (In dealing with these large red shifts we have to use formulas based on the theory of relativity. The calculation of distances to objects on the basis of their red shifts is not straightforward.) As we saw earlier, the radiation is coming at us from all directions, a fact which seems to confirm the Big Bang theory.

the ultimate question

If we assume, as most astronomers do, that the Big Bang theory correctly describes the development of the universe, we are still faced with perhaps the most important question of all. As we said

the universe and beyond

earlier, the universe "appeared" about 20 billion years ago. But *how* did it appear? Where did that radiation and matter originally come from? Can we explain how, at one moment, nothing existed in space, and at the next, the universe existed?

There must be an explanation, of course. Explanations have been proposed by philosophers and religious thinkers throughout all of human history. But so far, cosmologists have failed to find one. Although they believe that they have accurately described the beginning of the universe, they have no answer to the question of what came before the beginning. It is possible that no explanation will ever be found. But the search for an answer to this ultimate question will probably be the driving force behind scientific investigation throughout all of history to come.

The Trifid and Lagoon Nebulas in the constellation Sagittarius, seen against vast numbers of background stars belonging to our galaxy.

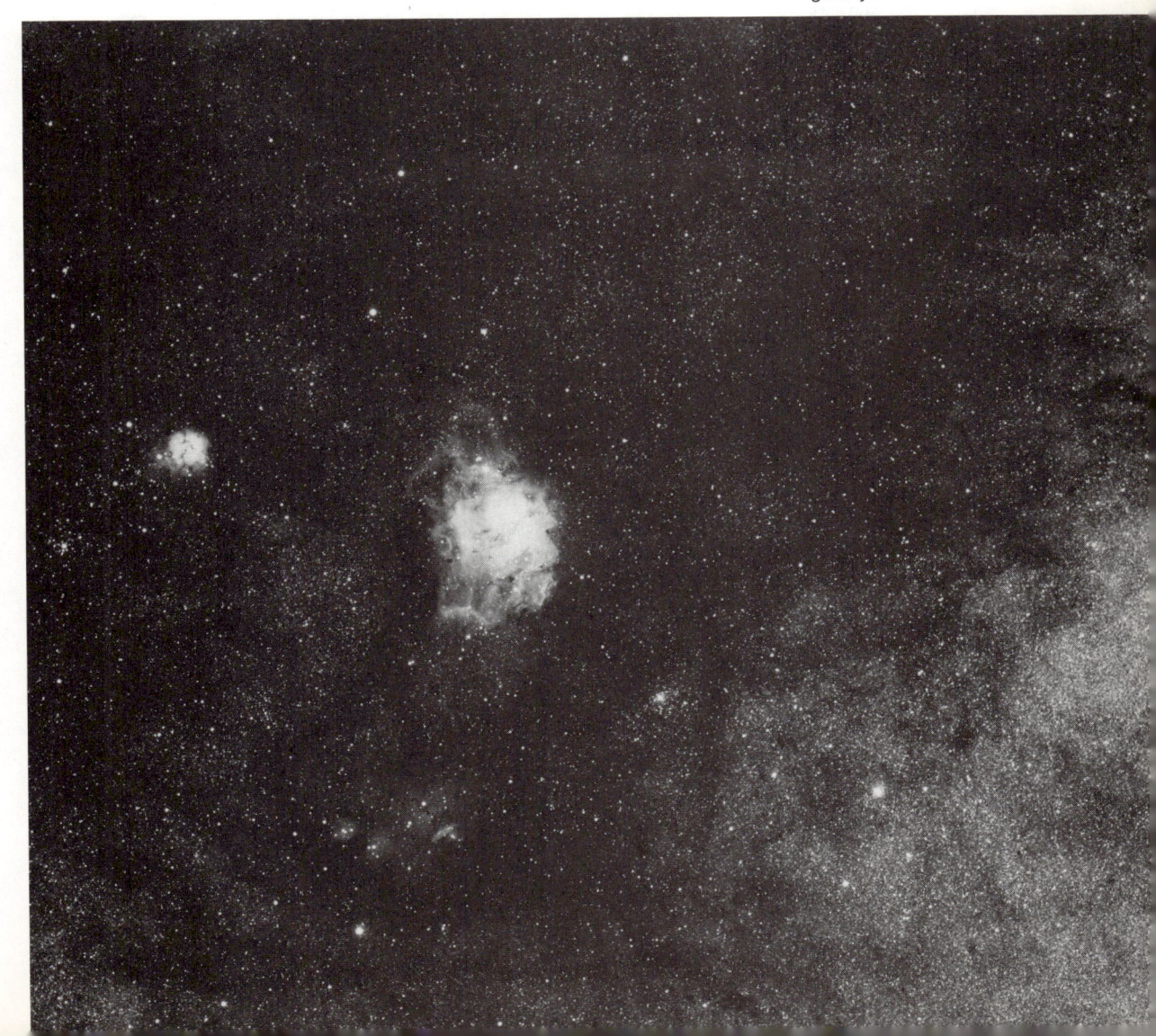

the universe and beyond

SUMMARY

The largest telescopes can photograph billions of stars, but only about 6,000 of these are visible to the naked eye.

The distance to a nearby star is calculated from its parallax. The brightness of a star depends on its distance and its luminosity. Brightness decreases n^2 times if the star's distance is increased n times.

The color of a star tells us its temperature. If obscuring matter alters the color, we can note which atoms are producing absorption lines in its spectrum, and calculate the star's temperature and composition from this.

If stars belong to binary systems we can calculate their masses. The luminosity of a star depends on its mass. Massive stars have short lifetimes because of their extremely high luminosities.

The Herzsprung–Russell diagram is used to plot luminosity against color. When stars are plotted they fall into a definite pattern. Most lie in a diagonal region of the diagram called the main sequence.

Since stars do not collapse under their own gravity they must have very high internal pressures, and thus extremely hot interiors. Only nuclear energy can explain the enormous quantity of energy radiated by stars. In their interiors, 4 hydrogen nuclei are converted into one helium nucleus. When a certain amount of helium has accumulated at the center, main-sequence stars evolve into red giants or supergiants.

Short-lived stars are always found near clouds of dust and gas, supporting the theory that they are formed when interstellar clouds collapse.

Giant stars are near the end of their lives. They can end their lives by becoming white dwarfs, neutron stars, or black holes. Neutron stars are also known as pulsars.

Galaxies contain billions of stars held together by gravitational attraction. Our galaxy, the Milky Way, contains both Pop I and Pop II stars, and star formation is continuing within the spiral arms. Galaxies may be elliptical, spiral, or irregular. Half of them occur in clusters. Our Local Group is a cluster containing about 17 members.

Cosmology is the study of the entire universe. Two rival cosmological theories are the Big Bang theory and the Steady State theory. The latter is now rejected by most astronomers.

Quasars appear star-like in photographs. But they are extremely distant objects that are hundreds of times more energetic than galaxies.

Radio waves are coming from all directions in space (the so-called 3K cosmic radiation). They are believed to be a result of the Big Bang that created the universe.

REVIEW QUESTIONS

the universe and beyond

1. What is the distance to a star whose parallax is $1/10''$? What are the limitations of the parallax method?

2. Two stars are the same size. One is red and the other is blue. Which one is more luminous?

3. How can we determine the mass of a star? How is its luminosity related to its mass?

4. How does the lifetime of a star that is ten times as massive as the sun compare with that of the sun?

5. Draw an H–R diagram. Indicate on it the main sequence, the sun, the red giants, and the white dwarfs.

6. Why do we believe that stars are powered by nuclear energy?

7. Describe several ways in which stars end their lives.

8. Stars A and B are white dwarfs. Star A is twice as massive as Star B. Which one is larger?

9. Is an open cluster or a globular cluster more likely to contain older stars?

10. What are neutron stars?

11. How were interstellar molecules discovered? Why were they not discovered earlier?

12. How are interstellar dust grains formed?

13. According to Hubble, what is the relationship between the distance to a galaxy and its red shift? If a galaxy is receding from us at 1700 km/sec, how far is it from us?

14. Compare the Big Bang theory with the Steady State theory.

15. What is the 3K cosmic radiation? Why does its presence support the Big Bang theory?

16. What are black holes? Why are they hard to detect?

17. What are quasars? How was the first quasar found?

18. Describe three different types of galaxies.

19. Describe the structure of our galaxy. Why is it called the Milky Way?

PART 6

ENVIRONMENT

CHAPTER SEVENTEEN

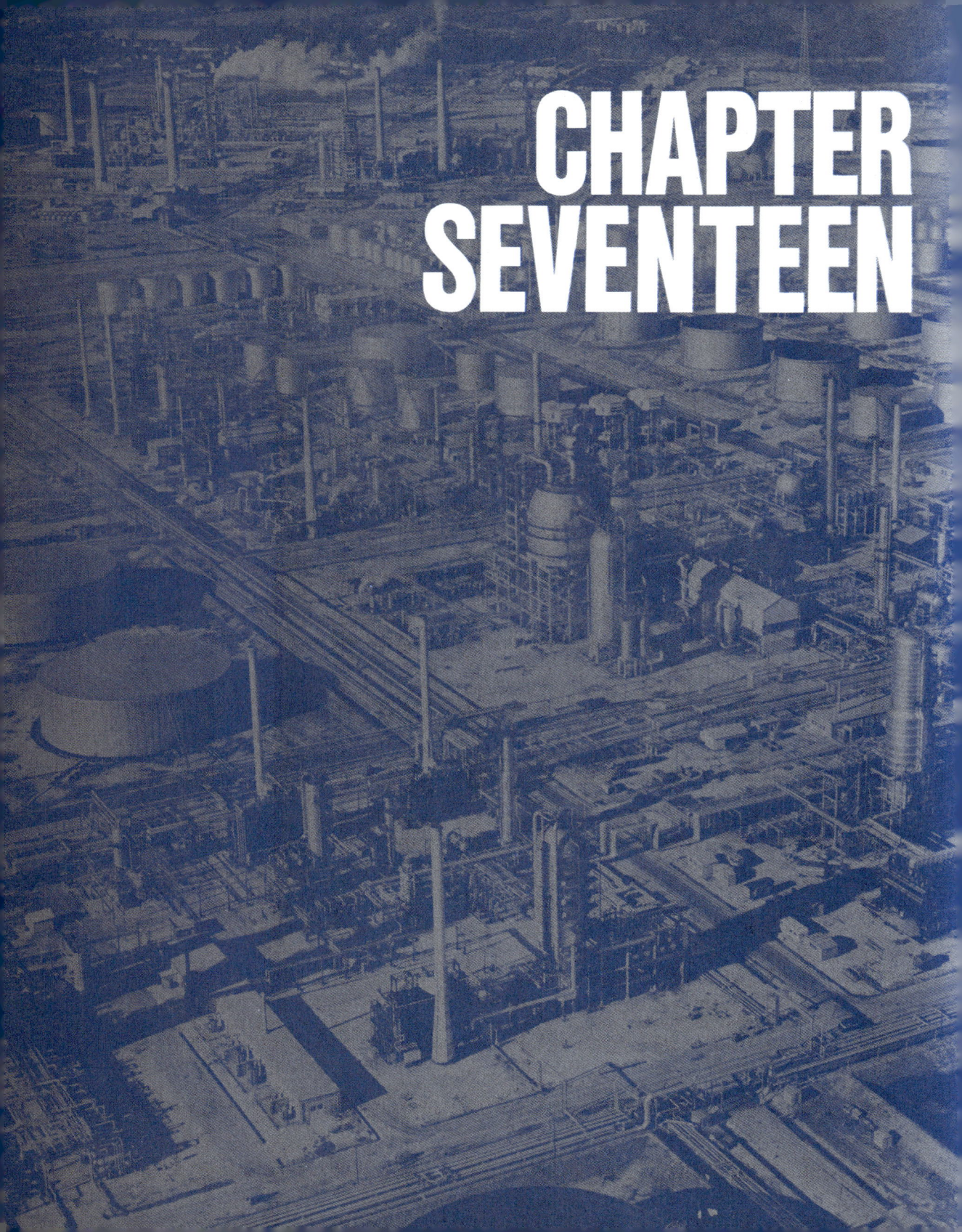

CURRENT ISSUES

Donora, Pennsylvania in 1948 would have been very low on anyone's list of beauty spots. It was a dirty, gritty industrial town of some twelve to fourteen thousand, crowded into a deep bend of the Monongahela River south of Pittsburgh. Most of its people depended for their livelihood on three huge mills — a steel mill, a wire coating plant, and a combination zinc smelter and sulfuric acid plant. Thick, acrid-smelling, multicolored smoke poured from the factory stacks night and day, and hung heavy in the air, trapped by the high bluffs that encircled the town on three sides. Because of the bluffs, and a range of hills on the fourth side, Donora rarely had much wind. Its weather was generally still and damp, and often foggy.

It was foggy on Friday, October 29, 1948. In fact, it had been foggy since Tuesday. The fog had mingled with the smoke from the mills until by now the air smelled and even tasted bad, and it was barely possible to see across the street. Bad smells were not in themselves unwelcome to Donora. They meant full operations at the mills, and hence prosperity for the town.

On this Friday, though, they were about to start meaning something else. At midafternoon an elderly man, an asthmatic, came to his doctor's office. He was having some trouble breathing and needed a shot to help him out. The doctor gave him the shot and thought nothing of it — asthmatics were always bothered by the fog. The man had scarcely left the office when another person staggered up the stairs, moaning and choking. Then the phone began to ring.

The phones kept on ringing for around thirty-six hours — the phones of doctors, the Red Cross, the police, the fire department, anyone who might be able to help hundreds of desperately ill people — vomiting, coughing, choking, spitting up blood, struggling desperately to breathe. In the small hours of

current issues

Saturday morning, another phone rang. It was the undertaker's. The first person had died.

On Sunday morning the mills finally shut down, though the management was still sure their emissions had nothing to do with the sickness. Sunday afternoon it began to rain, and the worst was over. On Tuesday, when most of the dead were buried, the weather was sunny and beautiful.

When news of the Donora "killer fog" got abroad, it caused a national shock and led to an intensive investigation. The final statistics showed that some 6,000 people had been ill. About 1,400 had sought help from the few overworked doctors in a town that had no hospital. Seventeen had died, or maybe it was twenty. How had it happened? Several ordinary factors—the circle of bluffs and hills, the normal seasonal fog, the daily flood of irritating chemicals and soot poured out by the factories—had combined with one unusual factor, a temperature inversion that pinned down the polluted air inside the town. The killer smog was, as we are now beginning to realize, only one of the unpleasant things an outraged earth can do to human beings who abuse it for too long.

a look at ecology

The physical and biological features of the earth form a variety of **systems.** For instance, a forest, including its special terrain, climate, and life forms, is one sort of system. An ocean, with its plant and animal life adapted to a marine environment, constitutes a different type of system. The earth is a vast network of neighboring systems, which can be viewed as independent entities. When we examine them more closely, however, we can see that all the systems are interconnected. The rain that falls on the forest runs off the land into streams and rivers, and finally into the sea, carrying with it accumulated materials. Winds carry evaporated water back from the sea to fall on the forest as rain and snow. In this way, systems are connected to each other by repeated patterns of activity, or **cycles.**

Ecology is the branch of natural science that studies the system of relationships which includes everything on earth. This science views our planet, or any given part of it that can be isolated, as a balanced interrelationship between energy and matter—a relationship in which a change in any part is followed by corresponding changes in other parts. Within the overall system of the earth are many smaller local systems, known as **ecosystems.**

Ecosystems are analyzed by the ecologist in terms of matter and energy. For practical purposes, the amount of matter on the earth is fixed. Minor additions or subtractions, such as the falling of mete-

current issues

orites or the departure of space probes, are negligible from the ecologist's point of view. But in regard to energy the situation is very different. Energy is constantly reaching the earth in the form of light and heat from the sun.

In this chapter, we will look at the ways energy and matter interact as they move through the energy-flow paths and the matter cycles of ecosystems. This interaction is going on all the time in the earth's natural systems. When it is altered, the system in which it occurs is altered also.

energy

The explosion of an atom or hydrogen bomb represents a controlled release of the energy locked in the nucleus of the atom.

Energy flows in a steady, one-way stream through our planet's ecosystems. Most of our energy comes from solar radiation, either directly, in the form of the immediate warmth and light of sunshine, or indirectly, as stored-up solar energy. There are other energy sources also, including the radiation from other stars, the gravitational pull of the moon, and the internal dynamics of the atom. But these are minor; the sun is by far the earth's most important source of energy.

There are several different ways to look at energy. Physicists take the most general and fundamental viewpoint, regarding matter and energy as abstractions. A more specialized view is taken by biologists, who consider energy from the standpoint of its flow through

current issues

an ecosystem. Still more specialized is the viewpoint of economists, who examine the uses of energy within human society. But for the purposes of earth science, we will have to combine these viewpoints.

Science makes a distinction between **kinetic** and **potential** energy. Kinetic energy, as we saw in chapter 11, is energy at work; potential energy is energy ready to do work. The calories in a piece of coal are potential energy until it is burned, when they become kinetic energy. One of the things we will be discussing is how energy is stored up as potential energy and released as kinetic energy.

A fundamental law of physical science states that molecules and atoms in all systems tend toward a condition of greater disorder in which the energy available for work **(free energy)** continually decreases as randomness **(entropy)** steadily increases. To put it simply, the entire universe is winding down. It is approaching a state in which no free energy will exist, and thus no power of action or motion.

The end result of this increase in entropy is ages away, and we need not take it into account for earth science. However, it does apply to this extent: the fresh supply of energy which our planet is continually receiving from the sun goes through an enormous number of transformations as it passes through the earth's ecosystems. We have various uses for it in its various forms. But energy always passes finally to a form in which we cannot use it. At the end of its one-way path, it becomes unavailable for our purposes. The wise use of energy depends upon how well we are able to exploit it in its many forms before entropy has put it beyond our reach.

energy pathways

Energy travels many different paths through the earth. For instance, the sun heats the surface of the earth, causing water to evaporate and the air to expand. This, in turn, produces winds and clouds in the atmosphere, and our weather patterns are the result.

Energy thus transformed can be used in various ways. For example, the energy of the wind can be harnessed by windmills and sailing ships. The energy of water evaporation is transformed into the kinetic energy of falling rain, and then of flowing streams, which can be harnessed by hydroelectric generators.

Living things seem to be an exception to the principle of entropy, which says that the universe is tending toward randomness. Life is anything but random. New living matter is being formed all the time, and even at the simplest level it reflects a high degree of organization. Molecules are intricately arranged to form cells, and each part of an individual cell is precisely structured to perform a particular function. In higher organisms cells, tissues, organs, and organ systems reach an extremely complex level of organization.

But this exception to the principle of entropy is only apparent. Biological systems are not exempt from the laws which govern nonliving systems. The only reason that cells can maintain order, and

current issues

hence life, is that they can extract energy from their environment and use it to perform work.

The ultimate source of energy for nearly all organisms is the sun. Green plants, algae, some protozoa, and several types of bacteria are able to capture radiant energy and use it to manufacture carbohydrates and other complex molecules, all of which store large amounts of energy in chemical bonds (see chapter 2). Organisms which can convert the energy of sunlight into the potential energy of chemical bonds are called producers or **autotrophs,** from the Greek words *autos* (self) and *trophe* (food). Organisms which are ultimately dependent on autotrophs for their food are called consumers or **heterotrophs,** from the Greek *hetero* (other) and *trophe*.

Several industries pour their wastes into the Hudson River. The resulting pollution is only all too apparent in this photograph.

All animals are heterotrophs. Although animals can bask in the sun for warmth, they cannot convert solar radiation into the chemical energy needed to drive cellular processes. Only green plants, algae, and some bacteria can do this, in the process called **photosynthesis.** To obtain the energy they need, animals must eat either plants or other animals. Plants thus form the only link between the living and nonliving communities, drawing water and minerals from the soil and combining them with sunlight and carbon dioxide from the air to make the food which supports animal life.

Plants may be eaten by small animals, such as insects, which in turn are eaten by larger ones, which are eaten in their turn by still larger animals. This sequence of eating and being eaten, with the resultant transfer of energy, is known as a **food chain.** When plants and animals die, their remains become food for scavengers and microorganisms. Their nutrients are thus returned to the food chain for reuse. Of course, this is a simplified way of stating what happens. There is so much interaction among organisms that in an ecosystem the food chain actually is a **food web,** involving the total pattern of

current issues

food relationships in a community.

At each step in the food chain, energy is transformed. It is sometimes useful to know how much energy the plants in a certain ecosystem are capturing. The term **primary productivity** refers to the amount of energy produced by photosynthesis in a given area in a given amount of time. **Secondary productivity** is the rate of transfer from producer to consumer levels in the food chain.

There is a great loss of energy as food passes up through the food chain. Typically, only about 0.1 percent to 2 percent of the sun's energy is used to produce food at the level of primary productivity. The various consumers at the different levels of secondary productivity use, on an average, only about 10 percent of the energy they consume. This helps us to understand why there are seldom more than four or five links in the food chain. Very little energy per square meter is available to carnivores at the fourth or fifth level, and they must be large and active, with an extensive range, if they are to survive.

The range required to support a human depends upon where the human stands in the food chain. Humans who eat diets rich in meat require more land to support them, other things being equal, because the animals which they eat pass on only about 10 percent of the energy they have consumed. Humans who eat a largely vegetarian diet occupy a far more efficient niche in the food chain than meat-eaters.

pollution

Pollution is the general term for a disruption of an ecosystem. Of course, ecosystems can be disrupted in a great many ways. The great geological upheavals of the earth during its early history brought about major changes in ecosystems. These movements are continuing at a very slow pace, and are accompanied by slow but perceptible environmental changes. The term *pollution* generally refers to the disruption of an ecosystem which results in harm to humans, or to other life forms on which humans depend for their wellbeing. In this chapter we shall consider chiefly pollution that is a result of human activity.

Today the increasing human population of the earth and the rapid advances in technology are bringing about ecological change at an abnormally fast rate. An obvious example is the transformation of the American West during the last century. What once was open prairie, traversed by huge herds of migrating buffalo, is now fenced farm and pasture land. The ecosystem has been altered. This alteration has made it possible for many more humans to be supplied with food and the other necessities of life. The settlement of the prairies is generally regarded as a beneficial rather than a harmful change.

current issues

Pollution of their fresh-water environment has killed these pond fish, either by direct poisoning of the water or by depletion of the water's oxygen content.

Ecosystems are exceedingly complex, and it often happens that a change produces some results that are helpful to human life, and others that are harmful. In our national debates, we try to calculate the possible gains and losses that may result from deliberate or accidental changes in the environment. This is the purpose of the environmental impact studies that are now required before certain kinds of programs can be approved by government authorities.

The agent of pollution is called a **pollutant.** It can be a material substance, a form of energy, or a complex interaction between these. No substance or form of energy is a pollutant in itself. It becomes a pollutant only when it disrupts an ecosystem in a way that produces harmful results. Here are some forms of pollution that earth scientists believe are especially dangerous to our planet right now.

air pollution

Most air pollution comes from smoke, but not everything in smoke is a pollutant. Water vapor, for example, is not. The polluting substances given off by combustion depend upon what is being burned and how it is being burned.

In the United States, the most important sources of air pollutants are automobiles and coal-burning electrical generators. The large, "gas-guzzling" cars which Americans traditionally have preferred burn more fuel than smaller models, and accordingly they produce more pollution.

The chief pollutants emitted from these sources are carbon monoxide, carbon dioxide, sulfur oxides, nitrogen oxides, hydrocarbons, and flyash, which is also called **particulate matter.** These sub-

current issues

In addition to the momentary discomforts it causes, such as coughing and watering eyes, low-lying smog over cities can have long-term damaging effects on health.

stances have different chemical properties, so thay cause different kinds of damage. Some are washed out of the atmosphere fairly rapidly by rain or snow, while others linger. Still others undergo chemical reactions in the atmosphere. A few air pollutants are not produced by combustion. These include pesticides and some types of radioactive residues which can be diffused through the air.

Air pollution is sometimes aggravated by local weather conditions. As we saw in chapter 12, a layer of warm air sometimes overlies a layer of colder air, forming a temperature inversion. The inversion imprisons the pollutants, which become concentrated. In 1952, such an inversion lasted for two weeks over London, and during that period the number of deaths there was 4,000 above normal.

The greatest health problems arising from air pollution seem to result from persistent exposure at levels that do not cause serious illness right away. Thousands of American communities are affected by air pollution of this sort—the perpetual smog over Los Angeles is a classic example. Residents of these areas may not feel the effects of breathing such air until years later. Statistical analysis shows a striking relationship between air pollution and the bronchitis death rate. Other problems correlated with air pollution include lung cancer, stomach cancer, heart disease, emphysema, and infant mortality.

groundwater pollution

Lakes, streams, underground aquifers and other in-shore waters may be carriers of pollutants. The worst sources of water pollution are industry, agriculture, and households. Manufacturing industries discharge organic matter, inorganic chemicals, and suspended solids; the chief offenders are the chemical, paper, food, and primary metal industries. Some of these materials may be **carcinogenic**—that is, they may be cancer-causing agents. One example is the PCBs, industrial waste products discharged until only recently in American waters. PCBs have been shown to cause cancer in laboratory animals.

Los Angeles, with its industry and its unusually high percentage of automobiles per capita, is the smog capital of the United States. The byproducts of internal combustion engines are a major component of smog.

Agricultural industry discharges a great deal of organic matter into water, largely in the form of animal wastes from feedlots. Agriculture is also the source of most of the pesticide residues and chemical fertilizers which pollute our water. Human wastes from households consist chiefly of organic matter and various suspended solids. But households also discharge a number of chemicals, primarily phosphate-containing detergents. In an effort to combat this problem, many communities now prohibit the sale of detergents which have a high phosphate content.

These discharges threaten us directly when they pollute our drinking, cooking, and recreational waters. They also threaten us indirectly, because these waters provide a home for the active and varied community of living things that is linked closely with life on land.

Much of the pollution in streams and rivers eventually reaches the sea, and pollutants from the air are deposited there as well. The oceans are also used as a dumping ground for raw sewage and sludge from coastal cities. Once in the oceans, the pollutants accumulate.

ocean pollution

current issues

There is no further dumping place, so the oceans cannot cleanse themselves as the air and fresh water do.

As the polluting chemicals accumulate in the sea, they begin to pass up the food chain and to be concentrated at its higher levels.

The beach at left has been heavily polluted by oil spilled from a passing ship. The long-term effects of such pollution are not yet known.

Oil spills pose a major threat to marine life. Birds, in particular, are often killed by accumulations of oil in their feathers which leave them unable to swim or fly.

One example is mercury, which is absorbed by small organisms, and stays in their systems. When those are eaten by bigger creatures, and those by still bigger ones, the mercury becomes more concentrated at every step. If human beings eat fish which contain heavy concentrations of mercury, they can suffer mercury poisoning. Even in very small amounts, mercury does irreparable damage to the nervous system. During the nineteenth century, hatmakers used mercury in their work. The brain damage that so many of them suffered gave rise to the expression, "mad as a hatter."

As recently as 1971, large quantities of swordfish and tuna were taken off the market in the United States, after tests demonstrated that they contained high levels of mercury. Even more tragic was the appearance of "Minamata disease" in Japan a few years ago. This disorder of the nervous system reduced many of its human victims to deformed, helpless cripples before its cause was finally traced to large amounts of mercury dumped into the ocean by cadmium manufacturing industries. The mercury had moved up through the food chain until it reached toxic levels in the fish which form an important part of the Japanese diet.

current issues

Oil spills from the huge tankers which supply petroleum to the industrial countries are another form of ocean pollution that is becoming increasingly serious. The oil washes up on beaches, killing sea birds and turning the sand into a sticky mess. It may also damage shellfish beds and the spawning grounds of other kinds of sea life. No one knows what long-term effects will result from this accumulation of oil in the oceans.

thermal pollution

The pollution we have been considering so far has been caused by the discharge of matter. Thermal pollution, however, is the result of the discharge of energy, in the form of heat. Many kinds of engines, such as the internal combustion engines of automobiles and steam-powered electrical generators, produce heat as a waste by-product. They must have cooling systems, such as a car's radiator, to disperse excess heat into the atmosphere.

Such heat discharge not only is a waste of energy, but can also have a polluting effect. Heavy traffic, for example, helps raise the air temperature in large cities. Electrical generators commonly use water for cooling systems and then return the heated water to the lake or river from which it was taken. This can raise the water's mean temperature enough that the fish living in it either die or are forced to migrate to cooler waters.

light pollution

Another growing problem is light pollution. The street lights and advertising signs in our large cities radiate so much light into the atmosphere that astronomers at such large observatories as Mount Palomar in California cannot see the fainter heavenly bodies they are

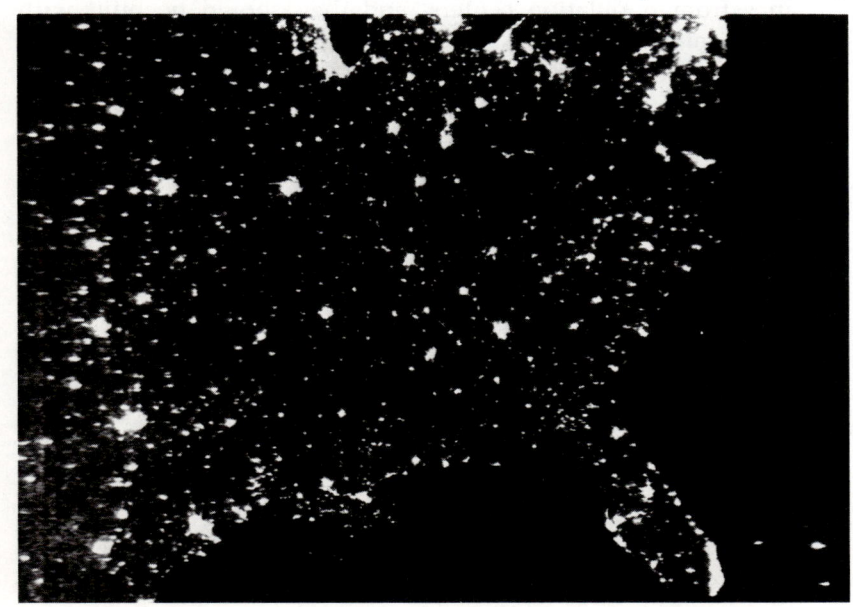

A satellite photograph of the United States, taken at night, shows that cities are a major source of light pollution.

current issues

radioactive wastes: a special problem

trying to observe. Some communities, including Tucson, Arizona, and Richland, Washington, have adopted regulations to keep artificial lights from brightening the night sky.

Pollution from atomic energy is a comparatively recent problem. Nuclear energy was not available until the development of the atomic bomb during World War II. Since then, it has been widely employed both for weapons and for peaceful purposes. Of course, the military use of nuclear weapons poses dangers that go beyond the problem of pollution. But even the peaceful use of atomic energy presents some difficult questions. What is an acceptable level of radioactivity in our environment? How can nuclear plants be prevented from dispersing radioactivity into the atmosphere? Where are the waste materials of such plants to be deposited? These are matters on which specialists disagree.

In view of this uncertainty, many people believe that the safest course is to proceed very slowly and cautiously with our development of nuclear energy. But others argue that going slowly has its own dangers. Our population and our economy are growing, they point out, and additional sources of energy are needed to meet their demands. One such source is nuclear energy. If it is not available, they say, we must burn ever-larger amounts of coal and oil to generate electricity, and the polluting effects of such combustion are proven. It is a case of a known danger on one hand balanced against an unknown danger on the other.

A compromise view is that the construction of nuclear power plants should proceed. But these should be located at a considerable distance from population centers, and they should be built with stringent safety standards. This course is being pursued in the United States; the debate over their safety is continuing.

Radioactive wastes from nuclear reactors present a special pollution problem. The energy from these reactors is badly needed, however, so efforts are being made to devise a safe method for storing the wastes.

current issues
resources

The earth is a generous provider of good things, but the supply is not endless. Furthermore, human demand for them threatens to increase indefinitely, for two reasons. First, the human population of the earth has been expanding at a high rate; in many countries the population has already reached the limit of the food supply, and only fam-

The United States offers a wide variety of beautiful scenic views. This is not one of them.

ine or some other disaster seems able to reduce the numbers. Second, the wealthier nations, with their industrial economies, are using the earth's riches in a very wasteful way. To find an example, you need look no farther than your own trash can. Each American discards about four pounds of solid wastes a day. There is no limit to the amount which careless humans can waste, but there is a very definite limit to how much the earth can supply for them to waste.

These things which our planet provides naturally are called our **resources.** Resources may be either material substances or forms of energy, and they may be **renewable** or **nonrenewable.** A nonrenewable resource is one which exists in a permanently limited amount, such as iron. There is a fixed amount of iron in the earth's crust, and no more will grow. A renewable resource is one which can replenish itself if allowed. An example is wood. For practical purposes, most of the material resources of our planet are fixed: they are

current issues

mineral resources

nonrenewable. But we are constantly receiving a fresh supply of energy, in the form of heat and light from the sun.

Metals are an absolute necessity to our industrial civilization. They form the skeletons of our skyscrapers and bridges; and almost all our vehicles, from the jumbo jet and the supertanker to the motor car and the bicycle, are made of them. As we produce and use more of these things, we are using increasing quantities of the nonrenewable metals.

The production of these goods also consumes large quantities of fuels. For the most part we use the fossil fuels—coal, oil, and natural gas. Strictly speaking, new fossil fuels can eventually be formed; but this takes geological ages, and we are now consuming in decades these resources which took millions of years to form. For practical purposes, therefore, fossil fuels must be regarded as nonrenewable.

It is predicted that at the current rate of growth in consumption, half of the known global reserves of industrial metals will be virtually gone within fifty years. These include tin, zinc, lead, copper, tungsten, manganese, aluminum, nickel, and chromium. None of them is expected to last a full century if industrial growth continues at its present pace. Already the richest iron ores of Minnesota, which were used to make steel in the blast furnaces of Pennsylvania, are running out. Now lower-grade ore must be mined to furnish iron, and the lower-grade ore is more expensive to process. The only reserves now remaining are those which are most difficult to extract from the earth. The cost of mining them can be expected to increase steadily.

The reserves of precious metals—gold, silver, and the platinum group—are vanishing even faster. We think of these metals as being primarily ornamental, but they have important industrial uses as well. At the current rate of consumption, all gold reserves should be exhausted in eleven years. If consumption increases at its present rate, the gold reserves will last only nine years.

The situation with fossil fuels is currently being debated. Some critics of the petroleum industry challenge the industry's estimates on reserves. But no one doubts that those reserves are limited. A probable estimate is that known global reserves of petroleum and natural gas may last 15 to 20 years. Those estimates are somewhat uncertain, however. New reserves may be discovered, and official estimates fluctuate. For instance, in 1975 the U.S. Geological Survey cut its estimate of American petroleum reserves in half.

The situation with coal is more favorable. Known global reserves are estimated at 5,000 billion tons, enough to last 2,300 years at the current rate of consumption. However, the rate of consumption is increasing quickly, because of the shortage and expense of oil and natural gas. If consumption continues to grow at this rapid rate, the global reserves of coal are only enough to last us for 111 years.

TURNING GARBAGE INTO FUEL

Each year, American cities produce a total of 140 million tons of garbage. Caught in the energy squeeze, more and more municipalities are doing with their trash what some European cities have been doing for years—converting it into energy, chiefly by burning it as fuel. Today, about thirty such trash-processing facilities are operating in this country, and about a hundred others are in the works or on the drawing board. Experts have estimated that if all major cities were to burn their trash as fuel, approximately 145 million barrels of fuel oil could be saved each year. What's more, the metal-reclamation facilities that are typically part of these plants could save about 20 percent of America's tin, 8 percent of its aluminum, and 7 percent of its iron.

One of the largest trash-burning and metal-recycling plants has recently begun operating in Baltimore, Maryland. This facility is designed to process a thousand tons of solid municipal waste per day. It is expected to save 15 million gallons of fuel per year and to reclaim about 70 tons of ferrous metals—iron and steel—each day.

Here's how the Baltimore plant works: first, trucks deposit municipal garbage at the facility, where it is shredded and then transferred to a kiln. In this 100-foot-long air-tight chamber, the trash is broken down by heating to temperatures up to 1,800 degrees Fahrenheit (1,000°C), a process known as pyrolysis.

Next, the trash is placed in a water tank to cool, and to allow the heavy residues to settle to the bottom. These heavy materials pass through a magnetic separator which extracts the ferrous metals. The remainder of the material, a glassy aggregate, is drawn off and used for road construction. The hot gases released during pyrolysis are transported to boilers which convert them into steam. The steam can then be used to heat or air-condition as many as half the buildings in the city's downtown area. Other similar trash-processing plants which are operating or planned include those at Riverside, California; New York City; and at a number of sites run by the Tennessee Valley Authority in Tennessee, Alabama, and Kentucky.

IN FOCUS:
is our technology destroying us?

It is the Year of Our Lord 2000. New York, Tokyo, and London sit under twenty feet of ocean. So do many of the world's other metropolises, including Washington, Buenos Aires, and Shanghai. In these cities, most of the old buildings have crumbled, and the waves are now assaulting the lower stories of the new skyscrapers. Panic has broken out among the populace who have fled the rising waters and are battling one another for dry space in inland regions. All told, more than 15 billion souls have become refugees, scavenging for food and a dry place to settle.

Many climatologists believe that this grim vision of the future could become a reality if we continue to heat up the earth with thermal pollution from our industries. The waste heat produced by power-hungry industrial nations could be transported to both poles by wind and ocean currents, causing the climate in these regions to grow warmer. Melting of the polar icecaps would raise the level of the oceans by as much as 200 feet. If power continues to be gobbled up at the present rate of growth — four to six percent annually — our planet could be awash by the turn of the century. Not all scientists agree with this gloomy scenario, but the mere fact that some consider it a possibility points up the importance of paying closer attention to the ways our technology may be inadvertently altering our climate.

One way we may be modifying the climate is by pumping more and more carbon dioxide (CO_2) into the atmosphere as we burn fossil fuels. The amount of carbon dioxide in the air has risen from an average of 290 parts per million (ppm) in 1850 to around 320 ppm at the beginning of the 1970s — a jump of ten percent. Atmospheric CO_2 is one agent behind the greenhouse effect, reflecting thermal radiation emitted by the earth back to its surface. The greater the CO_2 content of the atmosphere, the more radiation is reflected back to earth, and the higher the temperature becomes at the planet's surface. It has been calculated that a ten percent upsurge in carbon dioxide would produce a corresponding temperature rise of 0.3 degrees Celsius. By the year 2000, some say, the amount of atmospheric CO_2 will grow by 27 percent as more fossil fuels are burned, and temperatures around the world will rise an average of more than 0.7°C.

But the possibility of such a temperature surge may be offset by another development. Since 1940, global temperatures have been dropping, and now average some 0.3°C cooler. Some scientists believe that this drop has resulted because the air is carrying an increasing load of dust pollutants — also the product of industrial activities. Over some British cities during the mid-1950s, for instance, the amount of solar radiation reaching the earth was reduced as much as 50 percent by airborne dust particles. As dust blocks the incoming solar radiation, global temperatures fall.

This situation is complicated by still other factors. For one thing, the dust particles also reflect the outgoing thermal radiation back to earth; the amount reflected depends partly on the quantity and size of the particles. For another, the way these particles absorb and scatter light is not yet fully understood. And there is always the possibility that the effects of the carbon dioxide and the particulates may cancel one another out.

Another probable human-made villain in the climate modification picture is fluorocarbons—synthetic chemicals composed of fluorine, chlorine, and carbon. During the early 1970s a half million tons of these chemicals were released into the air, chiefly from aerosol spray cans. They ascended to the stratosphere, where ozone normally screens out incoming solar ultra-violet radiation. Ultra-violet can be harmful to life: if enough of it is allowed to reach the earth, the incidence of skin cancer and the rate at which living things mutate and evolve may increase dramatically. In recent years, some scientists believe, these fluorocarbons have wafted up to the stratosphere and destroyed some of the ozone there, permitting more of the deadly ultra-violet to reach the earth.

Another danger to the ozone layer is the supersonic transport, or SST. This huge aircraft cruises in the stratosphere, spewing out nitric oxide which interferes with the chemical processes that form ozone. The SST also releases water vapor, soot, and other waste products that tend to remain in the stratosphere for long periods of time. Large accumulations of water vapor in the air could blanket the earth with clouds, producing a greenhouse effect that would raise the temperature appreciably.

The jump in atmospheric CO_2, the increase in dust, the rise in fluorocarbons—which of them has changed our climate? Which will do the most damage in the future? The simple truth of the matter is that we don't know. Climate change can be brought about by so many complex and interrelated factors—winds, moisture, cloud cover, and ocean conditions, for example—that it is difficult to measure their effects accurately, let alone predict them. Another important consideration is that climates do change on their own, sometimes very radically—and they did so long before industrialization. It is conceivable, therefore, that the recent changes may reflect a natural trend that has been affected only slightly by the increased waste output of our technology.

Climatology is still an infant discipline, and much remains to be learned about climate and the many factors that affect it. We are fortunate, at least, that we now recognize climate as only one of the systems that keep Spaceship Earth going.

current issues

These figures probably do not tell the whole story. We can reasonably hope that some more reserves of minerals will be found. On the ocean bottom, for example, are mineral nodules that have a rich content of manganese, iron, nickel, cobalt, and other metals in trace amounts. The technology for exploiting such ocean-bottom resources now is being developed. However, even if substantial additional supplies of these resources should be discovered, it would not alter the fact that they are nonrenewable. We cannot afford to continue to waste the limited quantities that exist.

biogeochemical resources

As we have already noted, the minerals our global system are fixed in quantity, while energy is replaced and travels a one-way path through the environment. In this process, matter and energy interact. One such interaction is **nutrition,** in which living things employ energy to absorb nonliving matter for their life processes. The materials that living things absorb in this way are called **nutrients.**

Nutrients are constantly being recycled through the ecosystem. The hamburger you eat, for example, came from a steer that was nourished on grains and grasses; those plants, in turn, were nourished by soil partially composed of decayed organisms. About thirty to forty elements are necessary to life, and they are used again and again by successive generations of living things. The circuits which these elements follow between living organisms and the earth's water, air, soil, and rocks are referred to as **biogeochemical cycles.**

Often nutrients become trapped so that they can neither be reached nor used. This may occur through natural processes, but it can also be caused by human activity, which removes large amounts of material from the nutrient cycles. By such practices as logging, harvesting, and mining, humans withdraw vital elements from the nutrient supply in one area, and eventually deposit them in other ecosystems. These, in turn, may be severely unbalanced by the unnatural concentration of elements.

To take an example, phosphate rock is excavated at a current rate of about three million tons a year. The rock is treated with acid, to make it soluble, and then it is spread over fields as fertilizer. Rains eventually dissolve much of the phosphate, carrying it into ponds and lakes. There it nourishes a rapid growth of algae. Ordinarily, their growth would be limited by the normally low level of phosphates in fresh water.

Thus a new biogeochemical cycle is set in motion. The algae die, and nourish bacteria, which consume much of the oxygen in the water. Plants, fish and other organisms which also need that oxygen die for lack of it, and their decomposed forms nourish more bacteria. The cycle continues. Eventually the pond or lake which was once clear, fresh water stocked with a variety of life becomes a pool of stagnant, cloudy water containing little life except algae and bacteria. This process is called **eutrophication.** It represents another kind of waste of the earth's resources.

current issues

509

Spraying fields with pesticides allows more food to be grown on a smaller amount of land. At the same time, destruction of pests disturbs the ecological balance of the region.

conservation

Our extraordinary success at conquering our environment has, at the same time, created a distinct threat to our wellbeing, resulting from pollution and depletion of our planet's resources. **Conservation** includes all the steps being taken to reverse this. Besides continuing their struggle to set aside and preserve wilderness and green recreational areas for the growing population, conservationists now are championing such approaches as the reduction of wastes, recycling of waste materials, reclamation of land, and the search for new sources of energy.

The rising cost of raw materials such as iron and petroleum is expected to reduce waste. The large automobiles of America's past will probably become too expensive for most people. Very soon United States citizens may ride in smaller cars, as the Europeans and Japanese have been doing all along. These rising costs should also make it economically attractive to treat waste and discarded products so that their raw materials can be used again. The trash we are now disposing of in incinerators and landfills may soon be regarded as treasure. Household refuse, for instance, contains metals, glass, paper, food, wood, plastics, and other materials, which can be sorted out and put back to use. Old cars can be recycled as new steel, old paper as new paper, old wood as new wallboard.

Land reclamation is the restoration of terrain that has been damaged for human purposes by some disruption of the biogeochemical resources. The method of reclamation depends to some extent on how the land was originally damaged. Clear-cutting of forests

current issues

and overgrazing of range land are now prohibited on federal reserve lands, and are decreasing on private lands as owners realize how expensive they are in the long run. Contour plowing and earth-moving equipment are being used to halt erosion in such areas. Strip mining is now regulated by many states. Estuaries and salt marshes are being restored by the Army Corps of Engineers in Maryland, and along the Columbia River in Oregon.

All these conservation measures together, however, add up to only a small part of the effort needed to counter the destructive effects of our pollution and depletion of resources. Perhaps as those resources become ever more scarce, conservation will appear increasingly attractive to governments, businesses, and private citizens.

energy and economy

Human beings consume far more energy than is needed to sustain their bodily functions. An enormous amount of energy goes into producing and maintaining all the things we use in our industrial society. Even in subsistence economies, humans use additional energy. As long as a person's energy consumption depended upon the food he could eat, the rate of consumption was about 2,000 kilocalories a day. (A kilocalorie is 1,000 calories.) When humans learned to use fire, daily consumption rose to about 4,000 kilocalories. In primitive agricultural societies with some domestic animals, the rate rose to about 12,000 kilocalories.

At the height of the low-technology industrial revolution—between about 1850 and 1870 in England, Germany and the United States—the per capita daily consumption of energy reached about 70,000 kilocalories. By 1970, the figure in the United States alone was 230,000 kilocalories per person every day. The industrial regions of the world, with only 30 percent of the world's people, were using 80 percent of the world's energy. The United States, with six percent of the people, was consuming 35 percent.

The largest part of this dramatic increase in energy consumption was the steadily rising use of nonrenewable fossil fuels. These account for about 96 percent of the input into the United States energy economy. An increasing proportion of the oil and natural gas consumed in the United States is imported. Net imports of petroleum and petroleum products doubled between 1960 and 1970. In 1974, and again in 1975, energy consumption declined slightly in the United States before resuming its rise. But we can still make a general statement that the world's consumption of energy for industrial purposes is doubling approximately once per decade.

energy resources

It is clear that our use of nonrenewable fossil fuels cannot go on doubling decade after decade. Probably it cannot double even once more. Even if the rate of use does not increase, current supplies will soon be exhausted. Since fossil fuels cannot meet our demand for energy, what other sources of energy can we tap to supply the power requirements of a moderately industrialized world after the fossil fuels are gone?

Solar power could be directly used for such things as home heating units and electrical generators. Houses are now being built with such solar heating units, particularly in parts of the United States that receive a great deal of direct sunshine.

Tidal power can be harnessed by damming a bay or an estuary to hold back the tidal flow. One full-scale tidal-electric plant has been built, on the Rance estuary on the Channel Island coast of France. It has been in operation since 1966.

Geothermal power is obtained by extracting heat from the earth's interior. The difficulty with this is that over most of the earth, the heated region lies too deep to be easily reached—below the oceans or below the earth's thick crust. In Iceland, however, where the Mid-Atlantic Ridge comes above sea level, the crust is thin and heavily fractured, and geothermal heat is close to the surface. Here the water from hot springs is used to heat houses.

Nuclear power plants are the most widely developed of the alternative sources of energy. A number of these are in operation in the United States and elsewhere in the world, and others are planned or under construction. Their contribution to our power supply is expected to be significant. However, as we already have observed, nuclear generators pose serious pollution problems of their own.

energy in the future

From the standpoint of human history, the epoch of fossil fuels will be quite brief. Imagine that you are viewing a span of some 10,000 years, half before the present and half afterward. On such a scale, the complete cycle of the exploitation of the world's fossil fuels will be seen to have encompassed perhaps 1,300 years. The first 500 years will have seen the exhaustion of about 10 percent of the reserves, and the last 500 years will have seen the exhaustion of the last 10 percent. During the peak 300 years in the middle, 80 percent will have been used. Our lifetimes, of course, have fallen in those middle 300 years.

Even in the short run—during our own lifetimes—there will be dramatic changes in energy use. Some of these already are being adopted by forward-looking people. Solar energy and wood are being used to heat houses. Instead of having workers wear business suits and dresses in offices which must then be cooled to 70 degrees in the summer, some employers are having their workers wear clothing adapted to the climate. Americans are buying smaller cars, and commuters are forming car pools. Some consumers are cutting down on the amount of meat they eat.

Practices of this kind can be expected to become more widespread as the old, wasteful way of life becomes too expensive for most people. Other energy-saving ways of living are sure to be proposed.

current issues

SUMMARY

The earth is made up of a variety of systems. These, in turn, are divided into smaller ecosystems. Ecology is the science that studies relationships within and among the earth's systems. Energy flows in a one-way stream through our planet's ecosystems.

In physics, energy is defined as the ability to do work. Kinetic energy is energy performing work, whereas potential energy is ready to do work. Energy can be converted from one kind to another.

The sun is the ultimate source of energy for nearly all organisms. Those which can use it directly are called autotrophs. Organisms which depend on autotrophs for their food are called heterotrophs.

Primary productivity refers to the amount of energy produced by plant photosynthesis. Secondary productivity is the rate of transfer from producer to consumer levels in the food chain. A great deal of energy is lost as food passes up through the food chain.

Pollution refers to the disruption of an ecosystem which ultimately results in harm to humans. Air pollutants released as a byproduct of industry pose a distinct threat to health. Many pollutants released into groundwater may be carcinogenic, such as the PCBs, and may also destroy aquatic life. Pollutants that reach the ocean accumulate, pass up the food chain, and may do great damage to the consumers at the top of the chain. Thermal pollution can cause a rise in water temperature that makes streams and rivers uninhabitable by fish and other water life. Radioactive wastes pose a special problem; unless they can be safely stored, they present a threat to all life on earth.

Our natural resources are classified as either renewable or nonrenewable. Mineral resources, such as metals and the fossil fuels, are nonrenewable, and are rapidly being used up by industry. Renewable resources, such as lumber, can be replaced over the years by careful replanting and foresting.

Conservation includes all the steps being taken to reverse pollution and depletion of our resources. Some of the approaches are reduction of wastes, recycling of waste materials, reclamation of land, and the search for new energy sources. The rising cost of raw materials is also expected to reduce waste.

REVIEW QUESTIONS

1. Describe two different types of systems.
2. Name and describe the effects of several different types of pollutants.
3. Explain the difference between kinetic energy and potential energy.
4. Why is radioactive waste disposal a particularly difficult problem?
5. What is a biogeochemical cycle?
6. Name at least three sources of energy that might be used to replace fossil fuels.
7. Give an example of a pollutant that caused great damage after reaching the top of the food chain.
8. What are the three worst sources of water pollution?
9. Explain how air pollution can sometimes interact with local weather conditions.
10. Distinguish between primary productivity and secondary productivity.
11. Name some types of damage that result from oil spills.
12. Distinguish between renewable and nonrenewable resources, and list at least two examples of each.
13. List several approaches being taken by conservationists to preserve our resources.
14. How does plant life form a link between the living and non-living communities of the earth?
15. What is entropy?

terms and concepts

In this glossary, each entry is followed by a number in parenthesis. This is the number of the chapter in which the term was first introduced.

absorption line: a dark line in a spectrum indicating the absorption of the wavelength of a particular atom. **(16)**

abyssal plains: broad, nearly flat regions of the ocean bottom. **(14)**

air mass: a large body of air having distinct boundaries and characteristics. **(12)**

alkaline earth elements: a group of reactive elements which contain two more electrons than they require for stability. **(2)**

alkaline metals: a group of elements which are highly reactive because their atoms contain one more electron than they require for stability. **(2)**

amorphous mineral: a mineral whose atoms are not arranged in a definite order. **(2)**

anticline: an upward fold of the earth's crust. **(4)**

anticyclone: a system of winds blowing out from a high-pressure region. **(12)**

aquiclude: a layer of impermeable soil or rock. **(3)**

aquifer: a water-bearing layer of soil or rock. **(3)**

arête: a high ridge between adjacent cirques. **(3)**

asthenosphere: the region of upper mantle 60 to 400 km below the surface, consisting of plastic rock. **(3)**

atom: the smallest unit of an element that can exist. **(2)**

autotroph: an organism that uses photosynthesis to convert solar energy into the potential energy of chemical bonds. **(17)**

bar: a ridge of sand built up offshore, usually visible only at very low tide. **(14)**

barometer: an instrument used to measure atmospheric pressure. **(11)**

barrier island: a ridge of sediment built up offshore, usually rising above sea level. **(14)**

base level: the lowest level to which a stream can cut. **(3)**

batholith: a large solidified magma intrusion below a column of sedimentary rocks. **(6)**

bay barrier: a spit that blocks the mouth of a bay completely. **(14)**

bay-mouth bar: a spit that partially blocks the mouth of a bay. **(14)**

beach: the region between low-tide level and high-tide level. **(14)**

berm: the sandy terrace shoreward of high-tide line on a beach. **(14)**

big bang theory: the theory that the universe has been constantly expanding since its birth. **(16)**

binary stars: two stars which are in orbit around each other. **(16)**

biogeochemical cycle: the circuit followed by an element as it is used at successive levels in a system and then recycled. **(17)**

black hole: in astronomical theory, a neutron star which has collapsed and become so dense that light cannot escape from it. **(16)**

Bode's law: an equation that describes the relative positions of the planets. **(15)**

Bouguer anomaly: the difference between the standard gravity for a given location and the gravity that actually exists. **(4)**

Bowen's reaction series: the sequence in which crystallization of minerals occurs in a cooling magma. **(2)**

brightness: in astronomy, the amount of light that falls each second on 1 cm² of the earth's surface. **(16)**

caldera: a large depression formed when the walls of a volcanic crater collapse inward. **(6)**

cepheid variable: a star whose brightness varies in a cyclic pattern at regular intervals. **(16)**

chemical bond: a link between an atom and one or more other atoms. **(2)**

chondrite: a stony–iron meteorite that may be part of original solar-system material. **(15)**

chromosphere: the middle layer of the sun's atmosphere. **(15)**

circles of motion: the patterns of energy transmission within waves. **(13)**

cirque: the steep-walled basin in which a valley glacier originates. **(3)**

clastic rocks: rocks made up of fragments of pre-existing rocks. **(2)**

cleavage: a property possessed by some minerals, which causes them to break along predictable planes. **(2)**

coalescence: process in which medium-sized rain droplets sweep up smaller ones. **(12)**

comet: a chunk of frozen water and gases that travels around the sun in a elongated elliptical orbit. **(15)**

compound: a substance composed of two or more elements. **(2)**

condensation: formation of liquid water from water vapor. **(12)**

condensation nuclei: hygroscopic particles that promote condensation of water vapor. **(12)**

terms and concepts

conglomerate: a detrital sedimentary rock made up of large, rounded fragments. **(2)**

conservation law: a law which states that certain quantities, such as mass, remain constant. **(1)**

continental drift: the theory that continents have come together and separated several times during the earth's history. **(7)**

continental margin: the underwater area bordering a continent. The margin is divided into the continental rise, continental slope, and continental shelf. **(14)**

convection: the transfer of heat by a patterned flow of fluid. **(5)**

convergent boundary: a site at which a lithospheric plate is being subducted. **(8)**

core: the center of the earth, including the region deeper than 2,900 km below the surface. **(5)**

Coriolis effect: the deflection of moving objects to the right in the Northern Hemisphere and to the left in the Southern Hemisphere, caused by the earth's rotation. **(11)**

corona: the outer layer of the sun's atmosphere. **(15)**

Cosmological Principle: the principle which assumes that the universe is uniform. **(16)**

creep: the very slow downhill movement of soil or rock debris. **(3)**

crest: the high point of a wave. **(13)**

cross-bedding: situation in which one or more strata lie at an angle to the general orientation of the others. **(3)**

crust: the thin outer layer of the earth, probably about 40 km thick. **(5)**

cumulonimbus clouds: the distinctive clouds that accompany thunderstorms. **(12)**

cumulus clouds: billowy clouds produced by convective cooling. **(12)**

cyclone: a system of winds blowing in toward a low-pressure region. **(12)**

deflation: the process by which loose surface material is moved by wind. **(3)**

density: the amount of matter a material contains within a given volume of occupied space. **(2)**

desiccation: loss of water from a sediment as it is compacted. **(3)**

detrital sedimentary rock: sedimentary rock made up of fragments. **(2)**

divergent boundary: a site at which a lithospheric plate is being born. **(8)**

doldrums: belt of rising warm air around the equator, with only weak surface breezes. **(11)**

Doppler effect: a change in the apparent wavelength of light as it approaches or recedes from the viewer. **(16)**

ecology: the study of the relationship between living things and their environment. **(17)**

elasticity: the ability of a material to resume its original shape after a distorting stress has been removed. **(5)**

electron: fundamental atomic particle which carries a negative electrical charge. **(2)**

electrostatic force: the force of attraction between two objects carrying opposite electrical charges. **(2)**

element: the simplest substance that can be chemically isolated. At least 88 elements occur in nature. **(2)**

emergent coast: a coast formed by exposure of the former sea bottom. **(14)**

energy: in physics, the ability to perform work. Energy may be either potential or kinetic. **(17)**

entropy: a state of randomness in which more and more energy becomes unavailable for work. **(17)**

epicenter: point on earth's surface directly above the focus of an earthquake. **(5)**

equinox: a day on which the hours of darkness and of daylight are exactly equal. **(1)**

erosion: the removal, transport, and deposition of parts of the regolith by mechanical forces such as water and wind. **(3)**

estuary: a region at the mouth of a river where fresh and salt water mix. **(3)**

eutrophication: the destruction of a fresh-water environment by gradual loss of oxygen from the water. **(17)**

extrusive rocks: igneous rocks formed by cooling of magma at the earth's surface. **(2)**

fall line: the inland point at which a coastline was once located. **(14)**

fathom: nautical depth measurement equal to 6 feet, or about 2 meters. **(14)**

fault: a break in surface rocks caused by unequal stresses. **(4)**

fault block mountains: mountains formed by large-scale normal faulting. **(4)**

fault scarp: the steep face of the upthrown side of a fault. **(4)**

fault trace: the visible part of a fault at the surface. **(4)**

felsic minerals: minerals that contain no iron or magnesium. **(2)**

fjord: a flooded glacial valley. **(3)**

fluvioglacial outwash: material carried away from a glacier by flowing streams. **(3)**

focus: the sub-surface point at which an earthquake originates. **(5)**

foliation: the layered appearance of certain rocks which split easily along their plane of orientation. **(2)**

terms and concepts

food chain: a series of food-dependency relationships. (17)

food web: the total pattern of food relationships within a community. (17)

fossil: the remains or imprint of a plant or animal, usually preserved in sedimentary rock. (1)

freezing nuclei: particles that promote formation of ice crystals. (12)

frequency: the number of waves that pass a given point every second. (2)

front: the border between two air masses. (12)

fumarole: a vent in the ground that expels vapors and fumes. (6)

galaxy: an enormous group of stars which are gravitationally attracted to one another. (16)

geologic column: an orderly time sequence of rocks, from the oldest on the bottom to the youngest at the top. (9)

geyser: a hot spring that undergoes eruption in cycles or at set intervals. (6)

glacier: a large natural accumulation of ice which moves slowly. Glaciers can be classified as valley (alpine) or continental. (3)

globular cluster: a compact, spherical group of Pop II stars. (16)

graben: valley formed when a block drops between two normal faults. (4)

gravity: the force of attraction exerted by every object in the universe toward every other object. (1)

gravity sliding: the gradual downhill movement of a sedimentary stratum. (4)

greenhouse effect: a rise in atmospheric temperature caused by trapping of earth's radiation by CO_2 in the atmosphere. (11)

guyot: a seamount that has been eroded by waves, leaving a flat top. (14)

gyre: a large, slowly rotating region of the ocean. (13)

half-life: the length of time needed for half the atoms in a radioactive substance to decay. (9)

halogens: extremely reactive elements which possess one less electron than they require for stability. (2)

headland: a high ridge flanking a drowned river valley. (14)

heat equator: a region just north of the geographic equator where air is generally rising. (13)

Hertzsprung–Russell (H–R) diagram: a diagram in which the luminosity of a group of stars is plotted against their temperature. (16)

heterotroph: an organism that must get its energy either directly or indirectly from autotrophs. (17)

highlands: the older regions of the moon, covering most of its surface. (15)

hornblende: a silicate with an elaborate chain structure. Hornblende is also known as **amphibole.** (2)

horse latitudes: regions of sinking, hot, dry air about 30° on either side of the equator. (11)

horst: an elevation formed when a block is upthrown between two normal faults. (4)

Hubble's Law: law that expresses the relationship between the speed of a receding galaxy and its distance. (16)

humidity: the water content of the air. (11)

humus: the organic material contained in soil. (3)

hurricane: an intense small cyclone, usually no more than 300 miles in diameter, characterized by high winds and heavy rains. (12)

hydrocarbon compound: sedimentary deposit derived from organic material, such as peat, coal, or petroleum. (3)

hydrologic cycle: recurrent pattern of evaporation and condensation of water on the earth's surface. (12)

hydrolysis: a chemical weathering process in which a mineral combines chemically with water. (2)

hygroscopic: having the property of attracting water molecules. (12)

hypothesis: a tentative but unproven explanation of a natural phenomenon. (1)

igneous rocks: rocks formed by the cooling and solidification of magma. (2)

inclination: the tilt of a planet's axis with respect to the plane of its orbit. (1)

infra-red: a type of longwave radiation, not visible to the human eye. (2)

inert gases: a group of very unreactive elements which have extremely stable electron shells. (2)

inertia: the tendency of mass to resist change in motion. Unless a force acts on it, mass remains at rest or travels at a constant velocity in a straight line. (1)

insolation: incoming solar radiation. (11)

intrusive rock: an igneous rock formed by underground cooling of magma. (2)

inverse square law of gravity: the law which states that the gravitational force between two objects is measured by the product of their masses divided by the square of the distance between them. (1)

ion: an atom that has lost or gained one or more electrons. Ions carry either a positive or a negative electrical charge. (2)

ionic bond: a bond formed between two atoms whose opposite electrical charges attract them to each other. (2)

island arc: a chain of volcanic islands flanked by ocean-floor trenches. (8)

isobars: lines on a weather map that connect areas of equal barometric pressure. (12)

terms and concepts

isostasy: the equal distribution of pressure within the crust and subcrust of the earth. **(4)**

isotopes: atoms of the same element whose nuclei have different masses because they contain different numbers of neutrons. Isotopes of a given element have the same atomic number. **(2)**

Kelvin: a temperature scale in which absolute zero is put at zero degrees. The Kelvin system is generally used in astronomical temperature measurements. **(15)**

Kepler's first law: law of planetary motion which states that all planets in the solar system travel in elliptical orbits around the sun. **(1)**

Kepler's second law: law of planetary motion which states that the orbital velocity of a planet is faster when it is close to the sun than when it is farther away. **(1)**

Kepler's third law: law of planetary motion which states that the orbital period of a planet is related to the size of its orbit. **(1)**

kettle: a depression in a moraine formed by settling of debris where a block of ice melted. **(3)**

KREEP norite: a type of breccia found in the lunar highlands. **(15)**

laccolith: intrusive rock mass formed by magma solidifying between sedimentary strata. **(4)**

laminar flow: the smooth, rapid flow at the center of a stream. **(3)**

laterite: any of several tropical soils containing concentrations of iron oxides and hydrous aluminum oxide. **(3)**

leaching: the process in which the surface minerals of a rock are dissolved and carried away by rainwater. **(2)**

light, speed of: 300,000 km/sec, or about 186,000 mi/sec. **(16)**

lithification: the transformation of loose sediment into rock. **(3)**

lithosphere: the outer 60 km of the earth, including crust and part of the upper mantle. **(5)**

lithospheric plates: sections of the lithosphere that float on the asthenosphere. **(5)**

loess: a permeable, unlayered sediment formed from windblown silt. **(3)**

luminosity: a star's light output per second. **(16)**

mafic minerals: minerals containing a large proportion of iron and magnesium. **(2)**

magma: hot liquid rock that originates below the surface of the earth. **(2)**

main sequence (m-s): a diagonal band across an H–R diagram which includes about 90 percent of the stars. **(16)**

mantle: the thick middle layer between the earth's core and the crust. **(5)**

mare basalt: dark volcanic rock found in the moon's maria. **(15)**

maria (*sing.* mare): the lunar "seas" composed of dark basalt. **(15)**

marine terrace: an old beach either above or below present sea level. **(14)**

mass: the amount of matter contained in any object. **(1)**

mass wasting: land movement due to erosional activity over a large area. **(3)**

meander: a winding curve formed by a stream flowing through a floodplain. **(3)**

mesosphere: the topmost layer of the lower atmosphere. Also known as the **ionosphere. (11)**

metamorphic rocks: sedimentary or igneous rocks which have been transformed by intense heat and high pressure. **(2)**

metamorphism: the chemical and physical changes which produce metamorphic rock. **(2)**

meteor: a small mass of stone or metal from space that enters earth's atmosphere and burns up. **(15)**

meteorite: a meteor that reaches the earth's surface. **(15)**

Mid-Atlantic Rift: the deep ocean valley separating the two parallel ridges of the Mid-Atlantic Ridge. **(14)**

mineral: an inorganic substance of a definite chemical composition. **(2)**

mixed zone: the shallow surface layer of the ocean. **(13)**

molecule: the smallest recognizable part of a compound. Molecules are formed as the result of bonding between two or more atoms. **(2)**

monadnock: an isolated hill protruding from a peneplain. **(4)**

moraine: a pile of till dropped by a glacier. **(3)**

nanometer: a unit of light measurement equal to 1/100,000,000 of a centimeter. **(2)**

nappe: a recumbent fold. **(4)**

neutron: a subatomic particle which carries no electrical charge. **(2)**

neutron star: in astronomical theory, an extremely dense star formed by contraction of a white dwarf. **(16)**

Newton's law of universal gravitation: the law which states that every object in the universe exerts an attractive force on every other object. **(1)**

notch: a gap left in an uplifted mountain after a stream channel has been diverted. **(4)**

nuée ardente: cloud of incandescent gases and debris that may be released during volcanic eruption. **(6)**

oceanic ridge: a mountain range rising from the ocean floor. **(14)**

oceanic rise: a large, gently sloping region of the ocean floor. **(14)**

ocean trough: a deep trench which borders a continent. **(14)**

olivine: a greenish mineral with a single-tetrahedon structure. Olivine is the first mineral to crystallize in Bowen's reaction series. **(2)**

terms and concepts

orbit: the path traveled by an object that is revolving around another object. **(1)**

orogeny: the process of mountain building. **(4)**

oxidation: the tendency of certain elements to combine with oxygen. **(2)**

ozone: atmospheric gas which absorbs harmful incoming shortwave radiation. **(11)**

Pangaea: Wegener's name for the supercontinent before drift occurred. **(7)**

parallax: the difference in an object's apparent position when it is viewed from different angles. **(16)**

parsec: 30 trillion km, or about 3¼ ly. **(16)**

pedalfer: a type of soil containing aluminum and iron. **(3)**

pedocal: a type of soil containing large amounts of calcium carbonate. **(3)**

peneplanation: the erosion of a mountain region to a flat peneplain. **(4)**

Perfect Cosmological Principle: the principle which assumes that the universe is both uniform and unchanging. **(16)**

peridotite: an ultramafic rock believed to be the main component of the earth's mantle. **(2)**

petrology: the branch of geology concerned with the origin, composition, and classification of rocks. **(2)**

photosphere: the inner layer of the sun's atmosphere. **(15)**

photosynthesis: the process by which autotrophs convert solar energy to chemical energy. **(17)**

planetary motion, laws of: see **Kepler.** **(1)**

plate tectonics: the theory that deals with the mechanics of continental drift. **(8)**

Polar Zones: regions in which the sun does not set at least one day of the year. **(1)**

pollution: the harmful disruption of an ecosystem. **(17)**

Population I Star: a young star rich in heavy elements. **(16)**

Population II star: an old star containing few heavy elements. **(16)**

precipitation: a process in which the ions of a substance separate from water and re-combine into crystals. **(2)**

pressure: the force acting on a unit of area. **(11)**

pressure gradient: the difference in pressure between two points, divided by the distance between them. **(12)**

primary productivity: the amount of energy produced by photosynthesis in a given area during a given period. **(17)**

prime meridian: the reference point of longitude. The prime meridian is located at Greenwich, England. **(1)**

principle of superposition: the assumption that a low sedimentary stratum in a series is older than a higher stratum of the series. **(2)**

prominences: gaseous structures that extend outward from the sun's atmosphere. **(15)**

proton: a subatomic particle which carries a positive electrical charge. **(2)**

protostar: a forming star in which nuclear reactions have not yet begun. **(16)**

pulsar: in astronomical theory, a rotating neutron star with a spot from which radiation emerges. **(16)**

P-waves: primary seismic waves that transmit their energy longitudinally through solids, liquids, and gases. **(5)**

quasar: an enormously luminous astronomical object receding at extremely high velocity. **(16)**

radiocarbon: a radioactive isotope of carbon used in nuclear dating. Also known as carbon-14. **(9)**

red giant: a large, cool, red star, near the end of its life. **(16)**

red shift: a shift toward the red end of the spectrum observed in the light waves of receding stars. **(16)**

regolith: the loose surface material that lies above solid bedrock. **(3)**

rhyolitic rocks: fine-grained extrusive rocks composed largely of felsic minerals. **(2)**

rip current: a strong current flowing seaward from the shore. **(14)**

Roche limit: the closest point to which a satellite can approach its planet without being torn apart. **(15)**

salinity: the amount of dissolved salts contained in 1,000 g of water. **(13)**

saltation: process in which wind causes grains of sand to jump from place to place. **(3)**

saturation: condition of maximum concentration of water vapor. **(12)**

seamount: a volcanic peak that rises from the ocean floor. **(13)**

secondary productivity: the rate of energy transfer from producers to consumers within a food chain. **(17)**

sedimentary rocks: rocks formed by compaction and lithification of sediments. **(2)**

seismic wave: the wave generated by an earthquake or any other disturbance of the earth's crust. **(4)**

seismograph: an instrument for recording seismic waves. The record is called a seismogram. **(5)**

Seyfert galaxy: a galaxy with a small but extremely bright nucleus. **(16)**

sial: crustal material chiefly composed of silica and aluminum. **(4)**

silicates: the principal minerals found on earth. Silicates usually exist as the silicon–oxygen tetrahedon. **(2)**

sima: crustal material chiefly composed of silica and magnesium. **(4)**

soil: the portion of the regolith which contains organic material and can nourish plant life. **(3)**

soil horizon: any of three distinct soil layers which together make up the soil profile. **(3)**

solar wind: an outward flow of protons and electrons from the sun. **(15)**

specific gravity: the ratio of a material's density to the density of pure water. **(5)**

specific heat: the number of calories that must be added to or taken away from 1 g of a substance to change its temperature by 1°C. **(11)**

spectrum: a spread of colors, in order of wavelength, created when light is passed through a prism. **(16)**

spit: a sandbar built up above sea level and attached to the shore at one end. **(14)**

stalactite: stony projection from the roof of a limestone cave, formed by deposits of chemical limestone. **(2)**

stalagmite: stony projection from the floor of a limestone cave, formed by deposits of chemical limestone. **(2)**

steady state theory: the theory that the universe is infinite and unchanging, with matter being continuously created. **(16)**

storm surge: a sudden rise in sea level that precedes a hurricane. **(12)**

stratigraphy: the study of sedimentary strata. **(2)**

stratosphere: the middle layer of the lower atmosphere. **(11)**

stratum (pl. strata): a layer of sedimentary rock. **(2)**

stratus clouds: clouds with a layered appearance. **(12)**

streak: the powder left behind when a mineral is scraped across an unglazed porcelain plate. **(2)**

stream gradient: the change in elevation of a stream per unit of distance. **(3)**

striae: small scratches left in bedrock or in rocks carried by a glacier. **(3)**

submarine canyon: a steep valley cut across a continental shelf, probably formed by the action of turbidity currents. **(14)**

submergent coast: a coast that has been covered by water. **(14)**

sunspot: a dark, shallow depression in the sun's photosphere. **(15)**

supernova: the spectacular explosion of an aging star. **(16)**

surface waves: seismic waves that travel around the surface of the earth. Also known as **L-waves.** **(5)**

S-waves: secondary seismic waves that transmit their energy transversely through a solid medium only. **(5)**

swell: a wave that is no longer being built up by the wind. **(13)**

syncline: a downward fold. **(4)**

Temperate Zones: the two regions between the Polar Zones and the Tropical Zone, in which the sun never appears at zenith. **(1)**

temperature inversion: a stable atmospheric situation in which a cold air mass is trapped beneath a warm air mass. **(12)**

thermosphere: the lower layer of the upper atmosphere. **(11)**

thrust plane: contact point between folded sedimentary rock and the underlying crystalline bedrock. **(4)**

tidal current: the horizontal flow of water that accompanies a rising or falling tide. **(13)**

tidal flat: a coastal area that is submerged at high tide and exposed at low tide. **(13)**

till: the unsorted material dropped by a glacier. **(5)**

tornado: a very small, violent cyclone, the most severe of known storms. **(12)**

trade winds: winds in the horse latitudes that blow diagonally and westward toward the equator. **(11)**

Tropical Zone: a region surrounding the equator to the north and south, in which the sun reaches zenith at noon at least one day of the year. **(1)**

troposphere: the lowest layer of the atmosphere. **(11)**

trough: the low point of a wave. **(13)**

tsunami: a seismic sea wave resulting from an undersea earthquake. **(13)**

tundra soil: an immature soil found in arctic regions. **(3)**

turbidity current: a fast-moving underwater current heavily laden with sand, mud, and sediment. **(14)**

turbulent flow: the slow, eddying flow at the edges of a stream. **(3)**

ultra-violet: a type of shortwave radiation invisible to the human eye. **(2)**

unconformity: an interruption in the layering of sediments. **(9)**

uniformitarianism: a principle which assumes that the physical laws that govern the behavior of matter at present are identical to the laws that governed it in the past. **(1)**

upwelling: the upward movement of cool water from below the ocean surface. **(13)**

varve: a double-layered accumulation of glacial sediment. **(9)**

viscosity: the tendency of a liquid to resist flow. **(6)**

water table: the level below which the regolith is always saturated with water. **(3)**

wave: a disturbance or vibration moving progressively through a medium. **(13)**

wavelength: the distance between the crest of one wave and the crest of the next. **(2)**

wave period: the amount of time required for one complete wave to pass a fixed point. **(13)**

weight: a measurement of the force of gravitational attraction between an object and the closest large astronomical body. **(2)**

white dwarf: a very dense, small star formed by contraction of a red giant. **(16)**

X-rays: an extremely shortwave type of radiation. **(2)**

zenith: the position of the sun when it is directly overhead. **(1)**

index

Abrasion, 73
Absolute humidity, 309
Absolute zero, 402
Absorption line, 447
Abyssal plains, 369
Advective cooling, 310
Air cell, 302
Air masses, 313-315
Air pollution, 497-498
Alkaline earth elements, 32
Alkaline metals, 32
Altocumulus, 311
Altostratus, 311
Altus, 311
Amber, 234
Ammonites, 264
Amorphous minerals, 34
Amphibian, 257
Angiosperms, 265
Angular velocity, 299-300
Antarctic Circle, 17
Anticline, 105
Anticyclone, 321
Appalachians, 110-111
Aquiclude, 69
Aquifer, 68
Archaean rocks, 252-253
Arête, 78
Arctic Circle, 17
Aristotle, 8-9
Artesian wells, 69
Asteroids, 399
Asthenosphere, 131
Astronomical measurements, 400-402
Astronomical Unit (A.U.), 399
Atlantic Equatorial Countercurrent, 348
Atmosphere, 277-303
   energy budget, 291-293
   nature of, 278-281
   structure, 294-295, 298
Atmospheric gases, 279-281
Atomic number, 31
Atomic structure, 29-30
Atoms, 28-29
   chemical properties, 31-33
Autotrophs, 495

Autumnal equinox, 17
Azurite, 35

Backwash, 388
Balance point, 426
Barometer, 285-287
   aneroid, 286-287
Barred spirals, 471
Barrier islands, 389-390
Bars, 386
Basaltic rock, 42
Baseline, 400
Basins, 369
Bay barrier, 390
Bay-mouth bar, 390
Bays, 382
Beach, 380, 385-391
Beach face, 386
Belts, on Jupiter, 422
Berm, 386
Big Bang Theory, 481-482
Binary stars, 447
Biogenous debris, 372
Biogeochemical cycles, 508
Biogeochemical resources, 508
Black dwarfs, 454
Black Hills, 112-114
Black hole, 457, 458
Bode's Law, 409, 412
Bomb, 161
Bonding, covalent, 32-33
   ionic, 32
Bouguer anomaly, 100
Bowen, N. L., 38
Brahe, Tycho, 18
Breaker, 358
Breccia, 48

Cambrian Period, 254
Carbon-14, 239, 242
Carboniferous Period, 257-258
Carbon print, 234
Carcinogenic, 499
Cast, 234
Celsius, 8
Cementation, 86
Cenozoic Era, 266, 268-271
Cephalopods, 255
Cepheid variables, 471
Chart, 368

Chemical bond, 32-33
Chemical weathering, 43
Chondrichthytes, 256
Chondrites, 436
Chromosphere, 403
Chronology, 228-233
Circles of motion, 352
Cirque, 78
Cirrocumulus, 311
Cirrostratus, 311
Cirrus, 311
Civil time, 13
Clastic rock texture, 38
Cleavage, 36
Clouds, 310-312
Coalescence, 312
Coal, 88-89
Coast, 380
Coastal plain, 380-385
Cobble beach, 386
Cold front, 314
Comets, 399-400
Compounds, 28
Condensation, 310
Conglomerate, 48
Conservation, 509-510
Conservation laws, of energy, 21
  of mass, 21
Constellation, 442
Continental drift, 171-195
Continental margin, 376-379
Continental nucleus, 172
Continental rise, 376, 378
Continental shelf, 376, 379
Continental slope, 376, 377
Continuous reaction series, 40
Convection, 133-134
Convection cell hypothesis, 186-188, 190-191
Convergent boundary, 202
Copernicus, Nicolaus, 18
Core of earth, 126-128
Coriolis effect, 300, 302-303, 323, 347
Corona, 404
Coronagraph, 404
Cosmological Principle, 481
Cosmology, 476-485
Covalent bonding, 32-33
Crab Nebula, 456
Creep, 72
Crest, 351

Cretaceous Period, 265-266
Crust, 128-132
  in motion, 173-174
Crystalline minerals, 34
Crystalline rocks, 38
Crystals, 34-35
Cumulonimbus, 311
Cumulus, 311
Current, 343-350
Cycles, 492
Cyclone, 321

Dating, 237-239, 242
Davis, William Morris, 381
Decay, 29, 237
Deep zone, 341
Deflation, 72
Degrees, 11
Deimos, 420
Deltas, 65
Democritus, 29
Density, 33
Desiccation, 85
Detrital sedimentary rock, 46
Deuterium, 31
Devonian Period, 256-257
Diamond, 49
Diastrophism, 173
Diatoms, 264
Dike, 157
Discontinuity, 126
Discontinuous reaction series, 39
Diurnal tide, 362
Divergent boundary, 201
Dome landforms, 111-114
Domes, 431
Doppler effect, 464
Drumlins, 84
Dunes, 75
Dwarf ellipticals, 472

Earthquakes, 119-149
  famous, 140-149
  sensible, 135
  tectonic, 135
Echo sounder, 369
Eclipses, of moon, 428-429
Ecology, 492-493
Ecosystems, 492-493

Eddies, 354
Eddington, Arthur, 479
Edge, 403
Einstein, Albert, 21, 479
Elastic rebound theory, 136-138
Electrons, 29
Electron shells, 31
Elements, 28
Elliptical galaxies, 471-472
Emergent coast, 381
Emission nebula, 468
Energy, 21, 493-496
   and economy, 510
   resources, 510-511
English system, 7
Entropy, 494
Epicenter, 124
Equatorial calms, 299
Equatorial Currents, 347
Equinox, 17
Erosion, 59
Eskers, 83
Estuaries, 65
Euclidian geometry, 479
Eutrophication, 508
Exosphere, 298
Extrusive rock, 42

Faculae, 407
Fahrenheit, 8
Fall line, 384
FAMOUS project, 213-214, 217
Fathom, 368
Fault block mountains, 103-104
Faulting, 100-104
   transform, 205
   types of, 101-103
Feldspar, 40
Fertility, 56
Filtergram, 403
Fireballs, 435
Fissure eruption, 164
Flares, 407
Flatiron, 113
Flooding, 64-67
Florida Current, 348
Fog, 310-312
Folding, 104-107, 110-111
   types of, 106-107
Foliation, 49

Food chain, 495
Food web, 495
Fossil fuels, 87-89
Fossil magnetism, 134
Fossils, 11
   formation, 233-234
Fracture, 36
Fracture zone, 205
Free energy, 494
Full moon, 427
Fumarole, 166

Gabbroic rock, 42
Galactic violence, 474-476
Galaxy, 459-465, 468-476
   cluster, 473-474
   external, 470-472
   our own, 459-464
Galileo, 405, 408-409
Galton, Francis, 319-320
Gamma-rays, 31
Gas, behavior, 281-285
   natural, 89
Geological column, 229-231
Geology, 4
Geyser, 166
Giant ellipticals, 472
Gilbert, G. K., 111
Glaciers, 77-84
Globular clusters, 462
Gneiss, 50
Graben, 102
Grains of rocks, 37
Granitic rock, 42
Granitization, 154
Gravitational law, 20
Gravitational theory, 20
Gravity, 98-100
   and ocean, 345-346
   specific, 120
   standard, 99
Great Galaxy (M31), 459
Greenhouse effect, 292
Greenwich Mean Time (GMT), 12
Ground state, 30
Groundwater, 68-69
   pollution of, 498-499
Gulf Stream, 348
Guyots, 370
Gyre, 347

Hailstones, 313
Halite, 48
Hall, James, 110-111
Halo, 461
Halogens, 32
Hardness of minerals, 36
Headlands, 382
Heat equator, 348
Helium, 31
Hertzsprung–Russell (H-R) diagram, 449
Hess, H. H., 370
Heterotrophs, 495
Hexagonal crystals, 35
High, 321
Highlands, 430, 431-432
Hook, 390
Hornblende, 39
Horse Latitudes, 302
Horst, 102
Hubble's classification of galaxies, 471-472
Hubble's Law (H), 477-479, 483
Humidity, 279, 308-309
Humus, 56
Hurricane, 323-325
Hutton, James, 9-10, 236
Hydrocarbon compounds, 87-89
Hydrogen, 31
Hydrogenous debris, 372
Hydrologic cycle, 308
Hydrolysis, 43
Hydrophone, 369
Hypothesis, 7

Igneous rocks, 38-42
Incline of earth, 16
Inert gases, 31
Inertia, 20
Infra-red radiation, 31
Insolation, 289, 290
Interface, 126
International Date Line, 16
Interstellar dust, 468
Interstellar gas, 465
Interstellar matter, 465, 468-470
Interstellar molecules, 469-470
Intrusive rock, 42
Inverse square law of gravity, 20
Ion, 30
Ionic bonding, 32
Ionosphere, 298

Iron meteorites, 434
Irregular galaxies, 472
Island arc, 207-208
Isobar, 286
Isometric crystals, 34
Isostasy, 95-98
Isostatic adjustment, 98
Isotopes, 31

Jeffreys, Harold, 176
Jupiter, 422-423
Jovian planets, 412, 421-425
Jurassic Period, 264-265

Kames, 83
Karst regions, 71
Kelvin temperature scale, 402
Kepler, Johannes, 18, 20, 408-409
Kettles, 83
Kinetic energy, 283, 494
Kinetic theory of gases, 283-285
KREEP norite, 433

Laccolith, 111
Lagoon, 389
Laminar flow, 68
Landslide, 72
Langleys, 289
Laterites, 59
Latitude, 11-12
Lava, 158
  pillow, 164
Lavoisier, Antoine, 21
Law, 7
Leaching, 43
LGMs, 455
Light, nature of, 30-31
Lightning, 317-318
Light pollution, 501-502
Light year (ly), 444
Limestone, 48
Linear velocity, 299-300
Lithification, 85
Lithium, 31
Lithogenous debris, 372
Lithosphere, 130
  birth and death, 201-203
Lobachevskian geometry, 480
Local Group, 474
Loess, 76

Longitude, 11-12
Longshore current, 388
Longshore drifting, 389
Loran, 344
Low, 321
Lunar eclipse, 428
Lungfish, 257
Luster of minerals, 35
Lycopsida, 257
Lyell, Charles, 232-233, 236

Magellanic Clouds, 459
Magma, 38
   formation of, 154, 156
Magnetic declination, 192
Magnetic striping, 219
Magnetism, 132-134, 191-195
Main sequence (m-s), 449
Malachite, 35
Mantle, 128-132
Marble, 50
Mare basalt, 433
Maria, 430, 431
Marine terraces, 381
Mars, 417-420
Mass, 20, 21, 33
Mass wasting, 71
Maury, Matthew, 340
Maxwell, James Clerk, 21
Measurement systems, 7
Mechanical weathering, 43
Mercalli, Giuseppe, 138
Mercury, 412-413
Mercury poisoning, 500
Mesosphere, 298
Mesozoic Era, 259-261, 264-266
Metamorphic rocks, 38, 49-51
Metazoan, 253
Meteor craters, 436-437
Meteorites, 434, 436
Meteorology, 4
Meteors, 399, 434-437
Meteor showers, 435
Metric system, 7
Mica, 39
Microfossils, 234
Mid-Atlantic Rift, 372
Milky Way, 459
Millibar, 285
Mineralogy, 34

Mineral resources, 506, 508
Minerals, 33-36
Mixed tide, 362
Mixed zone, 341
Moho, 131
Mohs, Frederick, 36
Mold, 233
Molecules, 29
Monoclinic crystals, 35
Moon, 425-434
   eclipses, 428-429
   geology, 432-433
   missing atmosphere, 429
   motion, 426-427
   phases, 427-428
   surface, 430-432
Moraines, 80
Mountains, 94-116
   formation of, 173-174
Mud flat, 386
Murchison, Roderick, 231-232

Nanometers (nm), 31
Nappe, 107
Natural radioactive decay, 237
Neap tide, 362
Neptune, 424-425
Nereid, 425
Neutrons, 29
Neutron stars, 455
New England northeaster, 318-319
New moon, 427
Newton, Isaac, 20, 408-409
Nimbus, 311
Nonrenewable resources, 503
North Atlantic Drift, 347
North Pacific Drift, 347
North tropical air cell, 303
Nuclear dating, 237-239, 242
Nucleons, 29
Nutrients, 508
Nutrition, 508

Oblate spheroid, 11
Occluded front, 321
Ocean, floor, 182-183, 194-195, 368-373
   pollution of, 499-501
   ridges, 371
   rises, 369
   structure of, 340-343

troughs, 369
Oceanography, 4, 340
Oldland, 385
Olivine, 39
Open clusters, 463
Ordovician Period, 255
Organic chemistry, 33
Orogenic belts, 95
Orographic cooling, 310
Orthorhombic crystals, 34
Oxidation, 46

Paleozoic Era, 254-259
Parallax, 443
Parallels of latitude, 11
Parent rock, 57
Parsec (pc), 444
Partial eclipse, 428
Particulate matter, 497
Pascal, Blaise, 288-289
Peat, 88
Pebble beach, 386
Pedalfer, 59
Pedocals, 59
Peneplain, 105
Perfect Cosmological Principle, 481
Period, of planet, 20
Permafrost, 59
Permian Period, 258-259
Petroleum, 89
Petrology, 37
Phobos, 420
Photosphere, 403
Photosynthesis, 495
Phyllite, 50
Pinger, 369
Placoderms, 256
Planetesimals, 453
Planets, 408-425
Plate tectonics, 199-223
Pluto, 421
Plutonic rock, 42
Polar air cell, 314
Polar outbreak, 314
Polar Zones, 17
Pollution, 496-503
   air, 497-498
   groundwater, 498-499
   light, 501-502
   ocean, 499-501
   thermal, 501
Population I stars, 462
Population II stars, 462
Potential energy, 494
Precipitation, 312
Pressure, 282
Pressure gradient, 320
Pressure wave, 353
Prevailing westerlies, 303
Priestley, Joseph, 21
Primary productivity, 496
Prime meridian, 11-12
Prominences, 407
Proterozoic Era, 253
Protons, 29
Protostar, 453
Proust, Joseph Louis, 21
Psilopsida, 257
Pulsars, 455
Pyroxene, 39

Quaternary Period, 268-270
Quartz, 40
Quartzite, 50
Quasars, 482-484

Radar, 402
Radiation, infra-red, 31
   ultra-violet, 31
Radiative cooling, 310
Radioactive wastes, 502-503
Radio astronomy, 465
Radiocarbon dating, 239, 242
Radio telescopes, 465
Rayleigh waves, 142
Rays, 430
Reaction series, 39-40
Red giants, 450, 453-454
Red shift, 465
Regolith, 56
Reimanian geometry, 479-480
Relative humidity, 309
Relief features, 94
Renewable resources, 503
Resources, 503, 506-511
   biogeochemical, 508
   energy, 510-511
   mineral, 506, 508
Return stroke, 317
Rhyolitic rock, 42

Richter, Charles, 139
Rift, 132, 182
Rip current, 389
Roche limit, 412, 424
Rock controversy, 228-229
Rock cycle, 51
Rocks, defining and classifying, 36-51
   melting, 154
   naming, 231-233
Rosse, Lord, 470

Salinity, 343
Saltation, 73
Salt landforms, 114
Sand beach, 386
Sandstone, 48
Saturated air, 308
Saturn, 423-424
Scarps, 431
Schist, 50
Schmidt, Martin, 483
Scientific method, 6
Seamounts, 369
Secondary productivity, 496
Sedgwick, Adam, 231-232
Sediment, ocean, 372-373
Sedimentary rocks, 38, 43, 46, 48-49
Sedimentation, 46, 84-87
Seismic sea wave, 360-361
Seismography, 122-126
Seismology, 121-126
Semi-diurnal tide, 362
Seyfert galaxies, 475-476
Shadow zone, 126
Shake wave, 353
Shale, 48
Shallow-focus earthquakes, 205
Shear wave, 353
Shingle beach, 386
Shoreline, 380
Sial, 97
Silicate melt, 38-39
Silicates, 35
Sill, 156
Siltstone, 48
Silurian Period, 256
Sima, 97
Sinkholes, 70
Slate, 50
Sleet, 313

Smith, William, 229, 236
Soil, 56-59
   profile, 57
Solar energy, 293-294
Solar radiation, 291
Solar system, 397-437
Solstice, 16
Sonar, 132
Spacetime, 21, 480
Specific heat, 282
Spectrogram, 446
Spectrograph, 446
Spectrum, 446
Spiral galaxies, 471
Spiral nebulas, 470
Spit, 390
Spring tides, 362
Stalactites, 48-49
Standard gravity, 99
Standard time zones, 13
Star clusters, 461-462
Stars, 442-485
   brightness, 442-443
   colors, 445-446
   energy, 451
   formation, 452-453
   luminosity, 445
   masses, 447-448
   mass vs. luminosity, 449
   measuring distances, 443-444
   motion, 463-464
   observing, 442-450
   spectra, 446-447
   structure and evolution, 451-457
Static electricity, 317
Stationary front, 315
Steady State Theory, 482
Step leader, 317
Stony meteorites, 434-435
Storms, 321-325, 328
   and waves, 355, 358, 360-361
Storm surge, 325
Stratification, 86
Stratigraphy, 49
Stratocumulus, 311
Stratopause, 295
Stratosphere, 295, 298
Stratus, 311
Streak classification of minerals, 35
Stream gradient, 63

Stream system, 62, 64-66
Striae, 80
Submarine canyons, 378
Submergent coast, 381
Submersible, 213-214
Subsequent valley, 113
Subsoil, 57
Summer solstice, 16
Sun, 402-408
   energy source, 451-452
   outer layers, 403-404
   rotation, 404-405
Sunspots, 405-407
Supernova, 456
Supergiants, 450
Superposition principle, 49
Surface water, 62-63
Suture area, 211
Swells, 355
Syncline, 105
Systems, 492

Talus, 72, 81
Tarn, 78
Tectonics, 201
Temperate Zones, 18
Temperature, 282
Temperature scales, 8
Terrestrial planets, 412-421
Tertiary Period, 266, 268
Tetragonal crystals, 34
Texture of rock, 38
Texas norther, 319
Thecodonts, 260
Therapsids, 260
Thermal gradient, 121
Thermal pollution, 501
Thermosphere, 298
3K cosmic radiation, 484
Thrust plane, 107
Thunderhead, 316
Thunderstorms, 316-317
Tidal current, 363
Tidal flats, 363
Tides, 361-363
   types of, 362-363
Till, 80
Time zones, 13
Tornadoes, 325-328
Torricelli, Evangelista, 285-286

Trade winds, 303
Transducer, 369
Transform faults, 205
Transition zone, 341
Transponder, 215
Triangulation, 400
Triassic Period, 259-261, 264
Triclinic crystals, 35
Trilobite, 254
Tritium, 31
Triton, 425
Tropical depression, 325
Tropical Zone, 18
Tropopause, 295
Troposphere, 294-295
Trough, 352
Tsunami, 141, 360-361
Turbidity current, 378
Turbulent flow, 68
Twisters, 325

Ultramafic rock, 42
Unconformity, 231
Uniformitarianism, 10
Universal Time (UT), 12
Universe, 476
Uplift, 95-100
Upwelling, 350
Uranus, 424
Ussher, James, 4

Valleys, 79
Van Allen belts, 281
Varve, 237
Ventifacts, 73
Venus, 413-416
Vernal equinox, 17
Vine, Frederick J., 194
Viscosity, 156
Volcanic rock, 42
Volcano, 153-167
   types of, 157-158

Warm front, 315
Water gap, 106
Watershed, 62
Water table, 68
Wave frequency, 31, 352
Wave height, 352

Wavelength, 31
Wave period, 352
Waves, causes of, 354-355
   electromagnetic, 30
   ocean, 350-355, 358-361
   and storms, 355, 358, 360-361
Weathering, 43
Weather prediction, 328-333
Wegener, Alfred, 174-179, 200, 212
Weight, 33
Werner, Abraham, 9
West Wind Drift, 347
Whitecap, 355
White dwarfs, 450, 454

Wind, and ocean, 346
Wind arrow, 328
Wind erosion, 72-75
Wind gap, 106
Wind wave, 354
Winter solstice, 16
Wrinkle ridges, 431

X-rays, 31

Zenith, 16-17
Zodiac, 442
Zones, climatic, 18, 295
   in ocean, 341-342
   on Jupiter, 422

credits

Chapter 1. p. 2, New York City Public Library; 4, National Park Service; 5BL, Bell Labs; 5LR, National Oceanographic and Aeronautical Association; 6, National Aeronautics and Space Administration; 10, American Museum of Natural History; 15, NASA; 21, American Museum of Natural History; running head, NASA.

Chapter 2. p. 26, National Park Service; 32L, Morton Salt, Inc.; 34, Ward's; 36, Ward's; 39, American Museum of Natural History; 41, Bell Labs; 42, American Museum of Natural History; 45, Ward's; 48, American Museum of Natural History; 50 & 51, American Museum of Natural History.

Chapter 3. p. 54, Utah Tourist and Publicity Council; 63, U.S. Department of the Interior; 65T, Air Photo Div.: Energy, Mines & Resources of the Canadian Government; 65B, Wide World Photos, Inc.; 66, NASA; 70, Kentucky Department of Public Information; 71, Kentucky Department of Public Information; 73, Utah Travel Council; 74T, Norwegian National Tourist Office; 74B, 75, Utah Travel Council; 77, 78, 81, National Park Service, U.S. Department of the Interior; 82T & 82B, Norwegian National Tourist Office; 83, Prof. Black, Wisconsin Conservation Dep't., University of Wisconsin; 86, SCS photo by R. Brixrer NM-5063.

Chapter 4. p. 92, American Museum of Natural History; 94, NASA; 96, Norwegian National Tourist Office; 97, U.S. Coast Guard; 103, 104, 105L, 105R, 106, American Museum of Natural History; 109, National Publicity Studios, Lambton Quay, Wellington, N.Z.; 110, American Museum of Natural History.

Chapter 5. p. 118, Wide World Photos Inc.; 124, American Museum of Natural History; 137, Spence Aerials; 141T, 141B, 145, U.S. Department of the Interior; 146, 147, 148, 149, U.S. Geological Survey.

Chapter 6. p. 152, Hawaiian Visitors Bureau Photo; 156, American Museum of Natural History; 158, National Park Service; 159T, Wide World Photos; 159B, Brown Bros.; 161, NASA; 165, 167, National Publicity Studios, N.Z.

Chapter 7. p. 184T & 184B, Office of Naval Research; 195, Woods Hole Oceanographic Institute.

Chapter 8. p. 198, United States Navy; 206, CNEXO; 213, ESSA; 215, 216, USN; 218, CNEXO; 219, USN; 222, CNEXO.

Chapter 9. p. 226, 233, 234, American Museum of Natural History; 241, Bausch and Lomb; 244, American Museum of Natural History.

Chapter 10. p. 250, American Museum of Natural History; 253, Smithsonian Institute; 255T & 255B, American Museum of Natural History; 256, Courtesy of the Trustees of the British Museum; 257, T. Spencer for *Life;* 259, 260, Courtesy Yale Peabody Museum; 261, 263, 264T, 264B, 268, 269T, American Museum of Natural History; 269B, UPI; 270, American Museum of Natural History.

Chapter 11. p. 276, NASA; 297, National Park Service.

Chapter 12. p. 306, 309, 310, 311, 312, U.S. Department of Commerce, Weather Bureau; 314, NASA; 316, U.S. Dep't of Commerce; 317, ESSA; 320, NASA; 322, NOAA; 324, Wide World Photos; 327, NOAA; 328, © 1968 by New York Times Company, reprinted by permission.

Chapter 13. p. 338, American Museum of Natural History; 341, 344, 347, 350, 351, 355, NOAA; 357, U.S. Navy; 361, American Museum of Natural History; 363, Canadian Government—Office of Tourism; running head, U.S. Navy.

Chapter 14. p. 366, American Museum of Natural History; 369, NOAA; 373, U.S. Navy; 375, American Museum of Natural History; 376, U.S. Navy; 380, NOAA; 381, Canadian Government Travel Bureau; 390, NASA; 391T, American Museum of Natural History; 391B, U.S. Navy.

Chapter 15. p. 396, Hale Observatories; 399, Lick Observatory, University of California; 401BL, Cornell University, Office of Public Information; 401BR, NASA; 402L, 402R, 406, 408T, 408B, 411, Hale Observatories; 413T, 413R, 414B, NASA; 416, Arecibo Observatory, part of the National Astronomy and Ionosphere Center, operated by Cornell University under contract with the National Science Foundation; 416B, Novosti from Sovfoto; 417T & B, 418 all, 419 all, 420 all, 422, 423, 424, 429, 430, 431, 432, 433T & B, 435T, NASA; 435B, American Museum of Natural History; 436L, Dudley Observatory; 436R, NASA; 437, Yerkes Observatory, University of Chicago.

Chapter 16. p. 440, Hale Observatories; 442, American Museum of Natural History; 443, Griffith Observatory; 444TL, Yerkes Observatory, University of Chicago; 447, Hale Observatories; 448T, Yerkes Observatory, Univ. of Chicago; 453, Lick Observatory, Univ. of California; 455, Lick Observatory, Univ. of California; 455, Kitt Peak National Observatory; 456B, Hale Observatory; 459, Hale Observatories; 461, Yerkes Observatory, Univ. of Chicago; 462, 463TL, Kitt Peak National Observatory; 463TR, European Southern Observatory; 464, Hale Observatories; 465, Owens Valley Radio Observatory, California Institute of Technology; 467, NASA; 468T, Lick Observatory; 468L, 469, 470, 472 all, 473TL, Hale Observatories; 473TR, Lick Observatory; 474, 475T, 475BL, Hale Observatories; 475BR, Kitt Peak National Observatory; 476L & R, Hale Observatories; 479, American Museum of Natural History; 483 & 485, Hale Observatory.

Chapter 17. p. 490, Exxon; 493, ERDA; 497, USDA; 498, 499, 500L & R, UPI; 501, USAF; 502, Atomic Industrial Forum; 503 UPI; 505 British Airways; 509, USDA; running head, USDA.

Color Insert: Geology. g1. Rhodesia National Tourist Board; g2. Josef Muench; g3. National Park Service, Camp Verde, Arizona; g4B. F. M. Brown; g4T. Jerome Wyckoff; g5T. G. R. Roberts; g5B. Laurence Lowry; g6. Swissair Photo-AC Zurich; g7T. J. K. Snobble; g7B. Richard Harrington; g7BR. F. Martin Brown; g7BL. Fritz Coro for *Life;* g8TL. Dick Kent of Shostal Associates, Inc.; g8BR. John Bonte for Silver Burdett; g8TR. Ward's Natural Science Establishment, Inc.; g8L. Jerome Wyckoff; g8L. Steven C. Wilson.

Color Insert: Meteorology. m1. NASA; m2. E. Eliosofon for *Life;* m3. NASA; m4B. F. Martin Brown; m4T. ESSA; m5. A. Feininger for *Life;* m5TL. Neil Leifer for *Sports Illustrated;* m6T. Public Relations, Stone Mountain, Georgia; m6B. V. P. Hesslar; m7T. Arthur D. Little Inc.; m7B. Josef Muench; m8. Esther Henderson for Rapho Guillumette.

Color Insert: Oceanography. o1. Chuck Abbott for Rapho Guillumette; o2. George Hunter; o3. Wm. Keith of Keith Films; o4B. Laurence Lowry; o4T. G. R. Roberts; o5BR. Arabian American Oil Company; o5BL. Jerome Wyckoff; o6L. Ward's Natural Science Establishment, Inc.; o6T. Nova Scotia Information Service; o7T. Alfred T. Hughs; o7B. Mittet Foto AS; o8. Laurence Lowry.

Color Insert: Astronomy. a1. NASA; a2TR. Hale Observatories; a2L. W. S. Finsen; a3. NASA; a4. NASA; a5. NASA; a6. U.S. Naval Observatory; a7. Hale Observatories; a8. Hale Observatories.